Quantum Transport: Atom to Transistor

This book presents the conceptual framework underlying the atomistic theory of matter, emphasizing those aspects that relate to current flow. This includes some of the most advanced concepts of non-equilibrium quantum statistical mechanics. No prior acquaintance with quantum mechanics is assumed. Chapter 1 provides a description of quantum transport in elementary terms accessible to a beginner. The book then works its way from the hydrogen atom to nanostructures ending with a unified model for quantum transport. The final chapter summarizes the equations for quantum transport with illustrative examples showing how conductors evolve from the atomic to the ohmic regime as they get larger. Many numerical examples are used to provide concrete illustrations and the corresponding MATLAB codes can be downloaded from the web. Videostreamed lectures, keyed to specific sections of the book, are also available through the web. This book is primarily aimed at senior and graduate students.

Supriyo Datta is the Thomas Duncan Distinguished Professor in the School of Electrical and Computer Engineering at Purdue University. He shared the IEEE Cledo Brunetti award and the ASEE Terman award and is a Fellow of the IEEE, the American Physical Society (APS), and the Institute of Physics (IOP). He has written three other books including *Electronic Transport in Mesoscopic Systems*, also published by Cambridge University Press.

Quantum Transport: Atom to Transistor

Supriyo Datta
Purdue University

CAMBRIDGE
UNIVERSITY PRESS

University Printing House, Cambridge CB2 8BS, United Kingdom

One Liberty Plaza, 20th Floor, New York, NY 10006, USA

477 Williamstown Road, Port Melbourne, VIC 3207, Australia

314-321, 3rd Floor, Plot 3, Splendor Forum, Jasola District Centre, New Delhi - 110025, India

79 Anson Road, #06-04/06, Singapore 079906

Cambridge University Press is part of the University of Cambridge.

It furthers the University's mission by disseminating knowledge in the pursuit of education, learning and research at the highest international levels of excellence.

www.cambridge.org
Information on this title: www.cambridge.org/9781107632134

First published 2005
Reprinted with corrections 2006
5th printing 2011
First paperback edition 2013

A catalogue record for this publication is available from the British Library

ISBN 978-0-521-63145-7 Hardback
ISBN 978-1-107-63213-4 Paperback

Contents

* Asterisked sections indicate those that I use for undergraduate teaching.

Preface

The purpose of this book is to convey the conceptual framework that underlies the microscopic or atomistic theory of matter, emphasizing those aspects that relate to electronic properties, especially current flow. Even a hundred years ago the atomistic viewpoint was somewhat controversial and many renowned scientists of the day questioned the utility of postulating entities called atoms that no one could see.[1] What no one anticipated was that by the end of the twentieth century, scientists would actually be "seeing" and taking pictures of atoms and even building "nanostructures" engineered on a nanometer length scale.[2] The properties of such nanostructures cannot be modeled in terms of macroscopic concepts like mobility or diffusion. What we need is an atomic or microscopic viewpoint and that is what this book is about.

The microscopic theory of matter was largely developed in the course of the twentieth century following the advent of quantum mechanics and is gradually becoming an integral part of engineering disciplines, as we acquire the ability to engineer materials and devices on an atomic scale. It is finding use in such diverse areas as predicting the structure of new materials, their electrical and mechanical properties, and the rates of chemical reactions, to name just a few applications. In this book, however, I will focus on the flow of current through a nanostructure when a voltage is applied across it. This is a problem of great practical significance as electronic devices like transistors get downscaled to atomic dimensions. It is a rapidly evolving field of research and the specific examples I will use in this book may or may not be important twenty years from now. But the problem of current flow touches on some of the deepest issues of physics and the concepts we will discuss represent key fundamental concepts of quantum mechanics and non-equilibrium statistical mechanics that should be relevant to the analysis and design of nanoscale devices for many years into the future. This book is written very much in the spirit of a text-book that uses idealized examples to clarify general principles, rather than a research monograph that does justice to specific real-world issues.

[1] For an interesting description see Lindley (2001).

[2] The distance between two atoms is ~0.25 nm.

Describing the flow of current involves a lot more than just quantum mechanics – it requires an appreciation of some of the most advanced concepts of non-equilibrium statistical mechanics. Traditionally these topics are spread out over many physics/ chemistry courses that take many semesters to cover. My aim is to condense the essential concepts into a book that can be covered in a one-semester graduate course. I have also used a subset of this material to teach a senior-level undergraduate course. The only background I will assume is a knowledge of partial differential equations and matrix algebra including familiarity with MATLAB (or an equivalent mathematical software package).

The first chapter and the appendix are somewhat distinct from the rest of the book, but they have been included because I believe they should help the reader connect with the "big picture." The first chapter motivates the concepts covered in this book by laying out the factors that enter into a description of quantum transport in terms that are accessible to a beginner with no background in quantum mechanics. The appendix on the other hand is intended for the advanced reader and describes the same concepts using advanced formalism ("second quantization"). Both these chapters have been adapted from a longer article (Datta, 2004).

When I finished my last book, *Electronic Transport in Mesoscopic Systems* (*ETMS*) (Datta, 1995), I did not think I would want to write another. But *ETMS* was written in the early 1990s when quantum transport was a topic of interest mainly to physicists. Since then, electronic devices have been shrinking steadily to nanometer dimensions and quantum transport is fast becoming a topic of interest to electrical engineers as well. I owe it largely to my long-time friend and colleague Mark Lundstrom, that I was convinced to write this book with an engineering audience in mind. And this change in the intended audience (though I hope physicists too will find it useful) is reflected in my use of "q" rather than "e" to denote the electronic charge. However, I have not replaced "i" with "−j", since a Schrodinger equation with "$-j\partial\psi/\partial t$" just does not look right!

Anyway, this book has more substantial differences with *ETMS*. *ETMS* starts from the effective mass equation, assuming that readers had already seen it in a solid-state physics course. In this book, I spend Chapters 2 through 7 building up from the hydrogen atom to $E(k)$ diagrams and effective mass equations. Most importantly, *ETMS* was largely about low-bias conductance ("linear response") and its physical interpretation for small conductors, emphasizing the transmission formalism. In this book (Chapters 1, 8–11) I have stressed the full current–voltage characteristics and the importance of performing self-consistent calculations. I have tried to inject appropriate insights from the transmission formalism, like the Landauer formula and Buttiker probes, but the emphasis is on the non-equilibrium Green's function (NEGF) formalism which I believe provides a rigorous framework for the development of quantum device models that can be used to benchmark other simplified approaches. It bridges the gap between the fully coherent quantum transport models of mesoscopic physicists and the fully incoherent Boltzmann transport models of device physicists.

The NEGF formalism is usually described in the literature using advanced many-body formalism, but I have tried to make it accessible to a more general audience. In its simplest form, it reduces to a rate equation for a one-level system that I can teach undergraduates. And so in this book, I start with the "undergraduate" version in Chapter 1, develop it into the full matrix version, illustrate it with examples in Chapter 11, and provide a more formal justification using second quantization in the appendix. This book thus has a very different flavor from *ETMS*, which was primarily based on the transmission formalism with a brief mention of NEGF in the last chapter.

Another important distinction with *ETMS* is that in this book I have made significant use of MATLAB. I use many numerical examples to provide concrete illustrations and, for the readers' convenience, I have listed my MATLAB codes at the end of the book, which can also be downloaded from my website.[3] I strongly recommend that readers set up their own computer program on a personal computer to reproduce the results. This hands-on experience is needed to grasp such deep and diverse concepts in so short a time.

Additional problems designed to elaborate on the text material are posted on my website and I will be glad to share my solutions with interested readers. I plan to add more problems to this list and welcome readers to share problems of their own with the rest of the community. I will be happy to facilitate the process by adding links to relevant websites.

This book has grown out of a graduate course (and recently its undergraduate version) that I have been teaching for a number of years. The reader may find it useful to view the videostreamed course lectures, keyed to specific sections of this book, that are publicly available through the web, thanks to the Purdue University E-enterprise Center, the NSF Network for Computational Nanotechnology, and the NASA Institute for Nanoelectronics and Computing.

[3] http://dynamo.ecn.purdue.edu/~datta

Acknowledgements

There are far too many people to thank and so, rather than list them individually, let me simply say that over the years I feel fortunate to have interacted with so many brilliant students and outstanding colleagues at Purdue and elsewhere. To them, to my supportive family, and to all others who have helped me learn, I dedicate this book.

Symbols

Fundamental constants

q	electronic charge	1.602×10^{-19} C
h	Planck constant	6.626×10^{-34} J s
$\hbar$	$h/2\pi$	1.055×10^{-34} J s
m	free electron mass	9.11×10^{-31} kg
ε_0	permittivity of free space	8.854×10^{-12} F/m
a_0	Bohr radius, $4\pi\varepsilon_0\hbar^2/mq^2$	0.0529 nm
$G_0 = q^2/h$	conductance quantum	38.7×10^{-6} S (S $= 1/\Omega$ = A/V)
		$= 1/(25.8 \times 10^3\ \Omega)$

We will use rationalized MKS units throughout the book, with energy in electron-volts: 1 eV = 1.602×10^{-19} J.

E_0	$q^2/8\pi\varepsilon_0 a_0$	13.6 eV

Some of the other symbols used

		Units
I	current (external)	amperes (A)
J	current (internal)	amperes (A)
V	voltage	volts (V)
R	resistance	ohms (Ω = V/A)
G	conductance	siemens (S = A/V)
a	lattice constant	meters (m)
t_0	$\hbar^2/2m^*a^2$	electron-volts (eV)
t	time	seconds (s)
m^*	effective mass	kilograms (kg)
m_c	conduction band effective mass	kilograms (kg)
$\gamma_{1,2,3}$	Luttinger parameters	dimensionless
ε_r	relative permittivity	dimensionless
$\vec{F}$	electric field	V/m
L	channel length	m
S	cross-sectional area	m^2

$\vec{k}$	wavevector		/m
$\vec{v}$	velocity		m/s
n_S	electron density per unit area		/m^2
n_L	electron density per unit length		/m
N	number of electrons or number of photons		dimensionless
ρ	density matrix		dimensionless
ε	energy level		eV
H	Hamiltonian		eV
U	self-consistent potential		eV
E	energy		eV
μ	electrochemical potential		eV
$f(E)$	Fermi function		dimensionless
$n(E)$	electron density per unit energy		/eV
$D(E)$	density of states (DOS)		/eV
$A(E)$	spectral function		/eV
$G^n(E)$	(same as $-iG^<$) correlation function		/eV
$G^p(E)$	(same as $+iG^>$) hole correlation function		/eV
$G(E)$	Green's function (retarded)		/eV
$\overline{T}(E)$	transmission function		dimensionless
$T(E)$	transmission probability (<1)		dimensionless
$\gamma, \Gamma(E)$	broadening		eV
$\Sigma(E)$	self-energy (retarded)		eV
$\Sigma^{in}(E)$	(same as $-i\Sigma^<$) inscattering		eV
$\vartheta(E)$	unit step function	$\begin{cases} = 1, E > 0 \\ = 0, E < 0 \end{cases}$	dimensionless
$\delta(E)$	Dirac delta function		/eV
δ_{nm}	Kronecker delta	$\begin{cases} = 1, n = m \\ = 0, n \neq m \end{cases}$	dimensionless
+	Superscript to denote conjugate transpose		
T	Superscript to denote transpose		

1 Prologue: an atomistic view of electrical resistance

Let me start with a brief explanation since this is not a typical "prologue." For one it is too long, indeed as long as the average chapter. The reason for this is that I have a very broad objective in mind, namely to review *all* the relevant concepts needed to understand current flow through a very small object that has only one energy level in the energy range of interest. Remarkably enough, this can be done without invoking any significant background in quantum mechanics. What requires serious quantum mechanics is to understand where the energy levels come from and to describe large conductors with multiple energy levels. Before we get lost in these details (and we have the whole book for it!) it is useful to understand the factors that influence the current–voltage relation of a really small object.

This "bottom-up" view is different from the standard "top-down" approach to electrical resistance. We start in college by learning that the conductance G (inverse of the resistance) of a large macroscopic conductor is directly proportional to its cross-sectional area A and inversely proportional to its length L:

$$G = \sigma A/L \quad \text{(Ohm's law)}$$

where the conductivity σ is a material property of the conductor. Years later in graduate school we learn about the factors that determine the conductivity and if we stick around long enough we eventually talk about what happens when the conductor is so small that one cannot define its conductivity. I believe the reason for this "top-down" approach is historical. Till recently, no one was sure how to describe the conductance of a really small object, or if it even made sense to talk about the conductance of something really small. To measure the conductance of anything we need to attach two large contact pads to it, across which a battery can be connected. No one knew how to attach contact pads to a small molecule till the late twentieth century, and so no one knew what the conductance of a really small object was. But now that we are able to do so, the answers look fairly simple, except for unusual things like the Kondo effect that are seen only for a special range of parameters. Of course, it is quite likely that many new effects will be discovered as we experiment more on small conductors and the description presented here is certainly not intended to be the last word. But I think it should be the "first

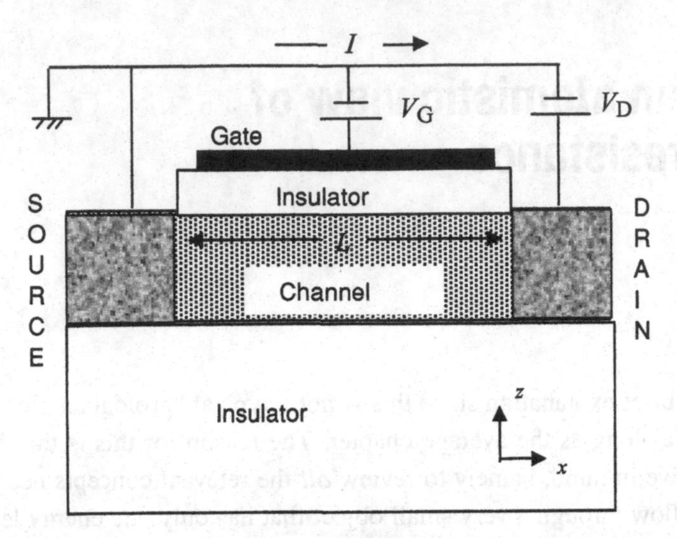

Fig. 1.1 Sketch of a nanoscale field effect transistor. The insulator should be thick enough to ensure that no current flows into the gate terminal, but thin enough to ensure that the gate voltage can control the electron density in the channel.

word" since the traditional top-down approach tends to obscure the simple physics of very small conductors.

The generic structure I will often use is a simple version of a "nanotransistor" consisting of a semiconducting channel separated by an insulator layer (typically silicon dioxide) from the metallic gate (Fig. 1.1). The regions marked source and drain are the two contact pads, which are assumed to be highly conducting. The resistance of the channel determines the current that flows from the source to the drain when a voltage V_D is applied between them. The voltage V_G on the gate is used to control the electron density in the channel and hence its resistance. Such a voltage-controlled resistor is the essence of any field effect transistor (FET) although the details differ from one version to another. The channel length L has been progressively reduced from ~10 μm in 1960 to ~0.1 μm in 2000, allowing circuit designers to pack $(100)^2 = 10\,000$ times more transistors (and hence that much more computing power) into a chip of given surface area. This increase in packing density is at the heart of the computer revolution. How much longer can the downscaling continue? No one really knows. However, one thing seems certain. Regardless of what form future electronic devices take, *we will have to learn how to model and describe the electronic properties of device structures that are engineered on an atomic scale.* The examples I will use in this book may or may not be important twenty years from now. But the problem of current flow touches on some of the deepest issues of physics related to the nature of "friction" on a microscopic scale and the emergence of irreversibility from reversible laws. The concepts we will discuss represent key fundamental concepts of quantum mechanics and non-equilibrium statistical mechanics that should be relevant to the analysis and design of nanoscale devices for many years into the future.

Outline: To model the flow of current, the first step is to draw an equilibrium energy level diagram and locate the electrochemical potential μ (also called the Fermi level or Fermi energy) set by the source and drain contacts (Section 1.1). Current flows when an external device such as a battery maintains the two contacts at different electrochemical potentials μ_1 and μ_2, driving the channel into a non-equilibrium state (Section 1.2). The current through a really small device with only one energy level in the range of interest is easily calculated and, as we might expect, depends on the quality of the contacts. But what is not obvious (and was not appreciated before the late 1980s) is that there is a maximum conductance for a channel with one level (in the energy range of interest), which is a fundamental constant related to the charge on an electron and Planck's constant:

$$G_0 \equiv q^2/h = 38.7\ \mu\text{S} = (25.8\ \text{k}\Omega)^{-1} \tag{1.1}$$

Actually small channels typically have two levels (one for up spin and one for down spin) at the same energy ("degenerate" levels) making the maximum conductance equal to $2G_0$. We can always measure conductances lower than this, if the contacts are bad. But the point is that there is an upper limit to the conductance that can be achieved even with the most perfect of contacts (Section 1.3). In Section 1.4, I will explain the important role played by charging and electrostatics in determining the shape of the current–voltage (*I–V*) characteristics, and how this aspect is coupled with the equations for quantum transport. Once this aspect has been incorporated we have all the basic physics needed to describe a one-level channel that is coupled "well" to the contacts. But if the channel is weakly coupled, there is some additional physics that I will discuss in Section 1.5. Finally, in Section 1.6, I will explain how the one-level description is extended to larger devices with multiple energy levels, eventually leading to Ohm's law. It is this extension to larger devices that requires the advanced concepts of quantum statistical mechanics that constitute the subject matter of the rest of this book.

1.1 Energy level diagram

Figure 1.1.1 shows the typical current–voltage characteristics for a well-designed transistor of the type shown in Fig. 1.1 having a width of 1 μm in the *y*-direction perpendicular to the plane of the paper. At low gate voltages, the transistor is in its off state, and very little current flows in response to a drain voltage V_D. Beyond a certain gate voltage, called the threshold voltage V_T, the transistor is turned on and the ON-current increases with increasing gate voltage V_G. For a fixed gate voltage, the current I increases at first with drain voltage, but it then tends to level off and saturate at a value referred to as the

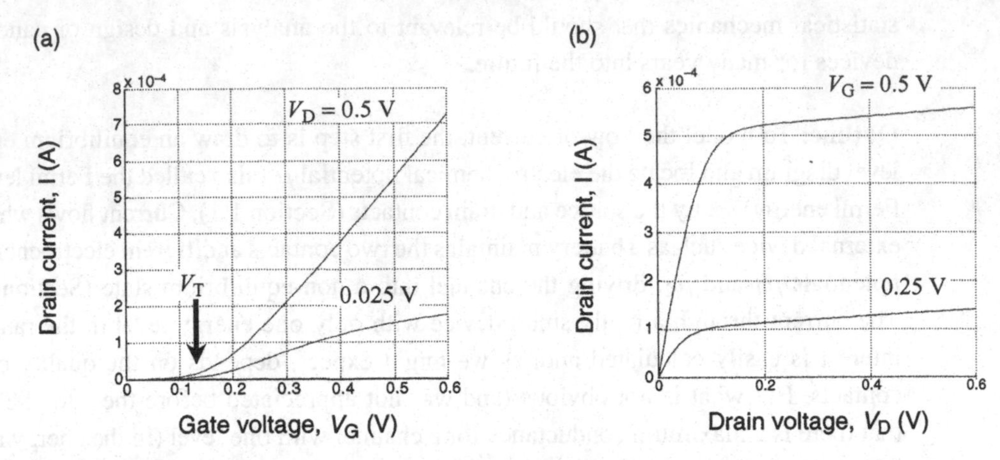

Fig. 1.1.1 (a) Drain current I as a function of the gate voltage V_G for different values of the drain voltage V_D. (b) Drain current as a function of the drain voltage for different values of the gate voltage.

ON-current. Let us start by trying to understand why the current increases when the gate voltage exceeds V_T (Fig. 1.1.1a).

The first step in understanding the operation of any inhomogeneous device structure (like the generic one shown in Fig. 1.1) is to draw an *equilibrium* energy level diagram (sometimes called a "band diagram") assuming that there is no voltage applied between the source and the drain. Electrons in a semiconductor occupy a set of energy levels that form bands as sketched in Fig. 1.1.2. Experimentally, one way to measure the occupied energy levels is to find the minimum energy of a photon required to knock an electron out into vacuum (photoemission (PE) experiments). We can describe the process symbolically as

$$S + h\nu \rightarrow S^+ + e^-$$

where "S" stands for the semiconductor device (or any material for that matter!).

The empty levels, of course, cannot be measured the same way since there is no electron to knock out. We need an inverse photoemission (IPE) experiment where an incident electron is absorbed with the emission of photons:

$$S + e^- \rightarrow S^- + h\nu$$

Other experiments like optical absorption also provide information regarding energy levels. All these experiments would be equivalent if electrons did not interact with each other and we could knock one electron around without affecting everything else around it. But in the real world subtle considerations are needed to relate the measured energies to those we use and we will discuss some of these issues in Chapter 3.

We will assume that the large contact regions (labeled source and drain in Fig. 1.1) have a continuous distribution of states. This is true if the contacts are metallic, but not

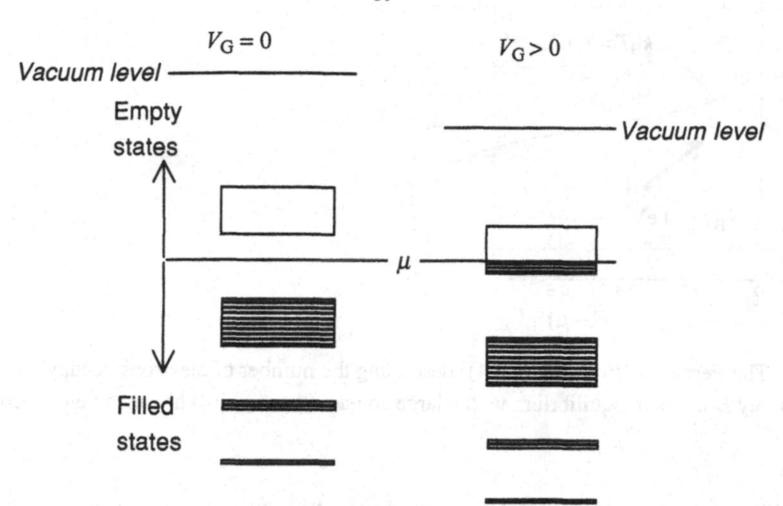

Fig. 1.1.2 Allowed energy levels that can be occupied by electrons in the active region of a device like the channel in Fig. 1.1. A positive gate voltage V_G moves the energy levels down while the electrochemical potential μ is fixed by the source and drain contacts, which are assumed to be in equilibrium with each other ($V_D = 0$).

exactly true of semiconducting contacts, and interesting effects like a decrease in the current with an increase in the voltage (sometimes referred to as negative differential resistance (NDR)) can arise as a result (see Exercise E.1.4); however, we will ignore this possibility in our discussion. The allowed states are occupied up to some energy μ (called the electrochemical potential) which too can be located using photoemission measurements. The work function is defined as the minimum energy of a photon needed to knock a photoelectron out of the metal and it tells us how far below the vacuum level μ is located.

Fermi function: If the source and drain regions are coupled to the channel (with V_D held at zero), then electrons will flow in and out of the device bringing them all in equilibrium with a common electrochemical potential, μ, just as two materials in equilibrium acquire a common temperature, T. In this equilibrium state, the average (over time) number of electrons in any energy level is typically not an integer, but is given by the Fermi function:

$$f_0(E - \mu) = \frac{1}{1 + \exp[(E - \mu)/k_B T]} \tag{1.1.1}$$

Energy levels far below μ are always full so that $f_0 = 1$, while energy levels far above μ are always empty with $f_0 = 0$. Energy levels within a few $k_B T$ of μ are occasionally empty and occasionally full so that the average number of electrons lies

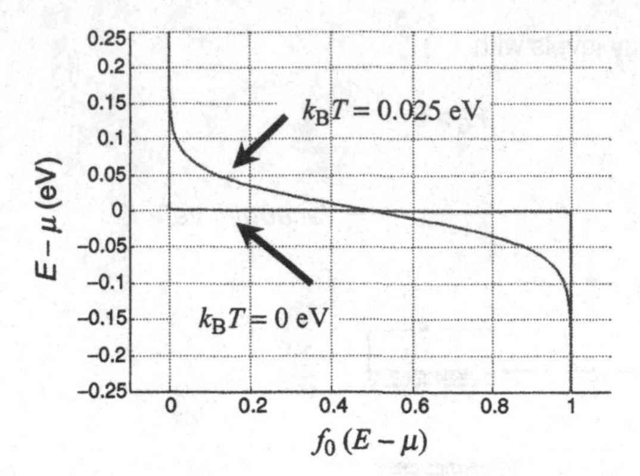

Fig. 1.1.3 The Fermi function (Eq. (1.1.1)) describing the number of electrons occupying a state with an energy E if it is in equilibrium with a large contact ("reservoir") having an electrochemical potential μ.

between 0 and 1: $0 \leq f_0 \leq 1$ (Fig. 1.1.3). Note that this number cannot exceed one because the exclusion principle forbids more than one electron per level.

n-type operation: A positive gate voltage V_G applied to the gate lowers the energy levels in the channel. However, the energy levels in the source and drain contacts are unchanged and hence the electrochemical potential μ (which must be the same everywhere) remains unaffected. As a result the energy levels move with respect to μ, driving μ into the empty band as shown in Fig. 1.1.2. This makes the channel more conductive and turns the transistor ON, since, as we will see in the next section, the current flow under bias depends on the number of energy levels available around $E = \mu$. The threshold gate voltage V_T needed to turn the transistor ON is thus determined by the energy difference between the equilibrium electrochemical potential μ and the lowest available empty state (Fig. 1.1.2) or what is called the conduction band edge.

p-type operation: Note that the number of electrons in the channel is not what determines the current flow. A negative gate voltage ($V_G < 0$), for example, reduces the number of electrons in the channel. Nevertheless the channel will become more conductive once the electrochemical potential is driven into the filled band as shown in Fig. 1.1.4, due to the availability of states (filled or otherwise) around $E = \mu$. This is an example of p-type or "hole" conduction as opposed to the example of n-type or electron conduction shown in Fig. 1.1.2. The point is that for current flow to occur, states are needed near $E = \mu$, but they need not be empty states. Filled states are just as good and it is not possible to tell from this experiment whether conduction is n-type (Fig. 1.1.2) or p-type (Fig. 1.1.4). This point should become clearer in Section 1.2 when we discuss why current flows in response to a voltage applied across the source and drain contacts.

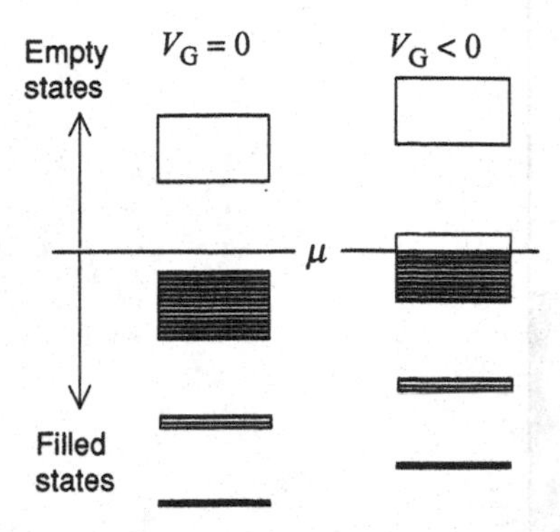

Fig. 1.1.4 Example of p-type or hole conduction. A negative gate voltage ($V_G < 0$) reduces the number of electrons in the channel. Nevertheless the channel will become more conductive once the electrochemical potential μ is driven into the filled band since conduction depends on the availability of states around $E = \mu$ and not on the total number of electrons.

Figures 1.1.2 and 1.1.4 suggest that the same device can be operated as an n-type or a p-type device simply by reversing the polarity of the gate voltage. This is true for short devices if the contacts have a continuous distribution of states as we have assumed. But in general this need not be so: for example, long devices can build up "depletion layers" near the contacts whose shape can be different for n- and p-type devices.

1.2 What makes electrons flow?

We have stated that conduction depends on the availability of states around $E = \mu$; it does not matter if they are empty or filled. To understand why, let us consider what makes electrons flow from the source to the drain. The battery lowers the energy levels in the drain contact with respect to the source contact (assuming V_D to be positive) and maintains them at distinct electrochemical potentials separated by qV_D

$$\mu_1 - \mu_2 = qV_D \tag{1.2.1}$$

giving rise to two different Fermi functions:

$$f_1(E) \equiv \frac{1}{1 + \exp[(E - \mu_1)/k_B T]} = f_0(E - \mu_1) \tag{1.2.2a}$$

$$f_2(E) \equiv \frac{1}{1 + \exp[(E - \mu_2)/k_B T]} = f_0(E - \mu_2) \tag{1.2.2b}$$

Each contact seeks to bring the channel into equilibrium with itself. The source keeps pumping electrons into it, hoping to establish equilibrium. But equilibrium is never

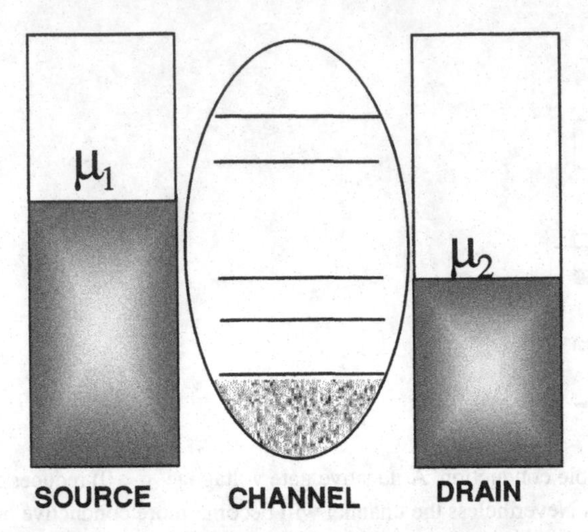

Fig. 1.2.1 A positive voltage V_D applied to the drain with respect to the source lowers the electrochemical potential at the drain: $\mu_2 = \mu_1 - qV_D$. Source and drain contacts now attempt to impose different Fermi distributions as shown, and the channel goes into a state intermediate between the two.

achieved as the drain keeps pulling electrons out in its bid to establish equilibrium with itself. The channel is thus forced into a balancing act between two reservoirs with different agendas and this sends it into a non-equilibrium state intermediate between what the source would like to see and what the drain would like to see (Fig. 1.2.1).

Rate equations for a one-level model: This balancing act is easy to see if we consider a simple one-level system, biased such that its energy ε lies between the electrochemical potentials in the two contacts (Fig. 1.2.2). Contact 1 would like to see $f_1(\varepsilon)$ electrons, while contact 2 would like to see $f_2(\varepsilon)$ electrons occupying the state where f_1 and f_2 are the source and drain Fermi functions defined in Eq. (1.2.2). The average number of electrons N at steady state will be something intermediate between $f_1(\varepsilon)$ and $f_2(\varepsilon)$. There is a net flux I_1 across the left junction that is proportional to $(f_1 - N)$, *dropping the argument* ε for clarity:

$$I_1 = \frac{q\gamma_1}{\hbar}(f_1 - N) \tag{1.2.3a}$$

where $-q$ is the charge per electron. Similarly the net flux I_2 across the right junction is proportional to $(f_2 - N)$ and can be written as

$$I_2 = \frac{q\gamma_2}{\hbar}(f_2 - N) \tag{1.2.3b}$$

We can interpret the rate constants $\gamma_1/\hbar$ and $\gamma_2/\hbar$ as the rates at which an electron placed initially in the level ε will escape into the source and drain contacts respectively. In principle, we could experimentally measure these quantities, which have the

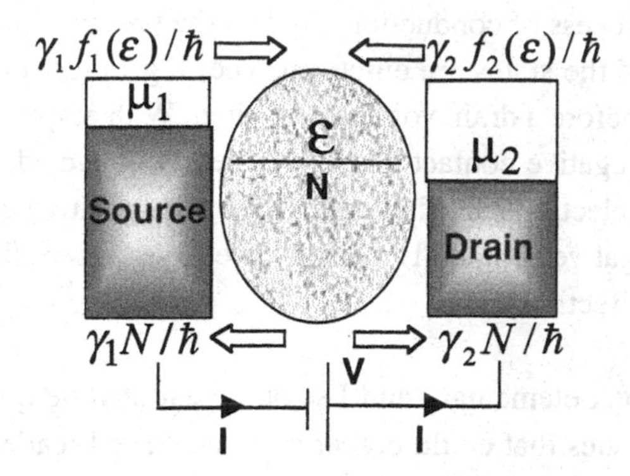

Fig. 1.2.2 Flux of electrons into and out of a one-level channel at the source and drain ends: simple rate equation picture.

dimension per second, so that γ_1 and γ_2 have the dimension of energy. At the end of this section I will say a few more words about the physics behind these equations. But for the moment, let us work out the consequences.

Current in a one-level model: At steady state there is no net flux into or out of the channel, $I_1 + I_2 = 0$, so that from Eqs. (1.2.3a, b) we obtain the reasonable result

$$N = \frac{\gamma_1 f_1 + \gamma_2 f_2}{\gamma_1 + \gamma_2} \tag{1.2.4}$$

that is, the occupation N is a weighted average of what contacts 1 and 2 would like to see. Substituting this result into Eq. (1.2.3a) or (1.2.3b) we obtain an expression for the steady-state current:

$$I = I_1 = -I_2 = \frac{q}{\hbar}\frac{\gamma_1\gamma_2}{\gamma_1 + \gamma_2}[f_1(\varepsilon) - f_2(\varepsilon)] \tag{1.2.5}$$

This is the current per spin. We should multiply it by two if there are two spin states with the same energy.

This simple result serves to illustrate certain basic facts about the process of current flow. Firstly, no current will flow if $f_1(\varepsilon) = f_2(\varepsilon)$. A level that is way below both electrochemical potentials μ_1 and μ_2 will have $f_1(\varepsilon) = f_2(\varepsilon) = 1$ and will not contribute to the current, just like a level that is way above both potentials μ_1 and μ_2 and has $f_1(\varepsilon) = f_2(\varepsilon) = 0$. It is only when the level lies within a few $k_B T$ of the potentials μ_1 and μ_2 that we have $f_1(\varepsilon) \neq f_2(\varepsilon)$ and a current flows. Current flow is thus the result of the "*difference in agenda*" between the contacts. Contact 1 keeps pumping in electrons striving to bring the number up from N to f_1, while contact 2 keeps pulling them out striving to bring it down to f_2. The net effect is a continuous transfer of electrons from contact 1 to 2 corresponding to a current I in the external circuit (Fig. 1.2.2). Note that the current is in a direction opposite to that of the flux of electrons, since electrons have negative charge.

It should now be clear why the process of conduction requires the presence of states around $E = \mu$. It does not matter if the states are empty (n-type, Fig. 1.1.2) or filled (p-type, Fig. 1.1.4) in equilibrium, before a drain voltage is applied. With empty states, electrons are first injected by the negative contact and subsequently collected by the positive contact. With filled states, electrons are first collected by the positive contact and subsequently refilled by the negative contact. Either way, we have current flowing in the external circuit in the same direction.

Inflow/outflow: Eqs. (1.2.3a, b) look elementary and I seldom hear anyone question them. But they hide many subtle issues that could bother more advanced readers and so I feel obliged to mention these issues briefly. I realize that I run the risk of confusing "satisfied" readers who may want to skip the rest of this section.

The right-hand sides of Eqs. (1.2.3a, b) can be interpreted as the difference between the influx and the outflux from the source and drain respectively (see Fig. 1.2.2). For example, consider the source. The outflux of $\gamma_1 N/\hbar$ is easy to justify since $\gamma_1/\hbar$ represents the rate at which an electron placed initially in the level ε will escape into the source contact. But the influx $\gamma_1 f_1/\hbar$ is harder to justify since there are many electrons in many states in the contacts, all seeking to fill up one state inside the channel and it is not obvious how to sum up the inflow from all these states. A convenient approach is to use a thermodynamic argument as follows. If the channel were in equilibrium with the source, there would be no net flux, so that the influx would equal the outflux. But the outflux under equilibrium conditions would equal $\gamma_1 f_1/\hbar$ since N would equal f_1. Under non-equilibrium conditions, N differs from f_1 but the influx remains unchanged since it depends only on the condition in the contacts which remains unchanged (note that the outflux does change giving a net current that we have calculated above).

"Pauli blocking"? Advanced readers may disagree with the statement I just made, namely that the influx "depends only on the condition in the contacts." Shouldn't the influx be reduced by the presence of electrons in the channel due to the exclusion principle ("Pauli blocking")? Specifically one could argue that the inflow and outflow (at the source contact) be identified respectively as

$$\gamma_1 f_1(1 - N) \quad \text{and} \quad \gamma_1 N(1 - f_1)$$

instead of

$$\gamma_1 f_1 \quad \text{and} \quad \gamma_1 N$$

as we have indicated in Fig. 1.2.2. It is easy to see that the net current given by the difference between inflow and outflow is the same in either case, so that the argument might appear "academic." What is not academic, however, is the level broadening that accompanies the process of coupling to the contacts, something we need to include in order to get quantitatively correct results (as we will see in the next section). I have chosen to define inflow and outflow in such a way that the outflow per electron

($\gamma_1 = \gamma_1 N/N$) is equal to the broadening (in addition to their difference being equal to the net current). Whether this broadening (due to the source) is γ_1 or $\gamma_1(1 - f_1)$ or something else is not an academic question. It can be shown that as long as energy relaxing or inelastic interactions are not involved in the inflow/outflow process, the broadening is γ_1, independent of the occupation factor f_1 in the contact. We will discuss this point a little further in Chapters 9 and 10, but a proper treatment requires advanced formalism as described in the appendix.

1.3 The quantum of conductance

Consider a device with a small voltage applied across it causing a splitting of the source and drain electrochemical potentials (Fig. 1.3.1a). We can write the current through this device from Eq. (1.2.5) and simplify it by assuming $\mu_1 > \varepsilon > \mu_2$ and the temperature is low enough that $f_1(\varepsilon) \equiv f_0(\varepsilon - \mu_1) \approx 1$ and $f_2(\varepsilon) \equiv f_0(\varepsilon - \mu_2) \approx 0$:

$$I = \frac{q}{\hbar}\frac{\gamma_1\gamma_2}{\gamma_1 + \gamma_2} = \frac{q\gamma_1}{2\hbar} \quad \text{if} \quad \gamma_2 = \gamma_1 \tag{1.3.1a}$$

This suggests that we could pump unlimited current through this one-level device by increasing γ_1 ($= \gamma_2$), that is by coupling it more and more strongly to the contacts. However, one of the seminal results of mesoscopic physics is that the maximum conductance of a one-level device is equal to G_0 (see Eq. (1.1)). What have we missed?

What we have missed is the broadening of the level that inevitably accompanies any process of coupling to it. This causes part of the energy level to spread outside the energy range between μ_1 and μ_2 where current flows. The actual current is then reduced below what we expect from Eq. (1.3.1a) by a factor $(\mu_1 - \mu_2)/C\gamma_1$ representing the fraction

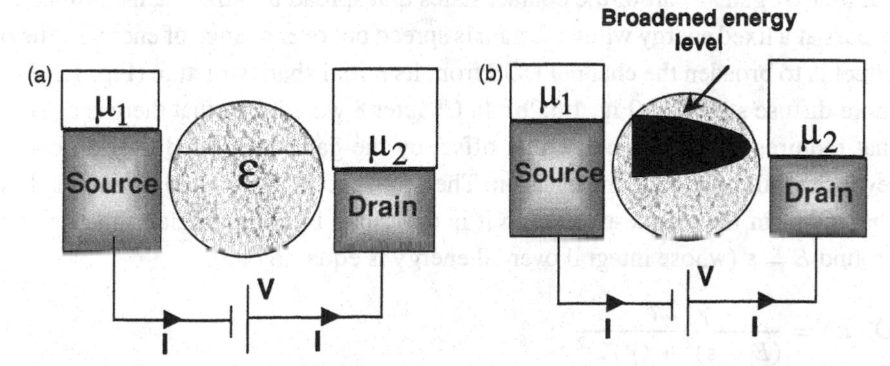

Fig. 1.3.1 (a) A channel with a small voltage applied across it causing a splitting of the source and drain electrochemical potentials $\mu_1 > \varepsilon > \mu_2$. (b) The process of coupling to the channel inevitably broadens the level, thereby spreading part of the energy level outside the energy range between μ_1 and μ_2 where current flows.

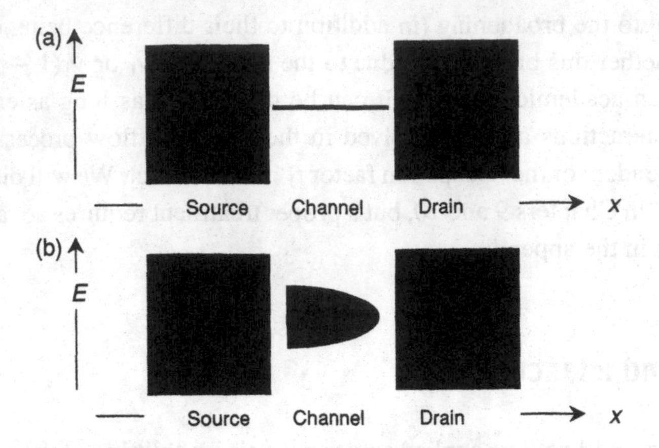

Fig. 1.3.2

of the level that lies in the window between μ_1 and μ_2, where $C\gamma_1$ is the effective width of the level, C being a numerical constant. Since $\mu_1 - \mu_2 = qV_D$, we see from Eq. (1.3.1b)

$$I = \frac{q\gamma_1}{2\hbar}\frac{qV_D}{C\gamma_1} \rightarrow G = \frac{I}{V_D} = \frac{q^2}{2C\hbar} \tag{1.3.1b}$$

that the conductance indeed approaches a constant value independent of the strength of the coupling ($\gamma_1 = \gamma_2$) to the contacts. We will now carry out this calculation a little more quantitatively so as to obtain a better estimate for C.

One way to understand this broadening is to note that, *before* we couple the channel to the source and the drain, the density of states (DOS) $D(E)$ looks something like Fig. 1.3.2a (dark indicates a high DOS). We have one sharp level in the channel and a continuous distribution of states in the source and drain contacts. On coupling, these states "spill over": the channel "loses" part of its state as it spreads into the contacts, but it also "gains" part of the contact states that spread into the channel. Since the loss occurs at a fixed energy while the gain is spread out over a range of energies, the overall effect is to broaden the channel DOS from its initial sharp structure (Fig. 1.3.2a) into a more diffuse structure (Fig. 1.3.2b). In Chapter 8 we will see that there is a "sum rule" that requires the loss to be exactly offset by the gain. Integrated over all energy, the level can still hold only one electron. The broadened DOS could in principle have any shape, but in the simplest situation it is described by a Lorentzian function centered around $E = \varepsilon$ (whose integral over all energy is equal to one):

$$D_\varepsilon(E) = \frac{\gamma/2\pi}{(E-\varepsilon)^2 + (\gamma/2)^2} \tag{1.3.2}$$

The initial delta function can be represented as the limiting case of $D_\varepsilon(E)$ as the broadening tends to zero: $\gamma \rightarrow 0$ (Fig. 1.3.3). The broadening γ is proportional to the strength of the coupling as we might expect. Indeed it turns out that $\gamma = \gamma_1 + \gamma_2$, where

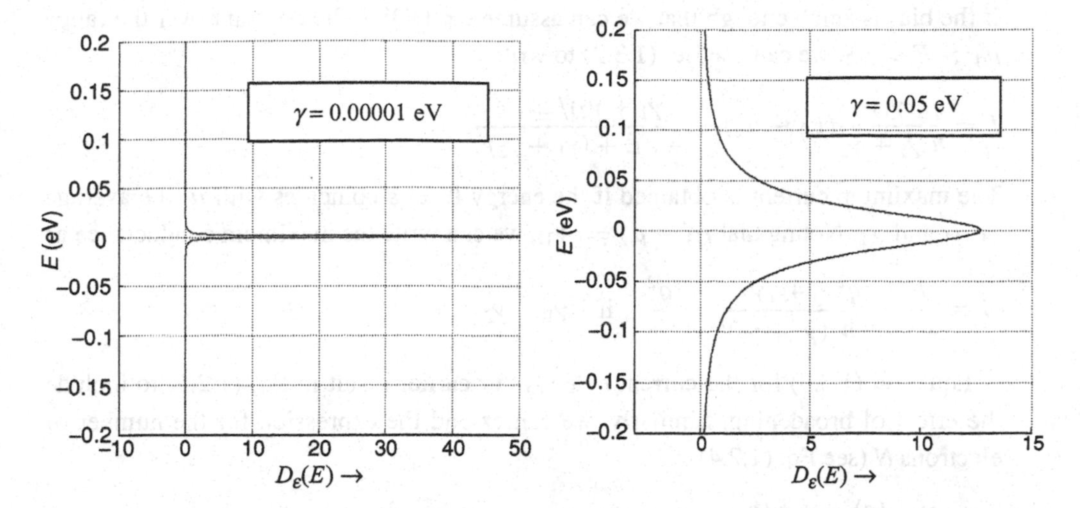

Fig. 1.3.3 An energy level at $E = \varepsilon$ is broadened into a continuous density of states $D_\varepsilon(E)$ by the process of coupling to the contacts. $D_\varepsilon(E)$ curves for two different values of coupling γ with $\varepsilon = 0$ eV are shown.

$\gamma_1/\hbar$ and $\gamma_2/\hbar$ are the escape rates introduced in Section 1.2. This will come out of our quantum mechanical treatment in Chapter 8, but at this stage we could rationalize it as a consequence of the "uncertainty principle" that requires the product of the lifetime ($= \hbar/\gamma$) of a state and its spread in energy (γ) to equal $\hbar$. Note that in general the lineshape need not be Lorentzian and this is usually reflected in an energy-dependent broadening $\gamma(E)$.

Anyway, the bottom line is that the coupling to the contacts broadens a single discrete energy level into a continuous density of states given by Eq. (1.3.2) and we can include this effect by modifying our expression for the current (Eq. (1.2.5))

$$I = \frac{q}{\hbar}\frac{\gamma_1\gamma_2}{\gamma_1+\gamma_2}[f_1(\varepsilon) - f_2(\varepsilon)]$$

to integrate (that is, sum) over a distribution of states, $D_\varepsilon(E)\,\mathrm{d}E$:

$$I = \frac{q}{\hbar}\int_{-\infty}^{+\infty} \mathrm{d}E\, D_\varepsilon(E)\frac{\gamma_1\gamma_2}{\gamma_1+\gamma_2}[f_1(E) - f_2(E)] \tag{1.3.3}$$

At low temperatures, we can write

$$\begin{aligned} f_1(E) - f_2(E) &= 1 \quad \text{if} \quad \mu_1 > E > \mu_2 \\ &= 0 \quad \text{otherwise} \end{aligned}$$

so that the current is given by

$$I = \frac{q}{\hbar}\frac{\gamma_1\gamma_2}{\gamma_1+\gamma_2}\int_{\mu_2}^{\mu_1} \mathrm{d}E\, D_\varepsilon(E)$$

If the bias is small enough that we can assume the DOS to be constant over the range $\mu_1 > E > \mu_2$, we can use Eq. (1.3.2) to write

$$I = \frac{q}{\hbar}\frac{\gamma_1\gamma_2}{\gamma_1+\gamma_2}(\mu_1-\mu_2)\frac{(\gamma_1+\gamma_2)/2\pi}{(\mu-\varepsilon)^2+(\gamma_1+\gamma_2)^2}$$

The maximum current is obtained if the energy level ε coincides with μ, the average of μ_1 and μ_2. Noting that $\mu_1 - \mu_2 = qV_D$, we can write the maximum conductance as

$$G \equiv \frac{I}{V_D} = \frac{q^2}{h}\frac{4\gamma_1\gamma_2}{(\gamma_1+\gamma_2)^2} = \frac{q^2}{h} \quad \text{if} \quad \gamma_1 = \gamma_2$$

Equation (1.3.3) for the current extends our earlier result in Eq. (1.2.5) to include the effect of broadening. Similarly, we can extend the expression for the number of electrons N (see Eq. (1.2.4))

$$N = \frac{\gamma_1 f_1(\varepsilon) + \gamma_2 f_2(\varepsilon)}{\gamma_1+\gamma_2}$$

to account for the broadened DOS:

$$N = \int_{-\infty}^{+\infty} \mathrm{d}E\, D_\varepsilon(E)\frac{\gamma_1 f_1(E) + \gamma_2 f_2(E)}{\gamma_1+\gamma_2} \tag{1.3.4}$$

1.4 Potential profile

Physicists often focus on the low-bias conductance ("linear response"), which is determined solely by the properties of the energy levels around the equilibrium electrochemical potential μ. What is not widely appreciated is that this is not enough if we are interested in the full current–voltage characteristics. It is then important to pay attention to the actual potential inside the channel in response to the voltages applied to the external electrodes (source, drain, and gate). To see this, consider a one-level channel with an equilibrium electrochemical potential μ located slightly above the energy level ε as shown in Fig. 1.4.1. When we apply a voltage between the source and drain, the electrochemical potentials separate by qV: $\mu_1 - \mu_2 = qV$. We know that a current flows (at low temperatures) only if the level ε lies between μ_1 and μ_2. Depending on how the energy level ε is affected by the applied voltage, we have different possibilities.

If we ignore the gate we might expect the potential in the channel to lie halfway between the source and the drain, $\varepsilon \rightarrow \varepsilon - (V/2)$, leading to Fig. 1.4.2 for positive and negative voltages (note that we are assuming the source potential to be held constant, relative to which the other potentials are changing). It is apparent that the energy level

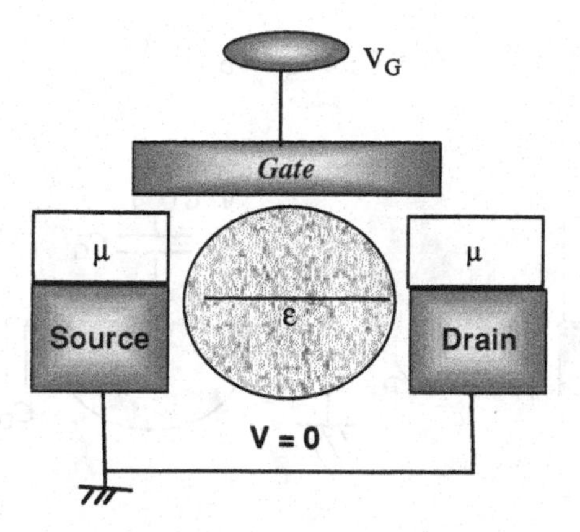

Fig. 1.4.1

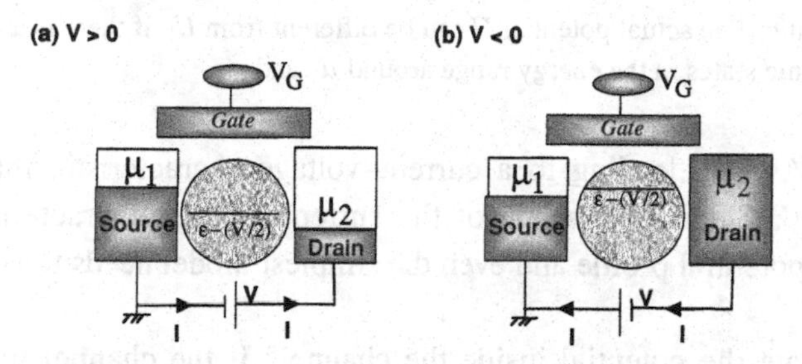

Fig. 1.4.2 Energy level diagram under (a) forward ($V > 0$) and (b) reverse ($V < 0$) bias, assuming that the channel potential lies halfway between the source and the drain.

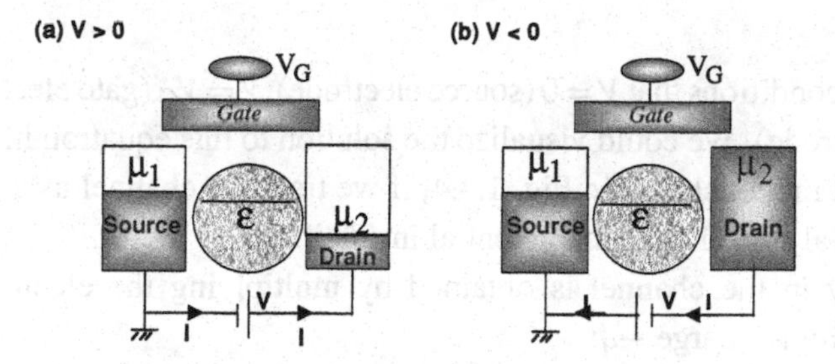

Fig. 1.4.3 Energy level diagram under (a) forward ($V > 0$) and (b) reverse ($V < 0$) bias assuming that the channel potential remains fixed with respect to the source.

lies halfway between μ_1 and μ_2 for either bias polarity ($V > 0$ or $V < 0$), leading to equal magnitudes of current for $+V$ and $-V$.

A different picture emerges if we assume that the gate is so closely coupled to the channel that the energy level follows the gate potential and is unaffected by the drain voltage or, in other words, ε remains fixed with respect to the source (Fig. 1.4.3). In this case the energy level lies between μ_1 and μ_2 for positive bias ($V > 0$) but

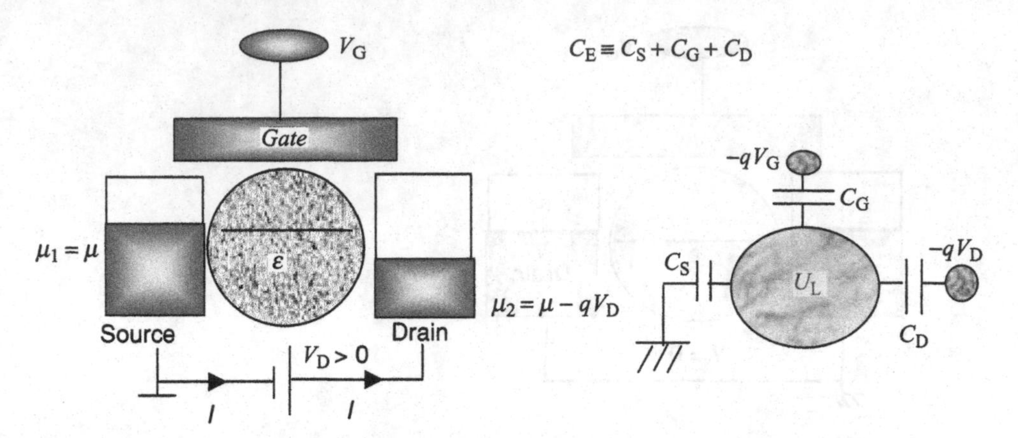

Fig. 1.4.4 A simple capacitive circuit model for the "Laplace" potential U_L of the active region in response to the external gate and drain voltages, V_G and V_D. The total capacitance is denoted C_E, where E stands for electrostatic. The actual potential U can be different from U_L if there is a significant density of electronic states in the energy range around μ_1 and μ_2.

not for negative bias ($V < 0$), leading to a current–voltage characteristic that can be very *asymmetric* in V. Clearly the shape of the current–voltage characteristic is affected strongly by the potential profile and even the simplest model needs to account for it.

So how do we calculate the potential inside the channel? If the channel were an insulator, we could solve Laplace's equation (ε_r is the relative permittivity, which could be spatially varying)

$$\vec{\nabla} \cdot (\varepsilon_r \vec{\nabla} V) = 0$$

subject to the boundary conditions that $V = 0$ (source electrode), $V = V_G$ (gate electrode), and $V = V_D$ (drain electrode). We could visualize the solution to this equation in terms of the capacitive circuit model shown in Fig. 1.4.4, if we treat the channel as a single point ignoring any spatial variation of the potential inside it.

The potential energy in the channel is obtained by multiplying the electrostatic potential V by the electronic charge $-q$:

$$U_L = \frac{C_G}{C_E}(-qV_G) + \frac{C_D}{C_E}(-qV_D) \tag{1.4.1a}$$

Here we have labeled the potential energy with a subscript L as a reminder that it is calculated from the Laplace equation ignoring any change in the electronic charge, which is justified if there are very few electronic states in the energy range around μ_1 and μ_2. Otherwise there is a change $\Delta\rho$ in the electron density in the channel and we need to solve the Poisson equation

$$\vec{\nabla} \cdot (\varepsilon_r \vec{\nabla} V) = -\Delta\rho/\varepsilon_0$$

for the potential. In terms of our capacitive circuit model, we could write the change in the charge as a sum of the charges on the three capacitors:

$$-q\Delta N = C_S V + C_G(V - V_G) + C_D(V - V_D)$$

so that the potential energy $U = -qV$ is given by the sum of the Laplace potential and an additional term proportional to the change in the number of electrons:

$$U = U_L + \frac{q^2}{C_E}\Delta N \qquad (1.4.1b)$$

The constant $q^2/C_E \equiv U_0$ tells us the change in the potential energy due to *one* extra electron and is called the single-electron charging energy, whose significance we will discuss further in the next section. The *change* ΔN in the number of electrons is calculated with respect to the reference number of electrons, originally in the channel, N_0, corresponding to which the energy level is believed to be located at ε.

Iterative procedure for self-consistent solution: For a small device, the effect of the potential U is to raise the DOS in energy and can be included in our expressions for the number of electrons N (Eq. (1.3.4)) and the current I (Eq. (1.3.3)) in a straightforward manner:

$$N = \int_{-\infty}^{+\infty} dE\, D_\varepsilon(E - U)\frac{\gamma_1 f_1(E) + \gamma_2 f_2(E)}{\gamma_1 + \gamma_2} \qquad (1.4.2)$$

$$I = \frac{q}{\hbar}\int_{-\infty}^{+\infty} dE\, D_\varepsilon(E - U)\frac{\gamma_1\gamma_2}{\gamma_1 + \gamma_2}[f_1(E) - f_2(E)] \qquad (1.4.3)$$

Equation (1.4.2) has a U appearing on its right-hand side, which in turn is a function of N through the electrostatic relation (Eq. (1.4.1b)). This requires a simultaneous or "self-consistent" solution of the two equations which is usually carried out using the iterative procedure depicted in Fig. 1.4.5.

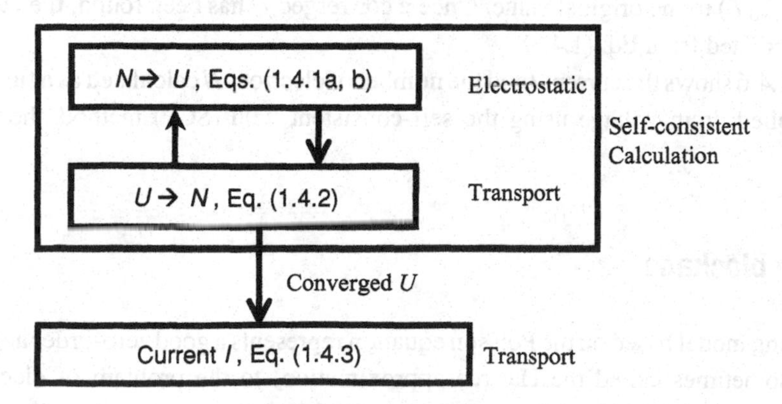

Fig. 1.4.5 Iterative procedure for calculating N and U self-consistently.

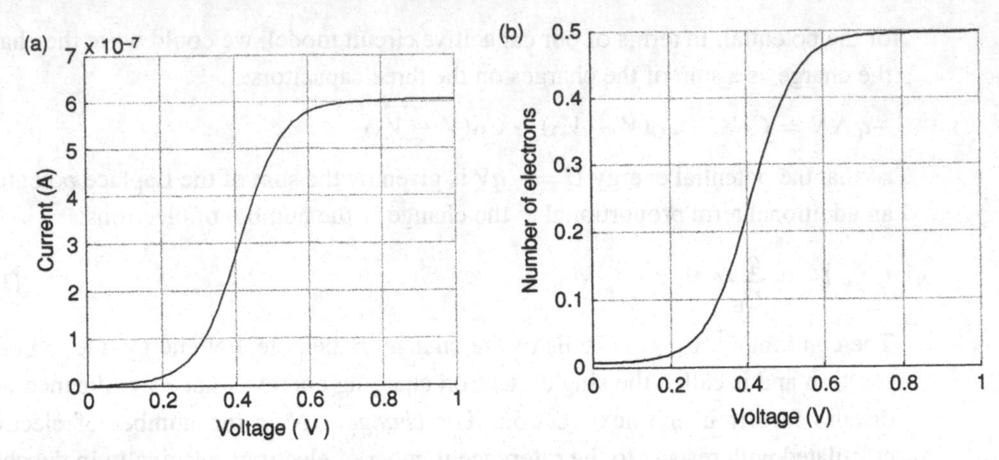

Fig. 1.4.6 (a) Current vs. voltage calculated using the SCF method (Fig. 1.4.5) with $\mu = 0$, $\varepsilon = 0.2\,\text{eV}$, $V_G = 0$, $k_BT = 0.025\,\text{eV}$, $U_0 = 0.025\,\text{eV}$, $C_D/C_E = 0.5$, and $\gamma_1 = \gamma_2 = 0.005\,\text{eV}$. (b) Number of electrons vs. voltage calculated using the SCF method (Fig. 1.4.5) with same parameters as in (a).

We start with an initial guess for U, calculate N from Eq. (1.4.2) with $D_\varepsilon(E)$ given by Eq. (1.3.2), calculate an appropriate U from Eq. (1.4.1b), with U_L given by Eq. (1.4.1a) and compare with our starting guess for U. If this new U is not sufficiently close to our original guess, we revise our guess using a suitable algorithm, say something like

$$U_{\text{new}} = U_{\text{old}} + \alpha(U_{\text{calc}} - U_{\text{old}}) \tag{1.4.4}$$

New guess — Old guess — Calculated

where α is a positive number (typically < 1) that is adjusted to be as large as possible without causing the solution to diverge (which is manifested as an increase in $U_{\text{calc}} - U_{\text{old}}$ from one iteration to the next). The iterative process has to be repeated till we find a U that yields an N that leads to a new U which is sufficiently close (say within a fraction of k_BT) to the original value. Once a converged U has been found, the current can be calculated from Eq. (1.4.3).

Figure 1.4.6 shows the current I and the number of electrons N calculated as a function of the applied drain voltage using the self-consistent field (SCF) method shown in Fig. 1.4.5.

1.5 Coulomb blockade

The charging model based on the Poisson equation represents a good zero-order approximation (sometimes called the Hartree approximation) to the problem of electron–electron interactions, but it is generally recognized that it tends to overestimate the

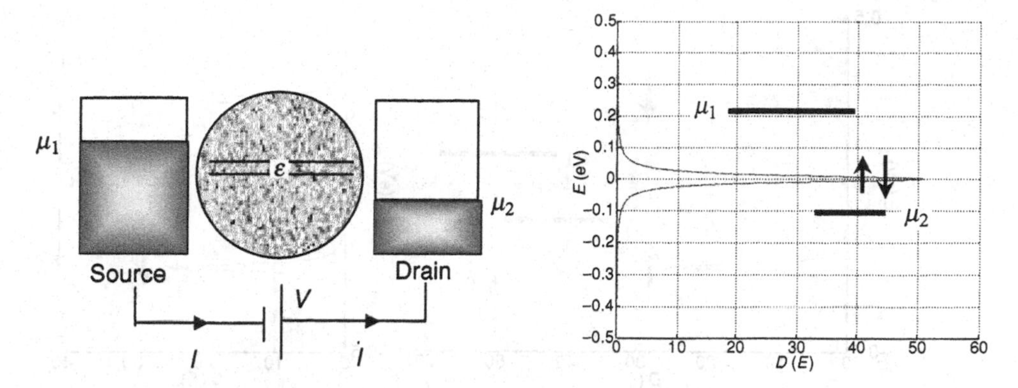

Fig. 1.5.1 A channel with two spin-degenerate levels containing one electron is expected to have an equilibrium electrochemical potential that lies in the center of its broadened density of states, so that current should flow easily under bias ($\gamma = 0.05$ eV).

effect and may need to be corrected (the so-called exchange and correlation effects). Discovering an appropriate function $U(N)$ (if there is one!) to replace our simple result (cf. Eq. (1.4.1b))

$$U(N) = q^2(N - N_0)/C_E$$

is arguably one of the central topics in many-electron physics and can in some cases give rise to profound effects like magnetism, which are largely outside the scope of this book. However, there is one aspect that I would like to mention right away, since it can affect our picture of current flow even for a simple one-level device and put it in the so-called Coulomb blockade or single-electron charging regime. Let me explain what this means.

Energy levels come in pairs, one up-spin and one down-spin, which are degenerate, that is they have the same energy. Usually this simply means that all our results have to be multiplied by two. Even the smallest device has two levels rather than one, and its maximum conductance will be twice the conductance quantum $G_0 \equiv q^2/h$ discussed earlier. The expressions for the number of electrons and the current should all be multiplied by two. However, there is a less trivial consequence that I would like to explain.

Consider a channel with two spin-degenerate levels (Fig. 1.5.1), containing one electron when neutral ($N_0 = 1$). We expect the broadened DOS to be twice our previous result (see Eq. (1.3.2))

$$D_\varepsilon(E) = 2\ (\text{for spin}) \times \frac{\gamma/2\pi}{(E - \varepsilon)^2 + (\gamma/2)^2} \tag{1.5.1}$$

where the total broadening is the sum of those due to each of the two contacts individually: $\gamma = \gamma_1 + \gamma_2$, as before. Since the available states are only half filled for a neutral

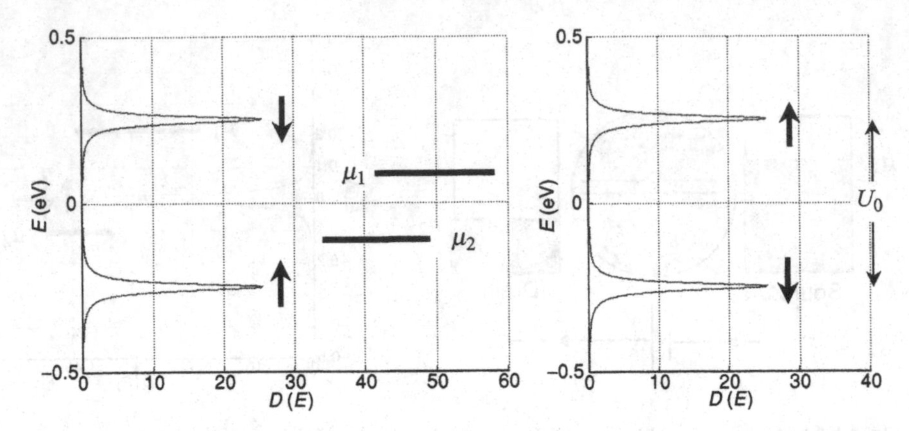

Fig. 1.5.2 Under certain conditions, the up-spin and down-spin density of states splits into two parts separated by the single-electron charging energy, U_0, instead of one single degenerate peak as shown in Fig. 1.5.1 ($\gamma = 0.05$ eV, $U_0 = 0.25$ eV).

channel, the electrochemical potential will lie exactly in the middle of the broadened DOS, so that we would expect a lot of current to flow when a bias is applied to split the electrochemical potentials in the source and drain as shown.

However, under certain conditions the DOS looks like one of the two possibilities shown in Fig. 1.5.2. The up-spin and the down-spin density of states splits into two parts separated by the single-electron charging energy

$$U_0 \equiv q^2/C_E \tag{1.5.2}$$

Very little current flows when we apply a small bias since there are hardly any states between μ_1 and μ_2 and this "Coulomb blockade" has been experimentally observed for systems where the charging energy U_0 exceeds the broadening γ.

It is hard to understand why the two peaks should separate based on the simple SCF picture. Two peaks with the same energy ("degenerate") should always remain degenerate as long as they feel the same self-consistent potential U. The point is that *no electron feels any potential due to itself*. Suppose the up-spin level gets filled first, causing the down-spin level to float up by U_0. But the up-spin level does not float up because it does not feel any self-interaction, leading to the picture shown on the left in Fig. 1.5.2. Of course, it is just as likely that the down-spin will fill up first leading to the picture on the right. In either case the DOS near μ is suppressed relative to the SCF picture (Fig. 1.5.1).

Describing the flow of current in this Coulomb blockade regime requires a very different point of view that we will not discuss in this book, except briefly in Section 3.4. But when do we have to worry about Coulomb blockade effects? Answer: only if U_0 exceeds both k_BT and γ $(= \gamma_1 + \gamma_2)$. Otherwise, the SCF method will give results that look much like those obtained from the correct treatment (see Fig. 3.4.3).

So what determines U_0? Answer: the extent of the electronic wavefunction. If we smear out one electron over the surface of a sphere of radius R, then we know from freshman physics that the potential of the sphere will be $q/4\pi\varepsilon_r\varepsilon_0 R$, so that the energy needed to put another electron on the sphere will be $q^2/4\pi\varepsilon_r\varepsilon_0 R \cong U_0$, which is ~0.025 eV if $R = 5$ nm and $\varepsilon_r = 10$. Levels with well-delocalized wavefunctions (large R) have a very small U_0 and the SCF method provides an acceptable description even at the lowest temperatures of interest. But if R is small, then the charging energy U_0 can exceed $k_B T$ and one could be in a regime dominated by single-electron charging effects that is not described well by the SCF method.

1.6 Towards Ohm's law

Now that we have discussed the basic physics of electrical conduction through small conductors, let us talk about the new factors that arise when we have large conductors. In describing electronic conduction through small conductors we can identify the following three regimes.

- *Self-consistent field (SCF) regime.* If $k_B T$ and/or γ is comparable to U_0, we can use the SCF method described in Section 1.4.
- *Coulomb blockade (CB) regime.* If U_0 is well in excess of both $k_B T$ and γ, the SCF method is not adequate. More correctly, one could use (under certain conditions) the multi-electron master equation that we will discuss in Section 3.4.
- *Intermediate regime.* If U_0 is comparable to the larger of $k_B T$ and γ, there is no simple approach: the SCF method does not do justice to the charging, while the master equation does not do justice to the broadening.

It is generally recognized that the intermediate regime can lead to novel physics that requires advanced concepts, even for the small conductors that we have been discussing. For example, experimentalists have seen evidence for the Kondo effect, which is reflected as an extra peak in the density of states around $E = \mu$ in addition to the two peaks (separated by U_0) that are shown in Fig. 1.5.2.

With large conductors too we can envision three regimes of transport that evolve out of these three regimes. We could view a large conductor as an array of unit cells as shown in Fig. 1.6.1. The inter-unit coupling energy t has an effect somewhat (but not exactly) similar to the broadening γ that we have associated with the contacts. If $t \geq U_0$, the overall conduction will be in the SCF regime and can be treated using an extension of the SCF method from Section 1.4. If $t \ll U_0$, it will be in the CB regime and can in principle be treated using the multi-electron master equation (to be discussed in Section 3.4), if quantum interference between energy levels can be neglected, perhaps due to level spacing or due to phase-breaking processes of the type to be discussed in Chapter 10. On the other hand, large conductors with $\gamma_s \ll t \leq U_0$ belong to an intermediate regime that presents major theoretical challenges, giving rise to intriguing

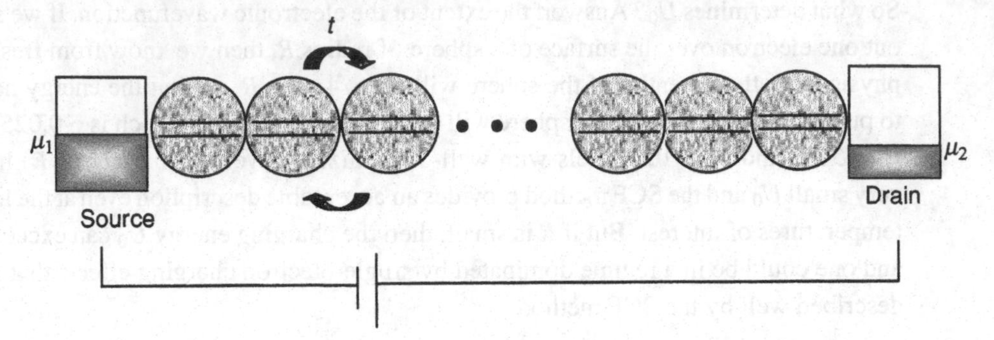

Fig. 1.6.1 A large conductor can be viewed as an array of unit cells. If the conductor is extended in the transverse plane, we should view each unit cell as representing an array of unit cells in the transverse direction.

possibilities. Indeed many believe that high-T_c superconductors (whose microscopic theory is still controversial) consist of unit cells whose coupling is delicately balanced on the borderline of the SCF and the CB regimes.

The more delocalized the electronic wavefunctions (large t), the more accurate the SCF description becomes and in this book I will focus on this regime. Basically I will try to explain how the simple one-level description from Section 1.4 is extended to larger conductors all the way to a nanotransistor, within the SCF picture that accounts for electron–electron interactions through an average potential $U(r)$ that one electron feels due to the other electrons.

Summary of results for one-level conductors: We have developed a model for current flow through a one-level device, starting with a simple discrete level (ε) in Section 1.2 and then extending it to include the broadening of the level into a Lorentzian density of states in Section 1.3

$$D_\varepsilon(E) = 2\,(\text{for spin}) \times \frac{\gamma/2\pi}{(E-\varepsilon)^2 + (\gamma/2)^2} \qquad \gamma \equiv \gamma_1 + \gamma_2 \tag{1.6.1}$$

and the self-consistent potential in Section 1.4

$$U = U_L + U_0(N - N_0) \tag{1.6.2}$$

$$U_L = \frac{C_G}{C_E}(-qV_G) + \frac{C_D}{C_E}(-qV_D)$$

$$U_0 = q^2/C_E \qquad C_E = C_G + C_S + C_D \tag{1.6.3}$$

The number of electrons N is given by (restricted SCF method)

$$N = \int_{-\infty}^{+\infty} \mathrm{d}E\, n(E)$$

where

$$n(E) = D(E - U)\left(\frac{\gamma_1}{\gamma} f_1(E) + \frac{\gamma_2}{\gamma} f_2(E)\right) \tag{1.6.4}$$

while the currents at the two terminals are given by

$$I_1 = \frac{q}{\hbar} \int_{-\infty}^{+\infty} \mathrm{d}E\, \gamma_1 [D(E - U) f_1(E) - n(E)] \tag{1.6.5a}$$

$$I_2 = \frac{q}{\hbar} \int_{-\infty}^{+\infty} \mathrm{d}E\, \gamma_2 [D(E - U) f_2(E) - n(E)] \tag{1.6.5b}$$

At steady state, the sum of the two currents is equated to zero to eliminate $n(E)$:

$$I = \frac{q}{h} \int_{-\infty}^{+\infty} \mathrm{d}E\, \overline{T}(E) [f_1(E) - f_2(E)]$$

where

$$\overline{T}(E) = D(E - U) 2\pi \gamma_1 \gamma_2 / \gamma \tag{1.6.6}$$

is called the *transmission* – a concept that plays a central role in the transmission formalism widely used in mesoscopic physics (see Section 9.4). Note that the Fermi functions f_1 and f_2 are given by

$$f_1(E) = f_0(E - \mu_1)$$
$$f_2(E) = f_0(E - \mu_2) \tag{1.6.7}$$

where $f_0(E) \equiv [1 + \exp(E/k_B T)]^{-1}$ and the electrochemical potentials in the source and drain contacts are given by

$$\mu_1 = \mu$$
$$\mu_2 = \mu - q V_D \tag{1.6.8}$$

μ being the equilibrium electrochemical potential.

Note that in Eqs. (1.6.4) through (1.6.6) I have used $D(E)$ instead of $D_\varepsilon(E)$ to denote the DOS. Let me explain why.

Large conductors – a heuristic approach: $D_\varepsilon(E)$ (see Eq. (1.6.1)) is intended to denote the DOS obtained by broadening a single discrete level ε, while $D(E)$ denotes the DOS in general for a multi-level conductor with many energy levels (Fig. 1.6.2).

If we make the rather cavalier assumption that all levels conduct independently, then we could use exactly the same equations as for the one-level device, replacing the

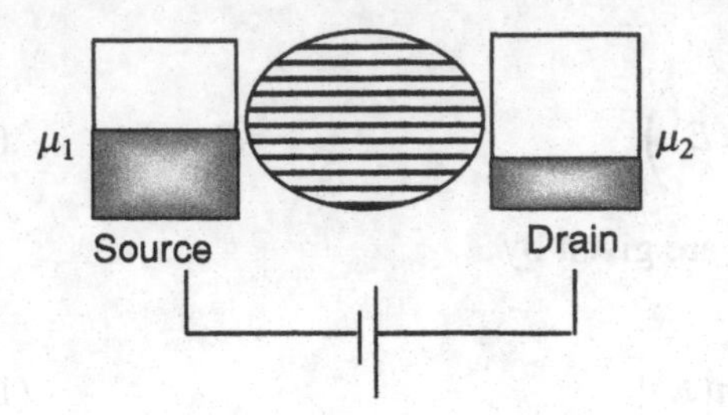

Fig. 1.6.2

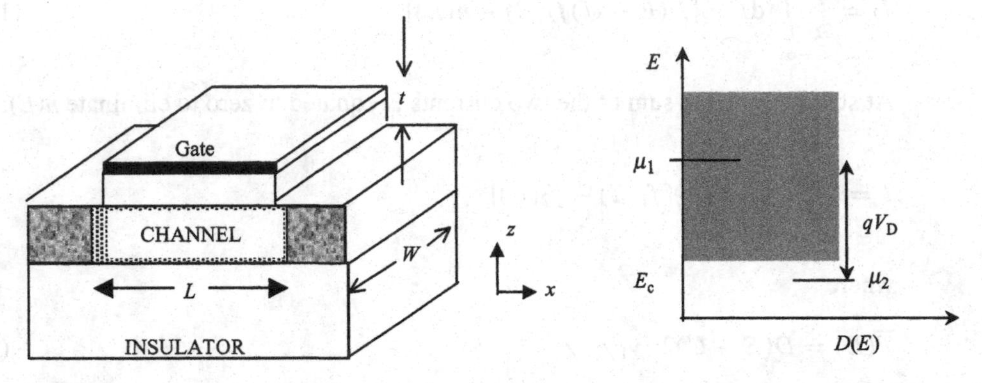

Fig. 1.6.3

one-level DOS $D_\varepsilon(E)$ in Eq. (1.6.1) with the total DOS $D(E)$. With this in mind, I will refer to Eqs. (1.6.4)–(1.6.6) as the *independent level model* for the current through a channel.

Nanotransistor – a simple model: As an example of this independent level model, let us model the nanotransistor shown in Fig. 1.1 by writing the DOS as (see Fig. 1.6.3, W is the width in the y-direction)

$$D(E) = m_c WL/\pi\hbar^2 \vartheta(E - E_c) \tag{1.6.9}$$

making use of a result that we will discuss in Chapter 6, namely that the DOS per unit area in a large two-dimensional (2D) conductor described by an effective mass m_c is equal to $m_c/\pi\hbar^2$, for energies greater than the energy E_c of the conduction band edge. The escape rates can be written down assuming that electrons are removed by the contact with a velocity v_R (somewhat like a "surface recombination velocity"):

$$\gamma_1 = \gamma_2 = \hbar v_R/L \tag{1.6.10}$$

The current–voltage relations shown in Fig. 1.1.1 were obtained using these model parameters: $m_c = 0.25m$, $C_G = 2\varepsilon_r\varepsilon_0 WL/t$, $C_S = C_D = 0.05C_G$, $W = 1$ μm,

$L = 10$ nm, insulator thickness $t = 1.5$ nm, $v_R = 10^7$ cm/s. At high drain voltages V_D the current saturates when μ_2 drops below E_c since there are no additional states to contribute to the current. Note that the gate capacitance C_G is much larger than the other capacitances, which helps to hold the channel potential fixed relative to the source as the drain voltage is increased (see Eq. (1.6.3)). Otherwise, the bottom of the channel density of states, E_c will "slip down" with respect to μ_1 when the drain voltage is applied, so that the current will not saturate. The essential feature of a well-designed transistor is that the gate is much closer to the channel than L, allowing it to hold the channel potential constant despite the voltage V_D on the drain.

I should mention that our present model ignores the profile of the potential along the length of the channel, treating it as a little box with a single potential U given by Eq. (1.6.2). Nonetheless the results (Fig. 1.1.1) are surprisingly close to experiments/realistic models, because the current in well-designed nanotransistors is controlled by a small region in the channel near the source whose length can be a small fraction of the actual length L. Luckily we do not need to pin down the precise value of this fraction, since the present model gives the same current independent of L.

Ohm's law? Would this independent level model lead to Ohm's law if we were to calculate the low-bias conductance of a large conductor of length L and cross-sectional area S? Since the current is proportional to the DOS, $D(E)$ (see Eq. (1.6.5)), which is proportional to the volume SL of the conductor, it might seem that the conductance $G \sim SL$. However, the coupling to the contacts decreases inversely with the length L of the conductor, since the longer a conductor is, the smaller is its coupling to the contact. While the DOS goes up with the volume, the coupling to the contact goes down as $1/L$, so that the conductance

$$G \sim SL/L = S$$

However, Ohm's law tells us that the conductance should scale as S/L; we are predicting that it should scale as S. The reason is that we are really modeling a *ballistic* conductor, where electrons propagate freely, the only resistance arising from the contacts. The conductance of such a conductor is indeed independent of its length. The ohmic length dependence of the conductance comes from scattering processes within the conductor that are not yet included in our thinking.

Without scattering, we could write the coupling $\gamma = \hbar v/L$, v being the velocity of an electron as it "bounces" back and forth between contacts. But in the presence of scatterers (defects, impurities etc.) an electron would not bounce ballistically with velocity v. Rather it would perform a random walk, taking a time $\tau \sim L^2/D$ ($D \equiv$ Diffusion constant, not DOS) to get from one contact to the other. Consequently the energy broadening $\gamma = \hbar/\tau \sim \hbar D/L^2$, instead of $\hbar v/L$. Since the DOS $\sim$ SL, this

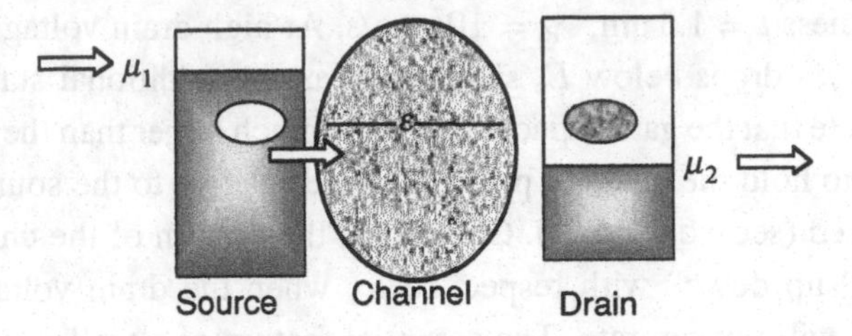

Fig. 1.6.4 When an electron goes from the source to the drain, an empty state or hole is left behind in the source, while an electron appears in the drain. Energy dissipating processes quickly take the electron down to μ_2 inside the drain and the hole up to μ_1 in the source. In our model we do not explicitly discuss these processes; we simply legislate that the contacts are maintained at equilibrium with the assumed electrochemical potentials.

leads to a conductance ~ S/L in keeping with Ohm's law. This seems like a simple argument, but it is important to note that the diffusive behavior we have invoked does not follow easily from a quantum mechanical description. Even if we were to refine our model appropriately to include "coherent" scatterers it would not lead to Ohm's law.

The full story requires us to include phase-breaking scattering processes that cause a change in the state of an external object. For example, if an electron gets deflected by a rigid (that is unchangeable) defect in the lattice, the scattering is said to be coherent. But if the electron transfers some energy to the atomic lattice causing it to start vibrating, that would constitute a phase-breaking or incoherent process.

Such incoherent scatterers are also needed to remove energy from the electrons and cause dissipation. For example, in this chapter we have developed a simple model that allows us to calculate the resistance R, but none of the associated Joule heat I^2R is dissipated in the channel; it is all dissipated in the contacts. This is evident if we consider what happens when an electron goes from the source to the drain (Fig. 1.6.4). An empty state or hole is left behind in the source at an energy lower than μ_1 while an electron appears in the drain at an energy higher than μ_2. Energy dissipating processes quickly take the electron down to μ_2 inside the drain and the hole up to μ_1 in the source. The overall effect is to take an electron from μ_1 in the source to μ_2 in the drain, and in our model the energy $(\mu_1 - \mu_2)$ is dissipated partly in the source and partly in the drain, but none in the channel. In the real world too there is experimental evidence that in nanoscale conductors, most of the heating occurs in the contacts outside the channel, allowing experimentalists to pump a lot more current through a small conductor without burning it up. But long conductors have significant incoherent scattering inside the channel and it is important to include it in our model.

The point is that the transition from ballistic conductors to Ohm's law has many subtleties that require a much deeper model for the flow of current than the independent level model (Eqs. (1.6.4)–(1.6.6)), although the latter can often provide an adequate description of short conductors. Let me now try to outline briefly the nature of this "deeper model" that we will develop in this book and illustrate with examples in Chapter 11.

Multi-level conductors – from numbers to matrices: The independent level model that we have developed in this chapter serves to identify the important concepts underlying the flow of current through a conductor, namely the location of the equilibrium *electrochemical potential* μ relative to the *density of states D(E)*, the *broadening* of the level $\gamma_{1,2}$ due to the coupling to contacts 1 and 2, the *self-consistent potential U* describing the effect of the external electrodes, and the change in the *number* of electrons *N*. In the general model for a multi-level conductor with *n* energy levels, each of these quantities is replaced by a corresponding matrix of size $(n \times n)$:

$\varepsilon \rightarrow [H]$	*Hamiltonian matrix*
$\gamma_{1,2} \rightarrow [\Gamma_{1,2}(E)]$	*Broadening matrix*
$2\pi D(E) \rightarrow [A(E)]$	*Spectral function*
$2\pi n(E) \rightarrow [G^{n}(E)]$	*Correlation function*
$U \rightarrow [U]$	*Self-consistent potential matrix*
$N \rightarrow [\rho] = \int (\mathrm{d}E/2\pi)[G^{n}(E)]$	*Density matrix*

Actually, the effect of the contacts is described by a "*self-energy*" *matrix*, $[\Sigma_{1,2}(E)]$, whose anti-Hermitian part is the broadening matrix: $\Gamma_{1,2} = \mathrm{i}[\Sigma_{1,2} - \Sigma_{1,2}^{+}]$. All quantities of interest can be calculated from these matrices. For example, in Section 1.2 we discussed the inflow/outflow of electrons from a one-level device. Figure 1.6.5 illustrates how these concepts are generalized in terms of these matrices. I should mention that in order to emphasize its similarity to the familiar concept of electron density, I have used $G^{n}(E)$ to denote what is usually written in the literature as $-\mathrm{i}G^{<}(E)$ following the non-equilibrium Green's function (NEGF) formalism pioneered by the works of Martin and Schwinger (1959), Kadanoff and Baym (1962) and Keldysh (1965).

Note that in the matrix model (Fig. 1.6.5b), I have added a third "contact" labeled "s-contact" representing scattering processes, without which we cannot make the transition to Ohm's law. Indeed it is only with the advent of mesoscopic physics in the 1980s that the importance of the contacts (Γ_1 and Γ_2) in interpreting experiments became widely recognized. Prior to that, it was common to ignore the contacts as minor experimental distractions and try to understand the physics of conduction in terms of the s-contact, though no one (to my knowledge) thought of scattering as a "contact" till Büttiker introduced the idea phenomenologically in the mid 1980s (see Büttiker, 1988; Datta, 1995). Subsequently, Datta (1989) showed from a microscopic model that

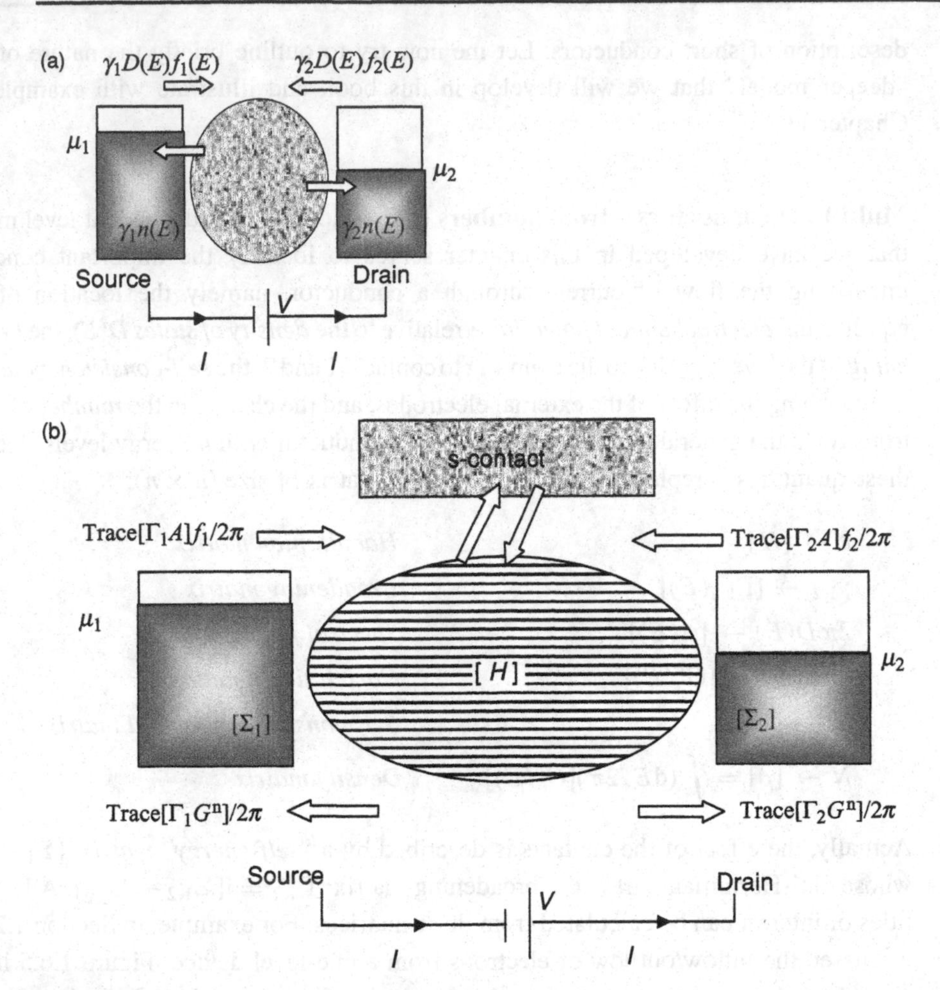

Fig. 1.6.5 From numbers to matrices: flux of electrons into and out of a device at the source and drain ends. (a) Simple result for independent level model, see Eqs. (1.6.4)–(1.6.6). (b) General matrix model, to be developed in this book. Without the "s-contact" this model is equivalent to Eq. (6) of Meir and Wingreen (1992). The "s-contact" distributed throughout the channel describes incoherent scattering processes (Datta, 1989). In general this "contact" cannot be described by a Fermi function, unlike the real contacts.

incoherent scattering processes act like a fictitious "contact" distributed throughout the channel that extracts and reinjects electrons. Like the real contacts, coupling to this "contact" too can be described by a broadening matrix Γ_s. However, unlike the real contacts, the scattering contact in general cannot be described by a Fermi function so that although the outflow is given by Trace[$\Gamma_s G^n/2\pi$], the inflow requires separate considerations that we will discuss in Chapter 10. The complete set of equations is summarized in Chapter 11.

The reader might wonder why the numbers become matrices, rather than just column vectors. For example, with one unit cell, we have an energy level ε. It seems reasonable that with many unit cells, we should talk about an energy level $\varepsilon(n)$ in each cell "n". But why do we need a matrix $H(m, n)$? This is a question that goes to the heart of quantum mechanics whereby all physical quantities are represented by matrices. We can find a representation that diagonalizes $[H]$ and in this representation we could write the energy eigenvalues as a column vector $\varepsilon(n)$. If all the other matrices were also approximately diagonal in this representation, then we could indeed work with column vectors with n elements rather than matrices with n^2 elements and that is what "semi-classical" methods commonly do. In general, no single representation will diagonalize all the matrices and a full quantum treatment is needed.

Figure 1.6.5b without the s-contact is often used to analyze small devices and in this form it is identical to the result obtained by Meir and Wingreen (1992, their Eq. (6)) following the method of Caroli *et al.* (1972) based on the NEGF formalism. In order to make this approach accessible to readers unfamiliar with advanced many-body physics, I will derive these results using elementary arguments. What we have derived in this chapter (Fig. 1.6.5a) could be viewed as a special case of this general formalism with all the matrices being (1×1) in size. Indeed if there is a representation that diagonalizes all the matrices, then the matrix model without the s-contact would follow quite simply from Fig. 1.6.5a. We could write down separate equations for the current through each diagonal element (or level) for this special representation, add them up and write the sum as a trace. The resulting equations would then be valid in any representation, since the trace is invariant under a change in basis. In general, however, the matrix model cannot be derived quite so simply since no single representation will diagonalize all the matrices. In Chapters 8–10, I have derived the full matrix model (Fig. 1.6.5b) using elementary quantum mechanics. In the appendix, I have provided a brief derivation of the same results using the language of second quantization, but here too I have tried to keep the discussion less "advanced" than the standard treatments available in the literature.

I should mention that the picture in Fig. 1.6.5 is not enough to calculate the current: additional equations are needed to determine the "density of states" $[A(E)]$ and the "electron density" $[G^n(E)]$. In our elementary model (Fig. 1.6.5a) we wrote down the density of states by "ansatz" (see Eq. (1.6.1)), but no separate equation was needed for the electron density which was evaluated by equating the currents (see derivation of Eq. (1.2.3) for a discrete level that was extended to obtain Eq. (1.6.4) for a broadened level). In the matrix model (Fig. 1.6.5b) too (without the s-contact), it was argued in Meir and Wingreen (1992) that $[G^n(\mathrm{E})]$ can be similarly eliminated if $[\Gamma_1]$ is equal to a constant times $[\Gamma_2]$. However, this can be true only for very short channels. Otherwise, the source end is distinct from the drain end, making $[\Gamma_1]$ a very different matrix from $[\Gamma_2]$ since they couple to different ends. We then need additional equations to determine both $[A(E)]$ and $[G^n(E)]$.

There is an enormous amount of physics behind all these matrices (both the diagonal and the off-diagonal elements) and we will introduce and discuss them in course of this book: the next five chapters are about $[H]$, Chapter 7 is about $[\rho]$, Chapter 8 is about $[\Sigma]$, Chapter 9 combines these concepts to obtain the inflow/outflow diagram shown in Fig. 1.6.5b, and Chapter 10 introduces the matrix Γ_s describing scattering to complete the model for dissipative quantum transport. Finally, in Chapter 11, we illustrate the full "machinery" using a series of examples chosen to depict the transition from ballistic transport to Ohm's law, or in other words, from the atom to the transistor.

After that rather long introduction, we are now ready to get on with the "details." We will start with the question of how we can write down the Hamiltonian $[H]$ for a given device, whose eigenvalues will tell us the energy levels. We will work our way from the hydrogen atom in Chapter 2 "up" to solids in Chapter 5 and then "down" to nanostructures in Chapter 6. Let us now start where quantum mechanics started, namely, with the hydrogen atom.

EXERCISES

E.1.1. Consider a channel with one spin-degenerate level assuming the following parameters: $\mu = 0$, $\varepsilon = 0.2\,\text{eV}$, $k_B T = 0.025\,\text{eV}$, $\gamma_1 = \gamma_2 = 0.005\,\text{eV}$. Calculate the current vs. drain voltage V_D assuming $V_G = 0$ with $U_L = -qV_D/2$ and $U_0 = 0.1\,\text{eV}$, $0.25\,\text{eV}$, using the SCF approach and compare with Fig. 1.4.6.

E.1.2. Calculate the current vs. gate and drain voltages for a nanotransistor as shown in Fig. 1.1.1 using the SCF equations summarized in Eqs. (1.6.2)–(1.6.7) with $D(E) = m_c WL/\pi\hbar^2$ and $\gamma_1 = \gamma_2 = \hbar v_R/L$ and the following parameters: $m_c = 0.25m$, $C_G = 2\varepsilon_r\varepsilon_0 WL/t$, $C_S = C_D = 0.05C_G$, $W = 1\,\mu m$, $L = 10\,\text{nm}$, insulator thickness, $t = 1.5\,\text{nm}$, $v_R = 10^7\,\text{cm/s}$.

E.1.3. Thermoelectric effect: In this chapter we have discussed the current that flows when a voltage is applied between the two contacts. In this case the current depends on the DOS near the Fermi energy and it does not matter whether the equilibrium Fermi energy μ_1 lies on (a) the lower end or (b) the upper end of the DOS:

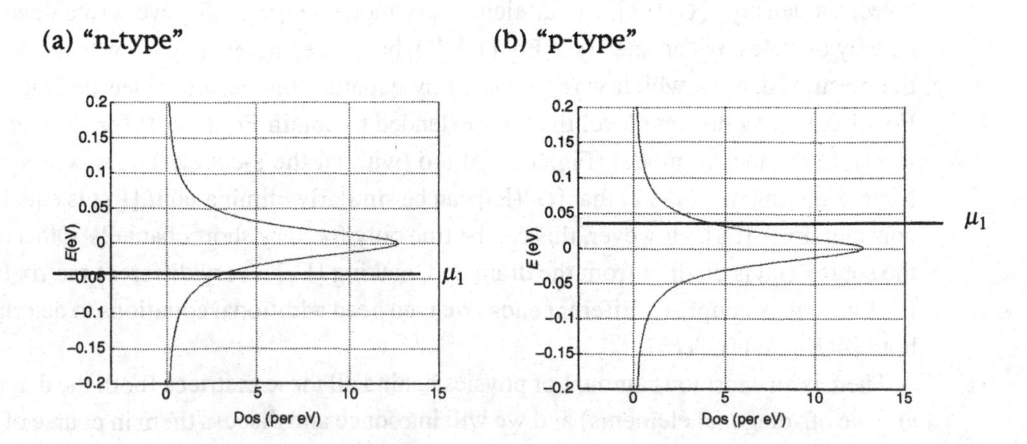

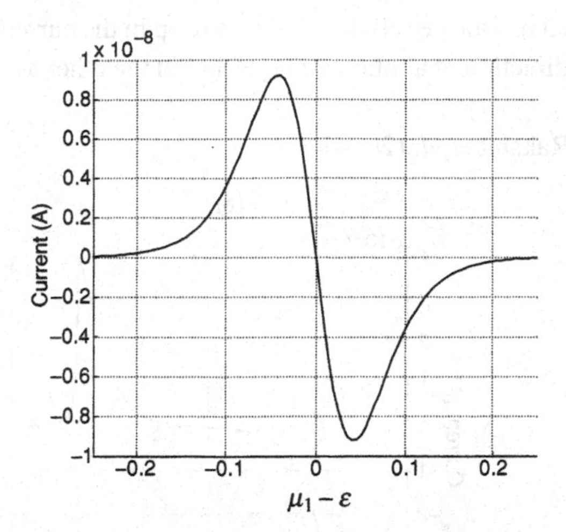

Fig. E.1.3 You should get a plot like this showing the reversal in the direction of the current from p-type ($\mu_1 < \varepsilon$) to n-type ($\mu_1 > \varepsilon$) samples.

However, if we simply heat up one contact relative to the other so that $T_1 > T_2$ (with no applied voltage) a thermoelectric current will flow whose direction will be different in case (a) and in case (b).

To see this, calculate the current from Eq. (1.6.6) with $U = 0$ (there is no need to perform a self-consistent solution), $V_D = 0$ and $V_G = 0$, and with $k_B T_1 = 0.026$ eV and $k_B T_2 = 0.025$ eV:

$$f_1(E) \equiv \left[1 + \exp\left(\frac{E - \mu_1}{k_B T_1}\right)\right]^{-1} \quad \text{and} \quad f_2(E) \equiv \left[1 + \exp\left(\frac{E - \mu_1}{k_B T_2}\right)\right]^{-1}$$

and plot it as a function of $(\mu_1 - \varepsilon)$ as the latter changes from -0.25 eV to $+0.25$ eV assuming $\gamma_1 = \gamma_2 = 0.05$ eV (Fig. E.1.3). This problem is motivated by Paulsson and Datta (2003).

E.1.4. Negative differential resistance: Figure 1.4.6a shows the current–voltage (I–V_D) characteristics calculated from a self-consistent solution of Eqs. (1.6.2)–(1.6.5) assuming

$$\varepsilon = 0.2\,\text{eV},\ k_B T = 0.025\,\text{eV},\ U_0 = 0.025\,\text{eV},\ V_G = 0,$$
$$\mu_1 = 0,\ \mu_2 = \mu_1 - qV_D,\ U_L = -qV_D/2$$

The broadening due to the two contacts γ_1 and γ_2 is assumed to be constant, equal to 0.005 eV.

Now suppose γ_1 is equal to 0.005 eV for $E > 0$, but is *zero* for $E < 0$ (γ_2 is still independent of energy and equal to 0.005 eV). Show that the current–voltage characteristics will now show negative differential resistance (NDR), that is, a drop in the current with an increase in the voltage, in one direction of applied voltage but not the other as shown in Fig. E.1.4.

This problem is motivated by Rakshit *et al.* (2004).

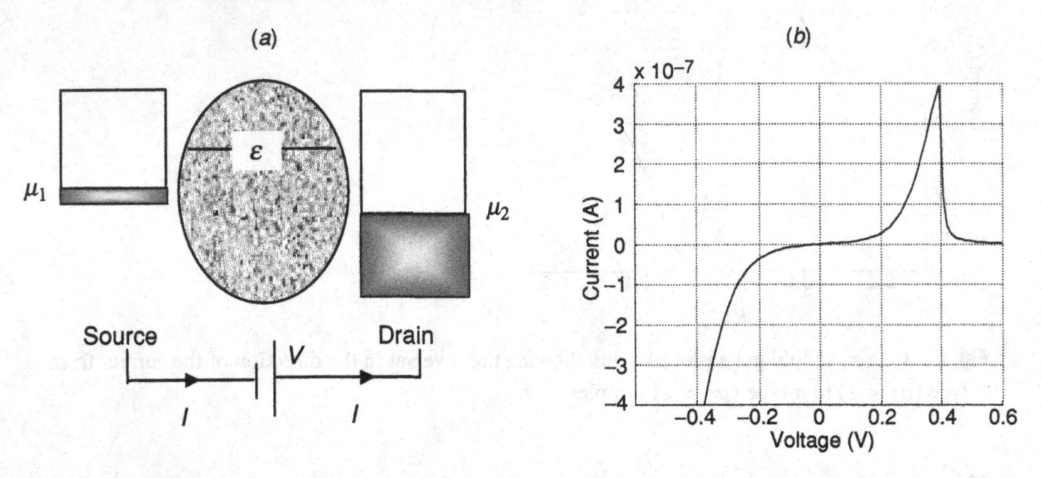

Fig. E.1.4

E.1.5. ***(Ohm's Law)*** The "bottom-up" view introduced in this chapter leads to the following expression for current (from eq. (1.6.6) with $\gamma_1 = \gamma_2$ and U = 0)

$$I = (q\gamma_1/2\hbar)\int_{-\infty}^{+\infty} dE\, D(E)\,(f_1(E) - f_2(E)) \tag{A}$$

The usual "top-down" view yields an expression for the current proportional to the cross-sectional area, S, the inverse length (1/L), the electron density n and the mobility $\tilde{\mu}$ (the tilde has been added to distinguish it from the electrochemical potential μ):

$$I = (S/L)n\tilde{\mu}(\mu_1 - \mu_2) \tag{B}$$

How do we get from A to B? Hint: (1) Relate the broadening to the diffusion constant $\tilde{D}$ through $\gamma_1 = 2\hbar\tilde{D}/L^2$ as mentioned in Section 1.6, (2) assume the electron system to be "non-degenerate" so that the Fermi functions are approximated by Boltzmann functions and assume the potential difference to be much less than k_BT, so that $f_1(E) - f_2(E) \approx e^{-(E-\mu_1)/k_BT}(\mu_1 - \mu_2)/k_BT$, and finally (3) make use of the Einstein relation $\tilde{D}/\tilde{\mu} = k_BT/q$ and the relation $nSL = \int_{-\infty}^{+\infty} dE\, D(E)e^{-(E-\mu_1)/k_BT}$.

2 Schrödinger equation

Our objective for the next few chapters is to learn how the Hamiltonian matrix $[H]$ for a given device structure (see Fig. 1.6.5) is written down. We start in this chapter with (1) the hydrogen atom (Section 2.1) and how it led scientists to the Schrödinger equation, (2) a simple approach called the finite difference method (Section 2.2) that can be used to convert this differential equation into a matrix equation, and (3) a few numerical examples (Section 2.3) showing how energy levels are calculated by diagonalizing the resulting Hamiltonian matrix.

2.1 Hydrogen atom

Early in the twentieth century scientists were trying to build a model for atoms which were known to consist of negative particles called electrons surrounding a positive nucleus. A simple model pictures the electron (of mass m and charge $-q$) as orbiting the nucleus (with charge Zq) at a radius r (Fig. 2.1.1) kept in place by electrostatic attraction, in much the same way that gravitational attraction keeps the planets in orbit around the Sun.

$$\frac{Zq^2}{4\pi\varepsilon_0 r^2} = \frac{mv^2}{r} \Rightarrow v = \sqrt{\frac{Zq^2}{4\pi\varepsilon_0 mr}} \tag{2.1.1}$$

Electrostatic force = Centripetal force

A faster electron describes an orbit with a smaller radius. The total energy of the electron is related to the radius of its orbit by the relation

$$E = -\frac{Zq^2}{4\pi\varepsilon_0 r} + \frac{mv^2}{2} = -\frac{Zq^2}{8\pi\varepsilon_0 r} \tag{2.1.2}$$

Potential energy + Kinetic energy = Total energy

However, it was soon realized that this simple viewpoint was inadequate since, according to classical electrodynamics, an orbiting electron should radiate electromagnetic waves like an antenna, lose energy continuously and spiral into the nucleus. Classically it is impossible to come up with a stable structure for such a system except with the

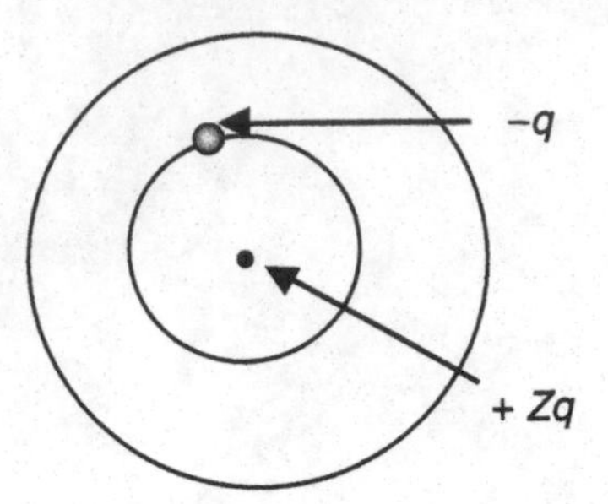

Fig. 2.1.1 Stationary orbits of an electron around a nucleus can be obtained by requiring their circumferences to be integer multiples of the de Broglie wavelength.

electron sitting right on top of the nucleus, in contradiction with experiment. It was apparent that a radical departure from classical physics was called for.

Bohr postulated that electrons could be described by stable orbits around the nucleus at specific distances from the nucleus corresponding to specific values of angular momenta. It was later realized that these distances could be determined by endowing the electrons with a wavelike character having a de Broglie wavelength equal to (h/mv), h being the Planck constant. One could then argue that the circumference of an orbit had to be an integer multiple of wavelengths in order to be stable:

$$2\pi r = n(h/mv) \tag{2.1.3}$$

Combining Eq. (2.1.3) with Eqs. (2.1.1) and (2.1.2) we obtain the radius and energy of stable orbits respectively:

$$r_n = (n^2/Z)a_0 \quad \text{(Bohr radius)} \tag{2.1.4}$$

$$\text{where } a_0 = 4\pi\varepsilon_0\hbar^2/mq^2 = 0.0529\,\text{nm} \tag{2.1.5}$$

$$E_n = -(Z^2/n^2)E_0 \tag{2.1.6a}$$

$$\text{where } E_0 = q^2/8\pi\varepsilon_0 a_0 = 13.6\,\text{eV} \quad \text{(1 Rydberg)} \tag{2.1.6b}$$

Once the electron is in its lowest energy orbit ($n = 1$) it cannot lose any more energy because there are no stationary orbits having lower energies available (Fig. 2.1.2a). If we heat up the atom, the electron is excited to higher stationary orbits (Fig. 2.1.2b). When it subsequently jumps down to lower energy states, it emits photons whose energy $h\nu$ corresponds to the energy difference between orbits m and n:

$$h\nu = E_m - E_n = E_0 Z^2 \left(\frac{1}{n^2} - \frac{1}{m^2}\right) \tag{2.1.7}$$

Experimentally it had been observed that the light emitted by a hydrogen atom indeed consisted of discrete frequencies that were described by this relation with integer values of n and m. This striking agreement with experiment suggested that there was some truth to this simple picture, generally known as the Bohr model.

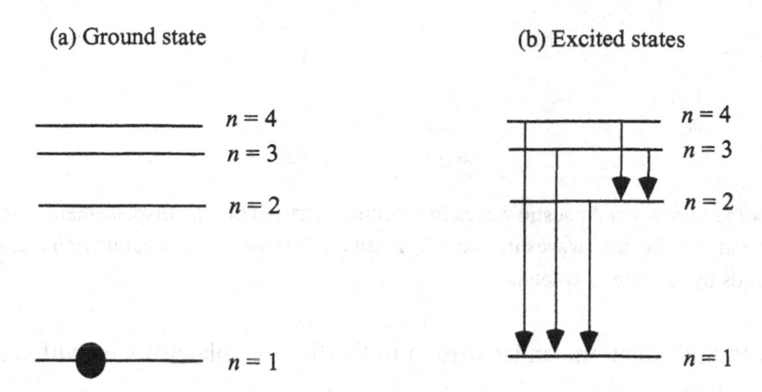

Fig. 2.1.2 (a) Left to itself, the electron relaxes to its lowest energy orbit ($n = 1$). (b) If we heat up the atom, the electron is excited to higher stationary orbits. When it subsequently jumps down to lower energy states, it emits photons whose energy $h\nu$ corresponds to the energy difference between the initial and final orbits.

The *Schrödinger equation* put this heuristic insight on a formal quantitative basis allowing one to calculate the energy levels for any confining potential $U(\vec{r})$.

$$i\hbar \frac{\partial \Psi}{\partial t} = \left(-\frac{\hbar^2}{2m} \nabla^2 + U(\vec{r}) \right) \Psi \tag{2.1.8}$$

How does this equation lead to discrete energy levels? Mathematically, one can show that if we assume a potential $U(\vec{r}) = -Zq^2/4\pi\varepsilon_0 r$ appropriate for a nucleus of charge $+Zq$, then the solutions to this equation can be labeled with three indices n, l and m

$$\Psi(\vec{r}, t) = \phi_{nlm}(\vec{r}) \exp(-\mathrm{i}E_n t/\hbar) \tag{2.1.9}$$

where the energy E_n depends only on the index n and is given by $E_n = -(Z^2/n^2)E_0$ in agreement with the heuristic result obtained earlier (see Eq. (2.1.6a)). The Schrödinger equation provides a formal wave equation for the electron not unlike the equation that describes, for example, an acoustic wave in a sound box . The energy E of the electron plays a role similar to that played by the frequency of the acoustic wave. It is well-known that a sound box resonates at specific frequencies determined by the size and shape of the box. Similarly an electron wave in an atomic box "resonates" at specific energies determined by the size and shape of the box as defined by the potential energy $U(\vec{r})$. Let us elaborate on this point a little further.

Waves in a box: To keep things simple let us consider the vibrations $u(x, t)$ of a one-dimensional (1D) string described by the 1D wave equation:

$$\frac{\partial^2 u}{\partial t^2} = v^2 \frac{\partial^2 u}{\partial x^2} \tag{2.1.10}$$

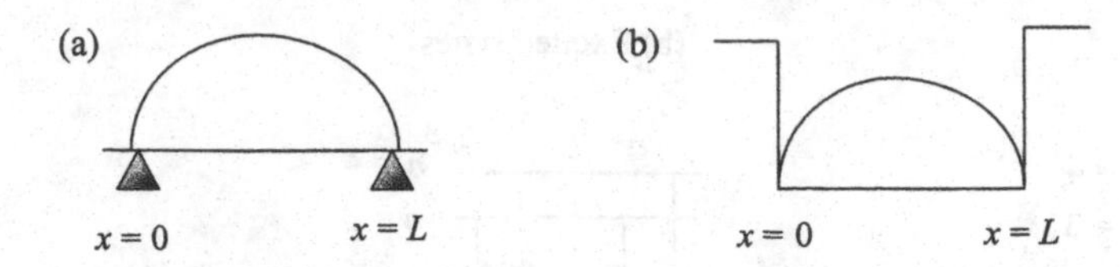

Fig. 2.1.3 Standing waves. (a) Acoustic waves in a "guitar" string with the displacement clamped to zero at either end. (b) Electron waves in a one-dimensional box with the wavefunction clamped to zero at both ends by an infinite potential.

The solutions to this equation can be written in the form of plane waves with a linear dispersion $\omega = \pm vk$:

$$u = u_0 \exp(ikx)\exp(-i\omega t) \Rightarrow \omega^2 = v^2k^2 \tag{2.1.11}$$

What happens if we clamp the two ends so that the displacement there is forced to be zero (Fig. 2.1.3)? We have to superpose solutions with $+k$ and $-k$ to obtain standing wave solutions. The allowed values of k are quantized leading to discrete resonant frequencies:

$$u = u_0 \sin(kx)\exp(-i\omega t) \Rightarrow k = n\pi/L \Rightarrow \omega = n\pi v/L \tag{2.1.12}$$

Well, it's the same way with the Schrödinger equation. If there is no confining potential ($U = 0$), we can write the solutions to the 1D Schrödinger equation:

$$i\hbar\frac{\partial\Psi}{\partial t} = -\frac{\hbar^2}{2m}\frac{\partial^2\Psi}{\partial x^2} \tag{2.1.13}$$

in the form of plane waves with a parabolic dispersion law $E = \hbar^2k^2/2m$:

$$\Psi = \Psi_0 \exp(ikx)\exp(-iEt/\hbar) \Rightarrow E = \hbar^2k^2/2m \tag{2.1.14}$$

If we fix the two ends we get standing waves with quantized k and resonant frequency:

$$\Psi = \Psi_0 \sin(kx)\exp(-iEt/\hbar) \Rightarrow k = n\pi/L$$
$$\Rightarrow E = \hbar^2\pi^2n^2/2mL^2 \tag{2.1.15}$$

Atomic "boxes" are of course defined by potentials $U(\vec{r})$ that are more complicated than the simple rectangular 1D potential shown in Fig. 2.1.2b, but the essential point is the same: anytime we confine a wave to a box, the frequency or energy is discretized because of the need for the wave to "fit" inside the box.

"Periodic" box: Another kind of box that we will often use is a ring (Fig. 2.1.4) where the end point at $x = L$ is connected back to the first point at $x = 0$ and there are no ends. Real boxes are seldom in this form but this idealization is often used since it simplifies the mathematics. The justification for this assumption is that if we are interested in the properties in the interior of the box, then what we assume at the ends (or surfaces) should

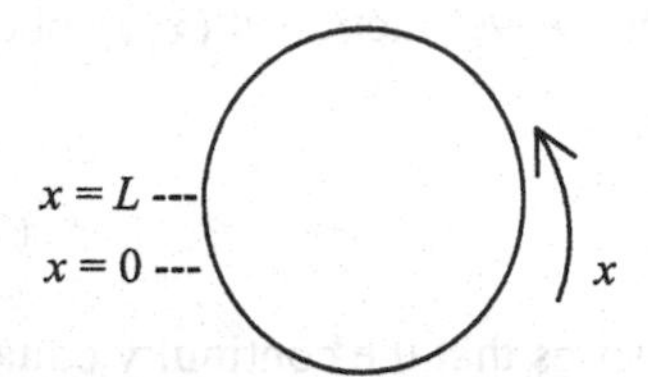

Fig. 2.1.4 Standing waves in a ring.

make no real difference and we could assume anything that makes our calculations simpler. However, this may not be a valid argument for "nanostructures" where the actual surface conditions can and do affect what an experimentalist measures.

Anyway, for a periodic box the eigenfunctions are given by (cf. Eq. (2.1.15))

$$\begin{aligned}\Psi &= \Psi_0 \sin(kx)\exp(-\mathrm{i}Et/\hbar)\\ \text{and } \Psi &= \Psi_0 \cos(kx)\exp(-\mathrm{i}Et/\hbar)\\ \text{with } k &= 2n\pi/L \Rightarrow E = 2\hbar^2\pi^2 n^2/mL^2 \end{aligned} \tag{2.1.16}$$

The values of k are spaced by $2\pi/L$ instead of π/L, so that there are half as many allowed values. But for each value of k there is a sine and a cosine function which have the same eigenvalue, so that the eigenvalues now come in pairs.

An important point to note is that whenever we have degenerate eigenstates, that is, two or more eigenfunctions with the same eigenvalue, any linear combination of these eigenfunctions is also an eigenfunction with the same eigenvalue. So we could just as well write the eigenstates as

$$\begin{aligned}\Psi &= \Psi_0 \exp(+\mathrm{i}kx)\exp(-\mathrm{i}Et/\hbar)\\ \text{and } \Psi &= \Psi_0 \exp(+\mathrm{i}kx)\exp(-\mathrm{i}Et/\hbar)\\ \text{with } k &= 2n\pi/L \Rightarrow E = 2\hbar^2\pi^2 n^2/mL^2 \end{aligned} \tag{2.1.17}$$

This is done quite commonly in analytical calculations and the first of these is viewed as the $+k$ state traveling in the positive x-direction while the second is viewed as the $-k$ state traveling in the negative x-direction.

Electron density and probability current density: An electron with a wavefunction $\Psi(x, t)$ has a probability of $\Psi^*\Psi\, \mathrm{d}V$ of being found in a volume $\mathrm{d}V$. When a number of electrons are present we could add up $\Psi^*\Psi$ for all the electrons to obtain the average electron density $n(x, t)$. What is the corresponding quantity we should sum to obtain the probability current density $J(x, t)$?

The appropriate expression for the probability current density

$$J = \frac{\mathrm{i}\hbar}{2m}\left(\Psi\frac{\partial\Psi^*}{\partial x} - \Psi^*\frac{\partial\Psi}{\partial x}\right) \tag{2.1.18}$$

is motivated by the observation that as long as the wavefunction $\Psi(x,t)$ obeys the Schrödinger equation, it can be shown that

$$\frac{\partial J}{\partial x} + \frac{\partial n}{\partial t} = 0 \tag{2.1.19}$$

if J is given by Eq. (2.1.18) and $n = \Psi^*\Psi$. This ensures that the continuity equation is satisfied regardless of the detailed dynamics of the wavefunction. The electrical current density is obtained by multiplying J by the charge $(-q)$ of an electron.

It is straightforward to check that the "$+k$" and "$-k$" states in Eq. (2.1.17) carry equal and opposite non-zero currents proportional to the electron density

$$J = (\hbar k/m)\,\Psi\Psi^* \tag{2.1.20}$$

suggesting that we associate $(\hbar k/m)$ with the velocity v of the electron (since we expect J to equal nv). However, this is true only for the plane wave functions in Eq. (2.1.17). The cosine and sine states in Eq. (2.1.16), for example, carry zero current. Indeed Eq. (2.1.18) will predict zero current for any *real* wavefunction.

2.2 Method of finite differences

The Schrödinger equation for a hydrogen atom can be solved analytically, but most other practical problems require a numerical solution. In this section I will describe one way of obtaining a numerical solution to the Schrödinger equation. Most numerical methods have one thing in common – they use some trick to convert the

	wavefunction $\Psi(\vec{r},t)$	*into a*	column vector $\{\psi(t)\}$
and the	differential operator H_{op}	*into a*	matrix $[H]$

so that the Schrödinger equation is converted from a

partial differential equation	*into a*	matrix equation
$\mathrm{i}\hbar\dfrac{\partial}{\partial t}\Psi(\vec{r},t) = H_{\text{op}}\Psi(\vec{r},t)$		$\mathrm{i}\hbar\dfrac{\mathrm{d}}{\mathrm{d}t}\{\psi(t)\} = [H]\{\psi(t)\}$

This conversion can be done in many ways, but the simplest one is to choose a discrete lattice. To see how this is done let us for simplicity consider just one dimension and discretize the position variable x into a lattice as shown in Fig. 2.2.1: $x_n = na$.

We can represent the wavefunction $\Psi(x,t)$ by a column vector $\{\psi_1(t)\ \psi_2(t)\ \cdots\ \cdots\}^{\mathrm{T}}$ ("T" denotes transpose) containing its values around each of the lattice points at time t. Suppressing the time variable t for clarity, we can write

$$\{\psi_1\quad \psi_2\quad \cdots\quad \cdots\} = \{\Psi(x_1)\quad \Psi(x_2)\quad \cdots\quad \cdots\}$$

This representation becomes exact only in the limit $a \to 0$, but as long as a is smaller than the spatial scale on which Ψ varies, we can expect it to be reasonably accurate.

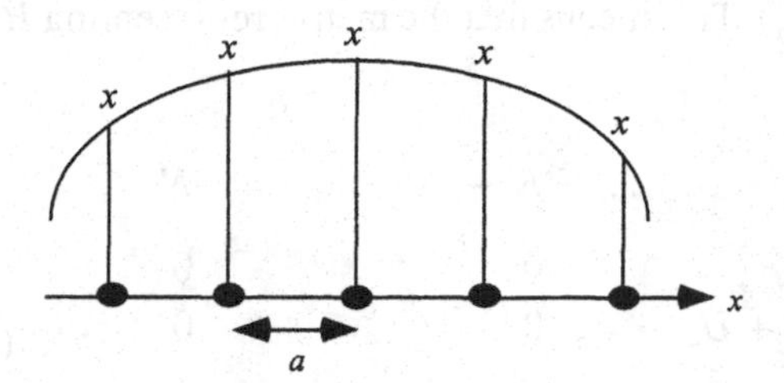

Fig. 2.2.1 A continuous function can be represented by its values at a set of points on a discrete lattice.

The next step is to obtain the matrix representing the Hamiltonian operator

$$H_{\mathrm{op}} \equiv -\frac{\hbar^2}{2m}\frac{\mathrm{d}^2}{\mathrm{d}x^2} + U(x)$$

Basically what we are doing is to turn a *differential* equation into a *difference* equation. There is a standard procedure for doing this – the finite difference technique:

$$\left(\frac{\partial^2 \Psi}{\partial x^2}\right)_{x=x_n} \rightarrow \frac{1}{a^2}\left[\Psi(x_{n+1}) - 2\Psi(x_n) + \Psi(x_{n-1})\right]$$

and

$$U(x)\,\Psi(x) \rightarrow U(x_n)\Psi(x_n)$$

This allows us to write (note: $t_0 \equiv \hbar^2/2ma^2$ and $U_n \equiv U(x_n)$)

$$\begin{aligned} i\hbar\,\frac{\mathrm{d}\psi_n}{\mathrm{d}t} &= \left[H_{\mathrm{op}}\,\psi\right]_{x=x_n} = (U_n + 2t_0)\,\psi_n - t_0\psi_{n-1} - t_0\psi_{n+1} \\ &= \sum_m \left[(U_n + 2t_0)\,\delta_{n,m} - t_0\delta_{n,m+1} - t_0\delta_{n,m-1}\right]\psi_m \end{aligned} \tag{2.2.1}$$

where $\delta_{n,m}$ is the Kronecker delta, which is one if $n = m$ and zero if $n \neq m$. We can write Eq. (2.2.1) as a matrix equation:

$$i\hbar\frac{\mathrm{d}}{\mathrm{d}t}\,\{\psi(t)\} = [H]\,\{\psi(t)\} \tag{2.2.2}$$

The elements of the Hamiltonian matrix are given by

$$H_{n,m} = [U_n + 2t_0]\,\delta_{n,m} - t_0\delta_{n,m+1} - t_0\delta_{n,m-1} \tag{2.2.3}$$

where $t_0 \equiv \hbar^2/2ma^2$ and $U_n \equiv U(x_n)$. This means that the matrix representing H looks like this

$$H = \begin{array}{c|ccccc} & 1 & 2 & \ldots & N-1 & N \\ \hline 1 & 2t_0 + U_1 & -t_0 & & 0 & 0 \\ 2 & -t_0 & 2t_0 + U_2 & & 0 & 0 \\ & \ldots & & \ldots & & \ldots \\ N-1 & 0 & 0 & & 2t_0 + U_{N-1} & -t_0 \\ N & 0 & 0 & & -t_0 & 2t_0 + U_N \end{array} \tag{2.2.4}$$

For a given potential function $U(x)$ it is straightforward to set up this matrix, once we have chosen an appropriate lattice spacing a.

Eigenvalues and eigenvectors: Now that we have converted the Schrödinger equation into a matrix equation (Eq. (2.2.2))

$$\mathrm{i}\hbar \frac{\mathrm{d}}{\mathrm{d}t}\{\psi(t)\} = [H]\{\psi(t)\}$$

how do we calculate $\{\psi(t)\}$ given some initial state $\{\psi(0)\}$? The standard procedure is to find the eigenvalues E_α and eigenvectors $\{\alpha\}$ of the matrix $[H]$:

$$[H]\{\alpha\} = E_\alpha \{\alpha\} \tag{2.2.5}$$

Making use of Eq. (2.2.5) it is easy to show that the wavefunction $\{\psi(t)\} = \mathrm{e}^{-\mathrm{i}E_\alpha t/\hbar}\{\alpha\}$ satisfies Eq. (2.2.2). Since Eq. (2.2.2) is linear, any superposition of such solutions

$$\{\psi(t)\} = \sum_\alpha C_\alpha \mathrm{e}^{-\mathrm{i}E_\alpha t/\hbar}\{\alpha\} \tag{2.2.6}$$

is also a solution. It can be shown that this form, Eq. (2.2.6), is "complete," that is, any solution to Eq. (2.2.2) can be written in this form. Given an initial state we can figure out the coefficients C_α. The wavefunction at subsequent times t is then given by Eq. (2.2.6). Later we will discuss how we can figure out the coefficients. For the moment we are just trying to make the point that the dynamics of the system are easy to visualize or describe in terms of the eigenvalues (which are the energy levels that we talked about earlier) and the corresponding eigenvectors (which are the wavefunctions associated with those levels) of $[H]$. That is why the first step in discussing any system is to write down the matrix $[H]$ and to find its eigenvalues and eigenvectors. This is easily done using any standard mathematical package like MATLAB as we will discuss in the next section.

2.3 Examples

Let us now look at a few examples to make sure we understand how to find the eigenenergies and eigenvectors numerically using the method of finite differences described in the last section. These examples are all simple enough to permit analytical solutions that we can use to compare and evaluate our numerical solutions. The advantage of the numerical procedure is that it can handle more complicated problems just as easily, even when no analytical solutions are available.

2.3.1 Particle in a box

Consider, first the "particle in a box" problem that we mentioned in Section 2.1. The potential is constant inside the box which is bounded by infinitely high walls at $x = 0$ and at $x = L$ (Fig. 2.3.1). The eigenstates $\phi_\alpha(x)$ are given by

$$\phi_\alpha(x) \sim \sin(k_\alpha x) \qquad \text{where } k_\alpha = \alpha\pi/L,\ \alpha = 1, 2, \ldots$$

and their energies are given by $E_\alpha = \hbar^2 k_\alpha^2/2m$.

We could solve this problem numerically by selecting a discrete lattice with 100 points and writing down a 100 × 100 matrix $[H]$ using Eq. (2.2.4) with all $U_n = 0$:

$$H = \begin{array}{c|cccc} & 1 & 2 & \ldots\ 99 & 100 \\ 1 & 2t_0 & -t_0 & 0 & 0 \\ 2 & -t_0 & 2t_0 & 0 & 0 \\ & \ldots & \ldots & \ldots & \\ 99 & 0 & 0 & 2t_0 & -t_0 \\ 100 & 0 & 0 & -t_0 & 2t_0 \end{array} \tag{2.3.1}$$

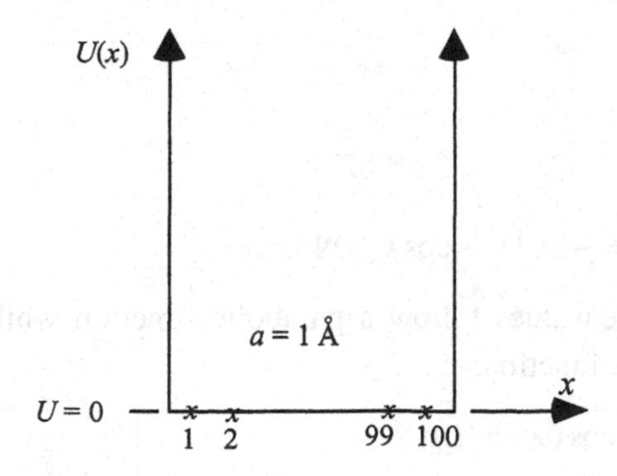

Fig. 2.3.1 Energy levels for a "particle in a box" are calculated using a discrete lattice of 100 points spaced by $a = 1$ Å.

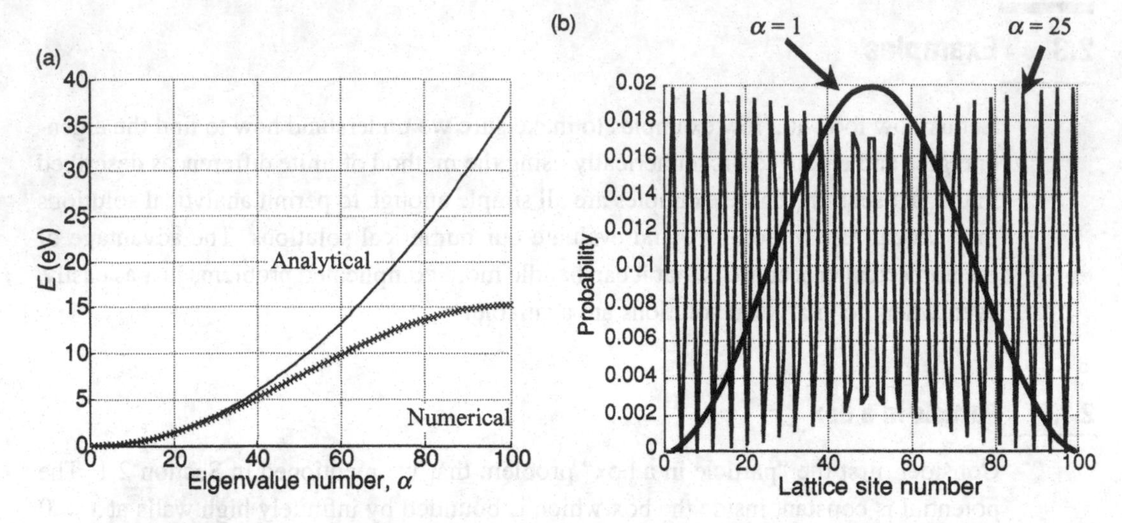

Fig. 2.3.2 (a) Numerical evaluation (see Fig. 2.3.1) yields 100 eigenvalues that follow the analytical result well for low energies but deviate at higher energies because the wavefunctions oscillate too rapidly. (b) Probability distribution (squared eigenfunction) for eigenvalues $\alpha = 1$ and $\alpha = 25$.

It is straightforward to set up this matrix and use any standard mathematical package like MATLAB to find the eigenvalues and the corresponding eigenvectors. We obtained 100 eigenvalues, which are plotted in Fig. 2.3.2a. They follow the analytical result $E_\alpha = \hbar^2\pi^2\alpha^2/2mL^2$, with $L = 101a$, fairly well at low energy, but deviate at higher energies because of the rapid oscillations in the wavefunction. Our finite difference approximation to the second derivative operator (note that $t_0 \equiv \hbar^2/2ma^2$)

$$-\frac{\hbar^2}{2m}\left(\frac{\partial^2\Psi}{\partial x^2}\right)_{x=x_n} \to -t_0\,[\Psi(x_{n+1}) - 2\Psi(x_n) + \Psi(x_{n-1})]$$

is accurate only if Ψ varies slowly enough on a length scale of a. Indeed if we put $\Psi \sim \sin(k_\alpha x)$ it is straightforward to show that

$$-\frac{\hbar^2}{2m}\left(\frac{\partial^2\Psi}{\partial x^2}\right)_{x=x_n} = t_0\,(k_\alpha a)^2\,\Psi(x_n)$$

while

$$-t_0\,[\Psi(x_{n+1}) - 2\Psi(x_n) + \Psi(x_{n-1})] = -2t_0(1 - \cos k_\alpha a)\Psi(x_n)$$

Since $k_\alpha = \alpha\pi/L$, the analytical eigenvalues follow a parabolic function while the numerical eigenvalues follow a cosine function:

$E_\alpha = t_0(\pi\alpha a/L)^2$	$E_\alpha = 2t_0[1 - \cos(\alpha\pi a/L)]$
Analytical result	Numerical result

The two are equivalent only if $k_\alpha a = \alpha\pi a/L \ll 1$ so that $\cos(k_\alpha a) \approx 1 - (k_\alpha^2 a^2/2)$.

Normalization: Figure 2.3.2b shows the eigenfunction squared corresponding to the eigenvalues $\alpha = 1$ and $\alpha = 25$. A word about the normalization of the wavefunctions: In analytical treatments, it is common to normalize wavefunctions such that

$$\int_{-\infty}^{+\infty} \mathrm{d}x \, |\phi_\alpha(x)|^2 = 1$$

Numerically, a normalized eigenvector satisfies the condition

$$\sum_{n=1}^{N} |\phi_\alpha(x_n)|^2 = 1$$

So when we compare numerical results with analytical results we should expect

$$\underbrace{|\phi_\alpha(x_n)|^2}_{\text{Numerical}} = \underbrace{|\phi_\alpha(x)|^2 a}_{\text{Analytical}} \tag{2.3.2}$$

where a is the lattice constant (see Fig. 2.3.1). For example, in the present case

$$\underbrace{|\phi_\alpha(x)|^2 = (2/L)\sin^2(k_\alpha x)}_{\text{Analytical}} \longrightarrow \underbrace{|\phi_\alpha(x_n)|^2 = (2a/L)\sin^2(k_\alpha x_n)}_{\text{Numerical}}$$

Since we used $a = 1$ Å and $L = 101$ Å, the numerical probability distribution should have a peak value of $2a/L \approx 0.02$ as shown in Fig. 2.3.2b.

Boundary conditions: One more point: Strictly speaking, the matrix $[H]$ is infinitely large, but in practice we always truncate it to a finite number, say N, of points. This means that at the two ends we are replacing (see Eq. (2.2.1))

$$-t_0\psi_0 + (2t_0 + U_1)\psi_1 - t_0\psi_2 \quad \textit{with} \quad (2t_0 + U_1)\psi_1 - t_0\psi_2$$

and

$$-t_0\psi_{N-1} + (2t_0 + U_N)\psi_N - t_0\psi_{N+1} \quad \textit{with} \quad -t_0\psi_{N-1} + (2t_0 + U_N)\psi_N$$

In effect we are setting ψ_0 and ψ_{N+1} equal to zero. This boundary condition is appropriate if the potential is infinitely large at point 0 and at point $N + 1$ as shown in Fig. 2.3.3. The actual value of the potential at the end points will not affect the results as long as the wavefunctions are essentially zero at those points anyway.

Another boundary condition that is often used is the *periodic boundary condition* where we assume that the last point is connected back to the first point so that there are no ends (Fig. 2.3.4). As we mentioned earlier (Fig. 2.1.4), the justification for this assumption is that if we are interested in the properties in the interior of a structure, then what we assume at the boundaries should make no real difference and we could assume anything to make our calculations simpler.

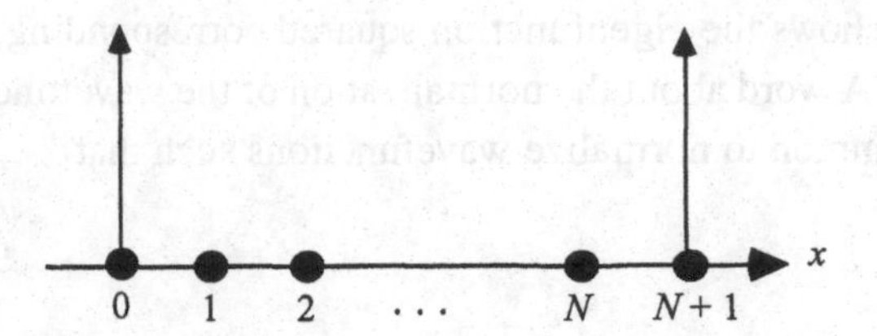

Fig. 2.3.3 The boundary condition $\psi_0 = 0$ and $\psi_{N+1} = 0$ can be used if we assume an infinitely large potential at points 0 and $N+1$.

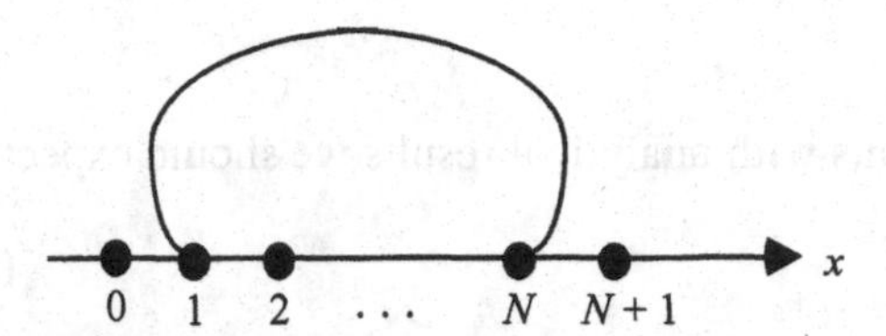

Fig. 2.3.4 Periodic boundary conditions assume that there are no "ends." Point N is connected back to point 1 as if the structure were in the form of a ring making $(N+1)$ equivalent to 1.

Mathematically, periodic boundary conditions are implemented by modifying the Hamiltonian to

$$
H = \begin{array}{c|ccccc}
 & 1 & 2 & \dots & 99 & 100 \\
\hline
1 & 2t_0 & -t_0 & & 0 & -t_0 \\
2 & -t_0 & 2t_0 & & 0 & 0 \\
 & \dots & \dots & & \dots & \\
99 & 0 & 0 & & 2t_0 & -t_0 \\
100 & -t_0 & 0 & & -t_0 & 2t_0
\end{array} \tag{2.3.3}
$$

Note that compared to the infinite wall boundary conditions (cf. Eq. (2.3.1)) the only change is in the elements $H(1, 100)$ and $H(100, 1)$. This does change the resulting eigenvalues and eigenvectors, but the change is imperceptible if the number of points is large. The eigenfunctions are now given by

$$\phi_\alpha(x) \sim \sin(k_\alpha x) \quad \text{and} \quad \cos(k_\alpha x)$$

where $k_\alpha = \alpha\, 2\pi/L$, $\alpha = 1, 2, \dots$ instead of

$$\phi_\alpha(x) \sim \sin(k_\alpha x)$$

where $k_\alpha = \alpha\, \pi/L$, $\alpha = 1, 2, \dots$

The values of $k\alpha$ are spaced by $2\pi/L$ instead of π/L, so that there are half as many allowed values. But for each value of k_α there is a sine and a cosine function which have the same eigenvalue, so that the eigenvalues now come in pairs as evident from Fig. 2.3.5.

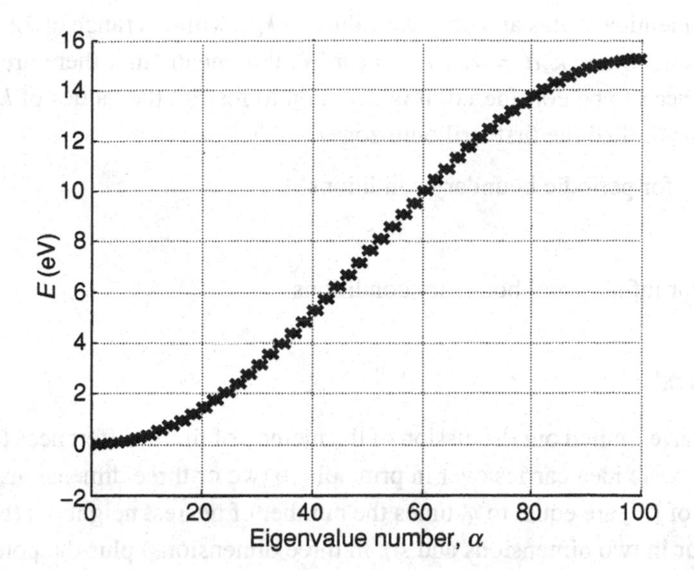

Fig. 2.3.5 Energy eigenvalues for a box of length 101 Å (same as Fig. 2.3.1) with periodic boundary conditions: the eigenvalues now come in pairs.

As we discussed earlier, instead of writing the eigenstates as

$$\cos(k_\alpha x) \quad \text{and} \quad \sin(k_\alpha x)$$

we could just as well write them as

$$e^{ik_\alpha x} = \cos(k_\alpha x) + i\sin(k_\alpha x) \quad \text{and} \quad e^{-ik_\alpha x} = \cos(k_\alpha x) - i\sin(k_\alpha x)$$

This is done quite commonly in analytical calculations, but numerical calculations will typically give the eigenvectors as $\cos(k_\alpha x)$ and $\sin(k_\alpha x)$. Both forms are equally correct though one may be more convenient than the other for certain calculations.

Number of eigenvalues: Another important point to note about the numerical solution is that it yields a finite number of eigenvalues (unlike the analytical solution for which the number is infinite). This is expected since a finite matrix can have only a finite number of eigenvalues, but one might wonder why we do not have an infinite number of E_α corresponding to an infinite number of $k_\alpha a = \alpha 2\pi a/L$, just as we have for the analytical result. The reason is that for a discrete lattice, the wavefunctions

$$\sin(k_\alpha\, x) \quad \text{and} \quad \sin([k_\alpha + (2\pi/a)]x)$$

represent the same state because at any lattice point $x_n = na$,

$$\sin(k_\alpha\, x_n) = \sin([k_\alpha + (2\pi/a)]x_n)$$

They are NOT equal between two lattice points and thus represent distinct states in a non-discrete representation. But once we adopt a discrete lattice, values of k_α differing by $2\pi/a$ represent identical states and only the values of $k_\alpha a$ within a range of 2π yield independent solutions. Since $k_\alpha a = \alpha\pi a/L = \alpha\pi/N$, this means that there are only N values of α that need to be considered. It is common to restrict the values of $k_\alpha a$ to the range (sometimes called the first Brillouin zone)

$$-\pi < k_\alpha a \le +\pi \quad \text{for periodic boundary conditions}$$

and

$$0 < k_\alpha a \le +\pi \quad \text{for infinite wall boundary conditions}$$

2.3.2 Particle in a 3D "box"

For simplicity we have limited our discussion of the method of finite differences to one dimension, but the basic idea carries over in principle to two or three dimensions. The diagonal elements of [H] are equal to t_0 times the number of nearest neighbors (two in one dimension, four in two dimensions and six in three dimensions) plus the potential $U(\vec{r})$ evaluated at the lattice site, while the off-diagonal elements are equal to $-t_0$ for neighboring sites on the lattice. That is, (ν is the number of nearest neighbors)

$$\begin{aligned} H_{nm} &= \nu t_0 && n = m \\ &= -t_0 && n, m \text{ are nearest neighbors} \\ &= 0 && \text{otherwise} \end{aligned} \tag{2.3.4}$$

However, we run into a practical difficulty in two or three dimensions. If we have lattice points spaced by 1 Å, then a one-dimensional problem with $L = 101$ Å requires a matrix [H] 100×100 in size. But in three dimensions this would require a matrix $10^6 \times 10^6$ in size. This means that in practice we are limited to very small problems. However, if the coordinates are separable then we can deal with three separate one-dimensional problems as opposed to one giant three-dimensional problem. This is possible if the potential can be separated into an x-, a y-, and a z-dependent part:

$$U(\vec{r}) = U_x(x) + U_y(y) + U_z(z) \tag{2.3.5}$$

The wavefunction can then be written in product form:

$$\Psi(\vec{r}) = X(x)Y(y)Z(z)$$

where each of the functions $X(x)$, $Y(y)$, and $Z(z)$ is obtained by solving a separate one-dimensional Schrödinger equation:

$$\begin{aligned} E_x X(x) &= \left(-\frac{\hbar^2}{2m}\frac{d^2}{dx^2} + U_x(x)\right) X(x) \\ E_y Y(y) &= \left(-\frac{\hbar^2}{2m}\frac{d^2}{dy^2} + U_y(y)\right) Y(y) \\ E_z Z(z) &= \left(-\frac{\hbar^2}{2m}\frac{d^2}{dz^2} + U_z(z)\right) Z(z) \end{aligned} \tag{2.3.6}$$

The total energy E is equal to the sum of the energies associated with each of the three dimensions: $E = E_x + E_y + E_z$.

Spherically symmetric potential: Some problems may not be separable in Cartesian coordinates but could be separable in cylindrical or spherical coordinates. For example, the potential in a hydrogen atom $U(\vec{r}) = -q^2/4\pi\varepsilon_0 r$ cannot be separated in (x, y, z). But it is separable in (r, θ, ϕ) and the wavefunction may be written in the form

$$\Psi(r,\theta,\phi) = [f(r)/r]\, Y_l^m(\theta,\phi) \tag{2.3.7}$$

where the radial wavefunction $f(r)$ is obtained by solving the radial Schrödinger equation:

$$Ef(r) = \left(-\frac{\hbar^2}{2m}\frac{d^2}{dr^2} + U(r) + \frac{l(l+1)\hbar^2}{2mr^2}\right) f(r) \tag{2.3.8}$$

Here $l = 0$ for s levels, $l = 1$ for p levels and so on. $Y_l^m(\theta, \phi)$ are the spherical harmonics given by

$$Y_0^0(\theta,\phi) = \sqrt{1/4\pi}$$

$$Y_1^0(\theta,\phi) = \sqrt{3/4\pi}\cos\theta$$

$$Y_1^{\pm 1}(\theta,\phi) = \pm\sqrt{3/8\pi}\sin\theta\, e^{\pm i\phi}$$

etc. Equation (2.3.8) can be solved numerically using the method of finite differences that we have described.

Normalization: Note that the overall wavefunctions are normalized such that

$$\int_0^\infty dr\, r^2 \int_0^\pi d\theta \sin\theta \int_0^{2\pi} d\phi\, |\Psi|^2 = 1$$

Since, from Eq. (2.3.7)

$$\Psi(r,\theta,\phi) = [f(r)/r] Y_l^m(\theta,\phi)$$

and the spherical harmonics are normalized such that

$$\int_0^\pi d\theta \sin\theta \int_0^{2\pi} d\phi \left|Y_l^m\right|^2 = 1$$

it is easy to see that the radial function $f(r)$ obeys the normalization condition

$$\int_0^\infty dr |f(r)|^2 = 1 \tag{2.3.9}$$

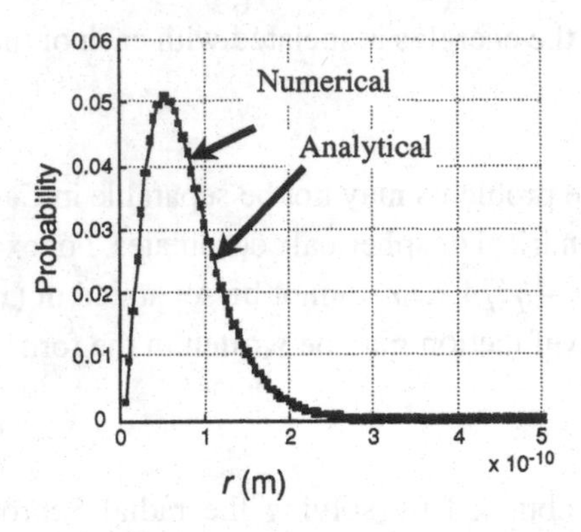

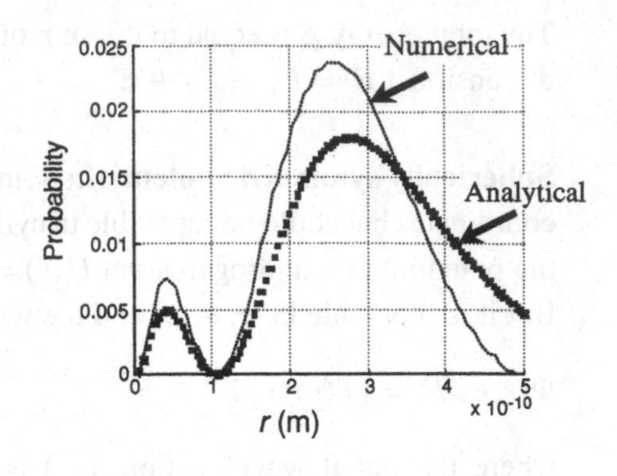

Fig. 2.3.6 Radial probability distribution $|f(r)|^2$ corresponding to the two lowest eigenvalues (−13.56 eV and −2.96 eV) for $l = 0$ (which correspond to the 1s and 2s levels). The dots show the analytical result (Eqs. (2.1.10a, b)) while the solid curve denotes the numerical result obtained using a lattice with 100 points spaced by $a = 0.05$ Å.

suggesting that we view $|f(r)|^2$ as a radial probability distribution function such that $|f(r)|^2\Delta r$ tells us the probability of finding the electron in the volume between r and $(r + \Delta r)$. Numerical results with a lattice spacing of a should be compared with the analytical values of $|f(r)|^2 a$. For example, for the 1s and 2s levels,

$$|f_{1s}|^2 a = \left(4ar^2/a_0^3\right) e^{-2r/a_0} \tag{2.3.10a}$$

$$|f_{2s}|^2 a = \left(ar^2/8a_0^3\right)\left(2 - \frac{r}{a_0}\right)^2 e^{-2r/2a_0} \tag{2.3.10b}$$

Numerical results: If we use a lattice with 100 points spaced by $a = 0.05$ Å then the two lowest eigenvalues with $l = 0$ (which correspond to the 1s and 2s levels) are

$E_{1s} = -13.56$ eV and $E_{2s} = -2.96$ eV

as compared with the analytical values (see Eq. (2.2.6)) $E_{1s} = -13.59$ eV and $E_{2s} = -3.4$ eV. The 1s level agrees well, but the 2s level is considerably off. The reason is easy to see if we plot the corresponding $|f(r)|^2$ and compare with the analytical results. It is evident from Fig. 2.3.6 that the 1s wavefunction matches well, but it is apparent that we do not have enough range for the 2s function. This can be fixed by choosing a larger lattice spacing, namely $a = 0.1$ Å. Figure 2.3.7 shows that the wavefunction now matches the analytical result quite well and the 2s eigenvalue is −3.39 eV, in good agreement with the analytical result. However, the 1s eigenvalue degrades slightly to −13.47 eV, because the wavefunction is not sampled frequently enough. We could improve the agreement for both 1s and 2s levels by using 200 points spaced by $a = 0.05$ Å, so that we would have both fine sampling and large range. But the calculation would then take longer since we would have to calculate the eigenvalues of a (200 × 200) matrix instead of a (100 × 100) matrix.

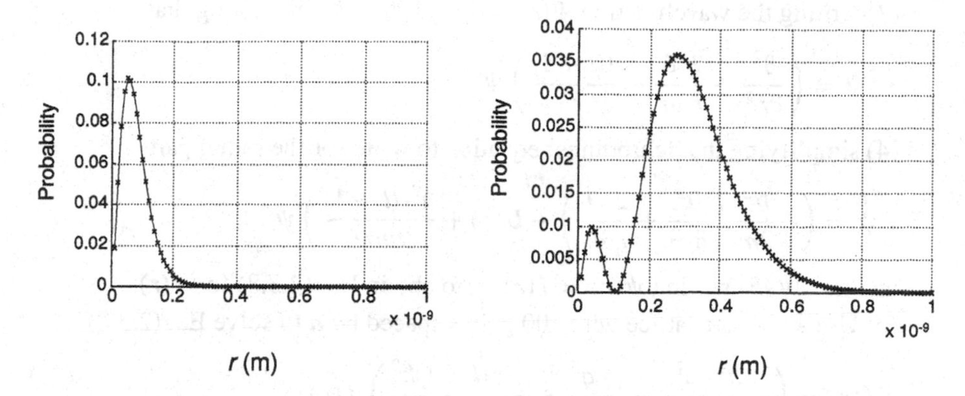

Fig. 2.3.7 Radial probability distribution $|f(r)|^2$ corresponding to the two lowest eigenvalues (−13.47 eV and −3.39 eV) for $l = 0$ (which correspond to the 1s and 2s levels). Solid line shows the analytical result (Eqs. (2.3.10a, b)) while the ×'s denote the numerical result obtained using a lattice with 100 points spaced by $a = 0.1$ Å.

This simple example illustrates the essential issues one has to consider in setting up the lattice for a numerical calculation. The lattice constant a has to be small enough to provide adequate sampling of the wavefunction while the size of the lattice has to be big enough to cover the entire range of the wavefunction. If it were essential to describe all the eigenstates accurately, our problem would be a hopeless one. Luckily, however, we usually need an accurate description of the eigenstates that lie within a certain range of energies and it is possible to optimize our matrix [H] so as to provide an accurate description over a desired range.

EXERCISES

E.2.1. (a) Use a discrete lattice with 100 points spaced by 1 Å to calculate the eigenenergies for a particle in a box with infinite walls and compare with $E_\alpha = \hbar^2\pi^2\alpha^2/2mL^2$ (cf. Fig. 2.3.2a). Plot the probability distribution (eigenfunction squared) for the eigenvalues $\alpha = 1$ and $\alpha = 50$ (cf. Fig. 2.3.2b). (b) Find the eigenvalues using periodic boundary conditions and compare with Fig. 2.3.5.

E.2.2. (a) Obtain the radial equation given in Eq. (2.3.8) by (1) writing the operator ∇^2 in the Schrödinger equation in spherical coordinates:

$$\nabla^2 \equiv \left(\frac{\partial^2}{\partial r^2} + \frac{2}{r}\frac{\partial}{\partial r}\right) + \frac{1}{r^2}\left(\frac{1}{\sin\theta}\frac{\partial}{\partial\theta}\left(\sin\theta\frac{\partial}{\partial\theta}\right) + \frac{1}{\sin^2\theta}\frac{\partial^2}{\partial\phi^2}\right)$$

(2) noting that the spherical harmonics $Y_l^m(\theta, \phi)$ are eigenfunctions of the angular part:

$$\left(\frac{1}{\sin\theta}\frac{\partial}{\partial\theta}\left(\sin\theta\frac{\partial}{\partial\theta}\right) + \frac{1}{\sin^2\theta}\frac{\partial^2}{\partial\phi^2}\right)Y_l^m = -l(l+1)Y_l^m$$

(3) writing the wavefunction $\Psi(r) = \Psi(r)\, Y_l^m(\theta, \phi)$ and noting that

$$\nabla^2 \Psi = \left(\frac{\partial^2}{\partial r^2} + \frac{2}{r}\frac{\partial}{\partial r} - \frac{l(l+1)}{r^2} \right) \Psi$$

(4) simplifying the Schrödinger equation to write for the radial part

$$E\psi = \left(-\frac{\hbar^2}{2m} \left(\frac{\partial^2}{\partial r^2} + \frac{2}{r}\frac{\partial}{\partial r} \right) + U(r) + \frac{\hbar^2 l(l+1)}{2mr^2} \right) \psi$$

and finally (5) writing $\psi(r) = f(r)/r$, to obtain Eq. (2.3.8) for $f(r)$.
(b) Use a discrete lattice with 100 points spaced by a to solve Eq. (2.3.8)

$$Ef(r) = \left(-\frac{\hbar^2}{2m}\frac{d^2}{dr^2} - \frac{q^2}{4\pi\varepsilon_0 r} + \frac{l(l+1)\hbar^2}{2mr^2} \right) f(r)$$

for the 1s and 2s energy levels of a hydrogen atom. Plot the corresponding radial probability distributions $|f(r)|^2$ and compare with the analytical results for (a) $a =$ 0.05 Å (cf. Fig. 2.3.6) and (b) $a = 0.1$ Å (cf. Fig. 2.3.7).

Strictly speaking one should replace the electron mass with the reduced mass to account for nuclear motion, but this is a small correction compared to our level of accuracy.

E.2.3. Use Eq. (2.1.18) to evaluate the current density associated with an electron having the wavefunction

$$\Psi(x, t) = (e^{+\gamma x} + a e^{-\gamma x}) e^{-iEt/\hbar}$$

assuming γ is (a) purely imaginary ($= i\beta$) and (b) purely real.

3 Self-consistent field

As we move from the hydrogen atom (one electron only) to multi-electron atoms, we are immediately faced with the issue of electron–electron interactions, which is at the heart of almost all the unsolved problems in our field. In this chapter I will explain (1) the self-consistent field (SCF) procedure (Section 3.1), which provides an approximate way to include electron–electron interactions into the Schrödinger equation, (2) the interpretation of the energy levels obtained from this so-called "one-electron" Schrödinger equation (Section 3.2), and (3) the energetic considerations underlying the process by which atoms "bond" to form molecules (Section 3.3). Finally, a supplementary section elaborates on the concepts of Section 3.2 for interested readers (Section 3.4).

3.1 The self-consistent field (SCF) procedure

One of the first successes of quantum theory after the interpretation of the hydrogen atom was to explain the periodic table of atoms by combining the energy levels obtained from the Schrödinger equation with the Pauli exclusion principle requiring that each level be occupied by no more than one electron. The energy eigenvalues of the Schrödinger equation for each value of l starting from $l = 0$ (see Eq. (2.3.8)) are numbered with integer values of n starting from $n = l + 1$. For any (n, l) there are $(2l + 1)$ levels with distinct angular wavefunctions (labeled with another index m), all of which have the same energy. For each (n, l, m) there a is an up-spin and a down-spin level making the number of degenerate levels equal to $2(2l + 1)$ for a given (n, l). The energy levels look something like Fig. 3.1.1.

The elements of the periodic table are arranged in order as the number of electrons increases by one from one atom to the next. Their electronic structure can be written as: hydrogen, $1s^1$; helium, $1s^2$; lithium, $1s^2 2s^1$; beryllium, $1s^2 2s^2$; boron, $1s^2 2s^2\ 2p^1$, etc., where the superscript indicates the number of electrons occupying a particular orbital.

How do we calculate the energy levels for a multi-electron atom? The time-independent Schrödinger equation

$$E_\alpha \Phi_\alpha(\vec{r}) = H_{\text{op}} \Phi_\alpha(\vec{r}) \quad \text{where} \quad H_{\text{op}} \equiv -\frac{\hbar^2}{2m}\nabla^2 + U(\vec{r})$$

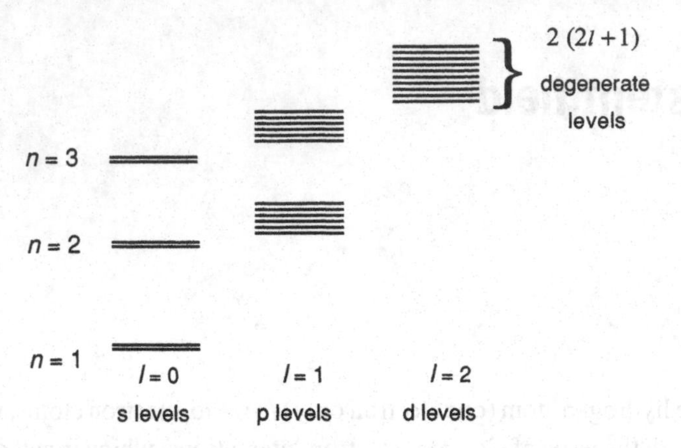

Fig. 3.1.1

provides a fairly accurate description of the observed spectra of all atoms, not just the hydrogen atom. However, multi-electron atoms involve electron–electron interactions that are included by adding a "self-consistent field (SCF)," $U_{SCF}(\vec{r})$, to the nuclear potential $U_{nuc}(\vec{r})$: $U(\vec{r}) = U_{nuc}(\vec{r}) + U_{SCF}(\vec{r})$, just as in Section 1.4 we added an extra potential to the Laplace potential U_L (see Eq. (1.4.1b)). The nuclear potential U_{nuc}, like U_L, is fixed, while U_{SCF} depends on the electronic wavefunctions and has to be calculated from a self-consistent iterative procedure. In this chapter we will describe this procedure and the associated conceptual issues.

Consider a helium atom consisting of two electrons bound to a nucleus with two positive charges $+2q$. What will the energy levels looks like? Our first guess would be simply to treat it just like a hydrogen atom except that the potential is

$$U(\vec{r}) = -2q^2/4\pi\varepsilon_0 r$$

instead of

$$U(\vec{r}) = -q^2/4\pi\varepsilon_0 r$$

If we solve the Schrödinger equation with $U(\vec{r}) = -Zq^2/4\pi\varepsilon_0 r$ we will obtain energy levels given by

$$E_n = -(Z^2/n^2)\,E_0 = -54.4\text{ eV}/n^2 \quad (Z = 2)$$

just as predicted by the simple Bohr model (see Eqs. (2.1.6a, b)). However, this does not compare well with experiment at all. For example, the ionization potential of helium is ~25 eV, which means that it takes a photon with an energy of at least 25 eV to ionize a helium atom:

$$\text{He} + h\nu \rightarrow \text{He}^+ + \text{e}^- \tag{3.1.1a}$$

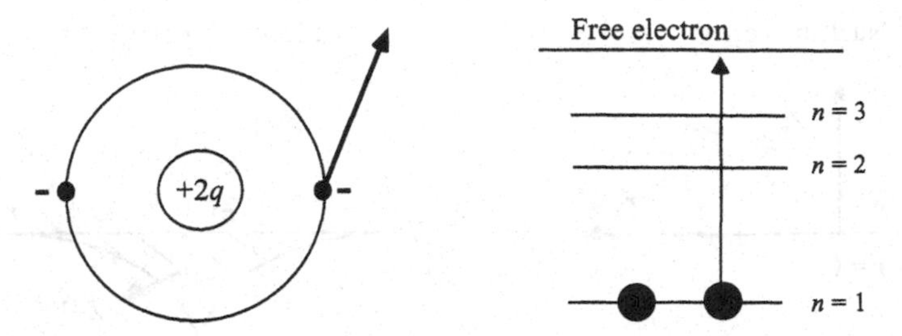

Fig. 3.1.2 Ionization of a neutral helium atom takes approximately 25 eV of energy, suggesting that the $n = 1$ level has an energy of -25 eV.

This suggests that the 1s level of a helium atom has an energy of -25 eV and not -54.4 eV as the simple argument would suggest. How could we be off by over 30 eV? It is because we did not account for the other electron in helium. If we were to measure the energy that it takes to remove the second electron from He^+

$$He^+ + h\nu \rightarrow He^{++} + e^- \tag{3.1.1b}$$

the result (known as the second ionization potential) is indeed close to 54.4 eV. But the (first) ionization potential is about 30 eV less, indicating that it takes 30 eV less energy to pull an electron out of a neutral helium atom than it takes to pull an electron out of a helium ion (He^+) that has already lost one electron. The reason is that an electron in a helium atom feels a repulsive force from the other electron, which effectively raises its energy by 30 eV and makes it easier for it to escape (Fig. 3.1.2).

In general, the ionization levels for multielectron atoms can be calculated approximately from the Schrödinger equation by adding to the nuclear potential $U_{nuc}(\vec{r})$, a self-consistent field $U_{SCF}(\vec{r})$ due to the other electrons (Fig. 3.1.3):

$$U(\vec{r}) = U_{nuc}(\vec{r}) + U_{SCF}(\vec{r}) \tag{3.1.2}$$

For all atoms, the nuclear potential arises from the nuclear charge of $+Zq$ located at the origin and is given by $U_{nuc}(\vec{r}) = -Zq^2/4\pi\varepsilon_0 r$. The self-consistent field arises from the other $(Z-1)$ electrons, since an electron does not feel any potential due to itself. In order to calculate the potential $U_{SCF}(\vec{r})$ we need the electronic charge which depends on the wavefunctions of the electron which in turn has to be calculated from the Schrödinger equation containing $U_{SCF}(\vec{r})$. This means that the calculation has to be done self-consistently as follows.

Step 1. Guess electronic potential $U_{SCF}(\vec{r})$.
Step 2. Find eigenfunctions and eigenvalues from Schrödinger equation.
Step 3. Calculate the electron density $n(\vec{r})$.
Step 4. Calculate the electronic potential $U_{SCF}(\vec{r})$.

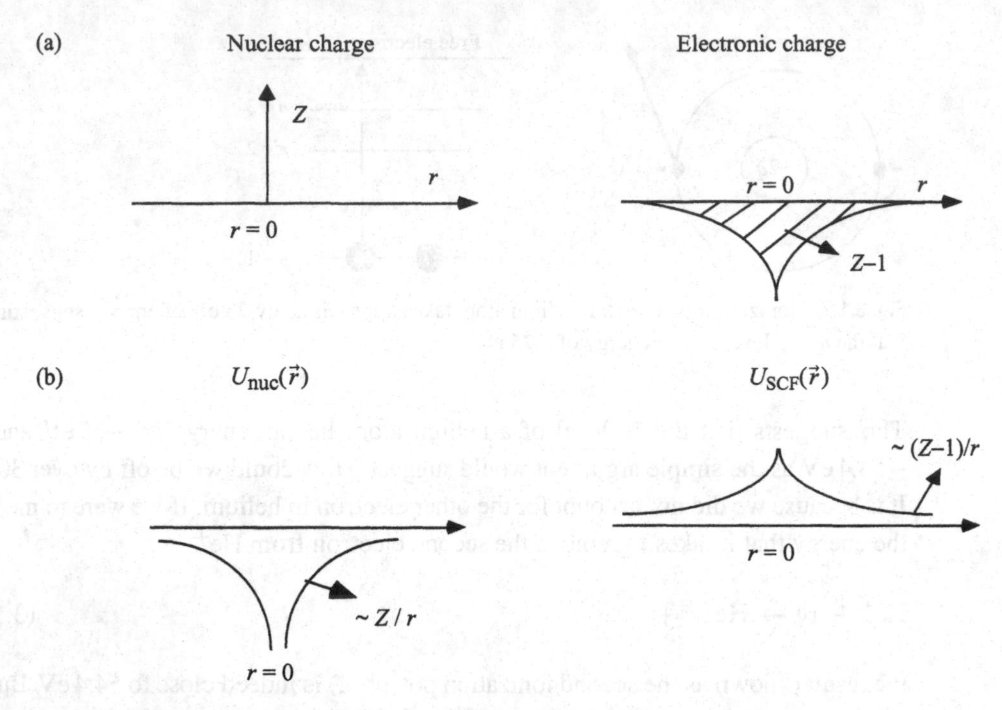

Fig. 3.1.3 Sketch of (a) the nuclear charge density and the electronic charge density; (b) potential energy felt by an additional electron due to the nucleus, $U_{nuc}(r)$, and the other electrons, $U_{SCF}(r)$. The latter has to be calculated self-consistently.

Step 5. If the new $U_{SCF}(\vec{r})$ is significantly different from last guess, update $U_{SCF}(\vec{r})$ and go back to Step 2. If the new $U_{SCF}(\vec{r})$ is within say 10 meV of the last guess, the result has converged and the calculation is complete.

For *Step 2* we can use essentially the same method as we used for the hydrogen atom, although an analytical solution is usually not possible. The potential $U_{SCF}(\vec{r})$ is in general not isotropic (which means independent of θ, ϕ) but for atoms it can be assumed to be isotropic without incurring any significant error. However, the dependence on r is quite complicated so that no analytical solution is possible. Numerically, however, it is just as easy to solve the Schrödinger equation with any $U(r)$ as it is to solve the hydrogen atom problem with $U(r) \sim 1/r$.

For *Step 3* we have to sum up the probability distributions for all the occupied eigenstates:

$$n(\vec{r}) = \sum_{\text{occ}\ \alpha} |\Phi_\alpha(\vec{r})|^2 = \sum_{\text{occ}\ n,l,m} \left|\frac{f_n(r)}{r}\right|^2 |Y_{lm}(\theta, \phi)|^2 \tag{3.1.3}$$

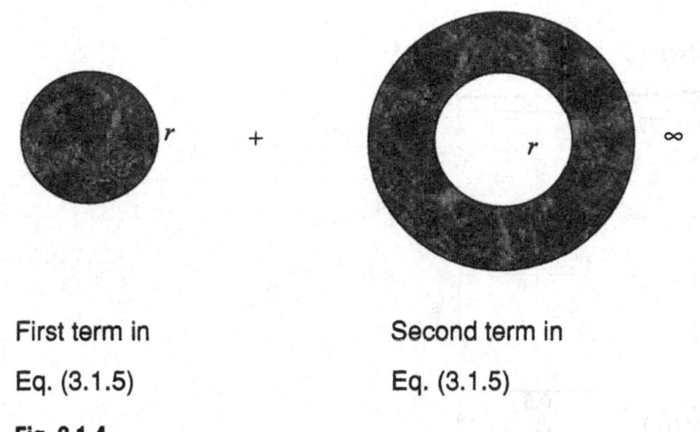

First term in Eq. (3.1.5) — Second term in Eq. (3.1.5)

Fig. 3.1.4

If we assume the charge distribution to be isotropic, we can write

$$\sigma(r) \equiv \int r^2 \sin\theta \, \mathrm{d}\theta \, \mathrm{d}\phi \, n(\vec{r}) = \sum_{\text{occ } n,l,m} |f_n(r)|^2 \tag{3.1.4}$$

For *Step 4* we can use straightforward electrostatics to show that

$$U_{\text{SCF}}(r) = \frac{Z-1}{Z}\left[\frac{q^2}{4\pi\varepsilon_0 r}\int_0^r \mathrm{d}r' \, \sigma(r') + \frac{q^2}{4\pi\varepsilon_0}\int_r^\infty \mathrm{d}r' \, \frac{\sigma(r')}{r'}\right] \tag{3.1.5}$$

The two terms in Eq. (3.1.5) arise from the contributions due to the charge within a sphere of radius r and that due to the charge outside of this sphere as shown in Fig. 3.1.4. The first term is the potential at r outside a sphere of charge that can be shown to be the same as if the entire charge were concentrated at the center of the sphere:

$$\frac{q^2}{4\pi\varepsilon_0 r}\int_0^r \mathrm{d}r' \, \sigma(r')$$

The second term is the potential at r inside a sphere of charge and can be shown to be the same as the potential at the center of the sphere (the potential is the same at all points inside the sphere since the electric field is zero)

$$\frac{q^2}{4\pi\varepsilon_0}\int_r^\infty \mathrm{d}r' \, \frac{\sigma(r')}{r'}$$

We obtain the total potential by adding the two components.

To understand the reason for the factor $(Z-1)/Z$ in Eq. (3.1.5), we note that the appropriate charge density for each eigenstate should exclude the eigenstate under consideration, since no electron feels any repulsion due to itself. For example, silicon has 14 electrons $1s^2\ 2s^2\ 2p^6\ 3s^2\ 3p^2$ and the self-consistent field includes all but one of

(a)

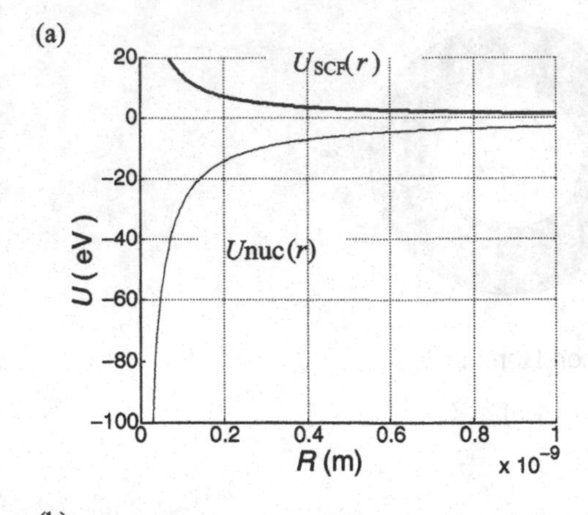

(b)

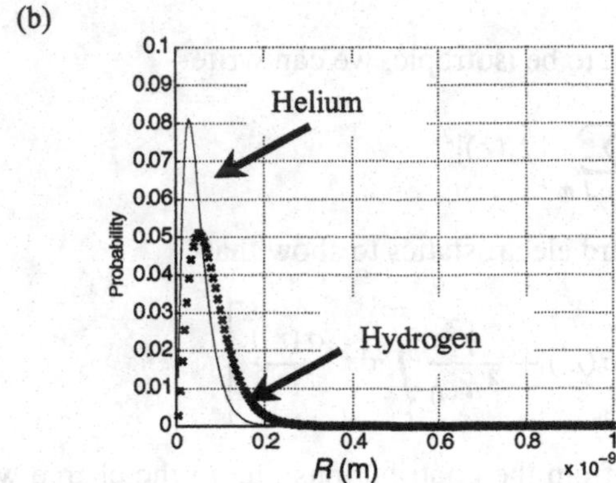

Fig. 3.1.5 Self-consistent field method applied to the helium atom. (a) Nuclear potential $U_{nuc}(r)$ and the self-consistent electronic potential $U_{SCF}(r)$. (b) Radial probability distribution for the 1s state in helium and hydrogen.

these electrons – for the 3p level we exclude the 3p electron, for the 3s level we exclude the 3s electron etc. However, it is more convenient to simply take the total charge density and scale it by the factor $(Z-1)/Z$. This preserves the spherical symmetry of the charge distribution and the difference is usually not significant. Note that the total electronic charge is equal to Z:

$$\int_0^\infty dr\, \sigma(r) = \sum_{\text{occ}\, n,l,m} 1 = Z \tag{3.1.6}$$

since the radial eigenfunctions are normalized: $\int_0^\infty dr\,|f_n(r)|^2 = 1$.

Helium atom: Figure 3.1.5 shows the potential profile and the probability distribution for the 1s state of helium obtained using the SCF method we have just described.

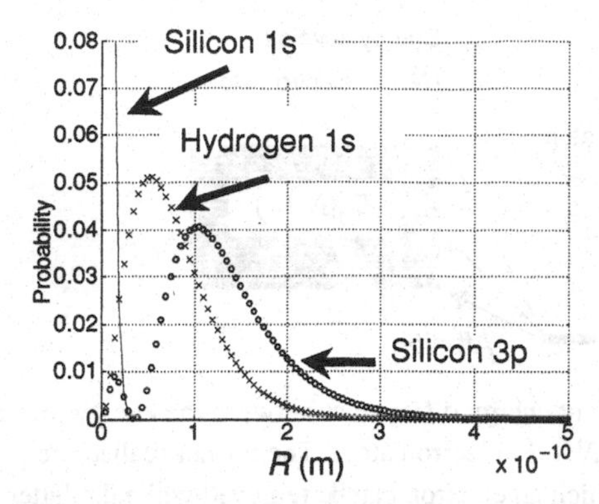

Fig. 3.1.6 Self-consistent field method applied to the silicon atom. The radial probability distributions for hydrogen 1s level and silicon 1s level and 3p level are shown.

Also shown for comparison is the 1s level of the hydrogen atom, discussed in the last chapter.

Silicon atom: Figure 3.1.6 shows the probability distribution for the 1s and 3p states of silicon obtained using the SCF method. Also shown for comparison is the 1s level of the hydrogen atom. Note that the silicon 1s state is very tightly confined relative to the 3p state or the hydrogen 1s state. This is typical of core states and explains why such states remain well-localized in solids, while the outer electrons (like 3p) are delocalized.

3.2 Relation to the multi-electron picture

Multi-electron Schrödinger equation: It is important to recognize that the SCF method is really an approximation that is widely used only because the correct method is virtually impossible to implement. For example, if we wish to calculate the eigenstates of a helium atom with two electrons we need to solve a two-electron Schrödinger equation of the form

$$E\Psi(\vec{r}_1, \vec{r}_2) = \left(-\frac{\hbar^2}{2m}\nabla^2 + U(\vec{r}_1) + U(\vec{r}_2) + U_{ee}(\vec{r}_1, \vec{r}_2)\right)\Psi(\vec{r}_1, \vec{r}_2) \qquad (3.2.1)$$

where $\vec{r}_1$ and $\vec{r}_2$ are the coordinates of the two electrons and U_{ee} is the potential energy due to their mutual repulsion: $U_{ee}(\vec{r}_1, \vec{r}_2) = e^2/4\pi\varepsilon_0|\vec{r}_1 - \vec{r}_2|$. This is more difficult to solve than the "one-electron" Schrödinger equation that we have been talking about, but it is not impossible. However, this approach quickly gets out of hand as we go to bigger atoms with many electrons and so is seldom implemented directly. But suppose we could actually calculate the energy levels of multi-electron atoms. How would we use our results (in principle, if not in practice) to construct a *one-electron energy level*

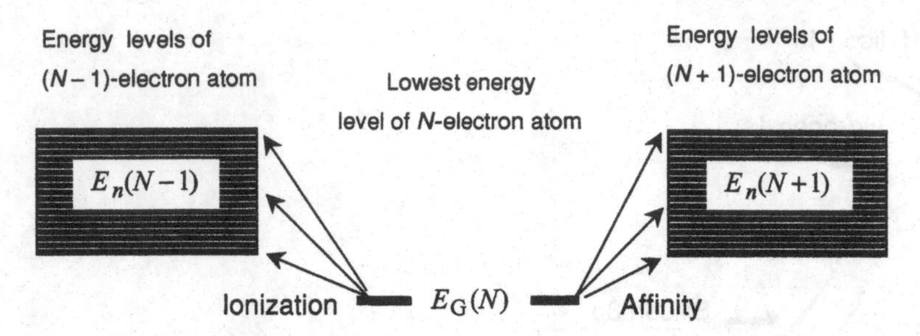

Fig. 3.2.1 One-electron energy levels represent energy differences between the energy levels of the N-electron atom and the $(N-1)$- or the $(N+1)$-electron atom. The former (called the ionization levels) are the filled states from which an electron can be removed while the latter (the affinity levels) are the empty states to which an electron can be added.

diagram like the ones we have been drawing? The answer depends on what we want our one-electron energy levels to tell us.

Ionization levels and affinity levels: Our interest is primarily in describing the flow of current, which involves inserting an electron and then taking it out or vice versa, as we discussed in Chapter 1. So we would want the one-electron energy levels to represent either the energies needed to take an electron out of the atom (ionization levels) or the energies involved in inserting an electron into the atom (affinity levels) (Fig. 3.2.1).

For the ionization levels, the one-electron energies ε_n represent the difference between the ground state energy $E_G(N)$ of the neutral N-electron atom and the nth energy level $E_n(N-1)$ of the positively ionized $(N-1)$-electron atom:

$$\varepsilon_n = E_G(N) - E_n(N-1) \tag{3.2.2a}$$

These ionization energy levels are measured by looking at the photon energy needed to ionize an electron in a particular level. Such photoemission experiments are very useful for probing the occupied energy levels of atoms, molecules, and solids. However, they only provide information about the occupied levels, like the 1s level of a helium atom or the valence band of a semiconductor. To probe unoccupied levels such as the 2s level of a helium atom or the conduction band of a semiconductor we need an inverse photoemission (IPE) experiment (see Fig. 3.2.2):

$$\text{He} + \text{e}^- \rightarrow \text{He}^- + h\nu$$

with which to measure the affinity of the atom for acquiring additional electrons. To calculate the affinity levels we should look at the difference between the ground state energy $E_G(N)$ and the nth energy level $E_n(N+1)$ of the negatively ionized $(N+1)$-electron atom:

$$\varepsilon_n = E_n(N+1) - E_G(N) \tag{3.2.2b}$$

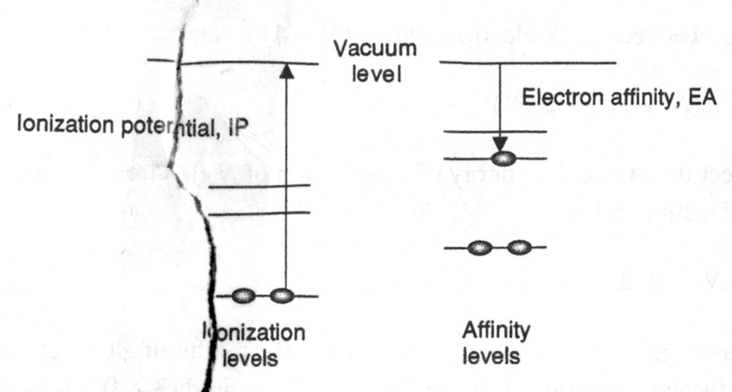

Fig. 3.2.2 The ionization levels include the repulsive potential from $Z-1$ electrons while the affinity levels include that of Z electrons, so that the latter is higher in energy by the single-electron charging energy U_0.

Note that if we want the energy levels to correspond to optical transitions then we should look at the difference between the ground state energy $E_G(N)$ and the nth energy level $E_n(N)$ of the N-electron atom, since visible light does not change the total number of electrons in the atom, just excites them to a higher energy:

$$\varepsilon_n = E_n(N) - E_G(N)$$

There is no *a priori* reason why the energy gap obtained from this calculation should correspond to the energy gap obtained from either the ionization or the affinity levels. In large solids (without significant excitonic effects) we are accustomed to assuming that the optical gap is equal to the gap between the valence and conduction bands, but this need not be true for small nanostructures.

Single-electron charging energy: As we have explained above, the straightforward approach for calculating the energy levels would be to calculate the energies $E_G(N)$ and $E_n(N \pm 1)$ from an N-electron and an $(N \pm 1)$-electron Schrödinger equation (cf. Eq. (3.2.1) which is a two-electron Schrödinger equation) respectively. This, however, is usually impossible and the only practical approach for large atoms, molecules, or solids is to include an effective potential $U_{SCF}(\vec{r})$ in the Schrödinger equation as we have been discussing.

How do we choose this effective potential? If we use $U_{ee}(N)$ to denote the total electron–electron interaction energy of an N-electron system then the appropriate U_{SCF} for the ionization levels is equal to the change in the interaction energy as we go from an N-electron to an $(N-1)$-electron atom:

$$[U_{SCF}]_{ionization} = U_{ee}(N) - U_{ee}(N-1) \quad (3.2.3a)$$

Similarly the appropriate U_{SCF} for the affinity levels is equal to the change in the interaction energy between an N-electron and an $(N+1)$-electron atom:

$$[U_{SCF}]_{affinity} = U_{ee}(N+1) - U_{ee}(N) \tag{3.2.3b}$$

The electron–electron interaction energy of a collection of N electrons is proportional to the number of distinct pairs:

$$U_{ee}(N) = U_0 N(N-1)/2 \tag{3.2.4}$$

where U_0 is the average interaction energy per pair, similar to the single-electron charging energy introduced in Section 1.4. From Eqs. (3.2.3a, b) and (3.2.4) it is easy to see that

$$[U_{SCF}]_{ionization} = U_0(N-1) \quad \text{while} \quad [U_{SCF}]_{affinity} = U_0 N \tag{3.2.5}$$

This means that to calculate the ionization levels of a Z-electron atom, we should use the potential due to $(Z-1)$ electrons (one electron for helium) as we did in the last section. But to calculate the affinity levels we should use the potential due to Z electrons (two electrons for helium). The energy levels we obtain from the first calculation are lower in energy than those obtained from the second calculation by the single-electron charging energy U_0.

As we discussed in Section 1.5, the single-electron charging energy U_0 depends on the degree of localization of the electronic wavefunction and can be several electron-volts in atoms. Even in nanostructures that are say 10 nm or less in dimension, it can be quite significant (that is, comparable to k_BT).

Typically one uses a single self-consistent potential

$$U_{SCF} = \partial U_{ee}/\partial N = U_0 N - (U_0/2) \tag{3.2.6}$$

for all levels so that the ionization levels are $(U_0/2)$ lower while the affinity levels are $(U_0/2)$ higher than the energy levels we calculate. One important consequence of this is that even if an SCF calculation gives energy levels that are very closely spaced compared to k_BT (see Fig. 3.2.3a), a structure may not conduct well, because the one-electron charging effects will create a "Coulomb gap" between the occupied and unoccupied levels (Fig. 3.2.3b). Of course, this is a significant effect only if the single-electron charging energy U_0 is larger than k_BT.

Hartree approximation: In large conductors (large R) U_0 is negligible and the distinction between Z and $(Z-1)$ can be ignored. The self-consistent potential for both ionization and affinity levels is essentially the same and the expression

$$U_{SCF} = \partial U_{ee}/\partial N \tag{3.2.7}$$

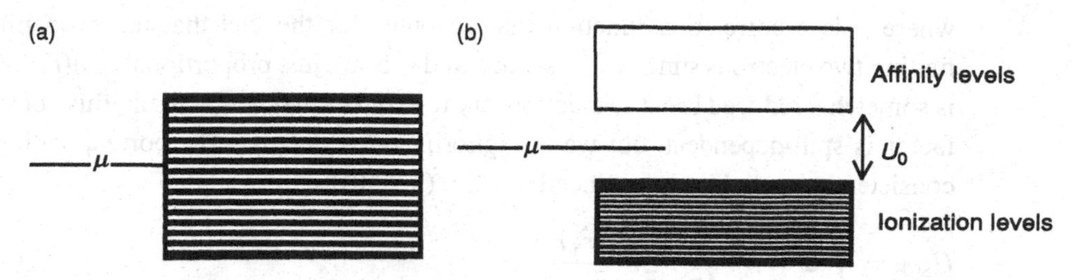

Fig. 3.2.3

can be generalized to obtain the standard expression used in density functional theory (DFT):

$$U_{\mathrm{SCF}}(\vec{r}) = \frac{\partial U_{\mathrm{ee}}}{\partial[n(\vec{r})]} \tag{3.2.8}$$

which tells us that the self-consistent potential at any point $\vec{r}$ is equal to the change in the electron–electron interaction energy due to an infinitesimal change in the number of electrons at the same point. If we use the standard expression for U_{ee} from classical electrostatics

$$U_{\mathrm{ee}} = \frac{1}{2}\int \mathrm{d}\vec{r} \int \mathrm{d}\vec{r}'\, \frac{q^2 n(\vec{r})n(\vec{r}')}{4\pi\varepsilon\,|\vec{r}-\vec{r}'|} \tag{3.2.9}$$

Equation (3.2.8) yields the Hartree approximation, $U_{\mathrm{H}}(\vec{r})$ for the self-consistent potential:

$$U_{\mathrm{H}}(\vec{r}) = \int \mathrm{d}\vec{r}'\, \frac{q^2 n(\vec{r}')}{4\pi\varepsilon\,|\vec{r}-\vec{r}'|} \tag{3.2.10}$$

which is a solution of the Poisson equation $-\nabla^2 U_{\mathrm{H}} = -q^2 n/\varepsilon$ in a homogeneous medium. Device problems often require us to incorporate complicated boundary conditions including different materials with different dielectric constants. It is then more convenient to solve a modified form of the Poisson equation that allows a spatially varying relative permittivity:

$$-\vec{\nabla}\cdot(\varepsilon_r \nabla U_{\mathrm{H})} = q^2 n/\varepsilon_0 \tag{3.2.11}$$

But for atoms, there is no complicated inhomogeneity to account for and it is more convenient to work with Eq. (3.2.10).

Correlation energy: The actual interaction energy is less than that predicted by Eq. (3.2.9) because electrons can correlate their motion so as to avoid each other – this correlation would be included in a many-electron picture but is missed in the one-particle picture. One way to include it is to write

$$U_{\mathrm{ee}} = \frac{1}{2}\int \mathrm{d}\vec{r} \int \mathrm{d}\vec{r}'\, \frac{e^2 n(\vec{r})n(\vec{r}')\,[1-g(\vec{r},\vec{r}')]}{4\pi\varepsilon\,|\vec{r}-\vec{r}'|}$$

where g is a correlation function that accounts for the fact that the probability of finding two electrons simultaneously at $\vec{r}$ and $\vec{r}'$ is not just proportional to $n(\vec{r})n(\vec{r}')$, but is somewhat reduced because electrons try to avoid each other (actually this correlation factor is spin-dependent, but we are ignoring such details). The corresponding self-consistent potential is also reduced (cf. Eq. (3.2.10)):

$$U_{\text{SCF}} = \int d\vec{r}' \frac{e^2 n(\vec{r}')\,[1 - g(\vec{r}, \vec{r}')]}{4\pi\varepsilon\,|\vec{r} - \vec{r}'|} \tag{3.2.12}$$

Much research has gone into estimating the function $g(\vec{r}, \vec{r}')$ (generally referred to as the exchange-correlation "hole").

The basic effect of the correlation energy is to add a *negative* term $U_{\text{xc}}(\vec{r})$ to the Hartree term $U_{\text{H}}(\vec{r})$ discussed above (cf. Eq. (3.2.10)):

$$U_{\text{SCF}}(\vec{r}) = U_{\text{H}}(\vec{r}) + U_{\text{xc}}(\vec{r}) \tag{3.2.13}$$

One simple approximation, called the local density approximation (LDA) expresses U_{xc} at a point in terms of the electron density at that point:

$$U_{\text{xc}}(\vec{r}) = -\frac{q^2}{4\pi\varepsilon_0} C[n(\vec{r})]^{1/3} \tag{3.2.14}$$

Here, C is a constant of order one. The physical basis for this approximation is that an individual electron introduced into a medium with a background electron density $n(r)$ will push other electrons in its neighborhood, creating a positive correlation "hole" around it. If we model this hole as a positive sphere of radius r_0 then we can estimate r_0 by requiring that the total charge within the sphere be equal in magnitude to that of an electron:

$$n(r)\, 4\pi r_0^3/3 = 1 \rightarrow r_0 = \frac{1}{C}\,[n(r)]^{-1/3}$$

C being a constant of order one. The potential in Eq. (3.2.14) can be viewed as the potential at the center of this positive charge contained in a sphere of radius r_0:

$$U_{\text{xc}}(\vec{r}) = -\frac{q^2}{4\pi\varepsilon_0\, r_0}$$

Much work has gone into the SCF theory and many sophisticated versions of Eq. (3.2.14) have been developed over the years. But it is really quite surprising that the one-electron picture with a suitable SCF often provides a reasonably accurate description of multi-electron systems. The fact that it works so well is not something that can be proved mathematically in any convincing way. Our confidence in the SCF method stems from the excellent agreement that has been obtained with experiment for virtually every atom in the periodic table (see Fig. 3.2.4). Almost all the work on the theory of electronic structure of atoms, molecules, and solids is based on this method and that is what we will be using.

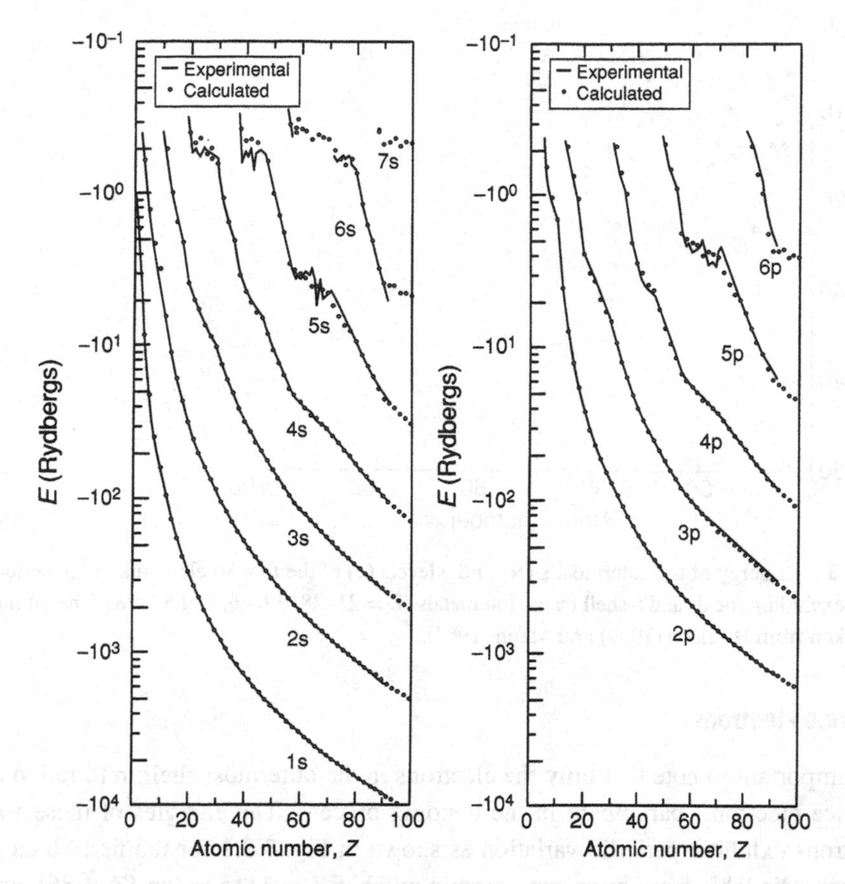

Fig. 3.2.4 Energy levels as a function of the atomic number calculated theoretically using a self-consistent field method. The results are in excellent agreement with experiment (adapted from Herman and Skillman (1963)). For a hydrogen atom, the s and p levels are degenerate (that is, they have the same energy). This is a consequence of the $\sim 1/r$ dependence of the nuclear potential. But this is not true of the self-consistent potential due to the electrons and, for multi-electron atoms, the s state has a lower energy than the p state.

3.3 Bonding

One of the first successes of quantum theory was to explain the structure of the periodic table of atoms by combining the energy levels obtained from the Schrödinger equation with the Pauli exclusion principle requiring that each level be occupied by no more than one electron. In Section 3.3.1 we will discuss the general trends, especially the periodic character of the energy levels of individual atoms. We will then discuss two bonding mechanisms (ionic (Section 3.3.2) and covalent (Section 3.3.3)) whereby a pair of atoms, A and B, can lower their overall energy by forming a molecule AB: $E(AB) < E(A) + E(B)$.

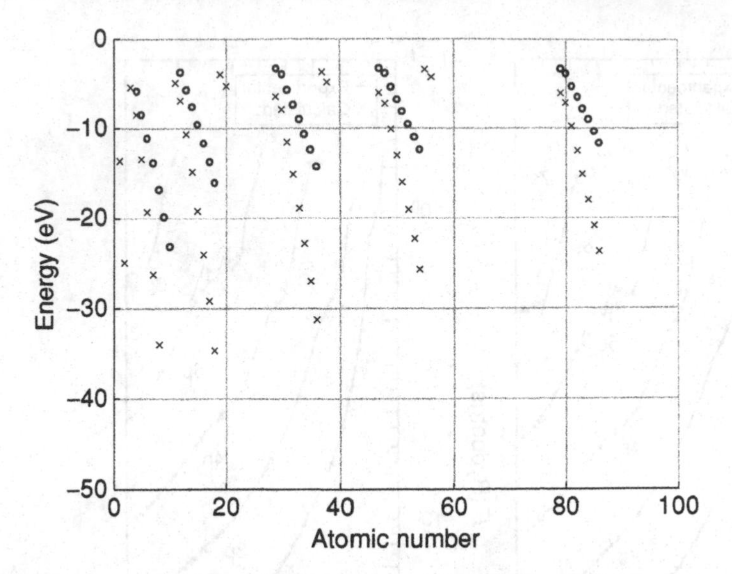

Fig. 3.3.1 Energy of the outermost s (×) and p levels (o) of the first 86 elements of the periodic table excluding the d- and f-shell transition metals ($Z = 21$–28, 39–46, and 57–78). The numbers are taken from Harrison (1999) and Mann (1967).

3.3.1 Valence electrons

It is important to note that only the electrons in the outermost shell, referred to as the valence electrons, participate in the bonding process. The energies of these valence electrons exhibit a periodic variation as shown in Fig. 3.3.1 for the first 86 atoms of the periodic table from hydrogen (atomic number $Z = 1$) to radon ($Z = 86$), excluding the d- and f-shell transition metals (see Table 3.3.1). The main point to notice is that the energies tend to go down as we go across a row of the periodic table from lithium (Li) to neon (Ne), increase abruptly as we step into the next row with sodium (Na) and then decrease as we go down the row to argon (Ar). This trend is shown by both the s and p levels and continues onto the higher rows. Indeed this periodic variation in the energy levels is at the heart of the periodic table of the elements.

3.3.2 Ionic bonds

Ionic bonds are typically formed between an atom to the left of the periodic table (like sodium, Na) and one on the right of the periodic table (like chlorine, Cl). The energy levels of Na and Cl look roughly as shown in Fig. 3.3.2. It seems natural for the 3s electron from Na to "spill over" into the 3p levels of Cl, thereby lowering the overall energy as shown. Indeed it seems "obvious" that the binding energy, E_{bin}, of NaCl would be

$$E_{\text{bin}} = E(\text{Na}) + E(\text{Cl}) - E(\text{Na}^+\text{Cl}^-) = 12.3 - 5.1 = 7.2\,\text{eV}.$$

Table 3.3.1 *First 86 atoms of the periodic table from hydrogen (atomic number Z = 1) to radon (Z = 86), excluding the d- and f-shell transition metals (Z = 21–28, 39–46, and 57–78)*

H (Z=1)										He (Z=2)
Li (Z=3)	Be (Z=4)				B (Z=5)	C (Z=6)	N (Z=7)	O (Z=8)	F (Z=9)	Ne (Z=10)
Na (Z=11)	Mg (Z=12)				Al (Z=13)	Si (Z=14)	P (Z=15)	S (Z=16)	Cl (Z=17)	Ar (Z=18)
K (Z=19)	Ca (Z=20)	...	Cu (Z=29)	Zn (Z=30)	Ga (Z=31)	Ge (Z=32)	As (Z=33)	Se (Z=34)	Br (Z=35)	Kr (Z=36)
Rb (Z=37)	Sr (Z=38)	...	Ag (Z=47)	Zn (Z=48)	In (Z=49)	Sn (Z=50)	Sb (Z=51)	Te (Z=52)	I (Z=53)	Xe (Z=54)
Cs (Z=55)	Ba (Z=56)	...	Au (Z=79)	Hg (Z=80)	Tl (Z=81)	Pb (Z=82)	B (Z=83)	Po (Z=84)	At (Z=85)	Rn (Z=86)

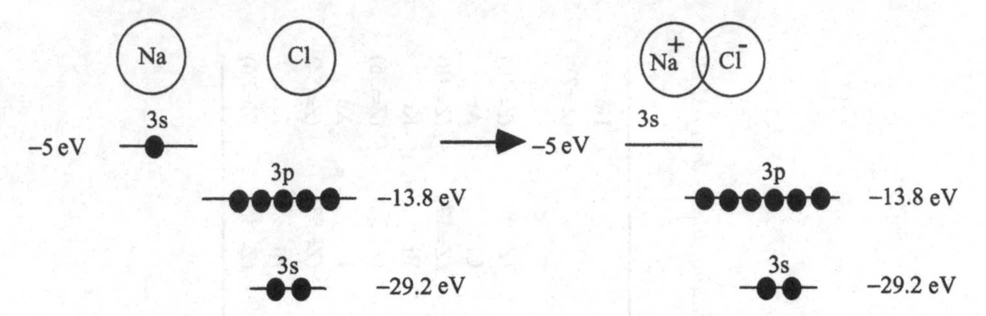

Fig. 3.3.2 Formation of Na^+Cl^- from individual Na and Cl atoms with a 3s electron from Na "spilling over" into the 3p levels of Cl thereby lowering the overall energy. This is only part of the story, since the overall energetics also includes the electrostatic energy stored in the microscopic capacitor formed by the two ions as explained in the text.

But this argument is incomplete because we also need to consider the change in the electrostatic energy due to the bonding. The correct binding energy is more like 4 eV.

The point is that the energy levels we have drawn here are all *ionization* levels. The energy needed to create a sodium ion is given by its ionization potential (IP)

$$E(\text{Na}^+) - E(\text{Na}) = \text{IP}(\text{Na}) = 5\,\text{eV} \tag{3.3.1a}$$

But the energy needed to create a chlorine ion is given by the *electron affinity* (EA) of Cl and this includes an extra charging energy U_0:

$$E(\text{Cl}) - E(\text{Cl}^-) = \text{EA}(\text{Cl}) = \text{IP}(\text{Cl}) - U_0 = 13.8\,\text{eV} - U_0 \tag{3.3.1b}$$

Combining Eqs. (3.3.1a) and (3.3.1b) we obtain

$$E(\text{Na}) + E(\text{Cl}) - E(\text{Na}^+) - E(\text{Cl}^-) = 8.8\,\text{eV} - U_0 \tag{3.3.2}$$

However, this is not the binding energy of NaCl. It gives us the energy gained in converting neutral Na and neutral Cl into a Na^+ and a Cl^- ion completely separated from each other. If we let a Na^+ and a Cl^- ion that are infinitely far apart come together to form a sodium chloride molecule, Na^+Cl^-, it will gain an energy U_0' in the process.

$$E(\text{Na}^+) + E(\text{Cl}^-) - E(\text{Na}^+\text{Cl}^-) = U_0'$$

so that the binding energy is given by

$$E_{\text{bin}} = E(\text{Na}) + E(\text{Cl}) - E(\text{Na}^+\text{Cl}^-) = 8.8\,\text{eV} - U_0 + U_0' \tag{3.3.3}$$

$U_0 - U_0'$ is approximately 5 eV, giving a binding energy of around 4 eV. The numerical details of this specific problem are not particularly important or even accurate. The main point I wish to make is that although the process of bonding by electron transfer may seem like a simple one where one electron "drops" off an atom into another with a lower energy level, the detailed energetics of the process require a more careful discussion. In general, care is needed when using one-electron energy level diagrams to discuss electron transfer on an atomic scale.

3.3.3 Covalent bonds

We have just seen how a lowering of energy comes about when we bring together an atom from the left of the periodic table (like sodium) and one from the right (like chlorine). The atoms on the right of the periodic table have lower electronic energy levels and are said to be more electronegative than those on the left. We would expect electrons to transfer from the higher energy levels in the former to the lower energy levels in the latter to form an ionic bond.

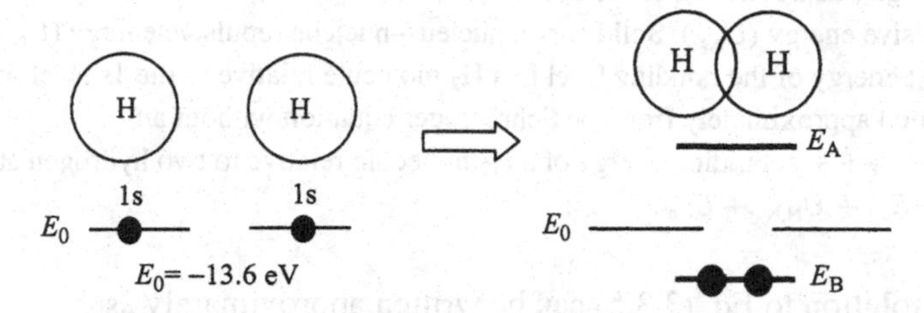

Fig. 3.3.3 Formation of H_2 from individual H atoms with a bonding level E_B and an anti-bonding level E_A.

However, this argument does not explain covalent bonds which involve atoms with roughly the same electronegativity. The process is a little more subtle. For example, it is hard to see why two identical hydrogen atoms would want to form a H_2 molecule, since no lowering of energy is achieved by transferring an electron from one atom to the other. What happens is that when the two atoms come close together the resulting energy levels split into a bonding level (E_B) and an anti-bonding level (E_A) as shown in Fig. 3.3.3. Both electrons occupy the bonding level which has an energy lower than that of an isolated hydrogen atom: $E_B < E_0$.

How do we calculate E_B? By solving the Schrödinger equation:

$$E_\alpha \Phi_\alpha(\vec{r}) = \left(-\frac{\hbar^2}{2m} \nabla^2 + U_N(\vec{r}) + U_{N'}(\vec{r}) + U_{SCF}(\vec{r}) \right) \Phi_\alpha(\vec{r}) \qquad (3.3.4)$$

where $U_N(r)$ and $U_{N'}(r)$ are the potentials due to the left and the right nuclei respectively and $U_{SCF}(r)$ is the potential that one electron feels due to the other. To keep things simple let us ignore $U_{SCF}(r)$ and calculate the electronic energy levels due to the nuclear potentials alone:

$$E_{\alpha 0} \Phi_{\alpha 0}(\vec{r}) = \left(-\frac{\hbar^2}{2m} \nabla^2 + U_N(\vec{r}) + U_{N'}(\vec{r}) \right) \Phi_{\alpha 0}(\vec{r}) \qquad (3.3.5)$$

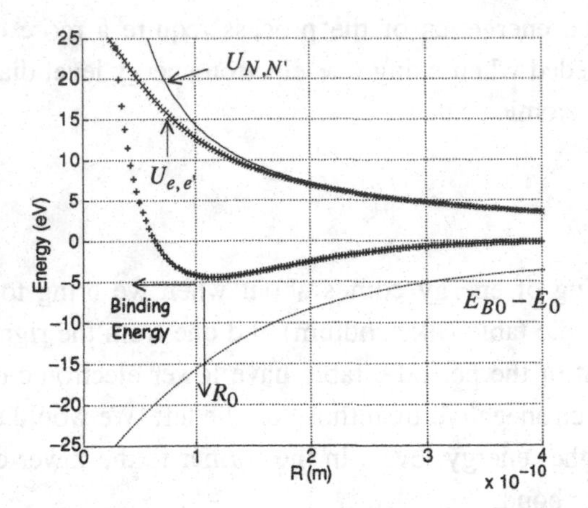

Fig. 3.3.4 Various energies as a function of the nuclear distance R. ×××, approximate electron–electron repulsive energy ($U_{e,e'}$). Solid curve, nucleus–nucleus repulsive energy ($U_{N,N'}$). Dashed curve, $E_{B0}-E_0$; energy of the bonding level in a H_2 molecule relative to the 1s level in a hydrogen atom calculated approximately from the Schrödinger equation without any self-consistent potential. ++++, binding energy of a H_2 molecule relative to two hydrogen atoms estimated from $2(E_{B0}-E_0)+U_{N,N'}+U_{e,e'}$.

The lowest energy solution to Eq. (3.3.5) can be written approximately as

$$E_{B0}=E_0+\frac{a+b}{1+s} \tag{3.3.6}$$

where (see Section 4.1)

$$a=-2E_0\frac{1-(1+\overline{R})e^{-2\overline{R}}}{\overline{R}}$$

$$b=-2E_0(1+\overline{R})\,e^{-\overline{R}}$$

$$s=e^{-\overline{R}}[1+\overline{R}+(\overline{R}^2/3)]$$

$$\overline{R}\equiv R/a_0$$

R being the center-to-center distance between the hydrogen atoms.

Let us now try to understand the competing forces that lead to covalent bonding. The dashed curve in Fig. 3.3.4 shows $E_{B0}-E_0$ versus the bond length R as given by Eq. (3.3.6). Experimentally, the bond length R for a H_2 molecule is 0.074 nm, indicating that the overall energy is a minimum for this value of R. Since the energy keeps decreasing as R is decreased, one might wonder why the two hydrogen atoms do not just sit on top of each other ($R=0$). To answer this question we need to calculate the overall energy which should include the electron–electron repulsion (note that $U_{SCF}(r)$

was left out from Eq. (3.3.6)) as well as the nucleus–nucleus repulsion. To understand the overall energetics let us consider the difference in energy between a hydrogen molecule (H_2) and two isolated hydrogen atoms (2H).

The energy required to assemble two separate hydrogen atoms from two protons (N, N′) and two electrons (e, e′) can be written as

$$E(2\text{H}) = U_{\text{e,N}} + U_{\text{e}',\text{N}'} = 2E_0 \tag{3.3.7a}$$

The energy required to assemble an H_2 molecule from two protons (N, N′) and two electrons (e, e′) can be written as

$$E(\text{H}_2) = U_{\text{N,N}'} + U_{\text{e,e}'} + U_{\text{e,N}} + U_{\text{e,N}'} + U_{\text{e}',\text{N}} + U_{\text{e}',\text{N}'} \tag{3.3.7b}$$

Equation (3.3.6) gives the quantum mechanical value of $(U_{\text{e,N}} + U_{\text{e,N}'})$ as well as $(U_{\text{e}',\text{N}} + U_{\text{e}',\text{N}'})$ as E_{B0}. Hence

$$E(\text{H}_2) = U_{\text{N,N}'} + U_{\text{e,e}'} + 2E_{\text{B0}} \tag{3.3.7c}$$

The binding energy is the energy it takes to make the hydrogen molecule dissociate into two hydrogen atoms and can be written as

$$E_{\text{bin}} = E(\text{H}_2) - E(2\text{H}) = 2(E_{\text{B0}} - E_0) + U_{\text{N,N}'} + U_{\text{e,e}'} \tag{3.3.8}$$

This is the quantity that ought to be a minimum at equilibrium and it consists of three separate terms. Eq. (3.3.6) gives us only the first term. The second term is easily written down since it is the electrostatic energy between the two nuclei, which are point charges:

$$U_{\text{N,N}'} = q^2/4\pi\varepsilon_0 R \tag{3.3.9a}$$

The electrostatic interaction between the two electrons should also look like $q^2/4\pi\varepsilon_0 R$ for large R, but should saturate to $\sim q^2/4\pi\varepsilon_0 a_0$ at short distances since the electronic charges are diffused over distances $\sim a_0$. Let us approximate it as

$$U_{\text{e,e}'} \cong q^2/4\pi\varepsilon_0\sqrt{R^2 + a_0^2} \tag{3.3.9b}$$

noting that this is just an oversimplified approximation to what is in general a very difficult quantum mechanical problem – indeed, electron–electron interactions represent the central outstanding problem in the quantum theory of matter.

The solid curve in Fig. 3.3.4 shows $U_{\text{N,N}'}$ (Eq. (3.3.9a)), while the ××× curve shows $U_{\text{e,e}'}$ (Eq. (3.3.9b)). The +++ curve shows the total binding energy estimated from Eq. (3.3.8). It has a minimum around 0.1 nm, which is not too far from the experimental bond length of 0.074 nm. Also the binding energy at this minimum is ~4.5 eV, very close to the actual experimental value. Despite the crudeness of the approximations used, the basic physics of bonding is illustrated fairly well by this example.

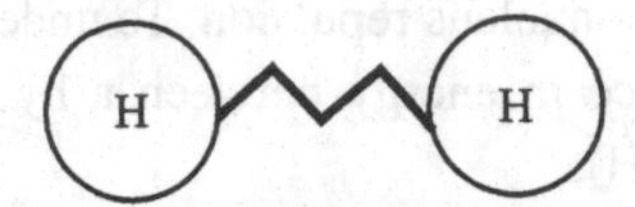

Fig. 3.3.5 A hydrogen molecule can be viewed as two masses connected by a spring.

Vibrational frequency: The shape of the binding energy vs. R curve suggests that we can visualize a hydrogen molecule as two masses connected by a spring (Fig. 3.3.5). An ideal spring with a spring constant K has a potential energy of the form $U(R) = K(R - R_0)^2/2$. The binding energy of the hydrogen molecule (see Fig. 3.3.4) can be approximated as $U(R) \cong U(R_0) + K(R - R_0)^2/2$, where the effective spring constant K is estimated from the curvature $[\mathrm{d}^2U/\mathrm{d}R^2]_{R=R_0}$. Indeed the vibrational frequency of the H–H bond can be estimated well from the resonant frequency $\sqrt{2K/M}$ of the mass and spring system where M is the mass of a hydrogen atom.

Ionization levels: As we have discussed, the energy levels of a multi-electron system usually denote the ionization levels, that is the energy it takes to strip an electron from the system. This means that in the present context the energy level E_B for a hydrogen molecule should represent

$$E_B = E(H_2) - E(H_2^+)$$

Since $E(H_2^+) = U_{N,N'} + U_{e',N} + U_{e',N'}$, we can write using Eq. (3.3.7b),

$$E_B = U_{e,e'} + U_{e,N} + U_{e,N'} = U_{e,e'} + E_{B0} \tag{3.3.10}$$

It is easy to check that for our model calculation (see Fig. 3.3.4) E_{B0} is nearly 15 eV below E_0, but E_B lies only about 4 eV below E_0. If we were to include a self-consistent field $U_{SCF}(r)$ in the Schrödinger equation, we would obtain the energy E_B which would be higher (less negative) than the non-interacting value of E_{B0} by the electron–electron interaction energy $U_{e,e'}$.

Binding energy: It is tempting to think that the binding energy is given by

$$E_{bin} = 2(E_B - E_0) + U_{N,N'}$$

since E_B includes the electron–electron interaction energy $U_{e,e'}$. However, it is easy to see from Eqs. (3.3.8) and (3.3.10) that the correct expression is

$$E_{bin} = 2(E_B - E_0) + (U_{N,N'} - U_{e,e'})$$

The point I am trying to make is that if we include the electron–electron interaction in our calculation of the energy level E_B then the overall energy of two electrons is NOT $2E_B$, for that would double-count the interaction energy between the two electrons. The correct energy is obtained by subtracting off this double-counted part: $2E_B - U_{e,e'}$.

3.4 Supplementary notes: multi-electron picture

As I mentioned in Section 3.2, the SCF method is widely used because the exact method based on a multi-electron picture is usually impossible to implement. However, it is possible to solve the multi-electron problem exactly if we are dealing with a small channel weakly coupled to its surroundings, like the one-level system discussed in Section 1.4. It is instructive to recalculate this one-level problem in the multi-electron picture and compare with the results obtained from the SCF method.

One-electron vs. multi-electron energy levels: If we have one spin-degenerate level with energy ε, the one-electron and multi-electron energy levels would look as shown in Fig. 3.4.1. Since each one-electron energy level can either be empty (0) or occupied (1), multi-electron states can be labeled in the form of binary numbers with number of digits equal to the number of one-particle states. N one-electron states thus give rise to 2^N multi-electron states, which quickly diverges as N increases, making a direct treatment impractical. That is why SCF methods are so widely used, even though they are only approximate.

Consider a system with two degenerate one-electron states (up-spin and down-spin) that can either be filled or empty. All other one-electron states are assumed not to change their occupation: those below remain filled while those above remain empty. Let us assume that the electron–electron interaction energy is given by

$$U_{ee}(N) = (U_0/2)N(N-1) \qquad \text{(same as Eq. (3.2.4))}$$

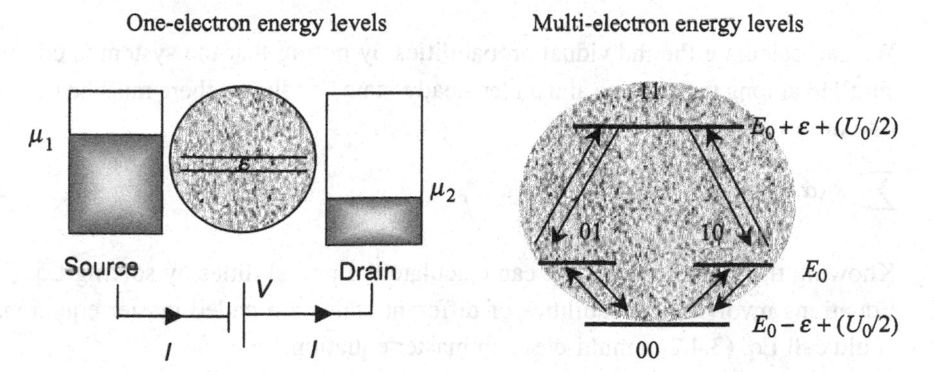

Fig. 3.4.1 One-electron vs. multi-electron energy levels in a channel with one spin-degenerate level having energy ε.

corresponding to a self-consistent potential (see Eq. (3.2.6))

$$\partial U_{ee}/\partial N = U_0 N - (U_0/2)$$

Suppose the number of electrons N_0 in the neutral state corresponds to having one of these states filled. The one-electron energy levels ε can be written as the sum of the "bare" levels $\tilde{\varepsilon}$ (obtained from a Schrödinger equation with just the nuclear potential, U_N) plus the self-consistent potential $[\partial U_{ee}/\partial N]_{N=N_0}$:

$$\varepsilon = \tilde{\varepsilon} + [\partial U_{ee}/\partial N]_{N=N_0} = \tilde{\varepsilon} + U_0 N_0 - (U_0/2)$$

Consider now the multi-electron picture. We have four available multi-electron states which we can designate as 00, 01, 10, and 11. In the neutral state, the system is in either the (10) or the (01) state whose total energy we denote as

$$E(10) = E(01) \equiv E_0$$

We can write the energies of the other multi-electron states as

$$\begin{aligned} E(11) &= E_0 + \tilde{\varepsilon} + U_{ee}(N_0 + 1) - U_{ee}(N_0) \\ &= E_0 + \tilde{\varepsilon} + U_0 N_0 = E_0 + \varepsilon + (U_0/2) \end{aligned}$$

and

$$\begin{aligned} E(00) &= E_0 - \tilde{\varepsilon} - U_{ee}(N_0) + U_{ee}(N_0 - 1) \\ &= E_0 - \tilde{\varepsilon} - U_0(N_0 - 1) = E_0 - \varepsilon + (U_0/2) \end{aligned}$$

Master equation: In the multi-electron picture, the overall system has different probabilities P_α of being in one of the 2^N possible states α and all the probabilities must add up to one:

$$\sum_\alpha P_\alpha = 1 \rightarrow P_{00} + P_{01} + P_{10} + P_{11} = 1 \tag{3.4.1}$$

We can calculate the individual probabilities by noting that the system is continually shuffled among these states and under steady-state conditions there must be no net flow into or out of any state:

$$\sum_\beta R(\alpha \rightarrow \beta)\, P_\alpha = \sum_\beta R(\beta \rightarrow \alpha) P_\beta \tag{3.4.2}$$

Knowing the rate constants, we can calculate the probabilities by solving Eq. (3.4.2). Equations involving probabilities of different states are called master equations. We could call Eq. (3.4.2) a multi-electron master equation.

The rate constants $R(\alpha \rightarrow \beta)$ can be written down assuming a specific model for the interaction with the surroundings. For example, if we assume that the interaction only involves the entry and exit of individual electrons from the source and drain contacts then for the 00 and 01 states the rate constants are given by

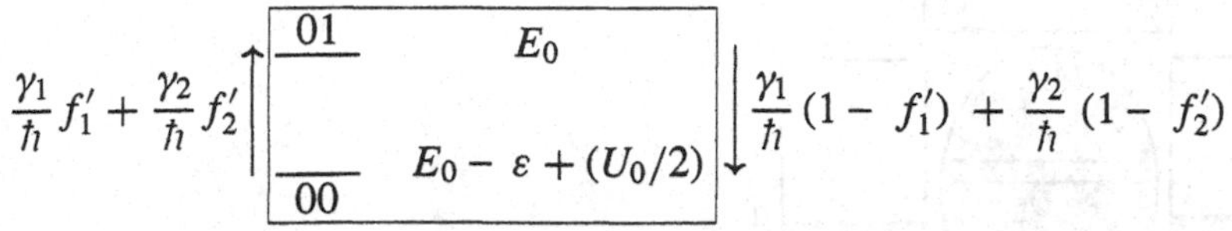

$$\frac{\gamma_1}{\hbar} f_1' + \frac{\gamma_2}{\hbar} f_2' \uparrow \quad \begin{array}{|ll|} \hline \underline{01} & E_0 \\ & \\ & E_0 - \varepsilon + (U_0/2) \\ \overline{00} & \\ \hline \end{array} \quad \downarrow \frac{\gamma_1}{\hbar}(1 - f_1') + \frac{\gamma_2}{\hbar}(1 - f_2')$$

where

$$f_1' \equiv f_0(\varepsilon_1 - \mu_1) \quad \text{and} \quad f_2' \equiv f_0(\varepsilon_1 - \mu_2)$$

tell us the availability of electrons with energy $\varepsilon_1 = \varepsilon - (U_0/2)$ in the source and drain contacts respectively. The entry rate is proportional to the available electrons, while the exit rate is proportional to the available empty states. The same picture applies to the flow between the 00 and the 10 states, assuming that up- and down-spin states are described by the same Fermi function in the contacts, as we would expect if each contact is locally in equilibrium.

Similarly we can write the rate constants for the flow between the 01 and the 11 states

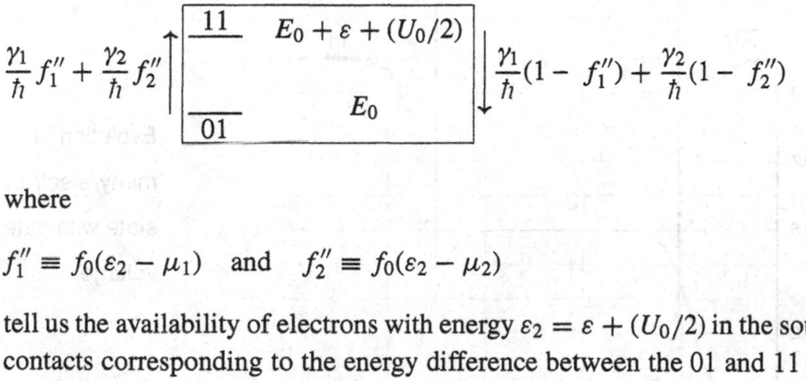

$$\frac{\gamma_1}{\hbar} f_1'' + \frac{\gamma_2}{\hbar} f_2'' \uparrow \quad \begin{array}{|ll|} \hline \underline{11} & E_0 + \varepsilon + (U_0/2) \\ & \\ & E_0 \\ \overline{01} & \\ \hline \end{array} \quad \downarrow \frac{\gamma_1}{\hbar}(1 - f_1'') + \frac{\gamma_2}{\hbar}(1 - f_2'')$$

where

$$f_1'' \equiv f_0(\varepsilon_2 - \mu_1) \quad \text{and} \quad f_2'' \equiv f_0(\varepsilon_2 - \mu_2)$$

tell us the availability of electrons with energy $\varepsilon_2 = \varepsilon + (U_0/2)$ in the source and drain contacts corresponding to the energy difference between the 01 and 11 states. This is larger than the energy difference ε between the 00 and 01 states because it takes more energy to add an electron when one electron is already present due to the interaction energy U_0.

Using these rate constants it is straightforward to show from Eq. (3.4.2) that

$$\frac{P_{10}}{P_{00}} = \frac{P_{01}}{P_{00}} = \frac{\gamma_1 f_1' + \gamma_2 f_2'}{\gamma_1(1 - f_1') + \gamma_2(1 - f_2')} \tag{3.4.3a}$$

and

$$\frac{P_{11}}{P_{10}} = \frac{P_{11}}{P_{01}} = \frac{\gamma_1 f_1'' + \gamma_2 f_2''}{\gamma_1(1 - f_1'') + \gamma_2(1 - f_2'')} \tag{3.4.3b}$$

Together with Eq. (3.4.1), this gives us all the individual probabilities. Figure 3.4.2 shows the evolution of these probabilities as the gate voltage V_G is increased

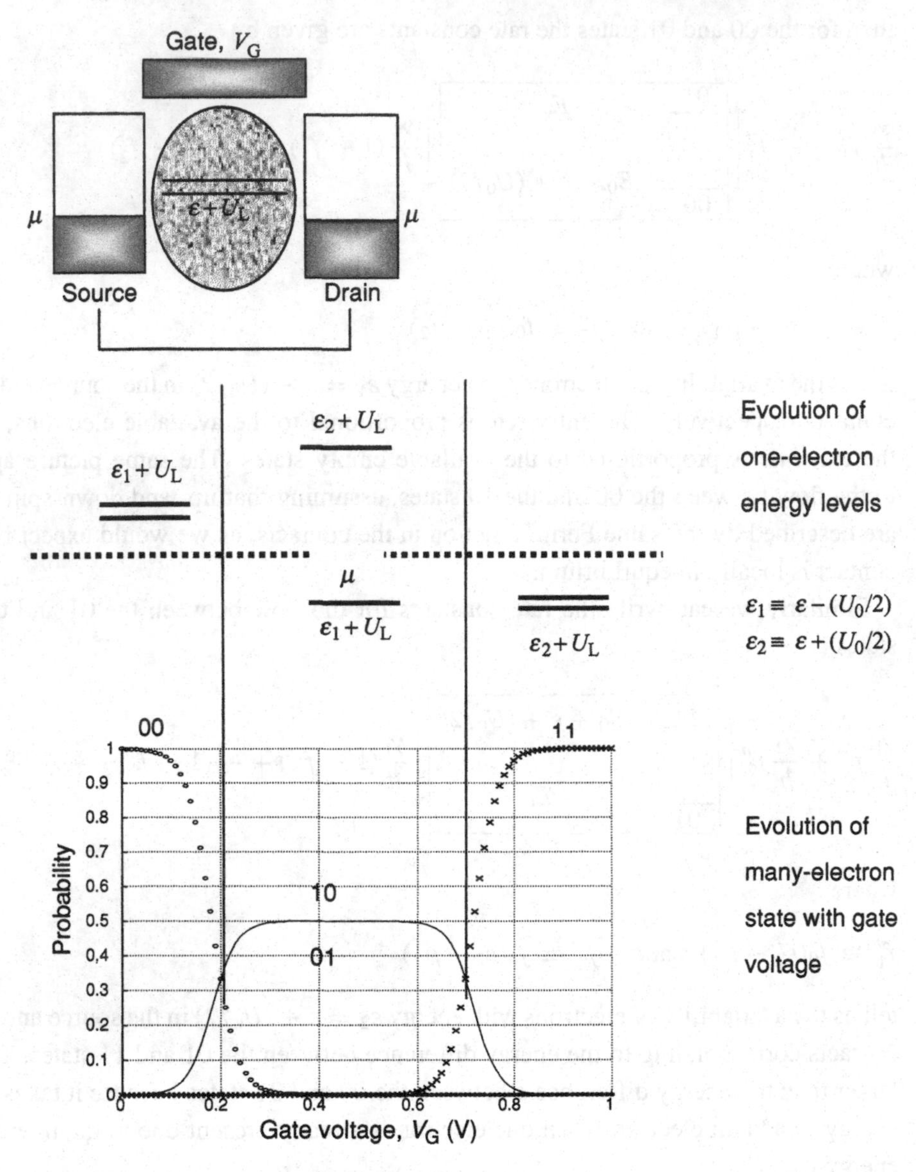

Fig. 3.4.2 Evolution of the energy levels of a channel with one spin-degenerate level as the gate voltage V_G is made more positive, holding the drain voltage V_D equal to zero. $\mu = 0$, $\varepsilon = 0.2$ eV, $k_B T = 0.025$ eV, $U_0 = 0.25$ eV, $U_L = -qV_G$. Lower plot shows the probabilities of finding the channel in one of its four states: P_{00} (∘), $P_{01} = P_{10}$ (solid) and P_{11} (×).

holding the drain voltage V_D equal to zero. The gate voltage shifts the one-electron level $\varepsilon \to \varepsilon + U_L$ (we have assumed $U_L = -qV_G$) and the probabilities are calculated from Eqs. (3.4.3a, b) and (3.4.1) noting that the Fermi functions are given by

$$f_1' = f_0(\varepsilon_1 + U_L - \mu_1),\ f_2' = f_0(\varepsilon_1 + U_L - \mu_2) \tag{3.4.4a}$$

$$f_1'' = f_0(\varepsilon_2 + U_L - \mu_1),\ f_2'' = f_0(\varepsilon_2 + U_L - \mu_2) \tag{3.4.4b}$$

The system starts out in the 00 state ($P_{00} = 1$), shifts to the 01 and 10 states ($P_{01} = P_{10} = 0.5$) once $\varepsilon_1 + U_L$ drops below μ, and finally goes into the 11 state ($P_{11} = 1$) when $\varepsilon_2 + U_L$ drops below μ.

Relation between the multi-electron picture and the one-electron levels: As I have emphasized in Section 3.2, one-electron energy levels represent *differences* between energy levels in the multi-electron picture corresponding to states that differ by *one electron*. Transitions involving the addition of one electron are called *affinity* levels while those corresponding to the removal of one electron are called *ionization* levels. For example (see Fig. 3.4.2), if the system is in the 00 state then there are two degenerate one-electron levels $\varepsilon_1 + U_L$ representing

$$\varepsilon_1 + U_L = E(10) - E(00) = E(01) - E(00) \qquad \textit{Affinity levels}$$

Once it is in the 10 state there are two one-electron levels

$$\varepsilon_1 + U_L = E(10) - E(00) \qquad \textit{Ionization level}$$

and $\varepsilon_2 + U_L = E(11) - E(10) \qquad$ *Affinity level*

In the 11 state there are two degenerate one-electron levels

$$\varepsilon_2 + U_L = E(11) - E(10) = E(11) - E(01) \qquad \textit{Ionization levels}$$

Affinity levels lie above μ, while ionization levels lie below μ as shown in Fig. 3.4.2. This is a very important general concept regarding the interpretation of the one-electron energy levels when dealing with complicated interacting objects. The occupied (or ionization) levels tell us the energy levels for removing an electron while the unoccupied (or affinity) levels tell us the energy levels for adding an extra electron. Indeed that is exactly how these levels are measured experimentally, the occupied levels by photoemission (PE) and the unoccupied levels by inverse photoemission (IPE) as mentioned in Section 1.1.

Law of equilibrium: Figure 3.4.2 represents an equilibrium calculation with both source and drain contacts having the same Fermi function: $f_1 = f_2$. Equilibrium problems do not really require the use of a master equation like Eq. (3.4.2). We can use the general principle of equilibrium statistical mechanics which states that the probability P_α that the system is in a multi-electron state α with energy E_α and N_α electrons is given by

$$P_\alpha = \frac{1}{Z} \exp[-(E_\alpha - \mu N_\alpha)/k_B T] \tag{3.4.5}$$

where the constant Z (called the partition function) is determined so as to ensure that the probabilities given by Eq. (3.4.5) for all states α add up to one:

$$Z = \sum_\alpha \exp[-(E_\alpha - \mu N_\alpha)/k_B T] \tag{3.4.6}$$

This is the central law of equilibrium statistical mechanics that is applicable to any system of particles (electrons, photons, atoms, etc.), interacting or otherwise (see for example, Chapter 1 of Feynman, 1972). The Fermi function is just a special case of this general relation that can be obtained by applying it to a system with just a single one-electron energy level, corresponding to two multi-electron states:

α	N_α	E_α	P_α
0	0	0	$1/Z$
1	1	ε	$(1/Z)\exp[(\mu-\varepsilon)/k_BT]$

so that $Z = 1 + \exp[(\mu - \varepsilon)/k_BT]$ and it is straightforward to show that the average number of electrons is equal to the Fermi function (Eq. (1.1.1)):

$$N = \sum_\alpha N_\alpha P_\alpha = P_1 = \frac{\exp[(\mu-\varepsilon)/k_BT]}{1+\exp[(\mu-\varepsilon)/k_BT]} = f_0\,[\varepsilon-\mu]$$

For multi-electron systems, we can use the Fermi function only if the electrons are not interacting. It is then justifiable to single out one level and treat it independently, ignoring the occupation of the other levels. The SCF method uses the Fermi function assuming that the energy of each level depends on the occupation of the other levels. But this is only approximate. The exact method is to abandon the Fermi function altogether and use Eq. (3.4.5) instead to calculate the probabilities of the different multi-particle states.

One well-known example of this is the fact that localized donor or acceptor levels (which have large charging energies U_0) in semiconductors at equilibrium are occupied according to a modified Fermi function (ν is the level degeneracy)

$$f = \frac{1}{1+(1/\nu)\exp\,[(\varepsilon-\mu)/k_BT]} \tag{3.4.7}$$

rather than the standard Fermi function (cf. Eq. (1.1.1)). We can easily derive this relation for two spin-degenerate levels ($\nu = 2$) if we assume that the charging energy U_0 is so large that the 11 state has zero probability. We can then write for the remaining states

α	N_α	E_α	P_α
00	0	0	$1/Z$
01	1	ε	$(1/Z)\exp[(\mu-\varepsilon)/k_BT]$
10	1	ε	$(1/Z)\exp[(\mu-\varepsilon)/k_BT]$

so that $Z = 1 + 2\exp[(\mu - \varepsilon)/k_BT]$ and the average number of electrons is given by

$$N = \sum_\alpha N_\alpha P_\alpha = P_{01} + P_{10} = \frac{2\exp[(\mu-\varepsilon)/k_BT]}{1+2\exp[(\mu-\varepsilon)/k_BT]}$$

$$= \frac{1}{1+(1/2)\exp[(\varepsilon-\mu)/k_BT]}$$

in agreement with Eq. (3.4.7). This result, known to every device engineer, could thus be viewed as a special case of the general result in Eq. (3.4.5).

Equation (3.4.5), however, can only be used to treat equilibrium problems. Our primary interest is in calculating the current under non-equilibrium conditions and that is one reason we have emphasized the master equation approach based on Eq. (3.4.2). For equilibrium problems, it gives the same answer. However, it also helps to bring out an important conceptual point. One often hears concerns that the law of equilibrium is a statistical one that can only be applied to large systems. But it is apparent from the master equation approach that the law of equilibrium (Eq. (3.4.5)) is not a property of the system. It is a property of the contacts or the "reservoir." The only assumptions we have made relate to the energy distribution of the electrons that come in from the contacts. As long as these "reservoirs" are simple, it does not matter how complicated or how small the "system" is.

Current calculation: Getting back to non-equilibrium problems, once we have solved the master equation for the individual probabilities, the source current can be obtained from

$$I_1 = -q \sum_{\beta} (\pm) R_1(\alpha \rightarrow \beta)\, P_\alpha$$

$(\pm)$: + if β has one more electron than α; − if β has one less electron than α

where R_1 represents the part of the total transition rate R associated with the source contact. In our present problem this reduces to evaluating the expression

$$\begin{aligned} I_1 = (-q/\hbar)\,(2\gamma_1 f_1' P_{00} &- \gamma_1(1 - f_1')(P_{01} + P_{10}) \\ &+ \gamma_1 f_1''(P_{01} + P_{10}) - 2\gamma_1(1 - f_1'')P_{11}) \end{aligned} \tag{3.4.8}$$

Figure 3.4.3 shows the current–drain voltage (I–V_D) characteristics calculated from the approach just described. The result is compared with a calculation based on the restricted SCF method described in Section 1.4. The SCF current–voltage characteristics look different from Fig. 1.4.6a because the self-consistent potential $U_0(N - N_0)$ has $N_0 = 1$ rather than zero and we have now included two spins. The two approaches agree well for $U_0 = 0.025$ eV, but differ appreciably for $U_0 = 0.25$ eV, showing evidence for Coulomb blockade or single-electron charging (see Exercise E.3.6).

The multi-electron master equation provides a suitable framework for the analysis of current flow in the Coulomb blockade regime where the single-electron charging energy U_0 is well in excess of the level broadening $\gamma_{1,2}$ and/or the thermal energy $k_B T$. We cannot use this method more generally for two reasons. Firstly, the

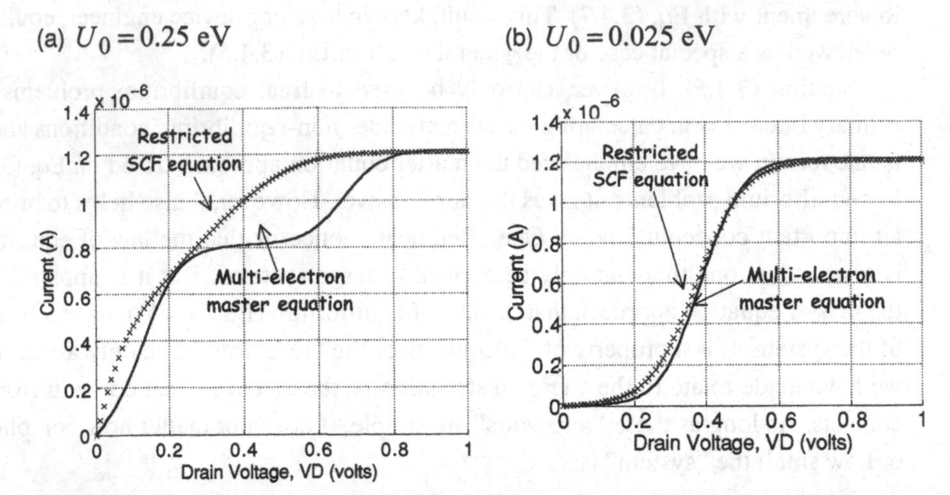

Fig. 3.4.3 Current vs. drain voltage V_D calculated assuming $V_G = 0$ with $\mu = 0$, $\varepsilon = 0.2$ eV, $k_BT = 0.025$ eV, $\gamma_1 = \gamma_2 = 0.005$ eV, $U_L = -qV_D / 2$. The two approaches (the SCF and the multi-electron master equation) agree well for $U_0 = 0.025$ eV, but differ appreciably for $U_0 = 0.25$ eV, showing evidence for Coulomb blockade or single-electron charging.

size of the problem increases exponentially and becomes prohibitive. Secondly, it is not clear how to incorporate broadening into this picture or any quantum interference between different energy levels. And so it remains a major challenge to provide a proper theoretical description of the intermediate transport regime $U_0 \sim \gamma_{1,2}, k_BT$: the regime where electronic motion is "strongly correlated" making a two-electron probability like $P(11)$ very different from the product of one-electron probabilities like $P(01)P(10)$. A lot of work has gone into trying to discover a suitable SCF within the one-electron picture that will capture the essential physics of correlation. For example, the self-consistent potential $U_{SCF} = U_0\Delta N$ we have used is the same for all energy levels or orbitals. One could use an "unrestricted" self-consistent field that is orbital-dependent such that the potential felt by level j excludes any self-interaction due to the number of electrons n_j in that level (see Exercise E.3.5(c)):

$$U_{SCF}(j) = U_0(\Delta N - \Delta n_j) \tag{3.4.9}$$

Such approaches can lead to better agreement with the results from the multi-electron picture but must be carefully evaluated, especially for non-equilibrium problems.

EXERCISES

E.3.1. Use the SCF method (only the Hartree term) to calculate the energy of the 1s level in a helium atom. (a) Plot the nuclear potential $U_N(r)$ and the self-consistent electronic potential $U_{SCF}(r)$ (cf. Fig. 3.1.4a). (b) Plot the wavefunction for the 1s level in helium and compare with that for the 1s level in hydrogen (cf. Fig. 3.1.4b).

E.3.2. Use the SCF method (only the Hartree term) to calculate the energies of the 3s and 3p levels in a silicon atom. Plot the wavefunction for the 1s and 3p levels in silicon and compare with that for the 1s level in hydrogen (cf. Fig. 3.1.4b).

E.3.3. Plot the approximate binding energy for a hydrogen molecule as a function of the hydrogen–hydrogen bond length, making use of Eqs. (3.3.6) and (3.3.9a, b) and compare with Fig. 3.3.4.

E.3.4: In Section 1.2 we obtained the following expression for the current through a single level

$$I = \frac{q}{h}\frac{\gamma_1\gamma_2}{\gamma_1+\gamma_2}\left[f_1(\varepsilon) - f_2(\varepsilon)\right]$$

and for the average number of electrons

$$N = \frac{\gamma_1 f_1 + \gamma_2 f_2}{\gamma_1+\gamma_2}$$

by writing a set of rate equations for a single one-electron energy level (without spin degeneracy). In the multi-electron picture we have two levels "0" and "1" corresponding to the one-electron level being empty or full respectively. Write down the appropriate rate equations in this picture and re-derive the expressions for "N" and "I".

E.3.5: Consider a channel with two spin-degenerate levels assuming the following parameters: $\mu = 0$, $\varepsilon = 0.2\,\text{eV}$, $k_B T = 0.025\,\text{eV}$, $\gamma_1 = \gamma_2 = 0.005\,\text{eV}$.

(a) Calculate the number of electrons vs. gate voltage V_G, with $V_D = 0$ and $U_L = -qV_G$, using (1) the multi-electron master equation and (2) a restricted SCF potential given by $U_{SCF} = U_0(N - N_0)$ with $N_0 = 1$. Use two different values of $U_0 = 0.025\,\text{eV}, 0.25\,\text{eV}$.

(b) Calculate the current vs. drain voltage V_D assuming $V_G = 0$ with $U_L = -qV_D/2$, using (1) the multi-electron master equation and (2) the restricted SCF potential given in (a). Use two different values of $U_0 = 0.025\,\text{eV}, 0.25\,\text{eV}$ and compare with Fig. 3.4.3.

(c) Repeat (a) and (b) with an unrestricted SCF potential (Eq. (3.4.9)) that excludes the self-interaction:

$$U_{SCF}(\uparrow) = U_0(\Delta N - \Delta n_\uparrow) = U_0(\Delta n_\downarrow) = U_0(n_\downarrow - 0.5)$$
$$U_{SCF}(\downarrow) = U_0(n_\uparrow - 0.5)$$

Note: The result may be different depending on whether the initial guess is symmetric, $U_{SCF}(\uparrow) = U_{SCF}(\downarrow)$ or not, $U_{SCF}(\uparrow) \neq U_{SCF}(\downarrow)$.

E.3.6: In Fig. 3.4.3a ($U_0 = 0.25\,\text{eV}$) the multi-electron approach yields two current plateaus: a lower one with $\varepsilon_2 + U_L > \mu_1 > \varepsilon_1 + U_L$ such that $f_1' \simeq 1, f_1'' \simeq 0$ and an upper one with $\mu_1 > \varepsilon_2 + U_L > \varepsilon_1 + U_L$, such that $f_1' \simeq 1, f_1'' \simeq 1$. In either case $f_2' \simeq 0, f_2'' \simeq 0$. Show from Eqs. (3.4.3) and (3.4.8) that the current at these plateaus is given by

$$\frac{2\gamma_1\gamma_2}{2\gamma_1 + \gamma_2} \quad \text{and} \quad \frac{2\gamma_1\gamma_2}{\gamma_1 + \gamma_2}$$

respectively.

4 Basis functions

We have seen that it is straightforward to calculate the energy levels for atoms using the SCF method, because the spherical symmetry effectively reduces it to a one-dimensional problem. Molecules, on the other hand, do not have this spherical symmetry and a more efficient approach is needed to make the problem numerically tractable. The concept of basis functions provides a convenient computational tool for solving the Schrödinger equation (or any differential equation for that matter). At the same time it is also a very important conceptual tool that is fundamental to the quantum mechanical viewpoint. In this chapter we attempt to convey both these aspects.

The basic idea is that the wavefunction can, in general, be expressed in terms of a set of basis functions, $u_m(\vec{r})$

$$\Phi(\vec{r}) = \sum_{m=1}^{M} \phi_m \, u_m(\vec{r})$$

We can then represent the wavefunction by a column vector consisting of the expansion coefficients

$$\Phi(\vec{r}) \rightarrow \{\phi_1 \quad \phi_2 \quad \cdots \quad \cdots \quad \phi_M\}^{\mathrm{T}}, \text{ 'T' denotes transpose}$$

In spirit, this is not too different from what we did in Chapter 2 where we represented the wavefunction by its values at different points on a discrete lattice:

$$\Phi(\vec{r}) \rightarrow \{\Phi(\vec{r}_1) \quad \Phi(\vec{r}_2) \quad \cdots \quad \cdots \quad \Phi(\vec{r}_M)\}^{\mathrm{T}}$$

However, the difference is that now we have the freedom to choose the basis functions $u_m(\vec{r})$: if we choose them to look much like our expected wavefunction, we can represent the wavefunction accurately with just a few terms, thereby reducing the size of the resulting matrix $[H]$ greatly. This makes the approach useful as a computational tool (similar in spirit to the concept of "shape functions" in the finite element method (Ramdas Ram-Mohan, 2002; White *et al.*, 1989)) as we illustrate with a simple example in Section 4.1.

But the concept of basis functions is far more general. One can view them as the coordinate axes in an abstract Hilbert space as described in Section 4.2 and we will illustrate the power and versatility of this viewpoint using the concept of the density matrix in Section 4.3 and further examples in Section 4.4.

4.1 Basis functions as a computational tool

The basic formulation can be stated fairly simply. We write the wavefunction in terms of any set of basis functions $u_m(\vec{r})$:

$$\Phi(\vec{r}) = \sum_m \phi_m \, u_m(\vec{r}) \tag{4.1.1}$$

and substitute it into the Schrödinger equation $E\Phi(\vec{r}) = H_{\mathrm{op}}\Phi(\vec{r})$ to obtain

$$E \sum_m \phi_m \, u_m(\vec{r}) = \sum_m \phi_m \, H_{\mathrm{op}} \, u_m(\vec{r})$$

Multiply both sides by $u_n^*(\vec{r})$ and integrate over all $\vec{r}$ to yield

$$E \sum_m S_{nm} \, \phi_m = \sum_m H_{nm} \, \phi_m$$

which can be written as a matrix equation

$$E[S]\{\phi\} = [H]\{\phi\} \tag{4.1.2}$$

where

$$S_{nm} = \int \mathrm{d}\vec{r} \, u_n^*(\vec{r}) u_m(\vec{r}) \tag{4.1.3a}$$

$$H_{nm} = \int \mathrm{d}\vec{r} \, u_n^*(\vec{r}) H_{\mathrm{op}} \, u_m(\vec{r}) \tag{4.1.3b}$$

To proceed further we have to evaluate the integrals and that is the most time-consuming step in the process. But once the matrix elements have been calculated, it is straightforward to obtain the eigenvalues E_α and eigenvectors $\{\Phi_\alpha\}$ of the matrix. The eigenfunctions can then be written down in "real space" by substituting the coefficients back into the original expansion in Eq. (4.1.1):

$$\Phi_\alpha(\vec{r}) = \frac{1}{\sqrt{Z_\alpha}} \sum_m \phi_{m\alpha} \, u_m(\vec{r}), \quad \Phi_\alpha^*(\vec{r}) = \frac{1}{\sqrt{Z_\alpha}} \sum_n \phi_{n\alpha}^* \, u_n^*(\vec{r}) \tag{4.1.4}$$

where Z_α is a constant chosen to ensure proper normalization:

$$1 = \int \mathrm{d}\vec{r} \, \Phi_\alpha^*(\vec{r}) \, \Phi_\alpha(\vec{r}) \rightarrow Z_\alpha = \sum_n \sum_m \phi_{n\alpha}^* \, \phi_{m\alpha} \, S_{nm} \tag{4.1.5}$$

Equations (4.1.1)–(4.1.5) summarize the basic mathematical relations involved in the use of basis functions.

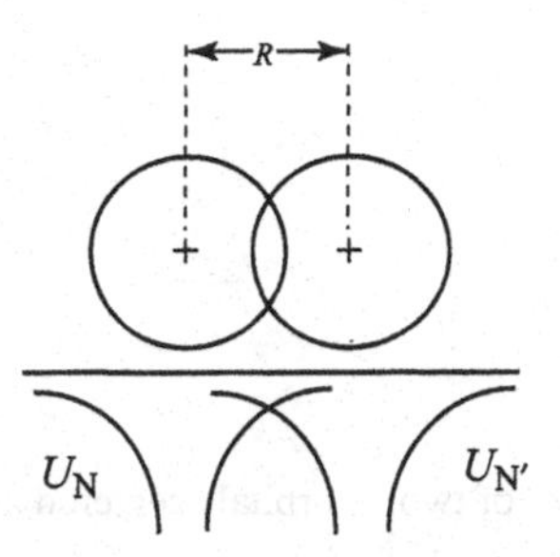

Fig. 4.1.1 U_N and $U_{N'}$ are the Coulombic potentials due to the left and right nuclei of a H_2 molecule respectively.

A specific example: To understand the underlying physics and how this works in practice let us look at a specific example. In Section 3.3.3 we stated that the lowest energy eigenvalue of the Schrödinger equation including the two nuclear potentials (Fig. 4.1.1) but excluding the self-consistent potential

$$E_{\alpha 0}\,\Phi_{\alpha 0}(\vec{r}) = \left(-\frac{\hbar^2}{2m}\nabla^2 + U_N(\vec{r}) + U_{N'}(\vec{r})\right)\Phi_{\alpha 0}(\vec{r}) \tag{4.1.6}$$

is approximately given by ($E_1 \equiv -E_0 = -13.6\,\text{eV}$)

$$E_{B0} = E_1 + \frac{a+b}{1+s} \tag{4.1.7}$$

where

$$a = -2E_0\,\frac{1-(1+R_0)e^{-2R_0}}{R_0} \qquad b = -2E_0(1+R_0)e^{-R_0}$$

$$s = e^{-R_0}\left[1 + R_0 + (R_0^2/3)\right] \qquad R_0 \equiv R/a_0$$

R being the center-to-center distance between the hydrogen atoms.

We will now use the concept of basis functions to show how this result is obtained from Eq. (4.1.6).

Note that the potential $U(\vec{r}) = U_N(\vec{r}) + U_{N'}(\vec{r})$ in Eq. (4.1.6) is not spherically symmetric, unlike the atomic potentials we discussed in Chapters 2 and 3. This means that we cannot simply solve the radial Schrödinger equation. In general, we have to solve the full three-dimensional Schrödinger equation, which is numerically quite challenging; the problem is made tractable by using basis functions to expand the wavefunction. In the present case we can use just two basis functions

$$\Phi_{\alpha 0}(\vec{r}) = \phi_L\,u_L(\vec{r}) + \phi_R\,u_R(\vec{r}) \tag{4.1.8}$$

where $u_L(\vec{r})$ and $u_R(\vec{r})$ represent a hydrogenic 1s orbital centered around the left and right nuclei respectively (see Fig. 4.1.2).

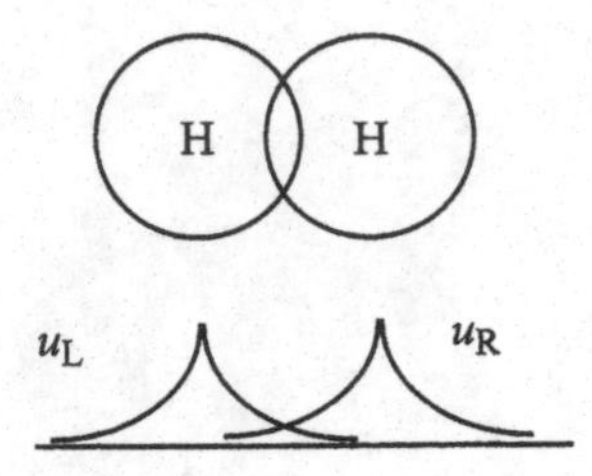

Fig. 4.1.2 A convenient basis set for the H_2 molecule consists of two 1s orbitals centered around the left and right nuclei respectively.

This means that

$$E_1\, u_L(\vec{r}) = \left(-\frac{\hbar^2}{2m}\nabla^2 + U_N(\vec{r})\right) u_L(\vec{r}) \tag{4.1.9a}$$

and

$$E_1 u_R(\vec{r}) = \left(-\frac{\hbar^2}{2m}\nabla^2 + U_{N'}(\vec{r})\right) u_R(\vec{r}) \tag{4.1.9b}$$

The ansatz in Eq. (4.1.8) is motivated by the observation that it clearly describes the eigenstates correctly if we move the two atoms far apart: the eigenstates are then given by

$$(\phi_L \quad \phi_R) = (1 \quad 0) \quad \text{and} \quad (\phi_L \quad \phi_R) = (0 \quad 1)$$

It seems reasonable to expect that if the bond length R is not too short (compared to the Bohr radius a_0) Eq. (4.1.8) will still provide a reasonably accurate description of the correct eigenstates with an appropriate choice of the coefficients ϕ_L and ϕ_R.

Since we have used only two functions u_L and u_R to express our wavefunction, the matrices [S] and [H] in Eqs. (4.1.2) are simple (2 × 2) matrices whose elements can be written down from Eqs. (4.1.3a, b) making use of Eqs. (4.1.9a, b):

$$S = \begin{bmatrix} 1 & s \\ s & 1 \end{bmatrix} \quad \text{and} \quad H = \begin{bmatrix} E_1 + a & E_1 s + b \\ E_1 s + b & E_1 + a \end{bmatrix} \tag{4.1.10}$$

where

$$s \equiv \int d\vec{r}\, u_L^*(\vec{r})\, u_R(\vec{r}) = \int d\vec{r}\, u_R^*(\vec{r})\, u_L(\vec{r}) \tag{4.1.11a}$$

$$a \equiv \int d(\vec{r})\, u_L^* \vec{r}\, U_{N'}(\vec{r})\, u_L(\vec{r}) = \int d\vec{r}\, u_R^*(\vec{r})\, U_N(\vec{r})\, u_R(\vec{r}) \tag{4.1.11b}$$

$$b \equiv \int d\vec{r}\, u_L^*(\vec{r})\, U_N(\vec{r})\, u_R(\vec{r}) = \int d\vec{r}\, u_L^*(\vec{r})\, U_{N'}(\vec{r})\, u_R(\vec{r})$$

$$= \int d\vec{r}\, u_R^*(\vec{r})\, U_N(\vec{r})\, u_L(\vec{r}) = \int d\vec{r}\, u_R^*(\vec{r})\, U_{N'}(\vec{r})\, u_L(\vec{r}) \tag{4.1.11c}$$

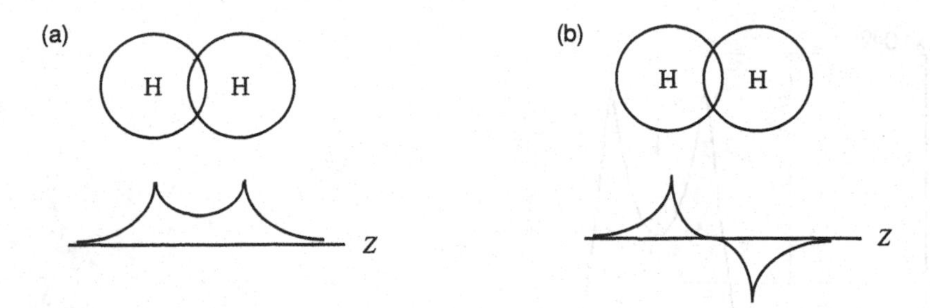

Fig. 4.1.3

Hence Eq. (4.1.2) becomes

$$E\begin{pmatrix}\phi_L \\ \phi_R\end{pmatrix} = \begin{bmatrix}1 & s \\ s & 1\end{bmatrix}^{-1} \begin{bmatrix}E_1 + a & E_1 s + b \\ E_1 s + b & E_1 + a\end{bmatrix}\begin{pmatrix}\phi_L \\ \phi_R\end{pmatrix} \tag{4.1.12}$$

from which it is straightforward to write down the two eigenvalues – the lower one is called the bonding level (B) and the higher one is called the anti-bonding level (A):

$$E_B = E_1 + \frac{a+b}{1+s} \quad \text{and} \quad E_A = E_1 + \frac{a-b}{1-s} \tag{4.1.13}$$

The quantities a, b, and s can be evaluated by plugging in the known basis functions $u_L(\vec{r})$, $u_R(\vec{r})$ and the nuclear potentials $U_N(\vec{r})$ and $U_{N'}(\vec{r})$ into Eqs. (4.1.11a, b, c). The integrals can be performed analytically to yield the results stated earlier in Eq. (4.1.7).

The wavefunctions corresponding to the bonding and anti-bonding levels are given by

$$(\phi_L \quad \phi_R)_B = (1 \quad 1) \quad \text{and} \quad (\phi_L \quad \phi_R)_A = (1 \quad -1)$$

which represent a symmetric (B) (Fig. 4.1.3a) and an antisymmetric (A) (Fig. 4.1.3b) combination of two 1s orbitals centered around the two nuclei respectively. Both electrons in a H_2 molecule occupy the symmetric or bonding state whose wavefunction can be written as

$$\Phi_{B0}(\vec{r}) = \frac{1}{\sqrt{Z}}\,[u_L(\vec{r}) + u_R(\vec{r})] \tag{4.1.14}$$

where

$$u_L(\vec{r}) = \frac{1}{\sqrt{\pi a_0^3}}\exp\left[\frac{-|\vec{r}-\vec{r}_L|}{a_0}\right] \qquad \vec{r}_L = -(R_0/2)\hat{z}$$

$$u_R(\vec{r}) = \frac{1}{\sqrt{\pi a_0^3}}\exp\left[\frac{-|\vec{r}-\vec{r}_R|}{a_0}\right] \qquad \vec{r}_R = +(R_0/2)\hat{z}$$

The constant Z has to be chosen to ensure correct normalization of the wavefunction:

$$1 = \int d\vec{r}\;\Phi^*_{B0}(\vec{r})\,\Phi_{B0}(\vec{r}) = \frac{2(1+s)}{Z} \rightarrow Z = 2(1+s)$$

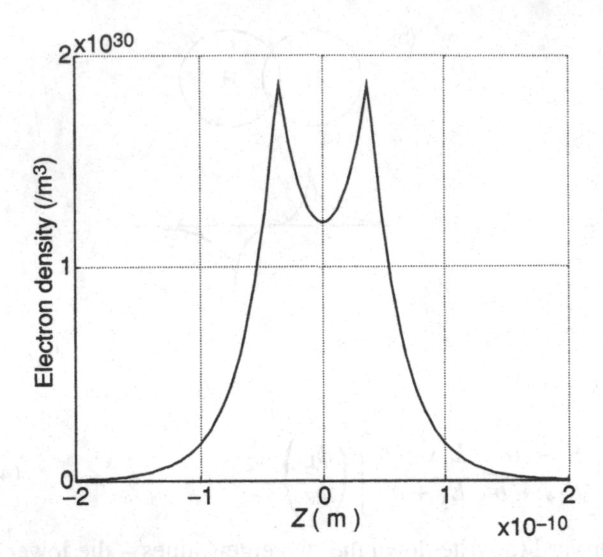

Fig. 4.1.4 Plot of electron density along the axis joining two hydrogen atoms assuming they are separated by the equilibrium bond distance of $R = 0.074$ nm.

The electron density $n(\vec{r})$ in a H_2 molecule is given by $|\Phi_{B0}(\vec{r})|^2$, multiplied by two since we have two electrons (one up-spin and one down-spin) with this wavefunction. Figure 4.1.4 shows a plot of the electron density along a line joining the two nuclei.

How can we get accurate results using just two basis functions? If we were to start from the Schrödinger equation and use a discrete lattice representation as we did in Chapters 2 and 3, we would need a fairly large number of basis functions per atom. For example if the lattice points are spaced by 0.5 Å and the size of an atom is 2.5 Å, then we need $5^3 = 125$ lattice points (each of which represents a basis function), since the problem is a three-dimensional one. What do we lose by using only one basis function instead of 125? The answer is that our results are accurate only over a limited range of energies.

To see this, suppose we were to use not just the 1s orbital as we did previously, but also the 2s, $2p_x$, $2p_y$, $2p_z$, 3s, $3p_x$, $3p_y$ and $3p_z$ orbitals (see Fig. 4.1.5). We argue that the lowest eigenstates will still be essentially made up of 1s wavefunctions and will involve negligible amounts of the other wavefunctions, so that fairly accurate results can be obtained with just one basis function per atom. The reason is that an off-diagonal matrix element M modifies the eigenstates of a matrix

$$\begin{bmatrix} E_1 & M \\ M & E_2 \end{bmatrix}$$

significantly only if it is comparable to the difference between the diagonal elements, that is, if $M \geq |E_1 - E_2|$. The diagonal elements are roughly equal to the energy levels

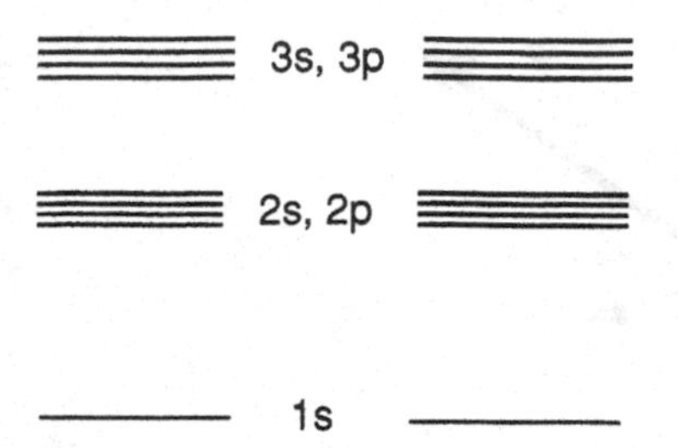

Fig. 4.1.5 Eigenstates of a H_2 molecule when the atoms are not too close. All these states could be used as basis functions for a more accurate treatment of the hydrogen molecule.

of the isolated atoms, so that $|E_1 - E_2|$ is $\sim$10 eV if we consider say the 1s and the 2s levels of a hydrogen atom. The off-diagonal element M depends on the proximity of the two atoms and for typical covalently bonded molecules and solids is $\sim$2 eV, which is smaller than $|E_1 - E_2|$. As a result the bonding level is primarily composed of 1s wavefunctions and our treatment based on a 2×2 matrix is fairly accurate. But a proper treatment of the higher energy levels would require more basis functions to be included.

***Ab initio* and semi-empirical models:** The concept of basis functions is widely used for *ab initio* calculations where the Schrödinger equation is solved directly including a self-consistent field. For large molecules or solids such calculations can be computationally quite intensive due to the large number of basis functions involved and the integrals that have to be evaluated to obtain the matrix elements. The integrals arising from the self-consistent field are particularly time consuming. For this reason, semi-empirical approaches are widely used where the matrix elements are adjusted through a combination of theory and experiment. Such semi-empirical approaches can be very useful if the parameters turn out to be "transferable," that is, if we can obtain them by fitting one set of observations and then use them to make predictions in other situations. For example, we could calculate suitable parameters to fit the known energy levels of an infinite solid and then use these parameters to calculate the energy levels in a finite nanostructure carved out of that solid.

4.2 Basis functions as a conceptual tool

We have mentioned that all practical methods for solving the energy levels of molecules and solids usually involve some sort of expansion in basis functions. However, the concept of basis functions is more than a computational tool. It represents an important conceptual tool for visualizing the physics and developing an intuition for what to expect. Indeed the concept of a wavefunction as a superposition of basis functions is central to the entire structure of quantum mechanics as we will try to explain next.

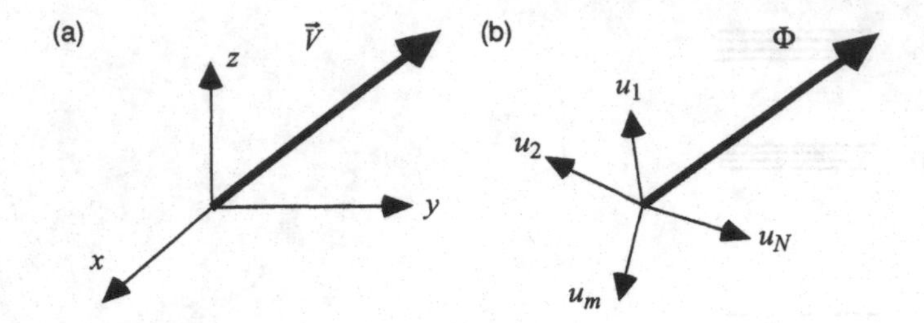

Fig. 4.2.1 (a) An ordinary vector $\vec{V}$ in three-dimensional space can be expressed in terms of its components along x, y, and z. (b) The state vector $\Phi(\vec{r})$ can be expressed in terms of its components along the basis functions $u_m(\vec{r})$.

Vector space vs. Hilbert space: It is useful to compare Eq. (4.1.1) with the expression for an ordinary three-dimensional vector $\vec{V}$ in terms of the three unit vectors $\hat{x}$, $\hat{y}$, and $\hat{z}$:

$$\vec{V} = V_x\hat{x} + V_y\hat{y} + V_z\hat{z} \quad \longleftrightarrow \quad \Phi = \phi_1 u_1 + \phi_2 u_2 + \phi_3 u_3 + \cdots$$

We can view the wavefunction as a state vector in an N-dimensional space called the Hilbert space, N being the total number of basis functions $u_m(\vec{r})$. The ϕ_ms in Eq. (4.1.1) are like the components of the state vector Φ, while the $u_m(\vec{r})$s are the associated unit vectors along the N coordinate axes. Choosing a different set of basis functions $u_m(\vec{r})$ is like choosing a different coordinate system: the components ϕ_m along the different axes all change, though the state vector remains the same. In principle, N is infinite, but in practice we can often get accurate results with a manageably finite value of N. We have tried to depict this analogy in Fig. 4.2.1 but it is difficult to do justice to an N-dimensional vector ($N > 3$) on two-dimensional paper. In the Dirac notation, which is very convenient and widely used, the state vector associated with wavefunction $\Phi(\vec{r})$ is denoted by a "ket" $|\Phi\rangle$ and the unit vectors associated with the basis functions $u_m(\vec{r})$ are also written as kets $|m\rangle$. In this notation the expansion in terms of basis functions (see Eq. (4.1.1)) is written as

$$\Phi(\vec{r}) = \sum_m \phi_m\, u_m(\vec{r}) \xrightarrow{\text{(Dirac notation)}} |\Phi\rangle = \sum_m \phi_m\, |m\rangle \tag{4.2.1}$$

Scalar product: A central concept in vector algebra is that of the scalar product:

$$\vec{A}\cdot\vec{B} = A_x B_x + A_y B_y + A_z B_z = \sum_{m=x,y,z} A_m B_m$$

The corresponding concept in Hilbert space is that of the overlap of any two functions $f(\vec{r})$ and $g(\vec{r})$:

$$\int \mathrm{d}\vec{r}\, f^*(\vec{r})\, g(\vec{r}) \xrightarrow{\text{(Dirac notation)}} \langle f|g\rangle$$

The similarity of the overlap integral to a scalar product can be seen by discretizing the integral:

$$\int d\vec{r}\, f^*(\vec{r})\, g(\vec{r}) \cong a^3 \sum_m f^*(\vec{r}_m)\, g(\vec{r}_m)$$

In the discrete lattice representation (see Fig. 2.2.1) the "component" f_m of $f(\vec{r})$ along $u_m(\vec{r})$ is given by $f(\vec{r}_m)$ just as A_m represents the component of the vector $\vec{A}$ along $\hat{m}$:

$$\int d\vec{r}\, f^*(\vec{r})\, g(\vec{r}) \cong a^3 \sum_m f_m^*\, g_m \quad \text{cf.} \quad \vec{A}\cdot\vec{B} = \sum_{m=x,y,z} A_m B_m$$

One difference here is that we take the complex conjugate of one of the functions (this is not important if we are dealing with real functions) which is represented by the "bra" $\langle f|$ as opposed to the "ket" $|g\rangle$. The scalar product is represented by juxtaposing a "bra" and a "ket" as in $\langle f|g\rangle$.

Orthogonality: Coordinate systems are said to be orthogonal if $\hat{n}\cdot\hat{m} = \delta_{nm}$, where the indices m and n stand for x, y, or z and δ_{nm} is the Kronecker delta which is defined as

$$\begin{aligned}\delta_{nm} &= 1 \quad \text{if } n = m \\ &= 0 \quad \text{if } n \neq m\end{aligned} \tag{4.2.2}$$

This is usually true (for example, $\hat{x}\cdot\hat{y} = \hat{y}\cdot\hat{z} = \hat{z}\cdot\hat{x} = 0$) but it is possible to work with non-orthogonal coordinate systems too. Similarly the basis functions $u_m(\vec{r})$ are said to be orthogonal if the following relation is satisfied:

$$\int d\vec{r}\, u_n^*(\vec{r})\, u_m(\vec{r}) = \delta_{nm} \xrightarrow{\text{(Dirac notation)}} \langle n|m\rangle = \delta_{nm} \tag{4.2.3a}$$

Note that the basis functions we used for the hydrogen molecule (see Fig. 4.1.2) are NON-orthogonal since

$$\int d\vec{r}\, u_L^*(\vec{r})\, u_R(\vec{r}) \equiv s = e^{-R_0}\left[1 + R_0 + (R_0^2/3)\right] \neq 0$$

In general

$$\int d\vec{r}\, u_n^*(\vec{r})\, u_m(\vec{r}) = S_{nm} \xrightarrow{\text{(Dirac notation)}} \langle n|m\rangle = S_{nm} \tag{4.2.3b}$$

Orthogonalization: Given a non-orthogonal set of basis functions $\{u_n(\vec{r})\}$, we can define another set

$$\tilde{u}_i(\vec{r}) = \sum_n [S^{-1/2}]_{ni}\, u_n(\vec{r}) \tag{4.2.4}$$

which will be orthogonal. This is shown as follows

$$\begin{aligned}\int d\vec{r}\,\tilde{u}_i^*(\vec{r})\,\tilde{u}_j(\vec{r}) &= \sum_n \sum_m [S^{-1/2}]_{in} S_{nm} [S^{-1/2}]_{mj} \\ &= [S^{-1/2} S\, S^{-1/2}]_{ij} \\ &= \delta_{ij}\end{aligned}$$

where we have made use of Eq. (4.2.3b). This means that if we use the new set $\{\tilde{u}_i(\vec{r})\}$ as our basis, then the overlap matrix $[S] = [I]$, where $[I]$ is the identity matrix which is a diagonal matrix with ones on the diagonal. This is a property of orthogonal basis functions which makes them conceptually easier to deal with.

Even if we start with a non-orthogonal basis, it is often convenient to orthogonalize it. What we might lose in the process is the local nature of the original basis which makes it convenient to visualize the physics. For example, the $\{u_n(\vec{r})\}$ we used for the hydrogen molecule were localized on the left and right hydrogen atoms respectively. But the orthogonalized basis $\{\tilde{u}_i(\vec{r})\}$ will be linear combinations of the two and thus less local than $\{u_n(\vec{r})\}$. As a rule, it is difficult to find basis functions that are both local and orthogonal. From here on we will generally assume that the basis functions we use are orthogonal.

Operators: An operator like H_{op} acting on a state vector changes it into a different state vector – we could say that it "rotates" the vector. With ordinary vectors we can represent a rotation by a matrix:

$$\begin{Bmatrix} A'_x \\ A'_y \end{Bmatrix} = \begin{bmatrix} R_{xx} & R_{xy} \\ R_{yx} & R_{yy} \end{bmatrix} \begin{Bmatrix} A_x \\ A_y \end{Bmatrix}$$

where for simplicity we have assumed a two-dimensional vector. How do we write down the matrix $[R]$ corresponding to an operator R_{op}? The general principle is the following: $R_{nm} = \hat{n} \cdot (R_{\mathrm{op}}\,\hat{m})$. For example, suppose we consider an operator that rotates a vector by an angle θ. We then obtain

$$\begin{aligned}R_{xx} &= \hat{x} \cdot (R_{\mathrm{op}}\,\hat{x}) = \hat{x} \cdot (\hat{x}\cos\theta + \hat{y}\sin\theta) = \cos\theta \\ R_{yx} &= \hat{y} \cdot (R_{\mathrm{op}}\,\hat{x}) = \hat{y} \cdot (\hat{x}\cos\theta + \hat{y}\sin\theta) = \sin\theta \\ R_{xy} &= \hat{x} \cdot (R_{\mathrm{op}}\,\hat{y}) = \hat{x} \cdot (-\hat{x}\sin\theta + \hat{y}\cos\theta) = -\sin\theta \\ R_{yy} &= \hat{y} \cdot (R_{\mathrm{op}}\,\hat{y}) = \hat{y} \cdot (-\hat{x}\sin\theta + \hat{y}\cos\theta) = \cos\theta\end{aligned}$$

The matrix representation for any operator A_{op} in Hilbert space is written using a similar prescription:

$$[A]_{nm} = \int d\vec{r}\,u_n^*(\vec{r})\,(A_{\mathrm{op}}\,u_m(\vec{r})) \xrightarrow{\text{(Dirac notation)}} [A]_{nm} = \langle n|A_{\mathrm{op}}\,m\rangle \tag{4.2.5}$$

Constant operator: What is the matrix representing a constant operator, one that simply multiplies a state vector by a constant C? In general, the answer is

$$[C]_{nm} = C \int d\vec{r}\, u_n^*(\vec{r})\, u_m(\vec{r}) = C[S]_{nm} \qquad (4.2.6)$$

which reduces to $C\,[I]$ for orthogonal bases.

Matrix representation of the Schrödinger equation: The matrix representation of the Schrödinger equation obtained in the last section (see Eqs. (4.1.2), (4.1.3a, b))

$$E\,\Phi(\vec{r}) = H_{\text{op}}\,\Phi(\vec{r}) \longrightarrow E[S]\{\phi\} = [H]\{\phi\} \qquad (4.2.7)$$

can now be understood in terms of the concepts described in this section. Like the rotation operator in vector space, any differential operator in Hilbert space has a matrix representation. Once we have chosen a set of basis functions, H_{op} becomes the matrix $[H]$ while the constant E becomes the matrix $E[S]$:

$$[S]_{nm} = \langle n|m\rangle \equiv \int d\vec{r}\, u_n^*(\vec{r})\, u_m(\vec{r}) \qquad (4.2.8a)$$

$$[H]_{nm} = \langle n|H_{\text{op}}\,m\rangle \equiv \int d\vec{r}\, u_n^*(\vec{r})\,(H_{\text{op}}\,u_m(\vec{r})) \qquad (4.2.8b)$$

We could orthogonalize the basis set following Eq. (4.2.5), so that in terms of the orthogonal basis $\{\tilde{u}_i(\vec{r})\}$, the Schrödinger equation has the form of a standard matrix eigenvalue equation:

$$E\{\tilde{\phi}\} = [\tilde{H}]\{\tilde{\phi}\}$$

where the matrix elements of $[\tilde{H}]$ are given by

$$[\tilde{H}]_{ij} = \int d\vec{r}\, \tilde{u}_i^*(\vec{r})\,(H_{\text{op}}\,\tilde{u}_j(\vec{r}))$$

Transformation of bases: Suppose we have expanded our wavefunction in one basis and would like to change to a different basis:

$$\Phi(\vec{r}) = \sum_m \phi_m\, u_m(\vec{r}) \longrightarrow \Phi(\vec{r}) = \sum_i \phi_i'\, u_i'(\vec{r}) \qquad (4.2.9)$$

Such a transformation can be described by a transformation matrix $[C]$ obtained by writing the new basis in terms of the old basis:

$$u_i'(\vec{r}) = \sum_m C_{mi}\, u_m(\vec{r}) \qquad (4.2.10)$$

From Eqs. (4.2.9) and (4.2.10) we can show that

$$\phi_m = \sum_i C_{mi}\,\phi_i' \longrightarrow \{\phi\} = [C]\{\phi'\} \qquad (4.2.11a)$$

Similarly we can show that any matrix $[A']$ in the new representation is related to the matrix $[A]$ in the old representation by

$$A'_{ji} = \sum_j \sum_i C^*_{nj} \, A_{nm} \, C_{mi} \longrightarrow [A'] = [C]^+[A][C] \tag{4.2.11b}$$

Unitary transformation: There is a special class of transformations which conserves the "norm" of a state vector, that is

$$\sum_m \phi^*_m \, \phi_m = \sum_i \phi'^*_i \, \phi'_i \longrightarrow \{\phi\}^+\{\phi\} = \{\phi'\}^+\{\phi'\} \tag{4.2.12}$$

Substituting for $\{\phi\}$ from Eq. (4.2.11a) into Eq. (4.2.12)

$$\{\phi'\}^+[C]^+[C]\{\phi'\} = \{\phi'\}^+\{\phi'\} \longrightarrow [C]^+[C] = I \tag{4.2.13}$$

A matrix $[C]$ that satisfies this condition (Eq. (4.2.13)) is said to be unitary and the corresponding transformation is called a unitary transformation.

Note that for a unitary transformation, $[C]^+ = [C]^{-1}$, allowing us to write an inverse transformation from Eq. (4.2.11b) simply as $[A] = [C][A'][C]^+$.

Hermitian operators: The matrix $[A]$ representing a Hermitian operator A_{op} is Hermitian (in any representation) which means that it is equal to its conjugate transpose $[A]^+$:

$$[A] = [A]^+, \quad \text{i.e. } A_{mn} = A^*_{nm} \tag{4.2.14}$$

If A_{op} is a function like $U(\vec{r})$ then it is easy to show that it will be Hermitian as long as it is real:

$$[U]^*_{mn} = \left[\int \mathrm{d}\vec{r} \, u^*_m(\vec{r}) \, U(\vec{r}) \, u_n(\vec{r}) \right]^* = [U]_{nm}$$

If A_{op} is a differential operator like $\mathrm{d}/\mathrm{d}x$ or $\mathrm{d}^2/\mathrm{d}x^2$ then it takes a little more work to check if it is Hermitian or not. An easier approach is to use the discrete lattice representation that we discussed in Chapter 2. Equation (2.3.1) shows the matrix representing $\mathrm{d}^2/\mathrm{d}x^2$ and it is clearly Hermitian in this representation. Also, it can be shown that a matrix that is Hermitian in one representation will remain Hermitian in any other representation. The Hamiltonian operator is Hermitian since it is a sum of Hermitian operators like $\partial^2/\partial x^2$, $\partial^2/\partial y^2$, and $U(\vec{r})$. An important requirement of quantum mechanics is that the eigenvalues corresponding to any operator A_{op} representing any observable must be real. This is ensured by requiring all such operators A_{op} to be Hermitian (not just the Hamiltonian operator H_{op} which represents the energy) since the eigenvalues of a Hermitian matrix are real.

Another useful property of a Hermitian matrix is that if we form a matrix [V] out of all the normalized eigenvectors

$$[V] = [\{V_1\}\{V_2\}\cdots]$$

then this matrix will be unitary, that is, $[V]^+[V] = [I]$. Such a unitary matrix can be used to transform all column vectors $\{\phi\}$ and matrices $[M]$ to a new basis that uses the eigenvectors as the basis:

$$\begin{aligned} \{\phi\}_{\text{new}} &= [V]^+\{\phi\}_{\text{old}} \longleftrightarrow \{\phi\}_{\text{old}} = [V]\{\phi\}_{\text{new}} \\ [M]_{\text{new}} &= [V]^+[M]_{\text{old}}[V] \longleftrightarrow [M]_{\text{old}} = [V][M]_{\text{new}}[V]^+ \end{aligned} \tag{4.2.15}$$

If [V] is the eigenvector matrix corresponding to a Hermitian matrix like [H], then the new representation of [H] will be diagonal with the eigenvalues E_m along the diagonal:

$$[H'] = [V]^+[H][V] = \begin{bmatrix} E_1 & 0 & 0 & 0 & \cdots \\ 0 & E_2 & 0 & 0 & \cdots \\ 0 & 0 & E_3 & 0 & \cdots \\ \cdots & \cdots & \cdots & \cdots & \cdots \end{bmatrix} \tag{4.2.16}$$

For this reason the process of finding eigenfunctions and eigenvalues is often referred to as diagonalization.

4.3 Equilibrium density matrix

The density matrix is one of the central concepts in statistical mechanics, which properly belongs in Chapter 7. The reason I am bringing it up in this chapter is that it provides an instructive example of the concept of basis functions. Let me start by briefly explaining what it means. In Chapter 3 we calculated the electron density, $n(\vec{r})$, in multi-electron atoms by summing up the probability densities of each occupied eigenstate α:

$$n(\vec{r}) = \sum_{\text{occ}\,\alpha} |\Phi_\alpha(\vec{r})|^2 \qquad \text{(see Eq. (3.1.3))}$$

This is true at low temperatures for closed systems having a fixed number of electrons that occupy the lowest available energy levels. In general, however, states can be partially occupied and in general the equilibrium electron density can be written as

$$n(\vec{r}) = \sum_\alpha f_0(\varepsilon_\alpha - \mu)|\Phi_\alpha(\vec{r})|^2 \tag{4.3.1}$$

where $f_0(E) \equiv [1 + \exp(E/k_BT)]^{-1}$ is the Fermi function (Fig. 1.1.3) whose value indicates the extent to which a particular state is occupied: "0" indicates unoccupied states, "1" indicates occupied states, while a value between 0 and 1 indicates the average occupancy of a state that is sometimes occupied and sometimes unoccupied.

Could we write a "wavefunction" $\Psi(\vec{r})$ for this multi-electron system such that its squared magnitude will give us the electron density $n(\vec{r})$? One possibility is to write it as

$$\Psi(\vec{r}) = \sum_{\alpha} C_\alpha \Phi_\alpha(\vec{r}) \tag{4.3.2}$$

where $|C_\alpha|^2 = f_0(\varepsilon_\alpha - \mu)$. But this is not quite right. If we square the magnitude of this multi-electron "wavefunction" we obtain

$$n(\vec{r}) = |\Psi(\vec{r})|^2 = \sum_{\alpha}\sum_{\beta} C_\alpha C_\beta^* \Phi_\alpha(\vec{r})\, \Phi_\beta(\vec{r})$$

which is equivalent to Eq. (4.3.1) if and only if,

$$|C_\alpha|^2 = f_0(\varepsilon_\alpha - \mu) \equiv f_\alpha, \qquad C_\alpha C_\beta^* = 0, \qquad \alpha \neq \beta \tag{4.3.3}$$

This is impossible if we view the coefficients C_α as ordinary numbers – in that case $C_\alpha C_\beta^*$ must equal $\sqrt{f_\alpha f_\beta}$ and cannot be zero unless both C_α and C_β are zero. If we wish to write the multi-electron wavefunction in the form shown in Eq. (4.3.2) we should view the coefficients C_α as stochastic numbers whose correlation coefficients are given by Eq. (4.3.3).

So instead of writing a wavefunction for multi-electron systems, it is common to write down a complete matrix $\rho(\alpha, \beta)$ indicating the correlation $C_\alpha C_\beta^*$ between every pair of coefficients. This matrix ρ is called the density matrix and in the eigenstate representation we can write its elements as (see Eq. (4.3.3))

$$\rho(\alpha, \beta) = f_\alpha\, \delta_{\alpha\beta} \tag{4.3.4}$$

where $\delta_{\alpha\beta}$ is the Kronecker delta defined as

$$\delta_{\alpha\beta} = \begin{cases} 1 & \text{if } \alpha = \beta \\ 0 & \text{if } \alpha \neq \beta \end{cases}$$

We can rewrite Eq. (4.3.1) for the electron density $n(\vec{r})$ in the form

$$n(\vec{r}) = \sum_{\alpha}\sum_{\beta} \rho(\alpha, \beta)\, \Phi_\alpha(\vec{r})\, \Phi_\beta^*(\vec{r}) \tag{4.3.5}$$

which can be generalized to define

$$\tilde{\rho}(\vec{r}, \vec{r}\,') = \sum_{\alpha}\sum_{\beta} \rho(\alpha, \beta)\, \Phi_\alpha(\vec{r})\, \Phi_\beta^*(\vec{r}\,') \tag{4.3.6}$$

such that the electron density $n(\vec{r})$ is given by its diagonal elements:

$$n(\vec{r}) = [\rho(\vec{r}, \vec{r}\,')]_{\vec{r}\,'=\vec{r}} \tag{4.3.7}$$

Now the point I want to make is that Eq. (4.3.6) *represents a unitary transformation from an eigenstate basis to a real space basis*. This is seen by noting that the transformation matrix [*V*] is obtained by writing each of the eigenstates (the old basis) as a column vector using the position (the new basis) representation:

$$[V]_{\vec{r},\alpha} = \Phi_\alpha(\vec{r})$$

and that this matrix is unitary: $V^{-1} = V^+$

$$\Rightarrow \quad [V^{-1}]_{\alpha,\vec{r}} = [V^+]_{\alpha,\vec{r}} = [V]^*_{\vec{r},\alpha} \Phi^*_\alpha(\vec{r})$$

so that Eq. (4.3.6) can be written in the form of a unitary transformation:

$$\tilde{\rho}(\vec{r},\vec{r}') = \sum_\alpha \sum_\beta V(\vec{r},\alpha)\rho(\alpha,\beta)V^+(\beta,\vec{r}') \Rightarrow \tilde{\rho} = V\rho V^+$$

This leads to a very powerful concept: The *density matrix ρ at equilibrium* can be written as the *Fermi function of the Hamiltonian matrix* (I is the identity matrix of the same size as H):

$$\rho = f_0(H - \mu I) \tag{4.3.8}$$

This is a general matrix relation that is valid *in any representation*. For example, if we use the eigenstates α of H as a basis then $[H]$ is a diagonal matrix:

$$[H] = \begin{bmatrix} \varepsilon_1 & 0 & 0 & \cdots \\ 0 & \varepsilon_2 & 0 & \cdots \\ 0 & 0 & \varepsilon_3 & \cdots \\ & \cdots & & \cdots \end{bmatrix}$$

and so is ρ:

$$\rho = \begin{bmatrix} f_0(\varepsilon_1 - \mu) & 0 & 0 & \cdots \\ 0 & f_0(\varepsilon_2 - \mu) & 0 & \cdots \\ 0 & 0 & f_0(\varepsilon_3 - \mu) & \cdots \\ \cdots & \cdots & \cdots & \cdots \end{bmatrix}$$

This is exactly what Eq. (4.3.4) tells us. But the point is that the relation given in Eq. (4.3.8) is valid, not just in the eigenstate representation, but in any representation. Given the matrix representation $[H]$, it takes just three commands in MATLAB to obtain the density matrix:

```
[V, D] = eig (H)
rho = 1./(1 + exp((diag(D) – mu)./kT))
rho = V * diag(rho) * V′
```

The first command calculates a diagonal matrix $[D]$ whose diagonal elements are the eigenvalues of $[H]$ and a matrix $[V]$ whose columns are the corresponding eigenvectors. In other words, $[D]$ is the Hamiltonian $[H]$ transformed to the eigenstate basis $\Rightarrow D = V^+HV$.

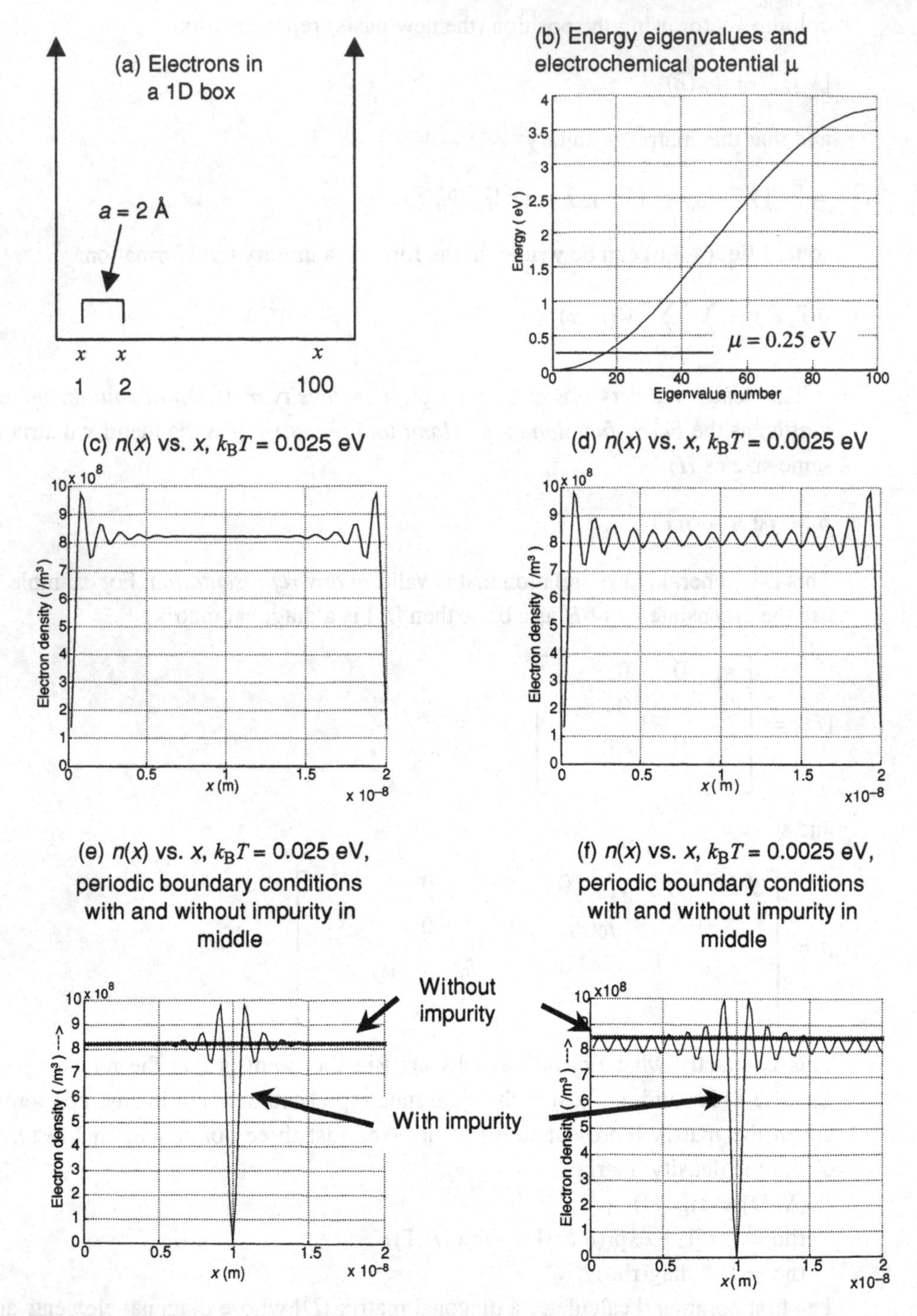

Fig. 4.3.1 Equilibrium electron density for a 1D box modeled with a discrete lattice of 100 points spaced by 2 Å.

The second command gives us the density matrix in the eigenstate representation, which is easy since in this representation both $[H]$ and $[\rho]$ are diagonal. The third command then transforms $[\rho]$ back to the original representation.

Figure 4.3.1 shows the equilibrium electron density for a 1D box modeled with a discrete lattice of 100 points spaced by 2 Å, with $\mu = 0.25$ eV. The Hamiltonian $[H]$ is a (100×100) matrix which can be set up following the prescription in Sections 2.2 and 2.3. The density matrix is then evaluated as described above and its diagonal elements give us the electron density $n(x)$ (times the lattice constant, a). Note the standing wave patterns in Figs. 4.3.1c, d which are absent when we use periodic boundary conditions (Figs. 4.3.1e, f). Figures 4.3.1e and f also show the standing wave patterns in the electron density when a large repulsive potential

$$U_0\delta[x - (L/2)] \text{ where } U_0 = 2 \text{ eV nm}$$

is included at the center of the box.

Note that the density matrix can look very different depending on what basis functions we use. In the eigenstate representation it is diagonal since the Hamiltonian is diagonal, but in the real-space lattice representation it has off-diagonal elements. In any basis m, the diagonal elements $\rho(m, m)$ tell us the number of electrons occupying the state m. In a real-space representation, the diagonal elements $\rho(\vec{r}, \vec{r})$ give us the electron density $n(\vec{r})$. The trace (sum of diagonal elements) of ρ, which is invariant in all representations, gives us the total number of electrons N:

$$N = \text{Trace}(\rho) \tag{4.3.9}$$

If we are only interested in the electron density, then the diagonal elements of the density matrix are all we need. But we cannot "throw away" the off-diagonal elements; they are needed to ensure that the matrix will transform correctly to another representation. Besides, depending on what we wish to calculate, we may need the off-diagonal elements too (see Section 4.4.1).

4.4 Supplementary notes

4.4.1 Density matrix

It is common in quantum mechanics to associate every observable A with an operator A_{op} for which we can find a matrix representation $[A]$ in any basis. The expectation value $\langle A \rangle$ for this observable (that is, the average value we expect to get in a series of measurements) is given by

$$\langle A \rangle = \int d\vec{r}\, \Psi^*(\vec{r}) A_{op} \Psi(\vec{r})$$

Substituting for the wavefunction in terms of the basis functions from Eq. (4.3.2), we can show that

$$\langle A\rangle = \sum_{\alpha}\sum_{\beta} C_\alpha C_\beta^* \int d\vec{r}\, \Phi_\beta^*(\vec{r})\, A_{\mathrm{op}}\, \Phi_\alpha(\vec{r})$$

so that

$$\langle A\rangle = \sum_{\alpha}\sum_{\beta} \rho_{\alpha\beta} A_{\beta\alpha} = \mathrm{Trace}[\rho A]$$

We could use this result to evaluate the expectation value of any quantity, even if the system is out of equilibrium, provided we know the density matrix. But what we have discussed here is the equilibrium density matrix. It is much harder to calculate the non-equilibrium density matrix, as we will discuss later in the book.

Plane wave vs. sine–cosine representations: Consider now a conductor of length L having just two plane wave (pw) states

$$\Psi_+(x) = \frac{1}{\sqrt{L}}\, e^{+ikx} \quad \text{and} \quad \Psi_-(x) = \frac{1}{\sqrt{L}}\, e^{-ikx}$$

The current operator in this basis is given by

$$[J_{\mathrm{op}}]_\pm = \frac{-q}{L}\begin{bmatrix} \hbar k/m & 0 \\ 0 & -\hbar k/m \end{bmatrix} \quad (\text{columns: } \text{“}+\text{”},\ \text{“}-\text{”})$$

and we could write the density matrix as

$$[\rho]_\pm = \begin{bmatrix} f_+ & 0 \\ 0 & f_- \end{bmatrix} \quad (\text{columns: } \text{“}+\text{”},\ \text{“}-\text{”})$$

where f_+ and f_- are the occupation probabilities for the two states. We wish to transform both these matrices from the "±" basis to a "cs" basis using cosine and sine states:

$$\Psi_c(x) = \sqrt{\frac{2}{L}}\cos kx \quad \text{and} \quad \Psi_s(x) = \sqrt{\frac{2}{L}}\sin kx$$

It is straightforward to write down the transformation matrix [V] whose columns represent the old basis (+, −) in terms of the new basis (c, s):

$$[V] = \frac{1}{\sqrt{2}}\begin{bmatrix} 1 & 1 \\ +i & -i \end{bmatrix}$$

so that in the "cs" representation

$$[\rho]_{\text{cs}} = [V]\,[\rho]_{\pm}\,[V]^{+} = \frac{1}{2}\begin{bmatrix} f_{+}+f_{-} & -\mathrm{i}(f_{+}-f_{-}) \\ -\mathrm{i}(f_{+}-f_{-}) & f_{+}+f_{-} \end{bmatrix} \quad \begin{matrix} \text{"c"} & \text{"s"} \end{matrix}$$

and

$$[J_{\text{op}}]_{\text{cs}} = [V][J_{\text{op}}]_{\pm}[V]^{+} = \begin{bmatrix} 0 & -\mathrm{i}\hbar k/mL \\ +\mathrm{i}\hbar k/mL & 0 \end{bmatrix} \quad \begin{matrix} \text{"c"} & \text{"s"} \end{matrix}$$

It is easy to check that the current $\langle J\rangle = \text{Trace}[\rho\, J_{\text{op}}]$ is the same in either representation:

$\langle J\rangle = (-q/L)(\hbar k/m)\,[f_{+}-f_{-}]$

This is expected since the trace is invariant under a unitary transformation and thus remains the same no matter which representation we use. But the point to note is that the current in the cosine–sine representation arises from the *off-diagonal* elements of the current operator and the density matrix, rather than the *diagonal* elements. The off-diagonal elements do not have an intuitive physical meaning like the diagonal elements. As long as the current is carried by the diagonal elements, we can use a semiclassical picture in terms of occupation probabilities. But if the "action" is in the off-diagonal elements then we need a more general quantum framework (I am indebted to A. W. Overhauser for suggesting this example).

4.4.2 Perturbation theory

Suppose we wish to find the energy levels of a hydrogen atom in the presence of an electric field F applied along the z-direction. Let us use the eigenstates 1s, 2s, $2p_x$, $2p_y$ and $2p_z$ as our basis set and write down the Hamiltonian matrix. If the field were absent the matrix would be diagonal:

$$[H_0] = \begin{bmatrix} E_1 & 0 & 0 & 0 & 0 \\ 0 & E_2 & 0 & 0 & 0 \\ 0 & 0 & E_2 & 0 & 0 \\ 0 & 0 & 0 & E_2 & 0 \\ 0 & 0 & 0 & 0 & E_2 \end{bmatrix} \quad (\text{columns: } 1s,\ 2s,\ 2p_x,\ 2p_y,\ 2p_z)$$

where $E_0 = 13.6$ eV, $E_1 = -E_0$, and $E_2 = -E_0/4$. The electric field leads to a matrix $[H_F]$ which has to be added to $[H]$. Its elements are given by

$$[H_F]_{nm} = qF\int_0^{\infty} \mathrm{d}r\,r^2 \int_0^{\pi} \sin\theta\,\mathrm{d}\theta \int_0^{2\pi} \mathrm{d}\phi\, u_n^*(\vec{r})\, r\cos\theta\, u_m(\vec{r})$$

Using the wavefunctions

$$u_{1s} = \sqrt{1/\pi a_0^3}\, e^{-r/a_0}$$

$$u_{2s} = \sqrt{1/32\pi a_0^3}\left(2 - \frac{r}{a_0}\right) e^{r/2a_0}$$

$$u_{2p_x} = \sqrt{1/16\pi a_0^3}\left(\frac{r}{a_0}\right) e^{-r/2a_0} \sin\theta \cos\phi$$

$$u_{2p_y} = \sqrt{1/16\pi a_0^3}\left(\frac{r}{a_0}\right) e^{-r/2a_0} \sin\theta \sin\phi$$

$$u_{2p_z} = \sqrt{1/16\pi a_0^3}\left(\frac{r}{a_0}\right) e^{-r/2a_0} \cos\theta$$

we can evaluate the integrals straightforwardly to show that

$$[H_F] = \begin{array}{c} \\ \end{array}\begin{array}{ccccc} 1s & 2s & 2p_x & 2p_y & 2p_z \end{array}$$
$$[H_F] = \begin{bmatrix} 0 & 0 & 0 & 0 & A \\ 0 & 0 & 0 & 0 & B \\ 0 & 0 & 0 & 0 & 0 \\ 0 & 0 & 0 & 0 & 0 \\ A & B & 0 & 0 & 0 \end{bmatrix}$$

where A and B are linear functions of the field:

$$A = (128\sqrt{2}/243)a_0 F, \quad B = -3a_0 F$$

Hence

$$\begin{array}{ccccc} 1s & 2s & 2p_z & 2p_x & 2p_y \end{array}$$
$$[H_0] + [H_F] = \begin{bmatrix} E_1 & 0 & A & 0 & 0 \\ 0 & E_2 & B & 0 & 0 \\ A & B & E_2 & 0 & 0 \\ 0 & 0 & 0 & E_2 & 0 \\ 0 & 0 & 0 & 0 & E_2 \end{bmatrix}$$

Note that we have relabeled the rows and columns to accentuate the fact that $2p_x$ and $2p_y$ levels are decoupled from the rest of the matrix and are *unaffected* by the field, while the 1s and 2s and $2p_z$ levels will be affected by the field.

Degenerate perturbation theory: If the field were absent we would have one eigenvalue E_1 and four degenerate eigenvalues E_2. How do these eigenvalues change as we turn up the field? As we have mentioned before, the eigenvalues are more or less equal to the diagonal values unless the off-diagonal term H_{mn} becomes comparable to the difference between the corresponding diagonal terms $(H_{mm} - H_{nn})$. This means that whenever we have two degenerate eigenvalues (that is, $H_{mm} - H_{nn} = 0$) even a small

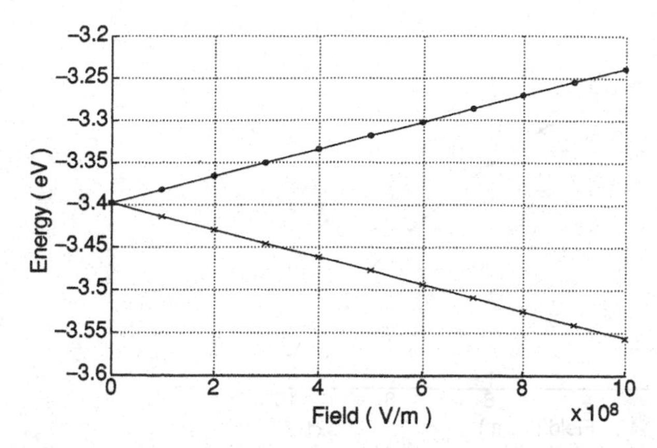

Fig. 4.4.1 Energy of 2s – 2p$_z$ levels due to an applied electric field F. The solid lines show the results obtained by direct diagonalization while o and × show perturbation theory results $E = E_2 \pm B$.

off-diagonal element H_{mn} has a significant effect. We thus expect the 2s and 2p$_z$ levels to be significantly affected by the field since they are degenerate to start with. We can get a very good approximation for these eigenvalues simply by looking at a subset of the $[H]$ matrix containing just these levels:

$$\begin{array}{cc} & \begin{array}{cc} \text{2s} & \text{2p}_z \end{array} \\ [H_0] + [H_F] = & \begin{bmatrix} E_2 & B \\ B & E_2 \end{bmatrix} \end{array}$$

It is easy to show that the eigenvalues are $E = E_2 \pm B$ and the corresponding eigenvectors are

$$|2\text{s}\rangle - |2\text{p}_z\rangle \quad \text{and} \quad |2\text{s}\rangle + |2\text{p}_z\rangle$$

This approximate approach (known as degenerate perturbation theory) describes the exact eigenvalues quite well (see Fig. 4.4.1) as long as the off-diagonal elements (like A) coupling these levels to the other levels are much smaller than the energy difference between these levels (like $E_2 - E_1$).

Non-degenerate perturbation theory: How is the 1s eigenvalue affected? Since there are no other degenerate levels the effect is much less and to first order one could simply ignore the rest of the matrix:

$$\begin{array}{cc} & \text{1s} \\ [H_0] + [H_F] = & [E_1] \end{array}$$

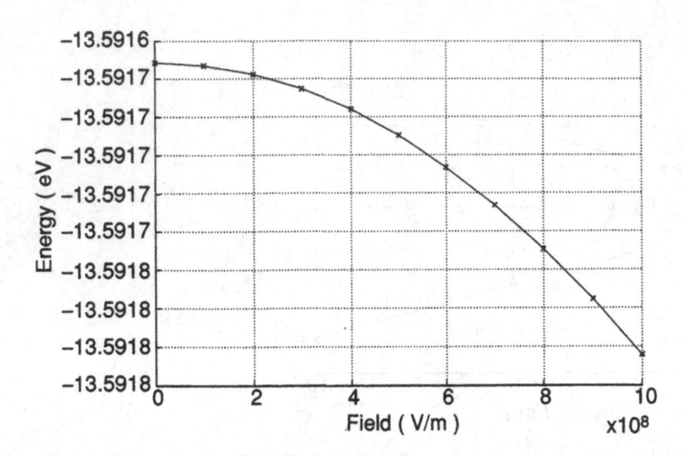

Fig. 4.4.2 Energy of 1s level due to an applied electric field F. The solid curve shows the results obtained by direct diagonalization while the crosses denote the perturbation theory results $E = E_1 + (A^2/(E_1 - E_2))$.

and argue that the eigenvalue remains E_1. We could do better by "renormalizing" the matrix as follows. Suppose we partition the $[H]$ matrix and write

$$[H]\{\phi\} = E\{\phi\} \longrightarrow \begin{bmatrix} H_{11} & H_{12} \\ H_{21} & H_{22} \end{bmatrix} \begin{Bmatrix} \phi_1 \\ \phi_2 \end{Bmatrix} = E \begin{Bmatrix} \phi_1 \\ \phi_2 \end{Bmatrix}$$

where $[H_{11}]$ denotes the part of the matrix we wish to keep (the 1s block in this case). It is easy to eliminate $\{\phi_2\}$ to obtain

$$[H']\{\phi_1\} = E\{\phi_1\}$$

where

$$[H'] = [H_{11}] + [H_{12}][EI - H_{22}]^{-1} H_{21}$$

I being an identity matrix of the same size as H_{22}. We haven't gained much if we still have to invert the matrix $EI - H_{22}$ including its off-diagonal elements. But to lowest order we can simply ignore the off-diagonal elements of H_{22} and write down the inverse by inspection. In the present case, this gives us

$$[H'] \approx E_1 + (0 \quad A) \begin{bmatrix} 1/(E - E_2) & 0 \\ 0 & 1/(E - E_2) \end{bmatrix} \begin{pmatrix} 0 \\ A \end{pmatrix} = E_1 + \frac{A^2}{E - E_2}$$

To lowest order, the eigenvalue E is approximately equal to E_1, so that

$$[H'] \approx E_1 + (A^2/(E_1 - E_2))$$

which shows that the correction to the eigenvalue is quadratic for non-degenerate states, rather than linear as it is for degenerate states. This approximate approach (known as non-degenerate perturbation theory) describes the exact eigenvalues quite well (see Fig. 4.4.2).

EXERCISES

E.4.1. Plot the electron density $n(x)$ in a hydrogen molecule along the axis joining the two hydrogen atoms assuming they are separated by the equilibrium bond distance of $R = 0.074$ nm and compare your result with Fig. 4.1.4.

E.4.2. Calculate the equilibrium electron density $n(x)$ in a one-dimensional box modeled with discrete lattice of 100 points spaced by 2 Å and compare with each of the results shown in Fig. 4.3.1c–f.

E.4.3. Consider a set of triangular basis functions on a one-dimensional discrete lattice with a lattice constant "a".

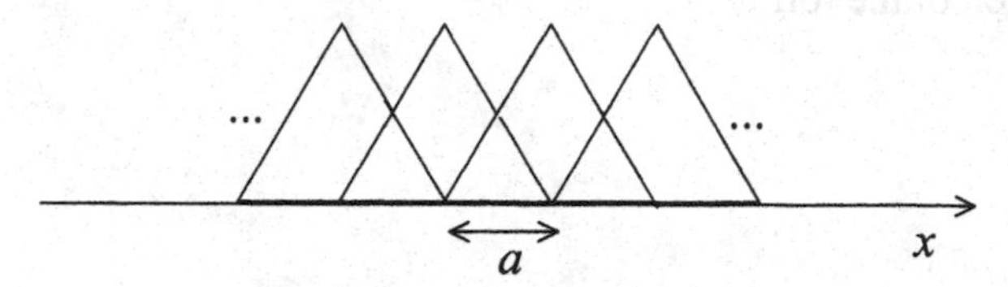

(a) Calculate the matrix elements of the overlap matrix [S] from Eq. (4.1.3a).

(b) Calculate the matrix elements of the Hamiltonian matrix [H] from Eq. (4.1.3b), assuming that $H_{op} = (\hbar^2/2m)\partial^2/\partial x^2$.

This is a special case of the finite element method often used in the literature. See for example, Ramdas Ram-Mohan (2002) and White *et al.* (1989). How does it compare with the finite difference method discussed in Section 2.2?

E.4.4. Find the eigenvalues of the matrix

$$\begin{bmatrix} E_1 & 0 & A & 0 & 0 \\ 0 & E_2 & B & 0 & 0 \\ A & B & E_2 & 0 & 0 \\ 0 & 0 & 0 & E_2 & 0 \\ 0 & 0 & 0 & 0 & E_2 \end{bmatrix}$$

where $E_1 = -E_0$, $E_2 = -E_0/4 (E_0 = 13.6\,\text{eV})$ and $A = (128\sqrt{2}/243)a_0 F$, $B = -3a_0 F$ and plot as a function of the electric field F. Compare with the perturbation theory results in Figs. 4.4.1 and 4.4.2.

5 Bandstructure

In the last chapter we saw how the atomic orbitals can be used as a basis to write down a matrix representation for the Hamiltonian operator, which can then be diagonalized to find the energy eigenvalues. In this chapter we will show how this approach can be used to calculate the energy eigenvalues for an infinite periodic solid. We will first use a few "toy" examples to show that the bandstructure can be calculated by solving a matrix eigenvalue equation of the form

$$E(\phi_0) = [h(\vec{k})](\phi_0)$$

where

$$[h(\vec{k})] = \sum_m [H_{nm}] \mathrm{e}^{\mathrm{i}\vec{k}\cdot(\vec{d}_m - \vec{d}_n)}$$

The matrix $[h(\vec{k})]$ is $(b \times b)$ in size, b being the number of basis orbitals per unit cell. The summation over m runs over all neighboring unit cells (including itself) with which cell n has any overlap (that is, for which H_{nm} is non-zero). The sum can be evaluated choosing any unit cell n and the result will be the same because of the periodicity of the lattice. The bandstructure can be plotted out by finding the eigenvalues of the $(b \times b)$ matrix $[h(\vec{k})]$ for each value of $\vec{k}$ and it will have b branches, one for each eigenvalue. This is the central result which we will first illustrate using toy examples (Section 5.1), then formulate generally for periodic solids (Section 5.2), and then use to discuss the bandstructure of 3D semiconductors (Section 5.3). We discuss spin–orbit coupling and its effect on the energy levels in semiconductors in Section 5.4. This is a relativistic effect whose proper treatment requires the Dirac equation as explained briefly in Section 5.5.

5.1 Toy examples

Let us start with a toy one-dimensional solid composed of N atoms (see Fig. 5.1.1). If we use one orbital per atom we can write down a $(N \times N)$ Hamiltonian matrix using one orbital per atom (the off-diagonal element has been labeled with a subscript "ss,"

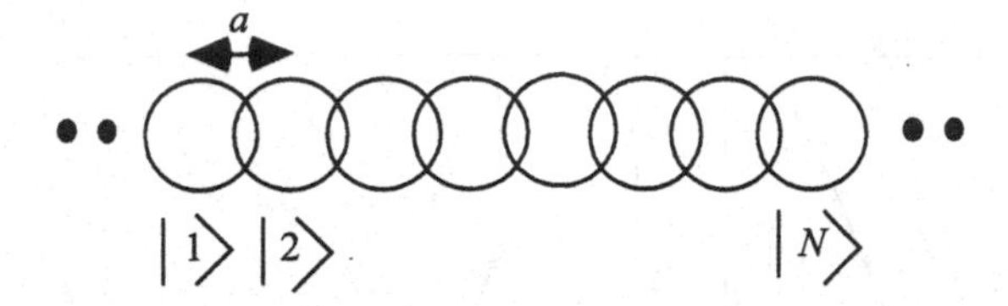

Fig. 5.1.1 A one-dimensional solid.

although the orbitals involved need not necessarily be s orbitals):

$$H = \begin{array}{c|ccccc} & |1\rangle & |2\rangle & \ldots & |N-1\rangle & |N\rangle \\ |1\rangle & E_0 & E_{ss} & & 0 & E_{ss} \\ |2\rangle & E_{ss} & E_0 & & 0 & 0 \\ & \cdots & \cdots & & \cdots & \\ |N-1\rangle & 0 & 0 & & E_0 & E_{ss} \\ |N\rangle & E_{ss} & 0 & & E_{ss} & E_0 \end{array} \quad (5.1.1)$$

We have used what is called the periodic boundary condition (PBC), namely, that the *N*th atom wraps around and overlaps the first atom as in a ring. This leads to non-zero values for the matrix elements $H_{1,N}$ and $H_{N,1}$ which would normally be zero if the solid were abruptly truncated. The PBC is usually not realistic, but if we are discussing the bulk properties of a large solid then the precise boundary condition at the surface does not matter and we are free to use whatever boundary conditions make the mathematics the simplest, which happens to be the PBC.

So what are the eigenvalues of the matrix [*H*] given in Eq. (5.1.1)? This is essentially the same matrix that we discussed in Chapter 1 in connection with the finite difference method. If we find the eigenvalues numerically we will find that they can all be written in the form (α is an integer)

$$E_\alpha = E_0 + 2E_{ss}\cos(k_\alpha a) \quad \text{where} \quad k_\alpha a = \alpha 2\pi/N \quad (5.1.2)$$

The values of $k_\alpha a$ run from $-\pi$ to $+\pi$ and are spaced by $2\pi/N$ as shown in Fig. 5.1.2. If *N* is large the eigenvalues are closely spaced (Fig. 5.1.2a); if *N* is small the eigenvalues are further apart (Fig. 5.1.2b).

Why is it that we can write down the eigenvalues of this matrix so simply? The reason is that because of its periodic nature, the matrix equation $E(\psi) = [H](\psi)$ consists of a set of *N* equations that are all identical in form and can all be written as ($n = 1, 2, \ldots N$)

$$E\psi_n = E_0\psi_n + E_{ss}\psi_{n-1} + E_{ss}\psi_{n+1} \quad (5.1.3)$$

This set of equations can be solved analytically by the ansatz:

$$\psi_n = \psi_0 e^{ikna} \quad (5.1.4)$$

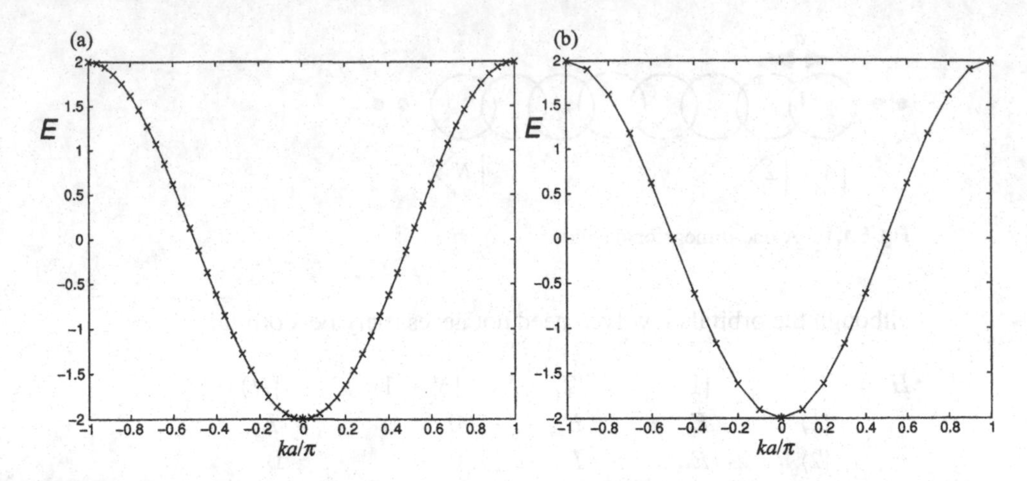

Fig. 5.1.2 The solid curves are plots of E vs. ka/π from Eq. (5.1.2) with $E_0 = 0$, $E_{ss} = -1$. The crosses denote the eigenvalues of the matrix in Eq. (5.1.1) with (a) $N = 100$ and (b) $N = 20$.

Substituting Eq. (5.1.4) into (5.1.3) and canceling the common factor exp($ikna$) we obtain

$$E\psi_0 = E_0\psi_0 + E_{ss}e^{-ika}\psi_0 + E_{ss}e^{ika}\psi_0$$

that is

$$E = E_0 + 2E_{ss}\cos(ka)$$

This shows us that a solution of the form shown in Eq. (5.1.4) will satisfy our set of equations for any value of k. But what restricts the number of eigenvalues to a finite number (as it must be for a finite-sized matrix)?

This is a result of two factors. Firstly, periodic boundary conditions require the wavefunction to be periodic with a period of Na and it is this finite lattice size that restricts the allowed values of k to the discrete set $k_\alpha a = \alpha 2\pi/N$ (see Eq. (5.1.2)). Secondly, values of ka differing by 2π do not represent distinct states on a discrete lattice. The wavefunctions

$$\exp(ik_\alpha x) \quad \text{and} \quad \exp(i[k_\alpha + (2\pi/a)]x)$$

represent the same state because at any lattice point $x_n = na$,

$$\exp(ik_\alpha x_n) = \exp(i[k_\alpha + (2\pi/a)]x_n)$$

They are NOT equal between two lattice points and thus represent distinct states in a continuous lattice. But once we adopt a discrete lattice, values of k_α differing by $2\pi/a$ represent identical states and only the values of $k_\alpha a$ within a range of 2π yield

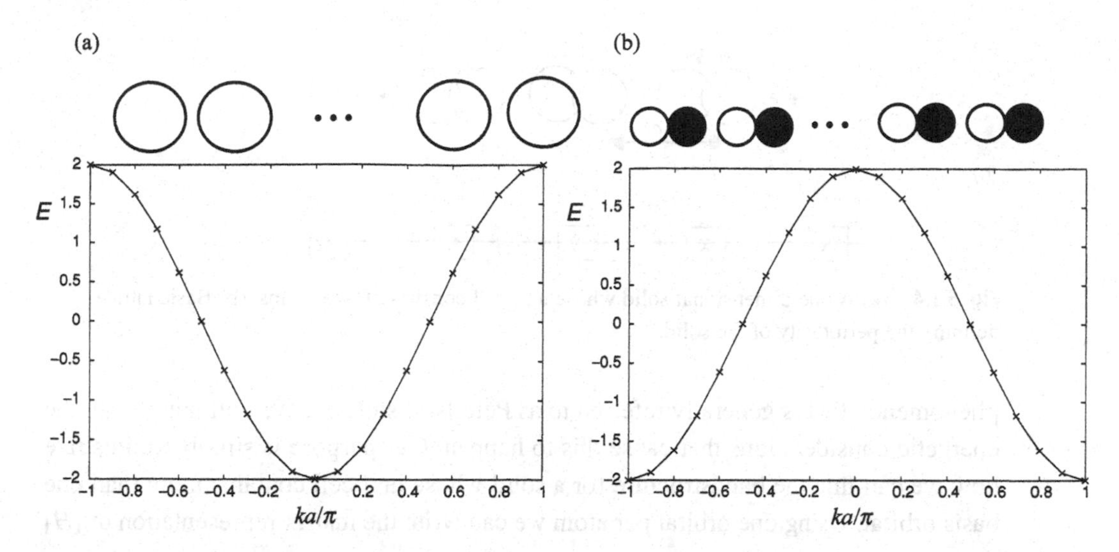

Fig. 5.1.3 As Fig. 5.1.2b with $E_0 = 0$ and (a) $E_{ss} = -1$ and (b) $E_{ss} = +1$.

independent solutions. In principle, any range of size 2π is acceptable, but it is common to restrict the values of $k_\alpha a$ to the range (sometimes called the first Brillouin zone)

$$-\pi \leq ka < +\pi \quad \text{for periodic boundary conditions} \tag{5.1.5}$$

It is interesting to note that the finite range of the lattice (Na) leads to a discreteness (in units of $2\pi/Na$) in the allowed values of k while the discreteness of the lattice (a) leads to a finite range of allowed k ($2\pi/a$). The number of allowed values of k

$$(2\pi/a)/(2\pi/Na) = N$$

is exactly the same as the number of points in the real space lattice. This ensures that the number of eigenvalues (which is equal to the number of allowed k values) is equal to the size of the matrix [H] (determined by the number of lattice points).

When do bands run downwards in *k*? In Fig. 5.1.2 we have assumed E_{ss} to be negative which is what we would find if we used, say, Eq. (4.1.11c) to evaluate it (note that the potentials U_L or U_R are negative) and the atomic orbitals were s orbitals. But if the atomic orbitals are p_x orbitals as shown in Fig. 5.1.3b then the sign of the overlap integral (E_{ss}) would be positive and the plot of $E(k)$ would run downwards in k as shown. Roughly speaking this is what happens in the valence band of common semiconductors which are formed primarily out of atomic p orbitals.

Lattice with a basis: Consider next a one-dimensional solid whose unit cell consists of two atoms as shown in Fig. 5.1.4. Actually one-dimensional structures like the one shown in Fig. 5.1.1 tend to distort spontaneously into the structure shown in Fig. 5.1.4 – a

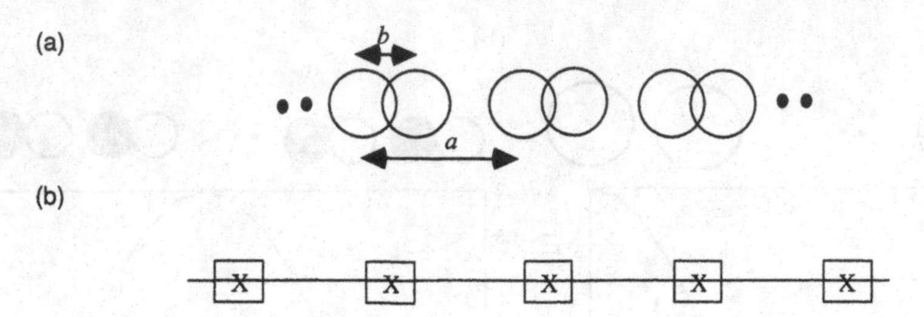

Fig. 5.1.4 (a) A one-dimensional solid whose unit cell consists of two atoms. (b) Basic lattice defining the periodicity of the solid.

phenomenon that is generally referred to as Peierls' distortion. We will not go into the energetic considerations that cause this to happen. Our purpose is simply to illustrate how we can find the bandstructure for a solid whose unit cell contains more than one basis orbital. Using one orbital per atom we can write the matrix representation of $[H]$ as

$$[H] = \begin{array}{c|ccccccc} & |1_A\rangle & |1_B\rangle & |2_A\rangle & |2_B\rangle & |3_A\rangle & |3_B\rangle & \dots \\ |1_A\rangle & E_0 & E_{ss} & 0 & 0 & 0 & 0 & \dots \\ |1_B\rangle & E_{ss} & E_0 & E'_{ss} & 0 & 0 & 0 & \dots \\ |2_A\rangle & 0 & E'_{ss} & E_0 & E_{ss} & 0 & 0 & \dots \\ |2_B\rangle & 0 & 0 & E_{ss} & E_0 & E'_{ss} & 0 & \dots \\ |3_A\rangle & 0 & 0 & 0 & E'_{ss} & E_0 & E_{ss} & \dots \\ |3_B\rangle & 0 & 0 & 0 & 0 & E_{ss} & E_0 & \dots \end{array} \tag{5.1.6}$$

Unlike the matrix in Eq. (5.1.1) there are two different overlap integrals E_{ss} and E'_{ss} appearing alternately. As such the ansatz in Eq. (5.1.4) cannot be used directly. But we could combine the elements of the matrix into (2×2) blocks and rewrite it in the form

$$[H] = \begin{array}{c|cccc} & |1\rangle & |2\rangle & |3\rangle & \dots \\ |1\rangle & H_{11} & H_{12} & 0 & \dots \\ |2\rangle & H_{21} & H_{22} & H_{23} & \dots \\ |3\rangle & 0 & H_{32} & H_{33} & \dots \end{array} \tag{5.1.7}$$

where

$$H_{nm} = \begin{bmatrix} E_0 & E_{ss} \\ E_{ss} & E_0 \end{bmatrix} \quad H_{n,n+1} = \begin{bmatrix} 0 & 0 \\ E'_{ss} & 0 \end{bmatrix} \quad H_{n,n-1} = \begin{bmatrix} 0 & E_{ss} \\ 0 & 0 \end{bmatrix}$$

The matrix in Eq. (5.1.7) is now periodic and we can write the matrix equation $E(\psi) = [H](\psi)$ in the form

$$E\phi_n = H_{nn}\phi_n + H_{n,n-1}\phi_{n-1} + H_{n,n+1}\phi_{n+1} \tag{5.1.8}$$

where ϕ_n represents a (2×1) column vector and the element H_{nm} is a (2×2) matrix.

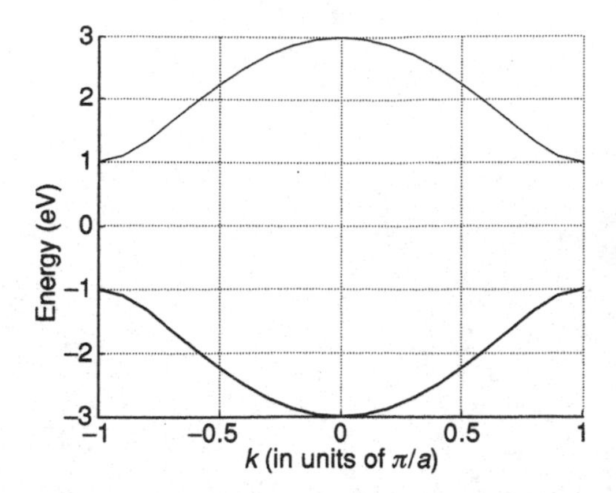

Fig. 5.1.5 Bandstructure for the "dimerized" one-dimensional solid shown in Fig. 5.1.4 plotted from Eq. (5.1.10) using $E_0 = 0$, $E_{ss} = 2$, $E'_{ss} = 1$.

We can solve this set of equations using the ansatz:

$$\phi_n = \phi_0 e^{ikna} \tag{5.1.9}$$

Substituting Eq. (5.1.9) into (5.1.8) and canceling the common factor exp[*ikna*] we obtain

$$E\phi_0 = H_{nn}\phi_0 + H_{n,n-1}e^{-ika}\phi_0 + H_{n,n+1}e^{ika}\phi_0$$

that is

$$E\{\phi_0\} = \begin{bmatrix} E_0 & E_{ss} + E'_{ss}e^{-ika} \\ E_{ss} + E'_{ss}e^{ika} & E_0 \end{bmatrix} \{\phi_0\}$$

We can now find the eigenvalues by setting the determinant to zero:

$$\det \begin{bmatrix} E_0 - E & E_{ss} + E'_{ss}e^{-ika} \\ E_{ss} + E'_{ss}e^{ika} & E_0 - E \end{bmatrix} = 0$$

that is

$$E = E_0 \pm \left(E_{ss}^2 + E_{ss}'^2 + 2E_{ss}E'_{ss}\cos(ka)\right)^{1/2} \tag{5.1.10}$$

Equation (5.1.10) gives us an $E(k)$ diagram with two branches as shown in Fig. 5.1.5.

5.2 General result

It is straightforward to generalize this procedure for calculating the bandstructure of any periodic solid with an arbitrary number of basis functions per unit cell. Consider any particular unit cell n (Fig. 5.2.1) connected to its neighboring unit cells m by a

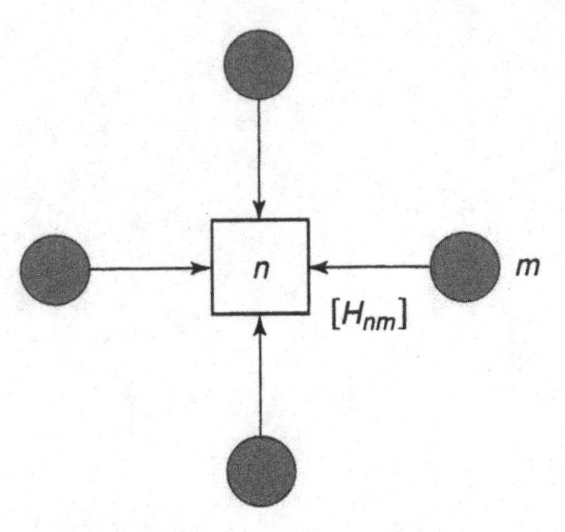

Fig. 5.2.1 Schematic picture showing a unit cell n connected to its neighboring unit cells m by a matrix $[H_{nm}]$ of size $(b \times b)$, b being the number of basis functions per unit cell. The configuration of neighbors will differ from one solid to another, but in a periodic solid the configuration is identical regardless of which n we choose.

matrix $[H_{nm}]$ of size $(b \times b)$, b being the number of basis functions per unit cell. We can write the overall matrix equation in the form

$$\sum_m [H_{nm}]\{\phi_m\} = E\{\phi_n\} \tag{5.2.1}$$

where $\{\phi_m\}$ is a $(b \times 1)$ column vector denoting the wavefunction in unit cell m.

The important insight is the observation that this set of equations can be solved by the ansatz

$$\{\phi_m\} = \{\phi_0\} \exp^{i\vec{k}\cdot\vec{d}_m} \tag{5.2.2}$$

provided Eq. (5.2.1) looks the same in every unit cell n. We could call this a discrete version of "Bloch's theorem" discussed in standard texts like Ashcroft and Mermin (1976). It is a consequence of the periodicity of the lattice and it ensures that when we substitute our ansatz Eq. (5.2.2) into Eq. (5.2.1) we obtain

$$E\{\phi_0\} = [h(\vec{k})]\{\phi_0\} \tag{5.2.3}$$

with

$$[h(\vec{k})] = \sum_m [H_{nm}] e^{i\vec{k}\cdot(\vec{d}_m - \vec{d}_n)} \tag{5.2.4}$$

independent of which unit cell n we use to evaluate the sum in Eq. (5.2.4). This is the *central result* underlying the bandstructure of periodic solids. The summation over m in Eq. (5.2.4) runs over all neighboring unit cells (including itself) with which cell n has any overlap (that is, for which H_{nm} is non-zero). The size of the matrix $[h(\vec{k})]$ is $(b \times b)$, b being the number of basis orbitals per unit cell. The bandstructure can be

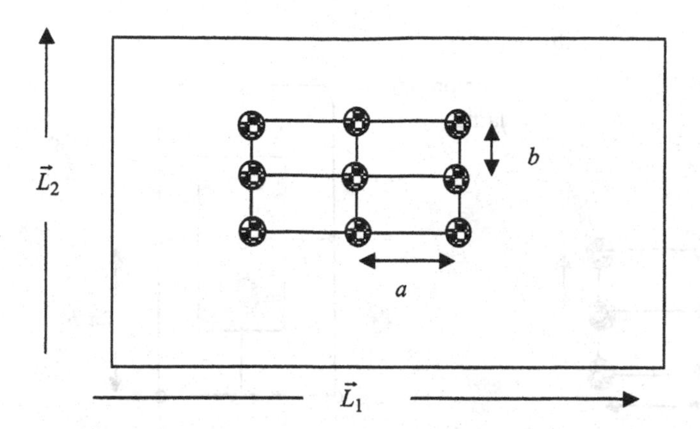

Fig. 5.2.2 A finite 2D rectangular lattice with M unit cells stacked along the x-direction and N unit cells stacked along the y-direction ($\vec{L}_1 = \hat{x}Ma$, $\vec{L}_2 = \hat{y}Nb$).

plotted by finding the eigenvalues of the $(b \times b)$ matrix $[h(\vec{k})]$ for each value of $\vec{k}$ and it will have b branches, one for each eigenvalue.

Allowed values of k: In connection with the 1D example, we explained how k has only a finite number of allowed values equal to the number of unit cells in the solid. To reiterate the basic result, the finite range of the lattice (Na) leads to a discreteness (in units of $2\pi/Na$) in the allowed values of k while the discreteness of the lattice (a) leads to a finite range of allowed $k(2\pi/a)$. How do we generalize this result beyond one dimension?

This is fairly straightforward if the solid forms a rectangular (or a cubic) lattice as shown in Fig. 5.2.2. In 2D the allowed values of $\vec{k}$ can be written as

$$[\vec{k}]_{mn} = \hat{x}(m\ 2\pi/Ma) + \hat{y}(n\ 2\pi/Nb) \tag{5.2.5}$$

where (m, n) are a pair of integers while M, N represent the number of unit cells stacked along the x- and y-directions respectively. This seems like a reasonable extension of the 1D result (cf. Eq. (5.1.2): $k_\alpha = \alpha(2\pi/Na)$). Formally we could derive Eq. (5.2.5) by writing ($\vec{L}_1 = \hat{x}Ma$, $\vec{L}_2 = \hat{y}Nb$)

$$\vec{k} \cdot \vec{L}_1 = m\ 2\pi \rightarrow k_x = m\ 2\pi/Ma$$

$$\vec{k} \cdot \vec{L}_2 = n\ 2\pi \rightarrow k_y = n\ 2\pi/Nb$$

Brillouin zone: Formally, the general procedure for constructing the Brillouin zone starts by constructing the reciprocal lattice (Fig. 5.2.3b) in k-space, which can be viewed as the Fourier transform of the direct lattice. In 1D we know that a set of impulses separated by a (Fig. 5.2.4a) has a Fourier transform consisting of a set of impulses separated

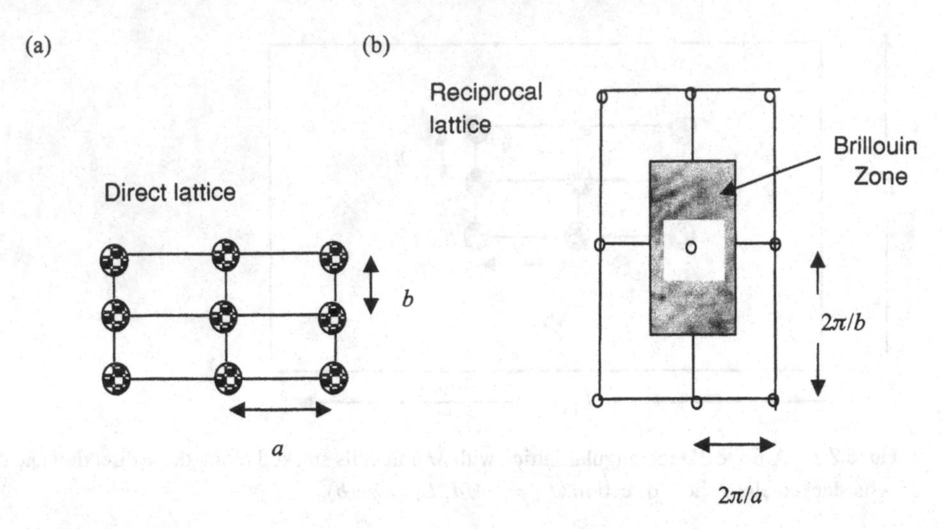

Fig. 5.2.3 (a) Rectangular lattice in real space. (b) Corresponding reciprocal lattice.

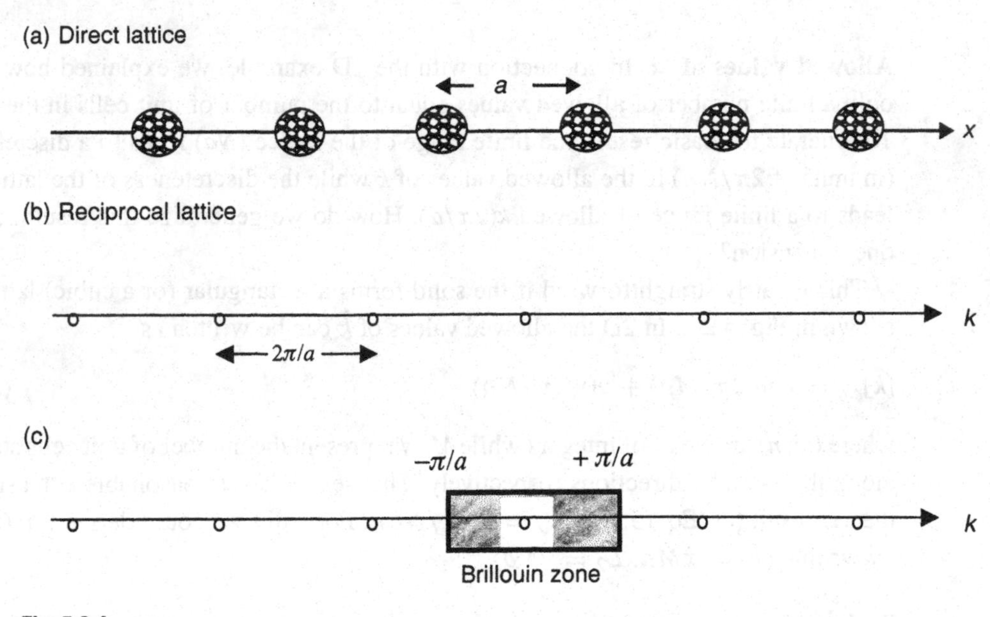

Fig. 5.2.4

by $2\pi/a$ (Fig. 5.2.4b). We could then construct the first Brillouin zone centered around $k = 0$ by connecting it to the neighboring points on the reciprocal lattice and drawing their bisectors (Fig. 5.2.4c). Similarly for a 2D rectangular lattice we can construct a reciprocal lattice and then obtain the first Brillouin zone by drawing perpendicular bisectors of the lines joining $\vec{k} = (0, 0)$ to the neighboring points on the reciprocal lattice.

The Brillouin zone obtained from this procedure defines the allowed range of values of $\vec{k}$

$$-\pi \leq k_x a < +\pi \quad \text{and} \quad -\pi \leq k_y b < +\pi \tag{5.2.6}$$

which agrees with what one might write down from a heuristic extension of Eq. (5.1.5).

Reciprocal lattice: In general, if the direct lattice is not rectangular or cubic, it is not possible to construct the reciprocal lattice quite so simply by inspection. We then need to adopt a more formal procedure as follows. We first note that any point on a direct lattice in 3D can be described by a set of three integers (m, n, p) such that

$$\vec{R} = m\vec{a}_1 + n\vec{a}_2 + p\vec{a}_3 \tag{5.2.7a}$$

where $\vec{a}_1, \vec{a}_2, \vec{a}_3$ are called the basis vectors of the lattice. The points on the reciprocal lattice can be written as

$$\vec{K} = M\vec{A}_1 + N\vec{A}_2 + P\vec{A}_3 \tag{5.2.7b}$$

where (M, N, P) are integers and $\vec{A}_1, \vec{A}_2, \vec{A}_3$ are determined such that

$$\vec{A}_j \cdot \vec{a}_i = 2\pi\,\delta_{ij} \tag{5.2.8}$$

δ_{ij} being the Kronecker delta function (equal to one if $i = j$, and equal to zero if $i \neq j$). Equation (5.2.8) can be satisfied by writing:

$$\vec{A}_1 = \frac{2\pi(\vec{a}_2 \times \vec{a}_3)}{\vec{a}_1 \cdot (\vec{a}_2 \times \vec{a}_3)} \qquad \vec{A}_2 = \frac{2\pi(\vec{a}_3 \times \vec{a}_1)}{\vec{a}_2 \cdot (\vec{a}_3 \times \vec{a}_1)} \qquad \vec{A}_3 = \frac{2\pi(\vec{a}_1 \times \vec{a}_2)}{\vec{a}_3 \cdot (\vec{a}_1 \times \vec{a}_2)} \tag{5.2.9}$$

It is easy to see that this formal procedure for constructing the reciprocal lattice leads to the lattice shown in Fig. 5.2.3b if we assume the real-space basis vectors to be $\vec{a}_1 = \hat{x}a$, $\vec{a}_2 = \hat{y}b$, $\vec{a}_3 = \hat{z}c$. Equation (5.2.9) then yields

$$\vec{A}_1 = \hat{x}(2\pi/a) \qquad \vec{A}_2 = \hat{y}(2\pi/b) \qquad \vec{A}_3 = \hat{z}(2\pi/c)$$

Using Eq. (5.2.7) we can now set up the reciprocal lattice shown in Fig. 5.2.3b. Of course, in this case we do not really need the formal procedure. The real value of the formal approach lies in handling non-rectangular lattices, as we will now illustrate with a 2D example.

A 2D example: The carbon atoms on the surface of a sheet of graphite (often called a graphene layer) are arranged in a hexagonal pattern as shown in Fig. 5.2.5a. It can be seen that the structure is not really periodic. Adjacent carbon atoms do not have identical environments. But if we lump two atoms together into a unit cell then the lattice of unit cells is periodic: every site has an identical environment (Fig. 5.2.5b).

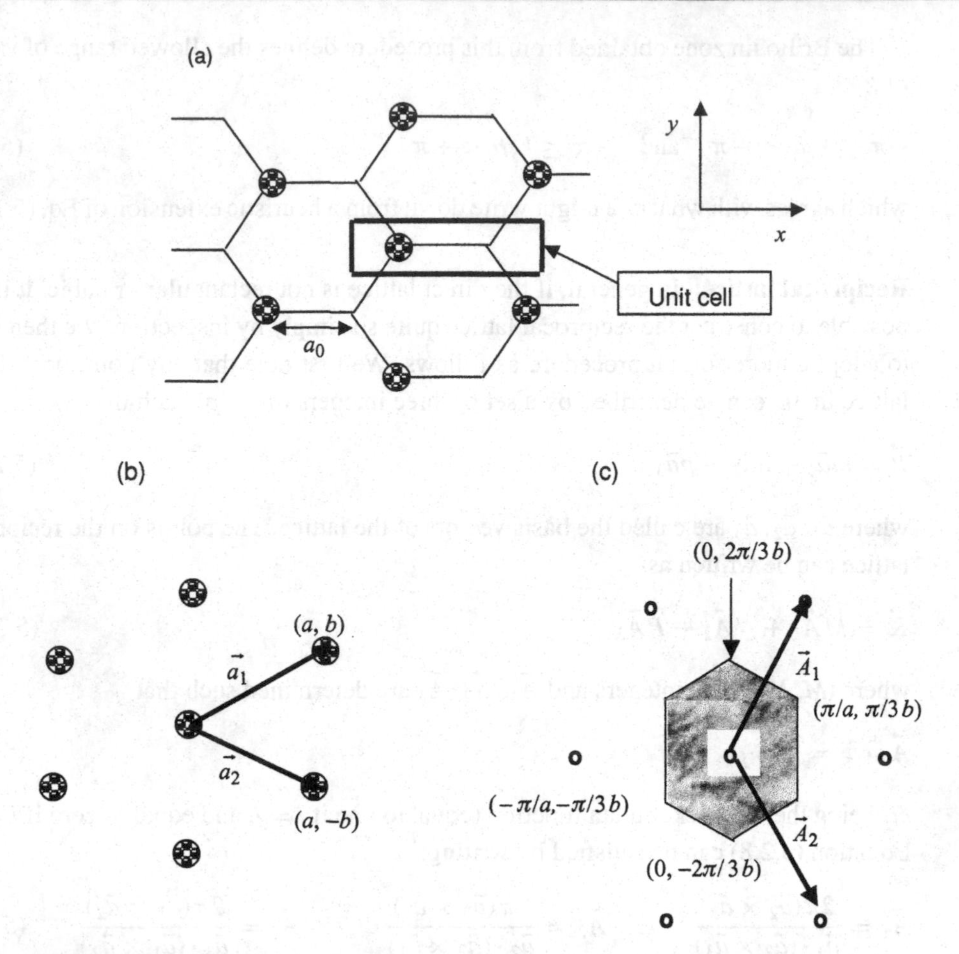

Fig. 5.2.5 (a) Arrangement of carbon atoms on the surface of graphite, showing the unit cell of two atoms. (b) Direct lattice showing the periodic arrangement of unit cells with basis vectors $\vec{a}_1$ and $\vec{a}_2$. (c) Reciprocal lattice with basis vectors $\vec{A}_1$ and $\vec{A}_2$ determined such that $\vec{A}_1 \cdot \vec{a}_1 = \vec{A}_2 \cdot \vec{a}_2 = 2\pi$ and $\vec{A}_1 \cdot \vec{a}_2 = \vec{A}_2 \cdot \vec{a}_1 = 0$. Also shown is the Brillouin zone (shaded) obtained by drawing the perpendicular bisectors of the lines joining the origin (0, 0) to the neighboring points on the reciprocal lattice.

Every point on this periodic lattice formed by the unit cells can be described by a set of integers (m, n, p) where

$$\vec{R} = m\vec{a}_1 + n\vec{a}_2 + p\vec{a}_3 \tag{5.2.10}$$

with

$$\vec{a}_1 = \hat{x}a + \hat{y}b \qquad \vec{a}_2 = \hat{x}a - \hat{y}b \qquad \vec{a}_3 = \hat{z}c$$

where

$$a \equiv 3a_0/2 \quad \text{and} \quad b \equiv \sqrt{3}a_0/2$$

Here c is the length of the unit cell along the z-axis, which will play no important role in this discussion since we will talk about the electronic states in the x–y plane assuming that different planes along the z-axis are isolated (which is not too far from the truth in real graphite). The points on the reciprocal lattice in the k_x–k_y plane are given by

$$\vec{K} = M\vec{A}_1 + N\vec{A}_2 \tag{5.2.11}$$

where (M, N) are integers and $\vec{A}_1, \vec{A}_2$ are determined from Eq. (5.2.9):

$$\vec{A}_1 = \frac{2\pi(\vec{a}_2 \times \hat{z})}{\vec{a}_1 \cdot (\vec{a}_2 \times \hat{z})} = \hat{x}\left(\frac{\pi}{a}\right) + \hat{y}\left(\frac{\pi}{b}\right)$$

$$\vec{A}_2 = \frac{2\pi(\hat{z} \times \vec{a}_1)}{\vec{a}_2 \cdot (\hat{z} \times \vec{a}_1)} = \hat{x}\left(\frac{\pi}{a}\right) - \hat{y}\left(\frac{\pi}{b}\right)$$

Using these basis vectors we can construct the reciprocal lattice shown in Fig. 5.2.5c. The Brillouin zone for the allowed k-vectors is then obtained by drawing the perpendicular bisectors of the lines joining the origin (0, 0) to the neighboring points on the reciprocal lattice.

The Brillouin zone tells us the range of k values while the actual discrete values of k have to be obtained from the finite size of the direct lattice, as explained following Eq. (5.2.5). But, for a given value of k how do we obtain the corresponding energy eigenvalues? Answer: from Eqs. (5.2.3) and (5.2.4). The size of the matrix $[h(\vec{k})]$ depends on the number of basis functions per unit cell. If we use the four valence orbitals of carbon (2s, $2p_x$, $2p_y$, $2p_z$) as our basis functions then we will have $4 \times 2 = 8$ basis functions per unit cell (since it contains two carbon atoms) and hence eight eigenvalues for each value of k.

It is found, however, for graphene that the levels involving 2s, $2p_x$, $2p_y$ orbitals are largely decoupled from those involving $2p_z$ orbitals; in other words, there are no matrix elements coupling these two subspaces. Moreover, the levels involving 2s, $2p_x$, $2p_y$ orbitals are either far below or far above the Fermi energy, so that the conduction and valence band levels right around the Fermi energy (which are responsible for electrical conduction) are essentially formed out of the $2p_z$ orbitals.

This means that the conduction and valence band states can be described quite well by a theory that uses only one orbital (the $2p_z$ orbital) per carbon atom resulting in a (2×2) matrix $[h(\vec{k})]$ that can be written down by summing over any unit cell and all its four neighboring unit cells (the matrix element is assumed equal to $-t$ between neighboring carbon atoms and zero otherwise):

$$[h(\vec{k})] = \begin{bmatrix} 0 & -t \\ -t & 0 \end{bmatrix} + \begin{bmatrix} 0 & -t\exp(i\vec{k}\cdot\vec{a}_1) \\ 0 & 0 \end{bmatrix} + \begin{bmatrix} 0 & -t\exp(i\vec{k}\cdot\vec{a}_2) \\ 0 & 0 \end{bmatrix}$$
$$+ \begin{bmatrix} 0 & 0 \\ -t\exp(-i\vec{k}\cdot\vec{a}_1) & 0 \end{bmatrix} + \begin{bmatrix} 0 & 0 \\ -t\exp(-i\vec{k}\cdot\vec{a}_2) & 0 \end{bmatrix}$$

Defining

$$h_0 \equiv -t(1 + e^{i\vec{k}\cdot\vec{a}_1} + e^{i\vec{k}\cdot\vec{a}_2}) = -t(1 + 2e^{ik_x a} \cos k_y b)$$

we can write

$$h(\vec{k}) = \begin{bmatrix} 0 & h_0 \\ h_0^* & 0 \end{bmatrix}$$

so that the eigenvalues are given by

$$E = \pm|h_0| = \pm t\sqrt{1 + 4\cos k_y b \cos k_x a + 4\cos^2 k_y b}$$

Note that we obtain two eigenvalues (one positive and one negative) for each value of $\vec{k}$ resulting in two branches in the $E(\vec{k})$ plot (cf. Fig. 5.1.5) – this is what we expect since we have two basis functions per unit cell. We will discuss the physics of this $E(\vec{k})$ relation (generally called the energy dispersion relation) in the next chapter when we discuss carbon nanotubes, which are basically graphite sheets rolled into cylinders. For the moment our main purpose is to illustrate the procedure for calculating the bandstructure using a 2D example that involves non-trivial features beyond the 1D examples from the last section and yet does not pose serious problems with visualization as the 3D example in the next section.

5.3 Common semiconductors

In this section, I will follow closely the treatment by Vogl *et al.* (1983). All the common semiconductors (like gallium arsenide) belong to the diamond structure which has a unit cell consisting of two atoms, a cation (like gallium) and an anion (like arsenic). For elemental semiconductors like silicon, both cationic and anionic sites are occupied by the same atom. For each atom we need to include at least four valence orbitals like 3s, $3p_x$, $3p_y$ and $3p_z$ for silicon. It is common to include the next higher orbital (4s for silicon) as well, giving rise to what is called the sp^3s^* model. In this model we have five orbitals per atom leading to ten basis orbitals per unit cell. Consequently the matrices $[h(\vec{k})]$ and $[H_{nm}]$ in Eq. (5.2.4) are each (10×10) in size. To perform the summation indicated in Eq. (5.2.4) we need to figure out how the nearest neighbors are located.

The diamond structure consists of two interpenetrating face-centered cubic (FCC) lattices. For example, if we look at GaAs, we find that the gallium atoms occupy the sites on an FCC lattice. The arsenic atoms occupy the sites of a different FCC lattice offset from the previous one by a quarter of the distance along the body diagonal – that is, the coordinates of this lattice can be obtained by adding $(\hat{x} + \hat{y} + \hat{z})a/4$ to those of the first one. If a gallium atom is located at the origin $(0\ 0\ 0)a/4$ then there will be an arsenic atom located at $(\hat{x} + \hat{y} + \hat{z})a/4$, which will be one of its nearest neighbors. Actually it will also have three more arsenic atoms as nearest neighbors. To see this consider

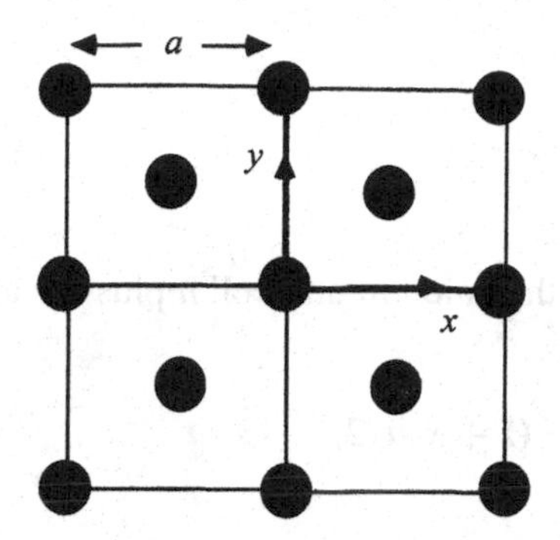

Fig. 5.3.1 x–y face of a face-centered cubic (FCC) lattice, showing the location of atoms.

where the nearest gallium atoms are located. There are four of them on the x–y face as shown in Fig. 5.3.1 whose coordinates can be written as $(\hat{x}+\hat{y})a/2$, $(-\hat{x}+\hat{y})a/2$, and $(-\hat{x}-\hat{y})a/2$.

The coordinates of the corresponding arsenic atoms are obtained by adding $(\hat{x}+\hat{y}+\hat{z})a/4$:

$$(3\hat{x}+3\hat{y}+\hat{z})a/4 \qquad (3\hat{x}-\hat{y}+\hat{z})a/4 \qquad (-\hat{x}+3\hat{y}+\hat{z})a/4 \qquad (-\hat{x}-\hat{y}+\hat{z})a/4$$

Of these the first three are too far away, but the fourth one is a nearest neighbor of the gallium atom at the origin. Similarly if we consider the neighboring gallium atoms on the y–z face and the z–x face we will find two more nearest neighbors, so that the gallium atom at the origin (0 0 0) has four nearest neighbor arsenic atoms located at

$$(\hat{x}+\hat{y}+\hat{z})a/4 \qquad (-\hat{x}-\hat{y}+\hat{z})a/4 \qquad (\hat{x}-\hat{y}-\hat{z})a/4 \qquad (-\hat{x}+\hat{y}-\hat{z})a/4$$

Every atom in a diamond lattice has four nearest neighbors of the opposite type (cation or anion) arranged in a tetrahedron.

To see how we perform the summation in Eq. (5.2.4) let us first consider just the s orbital for each atom. The matrices $[h(\vec{k})]$ and $[H_{nm}]$ in Eq. (5.2.4) are then each (2×2) in size. We can write $[H_{nm}]$ as

$$\begin{array}{ccc} & |s_a\rangle & |s_c\rangle \\ |s_a\rangle & E_{sa} & E_{ss} \\ |s_c\rangle & E_{ss} & E_{sc} \end{array} \tag{5.3.1a}$$

where E_{sa} and E_{sc} are the energies of the s orbitals for the anion and cation respectively, while E_{ss} represents the overlap integral between an s orbital on the anion and an s orbital on the cation. The anion in unit cell n overlaps with the cation in unit cell n plus the cations in three other unit cells m for which

$$\vec{d}_m-\vec{d}_n=(-\hat{x}-\hat{y})a/2 \qquad (-\hat{y}-\hat{z})a/2 \qquad (-\hat{z}-\hat{x})a/2$$

Each of these contributes a $[H_{nm}]$ of the form

	$\lvert m, s_{\mathrm{a}}\rangle$	$\lvert m, s_{\mathrm{c}}\rangle$
$\lvert n, s_{\mathrm{a}}\rangle$	0	E_{ss}
$\lvert n, s_{\mathrm{c}}\rangle$	0	0

(5.3.1b)

Similarly the cation in unit cell n overlaps with the anion in unit cell n plus the anions in three other unit cells m for which

$$\vec{d}_m - \vec{d}_n = (\hat{x} + \hat{y})a/2 \qquad (\hat{y} + \hat{z})a/2 \qquad (\hat{z} + \hat{x})a/2$$

Each of these contributes a $[H_{nm}]$ of the form

	$\lvert m, s_{\mathrm{a}}\rangle$	$\lvert m, s_{\mathrm{c}}\rangle$
$\lvert n, s_{\mathrm{a}}\rangle$	0	0
$\lvert n, s_{\mathrm{c}}\rangle$	E_{ss}	0

(5.3.1c)

Adding up all these contributions we obtain

$[h(\vec{k})] =$

	$\lvert s_{\mathrm{a}}\rangle$	$\lvert s_{\mathrm{c}}\rangle$
$\lvert s_{\mathrm{a}}\rangle$	E_{sa}	$4E_{\mathrm{ss}}g_0$
$\lvert s_{\mathrm{c}}\rangle$	$4E_{\mathrm{ss}}g_0^*$	E_{sc}

(5.3.2)

where

$$4g_0 \equiv 1 + \mathrm{e}^{-\mathrm{i}\vec{k}\cdot\vec{d}_1} + \mathrm{e}^{-\mathrm{i}\vec{k}\cdot\vec{d}_2} + \mathrm{e}^{-\mathrm{i}\vec{k}\cdot\vec{d}_3}$$

with

$$\vec{d}_1 \equiv (\hat{y} + \hat{z})a/2 \qquad \vec{d}_2 \equiv (\hat{z} + \hat{x})a/2 \qquad \vec{d}_3 \equiv (\hat{x} + \hat{y})a/2$$

To evaluate the full (10×10) matrix $[h(\vec{k})]$ including $\mathrm{sp^3s^*}$ levels we proceed similarly. The final result is

	$\lvert s_{\mathrm{a}}\rangle$	$\lvert s_{\mathrm{c}}\rangle$	$\lvert X_{\mathrm{a}}\rangle$	$\lvert Y_{\mathrm{a}}\rangle$	$\lvert Z_{\mathrm{a}}\rangle$	$\lvert X_{\mathrm{c}}\rangle$	$\lvert Y_{\mathrm{c}}\rangle$	$\lvert Z_{\mathrm{c}}\rangle$	$\lvert s_{\mathrm{a}}^*\rangle$	$\lvert s_{\mathrm{c}}^*\rangle$
$\lvert s_{\mathrm{a}}\rangle$	E_{sa}	$4E_{\mathrm{ss}}g_0$	0	0	0	$4E_{\mathrm{sapc}}g_1$	$4E_{\mathrm{sapc}}g_2$	$4E_{\mathrm{sapc}}g_3$	0	0
$\lvert s_{\mathrm{c}}\rangle$	$4E_{\mathrm{ss}}g_0^*$	E_{sc}	$4E_{\mathrm{pasc}}g_1^*$	$4E_{\mathrm{pasc}}g_2^*$	$4E_{\mathrm{pasc}}g_3^*$	0	0	0	0	0
$\lvert X_{\mathrm{a}}\rangle$	0	$4E_{\mathrm{pasc}}g_1$	E_{pa}	0	0	$4E_{xx}g_0$	$4E_{xy}g_3$	$4E_{xy}g_2$	0	$4E_{\mathrm{pas^*c}}g_1$
$\lvert Y_{\mathrm{a}}\rangle$	0	$4E_{\mathrm{pasc}}g_2$	0	E_{pa}	0	$4E_{xy}g_3$	$4E_{xx}g_0$	$4E_{xy}g_1$	0	$4E_{\mathrm{pas^*c}}g_2$
$\lvert Z_{\mathrm{a}}\rangle$	0	$4E_{\mathrm{pasc}}g_3$	0	0	E_{pa}	$4E_{xy}g_2$	$4E_{xy}g_1$	$4E_{xx}g_0$	0	$4E_{\mathrm{pas^*c}}g_3$
$\lvert X_{\mathrm{c}}\rangle$	$4E_{\mathrm{sapc}}g_1^*$	0	$4E_{xx}g_0^*$	$4E_{xy}g_3^*$	$4E_{xy}g_2^*$	E_{pc}	0	0	$4E_{\mathrm{s^*apc}}g_1^*$	0
$\lvert Y_{\mathrm{c}}\rangle$	$4E_{\mathrm{sapc}}g_2^*$	0	$4E_{xy}g_3^*$	$4E_{xx}g_0^*$	$4E_{xy}g_1^*$	0	$E_{\mathrm{pc}}0$	$4E_{\mathrm{s^*apc}}g_2^*$	0	
$\lvert Z_{\mathrm{c}}\rangle$	$4E_{\mathrm{sapc}}g_3^*$	0	$4E_{xy}g_2^*$	$4E_{xy}g_1^*$	$4E_{xx}g_0^*$	0	0	E_{pc}	$4E_{\mathrm{s^*apc}}g_3^*$	0
$\lvert s_{\mathrm{a}}^*\rangle$	0	0	0	0	0	$4E_{\mathrm{s^*apc}}g_1$	$4E_{\mathrm{s^*apc}}g_2$	$4E_{\mathrm{s^*apc}}g_3$	$E_{\mathrm{s^*a}}$	0
$\lvert s_{\mathrm{c}}^*\rangle$	0	0	$4E_{\mathrm{pas^*c}}g_1^*$	$4E_{\mathrm{pas^*c}}g_2^*$	$4E_{\mathrm{pas^*c}}g_3^*$	0	0	0	0	$E_{\mathrm{s^*c}}$

(5.3.3)

The factors g_1, g_2, and g_3 look much like the factor g_0 obtained above when discussing only the s orbitals:

$$4g_0 \equiv 1 + \mathrm{e}^{-\mathrm{i}\vec{k}\cdot\vec{d}_1} + \mathrm{e}^{-\mathrm{i}\vec{k}\cdot\vec{d}_2} + \mathrm{e}^{-\mathrm{i}\vec{k}\cdot\vec{d}_3} \tag{5.3.4a}$$

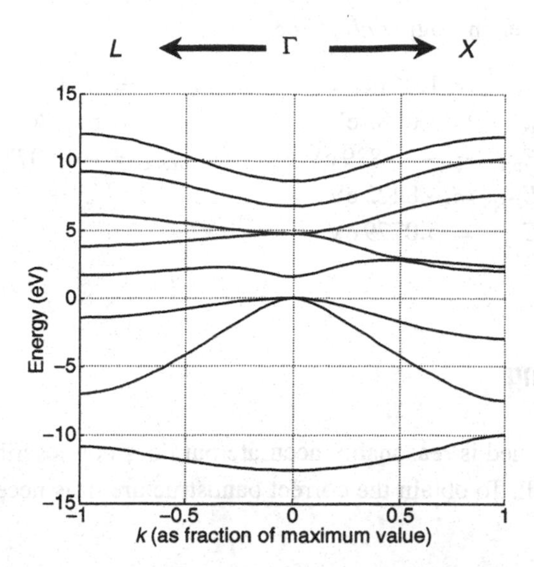

Fig. 5.3.2 $E(\vec{k})$ calculated by finding the eigenvalues of the matrix in Eq. (5.3.3) for each value of $\vec{k}$ along the Γ–X (that is, from $\vec{k} = 0$ to $\vec{k} = \vec{k} = \hat{x}\, 2\pi/a$) and Γ–L (that is, from $\vec{k} = 0$ to $\vec{k} = (\hat{x} + \hat{y} + \hat{z})\pi/a$) directions. The former is plotted along the positive axis and the latter along the negative axis. Only eight (rather than ten) lines are visible because some are degenerate.

However, the signs of some of the terms are negative:

$$4g_1 \equiv 1 + e^{-i\vec{k}\cdot\vec{d}_1} - e^{-i\vec{k}\cdot\vec{d}_2} - e^{-i\vec{k}\cdot\vec{d}_3} \tag{5.3.4b}$$

$$4g_2 \equiv 1 - e^{-i\vec{k}\cdot\vec{d}_1} + e^{-i\vec{k}\cdot\vec{d}_2} - e^{-i\vec{k}\cdot\vec{d}_3} \tag{5.3.4c}$$

$$4g_3 \equiv 1 - e^{-i\vec{k}\cdot\vec{d}_1} - e^{-i\vec{k}\cdot\vec{d}_2} + e^{-i\vec{k}\cdot\vec{d}_3} \tag{5.3.4d}$$

The negative signs arise because the wavefunction for p orbitals changes sign along one axis and so the overlap integral has different signs for different neighbors. This also affects the signs of the overlap integrals appearing in the expression for $[h(\vec{k})]$ in Eq. (5.2.4): the parameters E_{ss}, E_{pasc}, and E_{pas^*c} are negative, while the remaining parameters E_{sapc}, E_{xx}, E_{xy}, and E_{s^*apc} are positive. Note that the vectors

$$\vec{d}_1 \equiv (\hat{y} + \hat{z})a/2 \qquad \vec{d}_2 \equiv (\hat{z} + \hat{x})a/2 \qquad \vec{d}_3 \equiv (\hat{x} + \hat{y})a/2 \tag{5.3.5}$$

connect the cation in one unit cell to a cation in a neighboring cell (or an anion to an anion). Alternatively, we could define these vectors so as to connect the nearest neighbors – this has the effect of multiplying each of the factors g_0, g_1, g_2, and g_3 by a phase factor $\exp[i\vec{k}\cdot\vec{d}]$ where $\vec{d} = (\hat{x} + \hat{y} + \hat{z})a/4$. This is used by most authors like Vogl *et al.* (1983), but it makes no real difference to the result.

Figure 5.3.2 shows the bandstructure $E(\vec{k})$ calculated by finding the eigenvalues of the matrix in Eq. (5.3.3) for each value of $\vec{k}$ along the Γ–X and Γ–L directions. We have used the parameters for GaAs given in Vogl *et al.* (1983):

$$\begin{array}{llllll}
E_{\mathrm{sa}} & = -8.3431\,\mathrm{eV} & E_{\mathrm{pa}} & = 1.0414\,\mathrm{eV} & E_{\mathrm{s^*a}} & = 8.5914\,\mathrm{eV} \\
E_{\mathrm{sc}} & = -2.6569\,\mathrm{eV} & E_{\mathrm{pc}} & = 3.6686\,\mathrm{eV} & E_{\mathrm{s^*c}} & = 6.7386\,\mathrm{eV} \\
4E_{\mathrm{ss}} & = -6.4513\,\mathrm{eV} & 4E_{\mathrm{pasc}} & = -5.7839\,\mathrm{eV} & 4E_{\mathrm{pas^*c}} & = -4.8077\,\mathrm{eV} \\
4E_{\mathrm{sapc}} & = 4.48\,\mathrm{eV} & 4E_{\mathrm{s^*apc}} & = 4.8422\,\mathrm{eV} & & \\
4E_{xx} & = 1.9546\,\mathrm{eV} & 4E_{xy} & = 5.0779\,\mathrm{eV} & &
\end{array}$$

5.4 Effect of spin–orbit coupling

The bandstructure we have obtained is reasonably accurate but does not describe the top of the valence band very well. To obtain the correct bandstructure, it is necessary to include spin–orbit coupling.

Spinors: Let us first briefly explain how spin can be included explicitly in the Schrödinger equation. Usually we calculate the energy levels from the Schrödinger equation and fill them up with *two* electrons per level. More correctly we should view each level as two levels with the same energy and fill them up with *one* electron per level as required by the exclusion principle. How could we modify the Schrödinger equation so that each level becomes two levels with identical energies? The answer is simple. Replace

$$E(\psi) = [H_{\mathrm{op}}](\psi) \quad \text{with} \quad E\begin{pmatrix}\psi \\ \overline{\psi}\end{pmatrix} = \begin{bmatrix} H_{\mathrm{op}} & 0 \\ 0 & H_{\mathrm{op}} \end{bmatrix}\begin{pmatrix}\psi \\ \overline{\psi}\end{pmatrix} \tag{5.4.1}$$

where

$$H_{\mathrm{op}} = p^2/2m + U(\vec{r}) \quad (\vec{p} \equiv -i\hbar\vec{\nabla})$$

We interpret ψ as the up-spin component and $\overline{\psi}$ as the down-spin component of the electronic wavefunction. If we now choose a basis set to obtain a matrix representation, the matrix will be twice as big. For example if we were to use just the s orbital for each atom we would obtain a (4×4) matrix instead of the (2×2) matrix in Eq. (5.3.2):

$$\begin{array}{ccccc}
 & |s_{\mathrm{a}}\rangle & |s_{\mathrm{c}}\rangle & |\overline{s}_{\mathrm{a}}\rangle & |\overline{s}_{\mathrm{a}}\rangle \\
|s_{\mathrm{a}}\rangle & E_{\mathrm{sa}} & 4E_{\mathrm{ss}}g_0 & 0 & 0 \\
|s_{\mathrm{c}}\rangle & 4E_{\mathrm{ss}}g_0^* & E_{\mathrm{sc}} & 0 & 0 \\
|\overline{s}_{\mathrm{a}}\rangle & 0 & 0 & E_{\mathrm{sa}} & 4E_{\mathrm{ss}}g_0 \\
|\overline{s}_{\mathrm{c}}\rangle & 0 & 0 & 4E_{\mathrm{ss}}g_0^* & E_{\mathrm{sc}}
\end{array} \tag{5.4.2}$$

Similarly, with all ten orbitals included, the (10×10) matrix becomes a (20×20)

matrix:

$$[H_0(\vec{k})] = \begin{bmatrix} h(\vec{k}) & 0 \\ 0 & h(\vec{k}) \end{bmatrix} \tag{5.4.3}$$

where $[h(\vec{k})]$ is given by Eq. (5.3.3).

Spin–orbit coupling: If we were to calculate the bandstructure using Eq. (5.4.3) instead of Eq. (5.3.3) we would get exactly the same result, except that each line would have a second one right on top of it, which we would probably not even notice if a computer were plotting it out. But the reason we are doing this is that we want to add something called spin–orbit coupling to Eq. (5.4.3).

The Schrödinger equation is a non-relativistic equation. For electrons traveling at high velocities relativistic effects can become significant and we need to use the Dirac equation. Typically in solids the velocities are not high enough to require this, but the electric fields are very high near the nuclei of atoms leading to weak relativistic effects that can be accounted for by adding a spin–orbit correction H_{so} to the Schrödinger equation:

$$E\begin{pmatrix} \psi \\ \bar{\psi} \end{pmatrix} = [H_0]\begin{pmatrix} \psi \\ \bar{\psi} \end{pmatrix} + [H_{so}]\begin{pmatrix} \psi \\ \bar{\psi} \end{pmatrix} \tag{5.4.4}$$

where

$$H_0 = \begin{bmatrix} p^2/2m + U(\vec{r}) & 0 \\ 0 & p^2/2m + U(\vec{r}) \end{bmatrix} \tag{5.4.5}$$

and

$$H_{so} = \frac{q\hbar}{4m^2c^2}\begin{bmatrix} E_x p_y - E_y p_x & (E_y p_z - E_z p_y) - \mathrm{i}(E_z p_x - E_x p_z) \\ (E_y p_z - E_z p_y) - \mathrm{i}(E_z p_x - E_x p_z) & -(E_x p_y - E_y p_x) \end{bmatrix} \tag{5.4.6}$$

c being the velocity of light in vacuum. The spin–orbit Hamiltonian H_{so} is often written as

$$H_{so} = \frac{q\hbar}{4m^2c^2}\vec{\sigma}\cdot(\vec{E}\times\vec{p}) \tag{5.4.7}$$

where the Pauli spin matrices $\vec{\sigma}$ are defined as

$$\sigma_x = \begin{bmatrix} 0 & 1 \\ 1 & 0 \end{bmatrix} \quad \sigma_y = \begin{bmatrix} 0 & -\mathrm{i} \\ \mathrm{i} & 0 \end{bmatrix} \quad \sigma_z = \begin{bmatrix} 1 & 0 \\ 0 & -1 \end{bmatrix} \tag{5.4.8}$$

It is straightforward to show that the two expressions for the spin–orbit Hamiltonian H_{so} in Eqs. (5.4.7) and (5.4.6) are identical. I will not try to justify the origin of the spin–orbit term for this would take us too far afield into the Dirac equation, but the interested reader may find the brief discussion in Section 5.5 instructive.

Bandstructure with spin–orbit coupling: We already have the matrix representation for the non-spin–orbit part of the Hamiltonian, H_0 (Eq. (5.4.5)). It is given by Eq. (5.3.3). We now need to find a matrix representation for H_{so} and add it to H_0. Let us first see what we would do if we were to just use the s orbitals for each atom. Usually the spin–orbit matrix elements are significant only if both orbitals are centered on the same atom, so that we expect a matrix of the form

$$\begin{array}{c|cccc} & |s_a\rangle & |s_c\rangle & |\bar{s}_a\rangle & |\bar{s}_c\rangle \\ \hline |s_a\rangle & a_{11} & 0 & a_{12} & 0 \\ |s_c\rangle & 0 & c_{11} & 0 & c_{12} \\ |\bar{s}_a\rangle & a_{21} & 0 & a_{22} & 0 \\ |\bar{s}_c\rangle & 0 & c_{21} & 0 & c_{22} \end{array}$$

We would fill up the 11 elements of this matrix by taking the matrix elements of the 11 component of H_{so} (see Eq. (5.4.6)):

$$a_{11} = \langle s_a | E_x p_y - E_y p_x | s_a \rangle \qquad c_{11} = \langle s_c | E_x p_y - E_y p_x | s_c \rangle$$

To fill up the 12 elements of this matrix we take the matrix elements of the 12 component of H_{so} (see Eq. (5.4.6)):

$$a_{12} = \langle s_a | (E_y p_z - E_z p_y) - \mathrm{i}(E_z p_x - E_x p_z) | \bar{s}_a \rangle$$
$$c_{12} = \langle s_c | (E_y p_z - E_z p_y) - \mathrm{i}(E_z p_x - E_x p_z) | \bar{s}_c \rangle$$

Similarly we can go on with the 21 and the 22 components. As it turns out, all these matrix elements can be shown to be zero from symmetry arguments if we assume the potential $U(r)$ to be spherically symmetric as is reasonable for atomic potentials. The same is true for the s* orbitals as well. However, some of the matrix elements are non-zero when we consider the X, Y, and Z orbitals. These non-zero matrix elements can all be expressed in terms of a single number δ_a for the anionic orbitals:

$$\begin{array}{c|cccccc} & |X_a\rangle & |Y_a\rangle & |Z_a\rangle & |\overline{X}_a\rangle & |\overline{Y}_a\rangle & |\overline{Z}_a\rangle \\ \hline |X_a\rangle & 0 & -\mathrm{i}\delta_a & 0 & 0 & 0 & \delta_a \\ |Y_a\rangle & \mathrm{i}\delta_a & 0 & 0 & 0 & 0 & -\mathrm{i}\delta_a \\ |Z_a\rangle & 0 & 0 & 0 & -\delta_a & \mathrm{i}\delta_a & 0 \\ |\overline{X}_a\rangle & 0 & 0 & -\delta_a & 0 & \mathrm{i}\delta_a & 0 \\ |\overline{Y}_a\rangle & 0 & 0 & -\mathrm{i}\delta_a & -\mathrm{i}\delta_a & 0 & 0 \\ |\overline{Z}_a\rangle & \delta_a & \mathrm{i}\delta_a & 0 & 0 & 0 & 0 \end{array} \tag{5.4.9a}$$

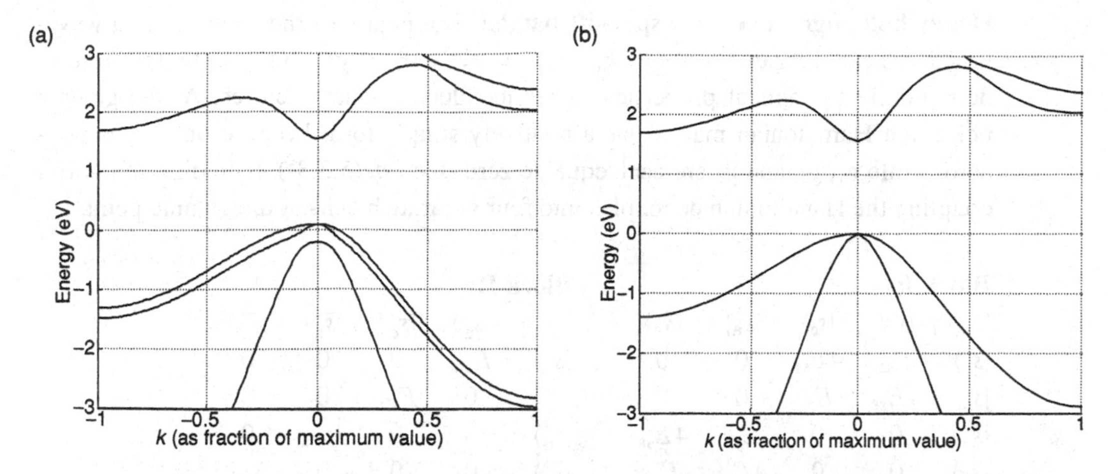

Fig. 5.4.1 Bandstructure of GaAs calculated (a) taking spin–orbit interaction into account (the Γ–X direction is plotted along the positive axis while the Γ–L direction is plotted along the negative axis) and (b) from Eq. (5.3.3) without adding the spin–orbit component.

and in terms of a single number δ_c for the cationic orbitals:

$$
\begin{array}{c|cccccc}
 & |X_c\rangle & |Y_c\rangle & |Z_c\rangle & |\overline{X}_c\rangle & |\overline{Y}_c\rangle & |\overline{Z}_c\rangle \\
|X_c\rangle & 0 & -i\delta_c & 0 & 0 & 0 & \delta_c \\
|Y_c\rangle & i\delta_c & 0 & 0 & 0 & 0 & -i\delta_c \\
|Z_c\rangle & 0 & 0 & 0 & -\delta_c & i\delta_c & 0 \\
|\overline{X}_c\rangle & 0 & 0 & -\delta_c & 0 & i\delta_c & 0 \\
|\overline{Y}_c\rangle & 0 & 0 & -i\delta_c & -i\delta_c & 0 & 0 \\
|\overline{Z}_c\rangle & \delta_c & i\delta_c & 0 & 0 & 0 & 0
\end{array}
\tag{5.4.9b}
$$

If we were to find the eigenvalues of either of these matrices we would obtain four eigenvalues equal to $+\delta_c$ (or $+\delta_a$) and two eigenvalues equal to $-2\delta_c$ (or $-2\delta_a$). The splitting between these two sets of levels is $3\delta_c$ (or $3\delta_a$) and is referred to as the spin–orbit splitting Δ_c (or Δ_a):

$$\Delta_c \text{ (or } \Delta_a) = 3\delta_c \text{ (or } 3\delta_a) \tag{5.4.10}$$

The spin–orbit splitting is well-known from both theory and experiment for all the atoms. For example, gallium has a spin–orbit splitting of 0.013 eV while that for arsenic is 0.38 eV. It is now straightforward to write down the full matrix representation for H_{so} making use of Eqs. (5.4.8) and (5.4.9), adding it to Eq. (5.3.3) and then calculating the bandstructure. For GaAs we obtain the result shown in Fig. 5.4.1a. For comparison, in Fig. 5.4.1b we show the results obtained directly from Eq. (5.3.3) without adding the spin–orbit part. This is basically the same plot obtained in the last section (see Fig. 5.3.2) except that the energy scale has been expanded to highlight the top of the valence band.

Heavy hole, light hole, and split-off bands: The nature of the valence band wavefunction near the gamma point ($k_x = k_y = k_z = 0$) plays a very important role in determining the optical properties of semiconductor nanostructures. At the gamma point, the Hamiltonian matrix has a relatively simple form because only g_0 is non-zero, while g_1, g_2, and g_3 are each equal to zero (see Eq. (5.3.3)). Including spin–orbit coupling the Hamiltonian decouples into four separate blocks at the gamma point:

Block I:

$$
\begin{array}{c|cccc}
 & |s_{\mathrm{a}}\rangle & |s_{\mathrm{c}}\rangle & |\overline{s}_{\mathrm{a}}\rangle & |\overline{s}_{\mathrm{c}}\rangle \\
\hline
|s_{\mathrm{a}}\rangle & E_{\mathrm{sa}} & 4E_{\mathrm{ss}} & 0 & 0 \\
|s_{\mathrm{c}}\rangle & 4E_{\mathrm{ss}} & E_{\mathrm{sc}} & 0 & 0 \\
|\overline{s}_{\mathrm{a}}\rangle & 0 & 0 & E_{\mathrm{sa}} & 4E_{\mathrm{ss}} \\
|\overline{s}_{\mathrm{c}}\rangle & 0 & 0 & 4E_{\mathrm{ss}} & E_{\mathrm{sc}}
\end{array}
$$

Block II:

$$
\begin{array}{c|cccc}
 & |s^*_{\mathrm{a}}\rangle & |s^*_{\mathrm{c}}\rangle & |\overline{s}^*_{\mathrm{a}}\rangle & |\overline{s}^*_{\mathrm{c}}\rangle \\
\hline
|s^*_{\mathrm{a}}\rangle & E_{\mathrm{s^*a}} & 0 & 0 & 0 \\
|s^*_{\mathrm{c}}\rangle & 0 & E_{\mathrm{s^*c}} & 0 & 0 \\
|\overline{s}^*_{\mathrm{a}}\rangle & 0 & 0 & E_{\mathrm{s^*a}} & 0 \\
|\overline{s}^*_{\mathrm{c}}\rangle & 0 & 0 & 0 & E_{\mathrm{s^*c}}
\end{array}
$$

Block III:

$$
\begin{array}{c|cccccc}
 & |X_{\mathrm{a}}\rangle & |Y_{\mathrm{a}}\rangle & |\overline{Z}_{\mathrm{a}}\rangle & |X_{\mathrm{c}}\rangle & |Y_{\mathrm{c}}\rangle & |\overline{Z}_{\mathrm{c}}\rangle \\
\hline
|X_{\mathrm{a}}\rangle & E_{\mathrm{pa}} & -\mathrm{i}\delta_{\mathrm{a}} & \delta_{\mathrm{a}} & 4E_{xx} & 0 & 0 \\
|Y_{\mathrm{a}}\rangle & \mathrm{i}\delta_{\mathrm{a}} & E_{\mathrm{pa}} & -\mathrm{i}\delta_{\mathrm{a}} & 0 & 4E_{xx} & 0 \\
|\overline{Z}_{\mathrm{a}}\rangle & \delta_{\mathrm{a}} & \mathrm{i}\delta_{\mathrm{a}} & E_{\mathrm{pa}} & 0 & 0 & 4E_{xx} \\
|X_{\mathrm{c}}\rangle & 4E_{xx} & 0 & 0 & E_{\mathrm{pc}} & -\mathrm{i}\delta_{\mathrm{c}} & \delta_{\mathrm{c}} \\
|Y_{\mathrm{c}}\rangle & 0 & 4E_{xx} & 0 & \mathrm{i}\delta_{\mathrm{c}} & E_{\mathrm{pc}} & -\mathrm{i}\delta_{\mathrm{c}} \\
|\overline{Z}_{\mathrm{c}}\rangle & 0 & 0 & 4E_{xx} & \delta_{\mathrm{c}} & \mathrm{i}\delta_{\mathrm{c}} & E_{\mathrm{pc}}
\end{array}
$$

Block IV:

$$
\begin{array}{c|cccccc}
 & |\overline{X}_{\mathrm{a}}\rangle & |\overline{Y}_{\mathrm{a}}\rangle & |Z_{\mathrm{a}}\rangle & |\overline{X}_{\mathrm{c}}\rangle & |\overline{Y}_{\mathrm{c}}\rangle & |Z_{\mathrm{c}}\rangle \\
\hline
|\overline{X}_{\mathrm{a}}\rangle & E_{\mathrm{pa}} & \mathrm{i}\delta_{\mathrm{a}} & -\delta_{\mathrm{a}} & 4E_{xx} & 0 & 0 \\
|\overline{Y}_{\mathrm{a}}\rangle & -\mathrm{i}\delta_{\mathrm{a}} & E_{\mathrm{pa}} & -\mathrm{i}\delta_{\mathrm{a}} & 0 & 4E_{xx} & 0 \\
|Z_{\mathrm{a}}\rangle & -\delta_{\mathrm{a}} & \mathrm{i}\delta_{\mathrm{a}} & E_{\mathrm{pa}} & 0 & 0 & 4E_{xx} \\
|\overline{X}_{\mathrm{c}}\rangle & 4E_{xx} & 0 & 0 & E_{\mathrm{pc}} & \mathrm{i}\delta_{\mathrm{c}} & -\delta_{\mathrm{c}} \\
|\overline{Y}_{\mathrm{c}}\rangle & 0 & 4E_{xx} & 0 & -\mathrm{i}\delta_{\mathrm{c}} & E_{\mathrm{pc}} & -\mathrm{i}\delta_{\mathrm{c}} \\
|Z_{\mathrm{c}}\rangle & 0 & 0 & 4E_{xx} & -\delta_{\mathrm{c}} & \mathrm{i}\delta_{\mathrm{c}} & E_{\mathrm{pc}}
\end{array}
$$

We can partially diagonalize Blocks III and IV by transforming to the heavy hole (HH), light hole (LH), and split-off (SO) basis using the transformation matrix

$$
[V] = \begin{array}{c|cccccc}
 & |\mathrm{HH_a}\rangle & |\mathrm{LH_a}\rangle & |\mathrm{SO_a}\rangle & |\mathrm{HH_c}\rangle & |\mathrm{LH_c}\rangle & |\mathrm{SO_c}\rangle \\
\hline
|X_{\mathrm{a}}\rangle & 1/\sqrt{2} & 1/\sqrt{6} & 1/\sqrt{3} & 0 & 0 & 0 \\
|Y_{\mathrm{a}}\rangle & \mathrm{i}/\sqrt{2} & -\mathrm{i}/\sqrt{6} & -\mathrm{i}/\sqrt{3} & 0 & 0 & 0 \\
|\overline{Z}_{\mathrm{a}}\rangle & 0 & \sqrt{2/3} & -1/\sqrt{3} & 0 & 0 & 0 \\
|X_{\mathrm{c}}\rangle & 0 & 0 & 0 & 1/\sqrt{2} & 1/\sqrt{6} & 1/\sqrt{3} \\
|Y_{\mathrm{c}}\rangle & 0 & 0 & 0 & \mathrm{i}/\sqrt{2} & -\mathrm{i}/\sqrt{6} & -\mathrm{i}/\sqrt{3} \\
|\overline{Z}_{\mathrm{c}}\rangle & 0 & 0 & 0 & 0 & \sqrt{2/3} & -1/\sqrt{3}
\end{array}
$$

and the usual rule for transformation, namely, $[H_{\text{new}}] = [V^{+}][H]_{\text{old}}[V]$. The transformed Hamiltonian for Block III looks like

$$\begin{array}{c|cccccc} & |\text{HH}_\text{a}\rangle & |\text{LH}_\text{a}\rangle & |\text{SO}_\text{a}\rangle & |\text{HH}_\text{c}\rangle & |\text{LH}_\text{c}\rangle & |\text{SO}_\text{c}\rangle \\ |\text{HH}_\text{a}\rangle & E_{\text{pa}}+\delta_\text{a} & 0 & 0 & 4E_{xx} & 0 & 0 \\ |\text{LH}_\text{a}\rangle & 0 & E_{\text{pa}}+\delta_\text{a} & 0 & 0 & 4E_{xx} & 0 \\ |\text{SO}_\text{a}\rangle & 0 & 0 & E_{\text{pa}}-2\delta_\text{a} & 0 & 0 & 4E_{xx} \\ |\text{HH}_\text{c}\rangle & 4E_{xx} & 0 & 0 & E_{\text{pc}}+\delta_\text{c} & 0 & 0 \\ |\text{LH}_\text{c}\rangle & 0 & 4E_{xx} & 0 & 0 & E_{\text{pc}}+\delta_\text{c} & 0 \\ |\text{SO}_\text{c}\rangle & 0 & 0 & 4E_{xx} & 0 & 0 & E_{\text{pc}}-2\delta_\text{c} \end{array}$$

Note how the three bands are neatly decoupled so that at the gamma point we can label the energy levels as HH, LH, and SO. As we move away from the gamma point, the bands are no longer decoupled and the eigenstates are represented by superpositions of HH, LH, and SO.

Similarly Block IV can be transformed using the transformation matrix

$$[\overline{V}] = \begin{array}{c|cccccc} & |\overline{\text{HH}_\text{a}}\rangle & |\overline{\text{LH}_\text{a}}\rangle & |\overline{\text{SO}_\text{a}}\rangle & |\overline{\text{HH}_\text{c}}\rangle & |\overline{\text{LH}_\text{c}}\rangle & |\overline{\text{SO}_\text{c}}\rangle \\ |\overline{X}_\text{a}\rangle & 1/\sqrt{2} & 1/\sqrt{6} & 1/\sqrt{3} & 0 & 0 & 0 \\ |\overline{Y}_\text{a}\rangle & -\text{i}/\sqrt{2} & \text{i}/\sqrt{6} & \text{i}/\sqrt{3} & 0 & 0 & 0 \\ |\overline{Z}_\text{a}\rangle & 0 & -\sqrt{2/3} & 1/\sqrt{3} & 0 & 0 & 0 \\ |\overline{X}_\text{c}\rangle & 0 & 0 & 0 & 1/\sqrt{2} & 1/\sqrt{6} & 1/\sqrt{3} \\ |\overline{Y}_\text{c}\rangle & 0 & 0 & 0 & -\text{i}/\sqrt{2} & \text{i}/\sqrt{6} & \text{i}/\sqrt{3} \\ |\overline{Z}_\text{c}\rangle & 0 & 0 & 0 & 0 & -\sqrt{2/3} & 1/\sqrt{3} \end{array}$$

to obtain

$$\begin{array}{c|cccccc} & |\overline{\text{HH}_\text{a}}\rangle & |\overline{\text{LH}_\text{a}}\rangle & |\overline{\text{SO}_\text{a}}\rangle & |\overline{\text{HH}_\text{c}}\rangle & |\overline{\text{LH}_\text{c}}\rangle & |\overline{\text{SO}_\text{c}}\rangle \\ |\overline{\text{HH}_\text{a}}\rangle & E_{\text{pa}}+\delta_\text{a} & 0 & 0 & 4E_{xx} & 0 & 0 \\ |\overline{\text{LH}_\text{a}}\rangle & 0 & E_{\text{pa}}+\delta_\text{a} & 0 & 0 & 4E_{xx} & 0 \\ |\overline{\text{SO}_\text{a}}\rangle & 0 & 0 & E_{\text{pa}}-2\delta_\text{a} & 0 & 0 & 4E_{xx} \\ |\overline{\text{HH}_\text{c}}\rangle & 4E_{xx} & 0 & 0 & E_{\text{pc}}+\delta_\text{c} & 0 & 0 \\ |\overline{\text{LH}_\text{c}}\rangle & 0 & 4E_{xx} & 0 & 0 & E_{\text{pc}}+\delta_\text{c} & 0 \\ |\overline{\text{SO}_\text{c}}\rangle & 0 & 0 & 4E_{xx} & 0 & 0 & E_{\text{pc}}-2\delta_\text{c} \end{array}$$

It is important to note that the eigenstates (which can be identified by looking at the columns of $[V]$ or $[\overline{V}]$) are not pure up-spin or pure down-spin states. However, we could view the lower block $[\overline{V}]$ as the spin-reversed counterpart of the upper block $[V]$ since it is straightforward to show that they are orthogonal, as we expect "up" and "down" spin states to be.

5.5 Supplementary notes: the Dirac equation

Relativistic electrons are described by the Dirac equation:

$$E\begin{pmatrix} \psi \\ \overline{\psi} \\ \phi \\ \overline{\phi} \end{pmatrix} = \begin{bmatrix} mc^2+U & 0 & cp_z & c(p_x-ip_y) \\ 0 & mc^2+U & c(p_x+ip_y) & -cp_z \\ cp_z & c(p_x-ip_y) & -mc^2+U & 0 \\ c(p_x+ip_y) & -cp_z & 0 & -mc^2+U \end{bmatrix}\begin{pmatrix} \psi \\ \overline{\psi} \\ \phi \\ \overline{\phi} \end{pmatrix}$$

which can be written compactly as

$$E\begin{Bmatrix} \Psi \\ \Phi \end{Bmatrix} = \begin{bmatrix} (mc^2+U)I & c\vec{\sigma}\cdot\vec{p} \\ c\vec{\sigma}\cdot\vec{p} & (-mc^2+U)I \end{bmatrix}\begin{Bmatrix} \Psi \\ \Phi \end{Bmatrix} \tag{5.5.1}$$

where

$$\Psi \equiv \begin{Bmatrix} \psi \\ \overline{\psi} \end{Bmatrix} \quad \text{and} \quad \Phi \equiv \begin{Bmatrix} \phi \\ \overline{\phi} \end{Bmatrix}$$

Assuming $U = 0$ and substituting a plane wave solution of the form

$$\begin{pmatrix} \Psi \\ \Phi \end{pmatrix} = \begin{pmatrix} \Psi \\ \Phi \end{pmatrix} e^{i\vec{k}\cdot\vec{r}}$$

we can show that the dispersion relation is given by

$$E(\vec{k}) = \pm\sqrt{m^2c^4 + c^2\hbar^2k^2}$$

which has two branches as shown in Fig. 5.5.1.

The negative branch is viewed as being completely filled even in vacuum. The separation between the two branches is $2mc^2$ which is approximately 1 MeV, well outside the range of energies encountered in solid-state experiments. In high-energy experiments electrons are excited out of the negative branch into the positive branch resulting in the creation of electron–positron pairs. But in common solid-state experiments energy exchanges are less than 10 eV and the negative branch provides an inert background. At energies around $E = mc^2$, we can do a binomial expansion of Eq. (5.5.1) to obtain the non-relativistic parabolic relation (apart from an additive constant equal to the relativistic rest energy mc^2):

$$E(\vec{k}) \approx mc^2 + (\hbar^2k^2/2m)$$

Relativistic corrections like the spin–orbit term are obtained by starting from the Dirac equation and eliminating the component Φ using approximate procedures valid at energies sufficiently small compared to mc^2.

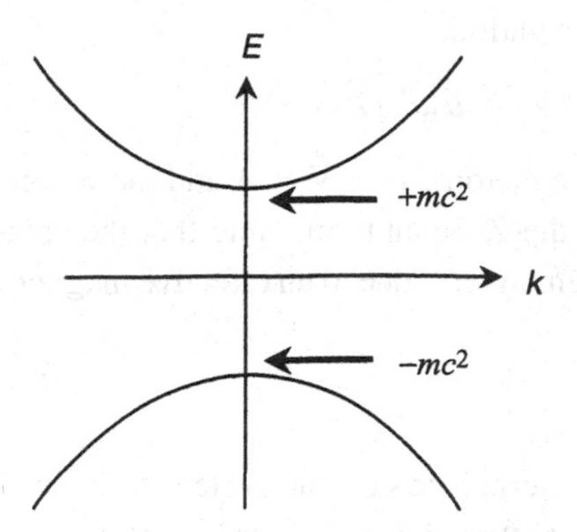

Fig. 5.5.1

Non-relativistic approximation to the Dirac equation: Starting from Eq. (5.5.1) we can show that

$$E\{\Psi\} = (mc^2 + U)\{\Psi\} + [c\vec{\sigma}\cdot\vec{p}]\left[\frac{1}{E + mc^2 - U}\right]^{-1}[c\vec{\sigma}\cdot\vec{p}]\{\Psi\}$$

Setting $E \approx mc^2$ on the right-hand side, we obtain the lowest order non-relativistic approximation

$$E\{\Psi\} = (mc^2 + U)\{\Psi\} + \frac{[\vec{\sigma}\cdot\vec{p}]^2}{2m}\{\Psi\} \tag{5.5.2}$$

which can be simplified to yield Eq. (5.4.1):

$$(E - mc^2)\{\Psi\} = \begin{bmatrix} U + p^2/2m & 0 \\ 0 & U + p^2/2m \end{bmatrix}\{\Psi\}$$

noting that

$$[\vec{\sigma}\cdot\vec{p}]^2 = \begin{bmatrix} p^2 & 0 \\ 0 & p^2 \end{bmatrix}$$

Effect of magnetic field: One question we will not discuss much in this book is the effect of a magnetic field on electron energy levels. The effect is incorporated into the Dirac equation by replacing $\vec{p}$ with $\vec{p} + q\vec{A}$:

$$E\begin{Bmatrix} \Psi \\ \Phi \end{Bmatrix} = \begin{bmatrix} (mc^2 + U)I & c\vec{\sigma}\cdot(\vec{p} + q\vec{A}) \\ c\vec{\sigma}\cdot(\vec{p} + q\vec{A}) & (-mc^2 + U)I \end{bmatrix}\begin{Bmatrix} \Psi \\ \Phi \end{Bmatrix}$$

As before (cf. Eq. (5.5.2)) we can obtain the lowest order non-relativistic approximation

$$E\{\Psi\} = (mc^2 + U)\{\Psi\} + \frac{[\vec{\sigma}\cdot(\vec{p} + q\vec{A})]^2}{2m}\{\Psi\}$$

which can be simplified to yield the Pauli equation:

$$(E - mc^2)\{\Psi\} = [U + (\vec{p} + q\vec{A})^2/2m][I]\{\Psi\} + \mu_B\vec{\sigma}\cdot\vec{B}\{\Psi\}$$

where $\mu_B \equiv q\hbar/2m$ (known as the Bohr magneton), $\vec{B} = \vec{\nabla}\times\vec{A}$ and the second term on the right-hand side ($\mu_B\vec{\sigma}\cdot B$) is called the Zeeman term. Note that the spin–orbit term in Eq. (5.4.7) can be viewed as the Zeeman term due to an effective magnetic field given by

$$B_{so} = (\vec{E}\times\vec{p})/2mc^2$$

Indeed, one way to rationalize the spin–orbit term is to say that an electron in an electric field "sees" this effective magnetic field due to "relativistic effects." To obtain the spin–orbit term directly from the Dirac equation it is necessary to go to the next higher order (see, for example, Sakurai, 1967).

EXERCISES

E.5.1. Set up the (2×2) matrix given in Eq. (5.1.10) for the one-dimensional dimerized toy solid and plot the $E(k)$ relation, cf. Fig. 5.1.5.

E.5.2. Set up the (10×10) matrix given in Eq. (5.3.3) using the parameters for GaAs given at the end of Section 5.3 and plot the dispersion relation $E(k_x, k_y, k_z)$ along Γ–X and Γ–L as shown in Fig. 5.3.2.

E.5.3. Set up the (20×20) matrix including the spin–orbit coupling as described in Section 5.4 for GaAs and plot E vs. k along Γ–X and Γ–L for GaAs ($\Delta_c = 0.013$ eV and $\Delta_a = 0.38$ eV) and compare with Fig. 5.4.1a.

E.5.4. Write down the eigenvalues of the matrix (a, b and c are constants)

$$\begin{bmatrix} a & b & c & 0 & c & b \\ b & a & b & c & 0 & c \\ c & b & a & b & c & 0 \\ 0 & c & b & a & b & c \\ c & 0 & c & b & a & b \\ b & c & 0 & c & b & a \end{bmatrix}$$

using the principles of bandstructure discussed in this chapter.

6 Subbands

In Chapter 5 we saw that the energy levels $E_b(\vec{k})$ in a periodic solid can labeled in terms of $\vec{k}$, with the number of branches b equal to the number of basis functions per unit cell. Strictly speaking, this requires us to assume periodic boundary conditions in all directions so that the periodicity is preserved everywhere even at the "ends." Real solids usually have "ends" where periodicity is lost, but this is commonly ignored as a surface effect that has no influence on bulk properties. The finite size of actual solids normally leads to no observable effects, but as we scale down the size of device structures, the discreteness of energy levels becomes comparable to the thermal energy $k_B T$ leading to experimentally observable effects. Our objective in this chapter is to describe the concept of subbands which is very useful in describing such "size quantization" effects. In Section 6.1 we will describe the effect of size quantization on the $E(\vec{k})$ relation using specific examples. We will then look at its effect on experimentally observable quantities, like the density of states (DOS), $D(E)$ in Section 6.2 and the number of subbands or modes, $M(E)$. In Section 6.3 we will see that the maximum conductance of a wire is proportional to the number of modes around the Fermi energy ($E = \mu$), the maximum conductance per mode being equal to the fundamental constant $G_0 \equiv q^2/h$ (Eq. (1.1)) as discussed in the introductory chapter. Finally in Section 6.4, I will discuss the matter of the appropriate velocity for an electron in a periodic solid. For free electrons, the wavefunction has the form of plane waves $\sim \exp(\pm\, ikx)$ and the corresponding velocity is $\hbar k/m$. Electrons in a periodic solid also have wavefunctions that can be labeled with a k, but they are not plane waves. So what is the quantity (if there is one!) that replaces $\hbar k/m$ and how do we obtain it from our knowledge of the bandstructure $E_b(\vec{k})$?

6.1 Quantum wells, wires, dots, and "nanotubes"

We saw in Chapter 5 that a good way to catalog the energy levels of a homogeneous periodic solid is in terms of the wavevector $\vec{k}$. How do we catalog the energy levels of a nanostructured device? As an example, consider the transistor structure discussed in Chapter 1, modified for convenience to include two gate electrodes symmetrically

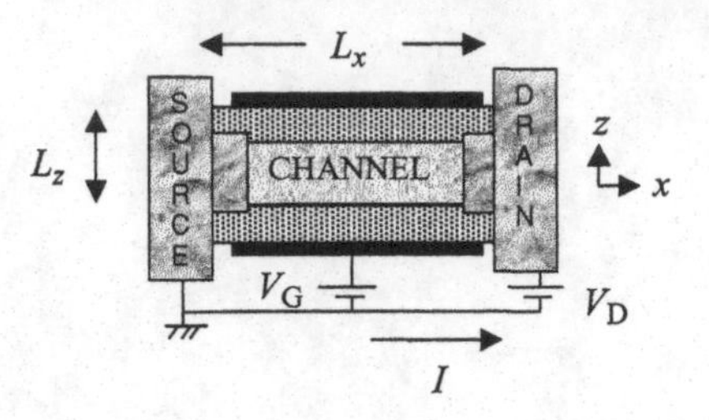

Fig. 6.1.1 Sketch of a dual-gate nanoscale field effect transistor (FET). The top gate is held at the same potential, V_G, as the bottom gate.

on either side of the channel (Fig. 6.1.1). The x-dimension (L) is getting quite short but since electrons can flow in and out at the contacts, one needs to enforce "open boundary conditions" which we will discuss in Chapter 8. But it is not too wrong to treat it as a closed system assuming periodic boundary conditions, at least in the absence of bias ($V_D = 0$). In the z-direction we have tight confinement leading to observable effects beyond what one might expect on the basis of periodic boundary conditions. The y-dimension (perpendicular to the plane of the paper) is typically a few microns and could be considered large enough to ignore surface effects, but as devices get smaller this may not be possible. So how do we label the energy levels of a structure like this?

In a homogeneous solid, electrons are free to move in all three directions and the energy levels can be classified as $E_b(k_x, k_y, k_z)$, where the subscript b refers to different bands. By contrast, the transistor structure shown in Fig. 6.1.1 represents a quantum well where the electrons are free to move only in the x–y plane. We could estimate the energy levels in this structure from our knowledge of the energy levels $E_b(k_x, k_y, k_z)$ of the homogeneous solid, by modeling the z-confinement as a ring of circumference L_z, so that the resulting periodic boundary conditions restrict the allowed values of k_z to a coarse lattice given by $k_z = p2\pi/L_z$:

$$E_{b,p}(k_x, k_y) \approx E_b(k_x, k_y, k_z = p2\pi/L_z)$$

where the additional subscript p can be called a *subband* index. This works quite well for ring-shaped structures like carbon nanotubes, but most low-dimensional structures involve more complicated confinement geometries and in general it takes considerably more work to compute the subband energy levels of a low-dimensional solid from the bulk bandstructure. For the transistor structure shown in Fig. 6.1.1 the insulator layers act like infinite potential walls (see Fig. 2.1.3b) and we can obtain fairly accurate estimates by assuming that the resulting box restricts the allowed values of k_z to a coarse lattice given by $k_z = p\pi/L_z$. The energy levels can then be classified as

$$E_{b,p}(k_x, k_y) \approx E_b(k_x, k_y, k_z = p\pi/L_z)$$

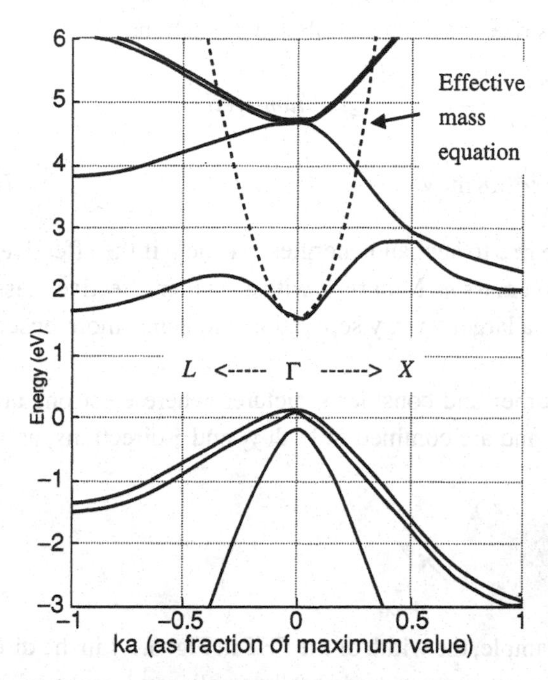

Fig. 6.1.2 Solid curves show the full bandstructure obtained from the sp^3s^* model described in Chapter 5. The dashed curve shows the dispersion obtained from a one-band effective mass model (Eq. (6.1.1)) with parameters adjusted for best fit: $E_c = 1.55$ eV and $m_c = 0.12m$ (m is the free electron mass). Actually the accepted value for the effective mass for GaAs is $0.07m$, but the sp^3s^* model parameters that we use are optimized to give the best fit over the entire band, and are not necessarily very accurate near the band edge.

As we have explained, this is only approximate, but the main point I am trying to make is that *quantum wells* have *1D subbands*, each having a *2D dispersion relation*, $E(k_x, k_y)$.

How small does the dimension L_z have to be in order for the structure to qualify as a quantum well? Answer: when it leads to experimentally observable effects. This requires that the discrete energy levels corresponding to the quantized values of $k_z = p\pi/L_z$ be less than or comparable to the thermal energy $k_B T$, since all observable effects tend to be smoothed out on this energy scale by the Fermi function. To obtain a "back-of-an-envelope" estimate, let us assume that the dispersion relation $E_b(\vec{k})$ in the energy range of interest is described by a parabolic relation with an effective mass, m_c:

$$E(\vec{k}) \approx E_c + \frac{\hbar^2\left(k_x^2 + k_y^2 + k_z^2\right)}{2m_c} \qquad \textit{Bulk solid} \qquad (6.1.1)$$

where E_c and m_c are constants that can be determined to obtain the best fit (see Fig. 6.1.2). These parameters are referred to as the conduction band edge and the conduction band effective mass respectively and are well-known for all common semiconductors.

The z-confinement then gives rise to subbands (labeled p) such that

$$E_p(k_x, k_y) \approx E_c + p^2 \varepsilon_z + \frac{\hbar^2 (k_x^2 + k_y^2)}{2m_c} \qquad \textit{Quantum well}$$

$$\varepsilon_z = \frac{\hbar^2 \pi^2}{2m_c L_z^2} = \frac{m}{m_c} \left(\frac{10\ \text{nm}}{L_z} \right)^2 \times 3.8\ \text{meV} \tag{6.1.2}$$

A layer 10 nm thick would give rise to subband energies ~4 meV if the effective mass m_c were equal to the free electron mass m. Materials with a smaller effective mass (like GaAs with $m_c = 0.07m$) lead to a larger energy separation and hence more observable effects of size quantization.

We could take this a step further and consider structures where electrons are free to move only in the x-direction and are confined in both y- and z-directions, as shown below.

This would be the case if, for example, the width of the FET in Fig. 6.1.1 in the direction perpendicular to the plane of the paper were made really small, say less than 100 nm. Such *quantum wires* have *2D subbands*, each having a *1D dispersion relation* that can be estimated by quantizing both the y- and z-components of the wavevector $\vec{k}$:

$$E_{n,p}(k_x) \approx E_c + n^2 \varepsilon_y + p^2 \varepsilon_z + \frac{\hbar^2 k_x^2}{2m_c} \qquad \textit{Quantum wire}$$

where ε_y is related to the y-dimension L_y by a relation similar to Eq. (6.1.2). Finally, one could consider structures that confine electrons in all three dimensions (as shown below) leading to discrete levels like atoms that can be estimated from

$$E_{m,n,p} \approx E_c + \frac{m^2 \hbar^2 \pi^2}{2m_c L_x^2} + \frac{n^2 \hbar^2 \pi^2}{2m_c L_y^2} + \frac{p^2 \hbar^2 \pi^2}{2m_c L_z^2} \qquad \textit{Quantum dot}$$

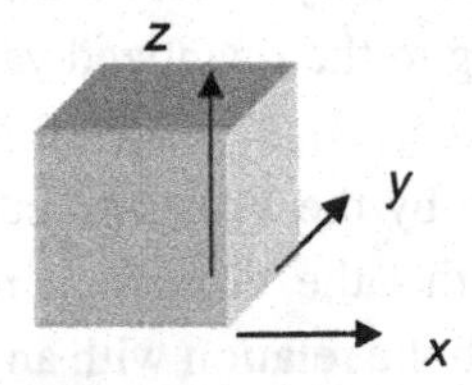

Such quantum dot structures are often referred to as artificial atoms.

Carbon nanotubes: Carbon nanotubes provide a very good example for illustrating the concept of subbands (Fig. 6.1.3). We saw in Section 5.2 that the energy levels of a sheet of graphite can be found by diagonalizing the (2 × 2) matrix

$$h(\vec{k}) = \begin{bmatrix} 0 & h_0 \\ h_0^* & 0 \end{bmatrix} \tag{6.1.3}$$

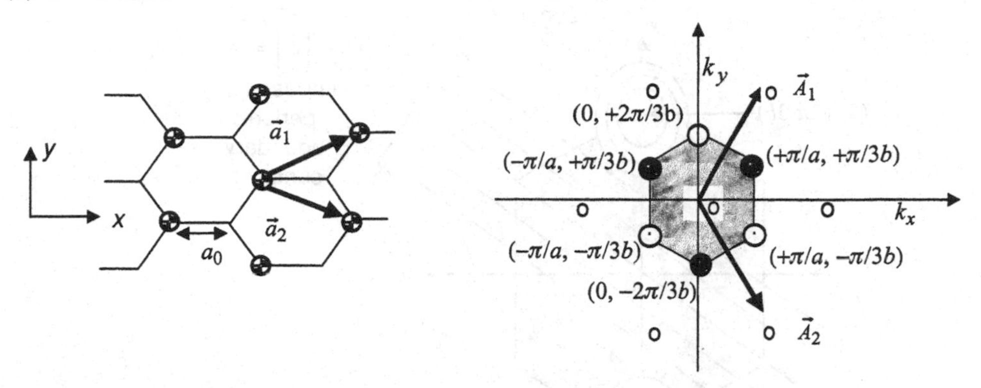

Fig. 6.1.3 (a) Arrangement of carbon atoms on the surface of graphite. (b) Reciprocal lattice showing Brillouin zone (shaded).

where

$$h_0 \equiv -t\left(1 + e^{i\vec{k}\cdot\vec{a}_1} + e^{i\vec{k}\cdot\vec{a}_2}\right) = -t\left(1 + 2e^{ik_x a}\cos k_y b\right)$$

The eigenvalues are given by

$$E = \pm|h_0| = \pm t\sqrt{1 + 4\cos k_y b \cos k_x a + 4\cos^2 k_y b} \tag{6.1.4}$$

Since each unit cell has two basis functions, the total number of states is equal to $2N$, N being the number of unit cells. Each carbon atom contributes one electron to the π-band, giving a total of $2N$ electrons that fill up exactly half the states. Since the energy levels are symmetrically disposed about $E = 0$, this means that all states with $E < 0$ are occupied while all states with $E > 0$ are empty, or equivalently one could say that the Fermi energy is located at $E = 0$.

Where in the k_x–k_y plane are these regions with $E = 0$ located? Answer: wherever $h_0(\vec{k}) = 0$. It is easy to see that this occurs at the six corners of the Brillouin zone:

$$e^{ik_x a}\cos k_y b = -1/2 \qquad k_x a = 0, \quad k_y b = \pm 2\pi/3$$
$$k_x a = \pi, \quad k_y b = \pm\pi/3$$

These six points are special as they provide the states right around the Fermi energy and thus determine the electronic properties. They can be put into two groups of three:

$$(k_x a, k_y b) = (0, -2\pi/3), \quad (-\pi, +\pi/3), \quad (+\pi, +\pi/3) \tag{6.1.5a}$$

$$(k_x a, k_y b) = (0, +2\pi/3), \quad (-\pi, -\pi/3), \quad (+\pi, -\pi/3) \tag{6.1.5b}$$

All three within a group are equivalent points since they differ by a reciprocal lattice vector. Each of the six points has *one-third* of a valley around it within the first Brillouin zone (shaded area in Fig. 6.1.3b). But we can translate these by appropriate reciprocal lattice vectors to form two *full* valleys around two of these points, one from

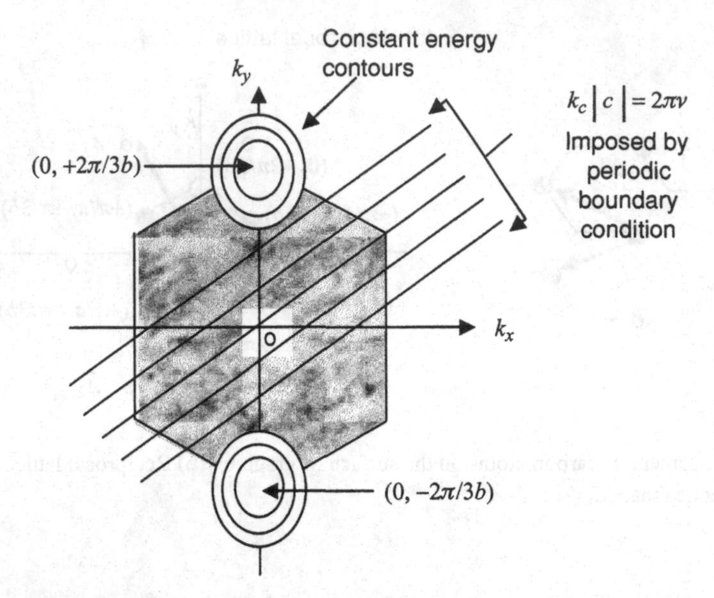

Fig. 6.1.4 Reciprocal lattice of graphite showing straight lines $k_c\,|c| = 2\pi\nu$ representing the constraint imposed by the nanotube periodic boundary conditions.

each group:

$$(k_x a, k_y b) = (0, \pm 2\pi/3)$$

Once a sheet of graphite is rolled up into a nanotube, the allowed values of k are constrained by the imposition of periodic boundary conditions along the circumferential direction. Note that this periodic boundary condition is a *real* one imposed by the physical structure, rather than a conceptual one used to facilitate the counting of states in a large structure whose exact boundary conditions are unimportant. Defining a circumferential vector

$$\vec{c} = m\vec{a}_1 + n\vec{a}_2 = \hat{x}(m+n)a + \hat{y}(m-n)b \tag{6.1.6}$$

that joins two equivalent points on the x–y plane that connect to each other on being rolled up, we can express the requirement of periodic boundary condition as

$$\vec{k}\cdot\vec{c} \equiv k_c|c| = k_x a(m+n) + k_y b(m-n) = 2\pi\nu \tag{6.1.7}$$

which defines a series of parallel lines, each corresponding to a different integer value for ν (Fig. 6.1.4). We can draw a one-dimensional dispersion relation along any of these lines, giving us a set of dispersion relations $E_\nu(k)$, one for each *subband* ν.

Whether the resulting subband dispersion relations will show an energy gap or not depends on whether one of the lines defined by Eq. (6.1.7) passes through the center of one of the valleys

$$(k_x a, k_y b) = (0, \pm 2\pi/3)$$

Fig. 6.1.5

where the energy levels lie at $E = 0$. It is easy to see from Eq. (6.1.7) that in order for a line to pass through $k_x a = 0$, $k_y b = 2\pi/3$ we must have

$$(m - n)/3 = \nu$$

Since ν is an integer this can only happen if $(m - n)$ is a multiple of three: nanotubes satisfying this condition are metallic.

Zigzag and armchair nanotubes: Consider a specific example: a nanotube with a circumferential vector along the y-direction, $\vec{c} = \hat{y}2bm$, which is called a zigzag nanotube because the edge (after rolling) looks zigzag (Fig. 6.1.5). The periodic boundary condition then requires the allowed values of k to lie parallel to the k_x-axis described by (the circumference is $2bm$)

$$k_y 2bm = 2\pi\nu \rightarrow k_y = \frac{2\pi}{3b}\frac{3\nu}{2m} \tag{6.1.8}$$

as shown in Fig. 6.1.6a.

Figure 6.1.7 shows the two "lowest" subbands corresponding to values of the subband index ν that give rise to the smallest gaps around $E = 0$. If $m = 66$ (i.e. a multiple of three), one of the subbands will pass through $(k_x a, k_y b) = (0, \pm 2\pi/3)$ and the dispersion relation for the lowest subbands look as shown in Fig. 6.1.7a, with no gap in the energy spectrum. But if $m = 65$ (not a multiple of three), then no subband will pass through $(k_x a, k_y b) = (0, \pm 2\pi/3)$ giving rise to a gap in the energy spectrum as shown in Fig. 6.1.7b.

A nanotube with a circumferential vector along the x-direction, $\vec{c} = \hat{x}2am$, is called an armchair nanotube because the edge (after rolling) looks like an armchair as shown in Fig. 6.1.8 (this requires some imagination!). The periodic boundary condition then requires the allowed values of k to lie parallel to the k_y-axis described by (the circumference is again $2bm$)

$$k_x 2am = 2\pi\nu \rightarrow k_x = \frac{2\pi\nu}{2ma} \tag{6.1.9}$$

as shown in Fig. 6.1.6b. The subband with $\nu = 0$ will always pass through the special point $(k_x a, k_y b) = (0, \pm 2\pi/3)$ giving rise to dispersion relations that look metallic (Fig. 6.1.7a) regardless of the value of m.

A useful approximation: Electrical conduction is determined by the states around the Fermi energy and so it is useful to develop an approximate relation that describes the regions of the E–k plot around $E = 0$. This can be done by replacing

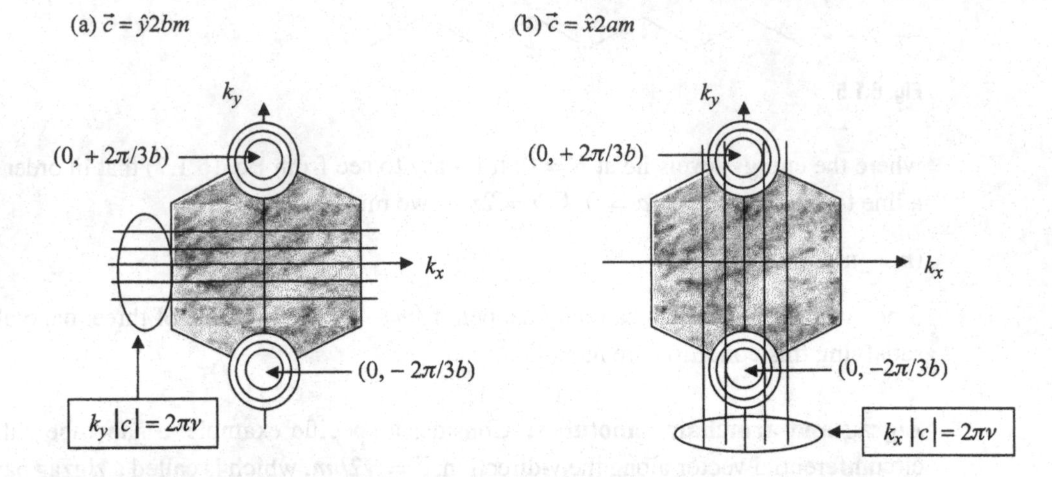

Fig. 6.1.6 (a) A zigzag nanotube obtained by rolling a sheet of graphite along the y-axis with $\vec{c} = \hat{y}2bm$ has its allowed k-values constrained to lie along a set of lines parallel to the k_x-axis as shown. One of the lines will pass through $(0, 2\pi/3b)$ only if m is a multiple of three. (b) An armchair nanotube obtained by rolling a graphite sheet along the x-axis with $\vec{c} = \hat{x}2am$ has its k-values constrained to lie along lines parallel to the k_y-axis as shown. One of the lines will always pass through $(0, 2\pi/3b)$ regardless of the value of m.

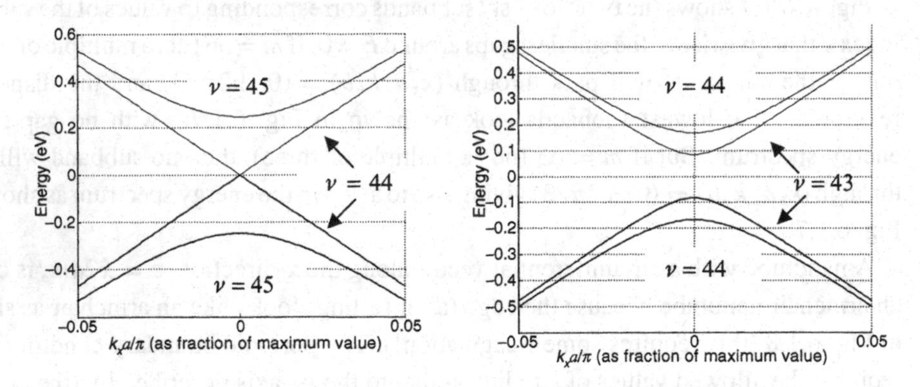

Fig. 6.1.7 Dispersion relation for the two "lowest" subbands of a zigzag nanotube with (a) $D = 5.09$ nm, $m = 66$ showing metallic character (no gap in energy spectrum) and (b) $D = 5.02$ nm, $m = 65$ showing semiconducting character (gap in energy spectrum).

the expression for $h_0(\vec{k}) = -t(1 + 2e^{ik_x a} \cos k_y b)$ with a Taylor expansion around $(k_x a, k_y b) = (0, \pm 2\pi/3)$ where the energy gap is zero (note that $h_0 = 0$ at these points):

$$h_0 \approx k_x \left[\frac{\partial h_0}{\partial k_x}\right]_{k_x a=0,\, k_y b=\pm 2\pi/3} + \left(k_y \mp \frac{2\pi}{3b}\right)\left[\frac{\partial h_0}{\partial k_y}\right]_{k_x a=0,\, k_y b=\pm 2\pi/3}$$

Fig. 6.1.8

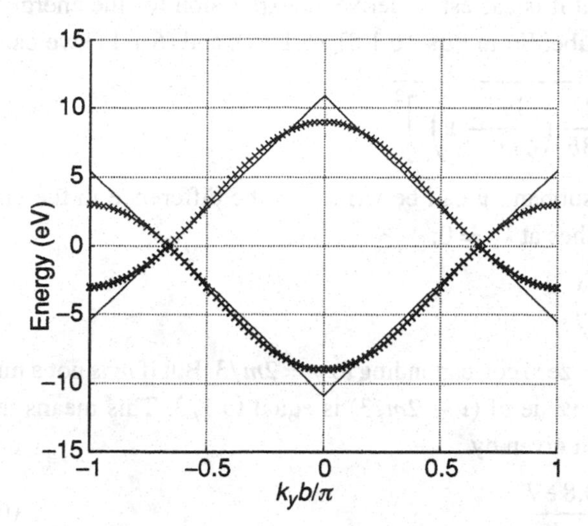

Fig. 6.1.9 Energy dispersion relation plotted as a function of k_yb along the line $k_xa = 0$. The solid curve is obtained from Eq. (6.1.11), while the crosses are obtained from Eq. (6.1.4).

It is straightforward to evaluate the partial derivatives:

$$\frac{\partial h_0}{\partial k_x} = \left[-2iat\, e^{ik_xa} \cos k_yb\right]_{k_xa=0,\ k_yb=\pm 2\pi/3} = iat = i3a_0t/2$$

$$\frac{\partial h_0}{\partial k_y} = \left[2bt\, e^{ik_xa} \sin k_yb\right]_{k_xa=0,\ k_yb=\pm 2\pi/3} = \pm bt\sqrt{3} = \pm 3a_0t/2$$

so that we can write

$$h_0(\vec{k}) \approx i\frac{3a_0t}{2}(k_x \mp i\beta_y)$$

where

$$\beta_y \equiv k_y \mp (2\pi/3b) \tag{6.1.10}$$

The corresponding energy dispersion relation (cf. Eq. (6.1.4)) can be written as

$$E(\vec{k}) = \pm|h_0| = \pm\frac{3ta_0}{2}\sqrt{k_x^2 + \beta_y^2} \tag{6.1.11}$$

This simplified approximate relation (obtained from a Taylor expansion of Eq. (6.1.4) around one of the two valleys) agrees with the exact relation fairly well over a wide range of energies, as is evident from Fig. 6.1.9. Within this approximation the constant-energy contours are circles isotropically disposed around the center of each valley, $(0, +2\pi/3b)$ or $(0, -2\pi/3b)$.

How large is the energy gap of a semiconducting nanotube? The answer is independent of the specific type of nanotube, as long as $(m - n)$ is not a multiple of three so that the gap is non-zero. But it is easiest to derive an expression for the energy gap if we consider a zigzag nanotube. From Eqs. (6.1.8), (6.1.10), and (6.1.11) we can write

$$E(k_x) = \pm\frac{3ta_0}{2}\sqrt{k_x^2 + \left[\frac{2\pi}{3b}\left(\frac{3\nu}{2m} - 1\right)\right]^2} \tag{6.1.12}$$

so that the energy gap for subband ν can be written as the difference in the energies between the + and − branches at $k_x = 0$:

$$E_{g,\nu} = 3ta_0\frac{2\pi}{2mb}\left(\nu - \frac{2m}{3}\right)$$

This has a minimum value of zero corresponding to $\nu = 2m/3$. But if m is not a multiple of three then the minimum value of $(\nu - 2m/3)$ is equal to 1/3. This means that the minimum energy gap is then given by

$$E_g = ta_0\frac{2\pi}{2mb} = \frac{2ta_0}{d} \approx \frac{0.8\,\text{eV}}{d} \tag{6.1.13}$$

where d is the diameter of the nanotube in nanometers, so that πd is equal to the circumference $2mb$.

6.2 Density of states

In the last section we discussed how size quantization effects modify the $E(\vec{k})$ relationship leading to the formation of subbands. However, it should be noted that such effects do not appear suddenly as the dimensions of a solid are reduced. It is not as if a bulk solid abruptly changes into a quantum well. The effect of reduced dimensions shows up gradually in experimental measurements and this can be appreciated by looking at the density of states (DOS), $D(E)$, which is reflected in conductance measurements.

The DOS tells us the number of energy eigenstates per unit energy range and it clearly depends on the $E(\vec{k})$ relationship. To be specific, let us assume for the moment that we are near one of the valleys in the conduction band where the energy levels can be described by a parabolic relation with some effective mass m_c:

$$E(\vec{k}) = E_c + \frac{\hbar^2 k^2}{2m_c} \tag{6.2.1}$$

What is the corresponding DOS if the vector $\vec{k}$ is constrained to one dimension (a quantum wire), two dimensions (a quantum well), or three dimensions (bulk solid)? The standard procedure for counting states is to assume a rectangular box of size

$L_x L_y L_z$ with periodic boundary conditions (see Fig. 2.1.4) in all three dimensions (cf. Eq. (2.1.17)):

$$k_x = \frac{2\pi}{L_x}\,\nu_x \qquad k_y = \frac{2\pi}{L_y}\,\nu_y \qquad k_z = \frac{2\pi}{L_z}\,\nu_z \tag{6.2.2}$$

where ν_x, ν_y, and ν_z are integers. We then assume that the box is so large that the allowed k-values are effectively continuous and we can replace any summations over these indices with integrals:

$$\sum_{k_x} \to \int_{-\infty}^{+\infty} \frac{\mathrm{d}k_x}{2\pi/L_x} \qquad \sum_{k_y} \to \int_{-\infty}^{+\infty} \frac{\mathrm{d}k_y}{2\pi/L_y} \qquad \sum_{k_z} \to \int_{-\infty}^{+\infty} \frac{\mathrm{d}k_z}{2\pi/L_z} \tag{6.2.3}$$

In other words, the allowed states in k-space are distributed with a density of $(L/2\pi)$ per unit k in each k-dimension. Hence the total number of allowed states $N(k)$, up to a certain maximum value k, is given by

$$\frac{L}{2\pi}2k = \frac{kL}{\pi} \qquad \text{1D with } L \equiv L_x$$

$$\frac{L_x L_y}{4\pi^2}\pi k^2 = \frac{k^2 S}{4\pi} \qquad \text{2D with } S \equiv L_x L_y$$

$$\frac{L_x L_y L_z}{8\pi^3}\frac{4\pi k^3}{3} = \frac{k^3 \Omega}{6\pi^2} \qquad \text{3D with } \Omega \equiv L_x L_y L_z$$

Using the dispersion relation Eq. (6.2.1) we can convert $N(k)$ into $N(E)$, which tells us the total number of states having an energy less than E. The derivative of this function gives us the density of states (DOS):

$$D(E) = \frac{\mathrm{d}}{\mathrm{d}E}N(E) \tag{6.2.4}$$

The results for one, two, and three dimensions are summarized in Table 6.2.1. It can be seen that a parabolic $E(\vec{k})$ relation (Eq. (6.2.1)) gives rise to a DOS that varies as $E^{-1/2}$ in 1D, E^0 (i.e. constant) in 2D, and $E^{1/2}$ in 3D. Note, however, that the 1D or 2D results do not give us the total DOS for a quantum wire or a quantum well. They give us the DOS *only for one subband* of a quantum wire or a quantum well. We have to sum over all subbands to obtain the full DOS. For example, if a quantum well has subbands p given by (see Eq. (6.1.1))

$$E_p(k_x, k_y) \approx E_c + p^2\varepsilon_z + \frac{\hbar^2\left(k_x^2 + k_y^2\right)}{2m_c}$$

then the DOS would look like a superposition of many 2D DOS:

$$D(E) = \frac{m_c S}{2\pi\hbar^2}\sum_p \vartheta(E - E_c - p^2\varepsilon_z) \tag{6.2.5}$$

Table 6.2.1 *Summary of DOS calculation in 1D, 2D, and 3D for parabolic isotropic dispersion relation (Eq. (6.2.1)). Plots assume $m_c = m$ (free electron mass) and results are for one spin only*

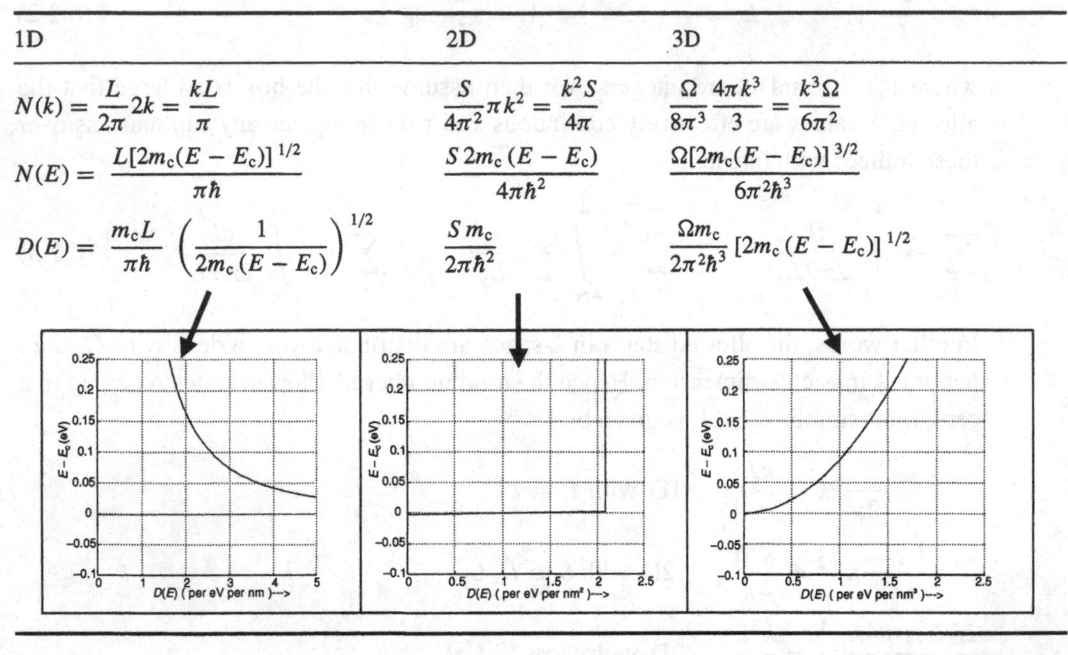

	1D	2D	3D
$N(k) =$	$\frac{L}{2\pi}2k = \frac{kL}{\pi}$	$\frac{S}{4\pi^2}\pi k^2 = \frac{k^2 S}{4\pi}$	$\frac{\Omega}{8\pi^3}\frac{4\pi k^3}{3} = \frac{k^3\Omega}{6\pi^2}$
$N(E) =$	$\frac{L[2m_c(E-E_c)]^{1/2}}{\pi\hbar}$	$\frac{S\,2m_c\,(E-E_c)}{4\pi\hbar^2}$	$\frac{\Omega[2m_c(E-E_c)]^{3/2}}{6\pi^2\hbar^3}$
$D(E) =$	$\frac{m_c L}{\pi\hbar}\left(\frac{1}{2m_c\,(E-E_c)}\right)^{1/2}$	$\frac{S\,m_c}{2\pi\hbar^2}$	$\frac{\Omega m_c}{2\pi^2\hbar^3}[2m_c\,(E-E_c)]^{1/2}$

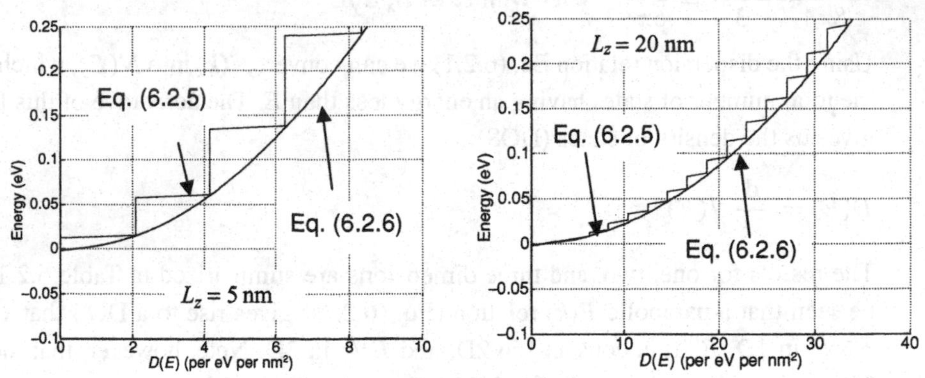

Fig. 6.2.1 Density of states $D(E)$ for a 2D box calculated from Eq. (6.2.5) (2D with quantized subbands) and compared with that obtained from the 3D relation in Eq. (6.2.6). As the width L_z of the box is increased from 5 to 20 nm, the DOS approaches the 3D result (the conduction band effective mass m_c is assumed equal to the free electron mass m).

Figure 6.2.1 shows the density of states calculated from Eq. (6.2.5) with $\varepsilon_z = \pi^2\hbar^2/2m_c L_z^2$ as given in Eq. (6.1.2). It is apparent that as the width L_z is increased from 5 to 20 nm, the DOS approaches the result obtained from the 3D DOS with the volume Ω set equal to SL_z:

$$D_{3D}(E) = \frac{Sm_c}{2\pi^2\hbar^3}\,[2m_c\,(E - E_c)]^{1/2}\,L_z \tag{6.2.6}$$

Table 6.2.2 *Summary of DOS per spin per valley*

	Graphite Eq. (6.2.7)	Zigzag nanotube Eq. (6.2.8)
$N(k) =$	$\frac{S}{4\pi}k^2$	$\frac{L}{\pi}k_x$
$N(E) =$	$\frac{SE^2}{4\pi a^2 t^2}$	$\sum_\nu \frac{L}{\pi at}\sqrt{E^2 - E_\nu^2}$
$D(E) =$	$\frac{SE}{2\pi a^2 t^2}$	$\sum_\nu \frac{L}{\pi at}\frac{E}{\sqrt{E^2 - E_\nu^2}}$

From graphite to a nanotube : The evolution of the DOS from a sheet of graphite to a nanotube also provides an instructive example of how size quantization effects arise as the dimensions are reduced. For a sheet of graphite we can approximate the $E(\vec{k})$ relation for each of the two valleys centered at $(k_x a, k_y b) = (0, \pm 2\pi/3)$ as (see Eq. (6.1.11), $a = 3a_0/2$)

$$E(\vec{k}) = \pm \frac{3ta_0}{2}\sqrt{k_x^2 + \beta_y^2} = \pm ta|\vec{k}| \tag{6.2.7}$$

As we have seen, the energy subbands for a zigzag nanotube are given by (see Eq. (6.1.12))

$$E(k_x) = \pm ta\sqrt{k_\nu^2 + k_x^2} = \pm\sqrt{E_\nu^2 + (tak_x)^2} \tag{6.2.8}$$

where

$$k_\nu \equiv \frac{2\pi}{3b}\left(\frac{3\nu}{2m} - 1\right) \quad \text{and} \quad E_\nu \equiv tak_\nu$$

In calculating $D(E)$ the steps are similar to those shown in Table 6.2.1, though the details are somewhat different because the dispersion relation is different (Eqs. (6.2.7), (6.2.8)), as summarized in Table 6.2.2. Figure 6.2.2 compares the density of states for a zigzag nanotube of length L and diameter d (note: circumference $= \pi d = 2mb$):

$$D_{\mathrm{ZNT}}(E) = \sum_\nu \frac{2L}{\pi at}\frac{E}{\sqrt{E^2 - E_\nu^2}} \quad \text{with} \quad E_\nu \equiv \frac{2at}{d}\left(\nu - \frac{2m}{3}\right) \tag{6.2.9}$$

with the density of states for a graphite sheet of area πLd:

$$D_{\mathrm{G}}(E) = \frac{Ld}{a^2t^2}|E| \tag{6.2.10}$$

for a nanotube with $m = 200$ (corresponding to $d = 15.4$ nm) and a nanotube with $m = 800$ ($d = 61.7$ nm).

It is apparent that the smaller nanotube has a DOS that is distinctly different from graphite. But the larger nanotube is less distinguishable, especially if we recall that experimental observations are typically convolved with the thermal broadening function

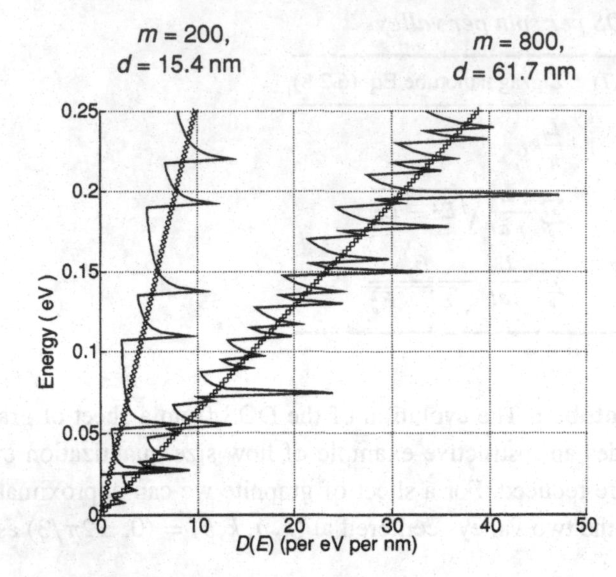

Fig. 6.2.2 Density of states $D(E)$ for a zigzag nanotube calculated from Eq. (6.2.9) (solid curves) compared with that obtained from the result for graphite (Eq. (6.2.10), crosses).

which has a width of $\sim k_B T \sim 0.026$ eV at room temperature (to be discussed later; see Fig. 7.3.4).

It is easy to see analytically that the DOS for zigzag nanotubes with the summation index ν replaced by an integral (cf. Eq. (6.2.9))

$$D_{\mathrm{ZNT}}(E) \approx \int 2\mathrm{d}\nu \, \frac{2L}{\pi a t} \, \frac{|E|}{\sqrt{E^2 - E_\nu^2}} \qquad \text{with} \quad \mathrm{d}E_\nu \equiv \frac{2at}{d} \, \mathrm{d}\nu$$

$$= \int_0^E \mathrm{d}E_\nu \frac{2Ld}{\pi a^2 t^2} \frac{|E|}{\sqrt{E^2 - E_\nu^2}} = \frac{Ld}{a^2 t^2} |E|$$

becomes identical to the DOS for graphite (cf. Eq. (6.2.10)).

Anisotropic dispersion relation: We have used a few examples to illustrate the procedure for converting an $E(\vec{k})$ relation to a density of states $D(E)$. This procedure (see Tables 6.2.1 and 6.2.2) works well when the dispersion relation $E(\vec{k})$ is isotropic. However, the details become more complicated if the relation is anisotropic. For example, silicon has six separate conduction band valleys, each of which is *ellipsoidal*:

$$E = E_c + \frac{\hbar^2 k_x^2}{2m_{xx}} + \frac{\hbar^2 k_y^2}{2m_{yy}} + \frac{\hbar^2 k_z^2}{2m_{zz}} \tag{6.2.11}$$

For a given energy E, the constant energy contour (Fig. 6.2.3) looks like an ellipsoid whose major axes are given by $\sqrt{2m_{xx}(E - E_c)}/\hbar$, $\sqrt{2m_{yy}(E - E_c)}/\hbar$,

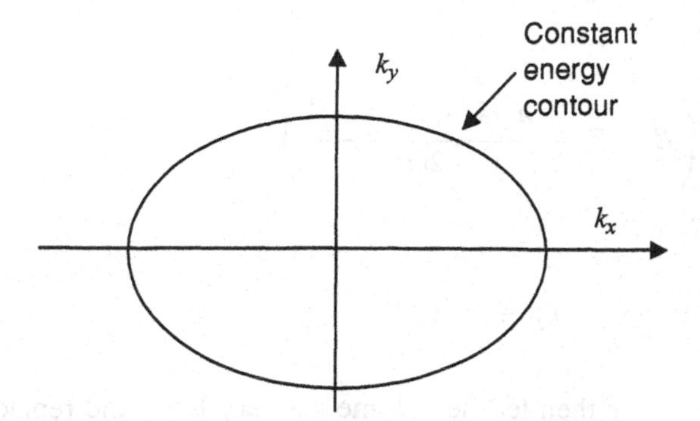

Fig. 6.2.3

and $\sqrt{2m_{zz}(E - E_c)}/\hbar$. The volume of this ellipsoid is

$$\frac{4\pi}{3}\frac{\sqrt{2m_{xx}(E - E_c)}}{\hbar}\frac{\sqrt{2m_{yy}(E - E_c)}}{\hbar}\frac{\sqrt{2m_{zz}(E - E_c)}}{\hbar}$$

so that

$$N(E) = \frac{\Omega}{8\pi^3}\frac{4\pi}{3}\frac{\sqrt{2m_{xx}(E - E_c)}}{\hbar}\frac{\sqrt{2m_{yy}(E - E_c)}}{\hbar}\frac{\sqrt{2m_{zz}(E - E_c)}}{\hbar}$$

and

$$D(E) \equiv \frac{d}{dE}N(E) = \frac{\Omega}{2\pi^2\hbar^3}\sqrt{2m_{xx}m_{yy}m_{zz}(E - E_c)} \tag{6.2.12}$$

which reduces to the earlier result (see Table 6.2.1) if the mass is isotropic.

Formal expression for DOS: In general if we have a system with eigenstates labeled by an index α, then we can write the total number of states, $N_T(E)$ with energy less than E as

$$N_T(E) = \sum_\alpha \vartheta(E - \varepsilon_\alpha) \tag{6.2.13}$$

where $\vartheta(E)$ denotes the unit step function which is equal to zero for $E < 0$, and equal to one for $E > 0$. The derivative of the unit step function is a delta function, so that

$$D(E) \equiv \frac{d}{dE}N_T(E) = \sum_\alpha \delta(E - \varepsilon_\alpha) \tag{6.2.14}$$

This expression represents a sequence of spikes rather than a continuous function. Formally we can obtain a continuous DOS from Eq. (6.2.14) by letting the size of the system get very large and replacing the summation by an integral as we have been doing. For example, if

$$E(\vec{k}) = E_c + \frac{\hbar^2k^2}{2m^*}$$

then $D(E) = \sum_{\vec{k}} \delta(E - \varepsilon_{\vec{k}})$

$$= \sum_{k_x}\sum_{k_y}\sum_{k_z} \delta\left(E - E_c - \frac{\hbar^2(k_x^2 + k_y^2 + k_z^2)}{2m^*}\right)$$

where

$$k_x = \frac{2\pi}{L_x}\nu_x \qquad k_y = \frac{2\pi}{L_y}\nu_y \qquad k_z = \frac{2\pi}{L_z}\nu_z$$

and ν_x, ν_y, and ν_z are integers. We then let the volume get very large and replace the summations over these indices with integrals:

$$\sum_{k_x} \to \int_{-\infty}^{+\infty} \frac{dk_x}{2\pi/L_x} \qquad \sum_{k_y} \to \int_{-\infty}^{+\infty} \frac{dk_y}{2\pi/L_y} \qquad \sum_{k_z} \to \int_{-\infty}^{+\infty} \frac{dk_z}{2\pi/L_z}$$

$$D(E) = \frac{\Omega}{8\pi^3}\int_0^{+\infty} dk_x\, dk_y\, dk_z\, \delta\left(E - E_c - \frac{\hbar^2 k^2}{2m^*}\right)$$

$$= \frac{\Omega}{8\pi^3}\int_0^{+\infty} 4\pi k^2\, dk\, \delta\left(E - E_c - \frac{\hbar^2 k^2}{2m^*}\right)$$

$$= \frac{\Omega}{2\pi^2}\int_0^{+\infty} \frac{m^* dE}{\hbar^2}\frac{\sqrt{2m^*(E - E_c)}}{\hbar}\delta\left(E - E_c - \frac{\hbar^2 k^2}{2m^*}\right)$$

$$= \frac{\Omega m^*}{2\pi^2\hbar^3}[2m^*(E - E_c)]^{1/2}$$

as we obtained earlier (cf. 3D DOS in Table 6.2.1). This procedure is mathematically a little more involved than the previous procedure and requires integrals over delta functions.

The real value of the formal expression in Eq. (6.2.14) is that it is generally valid regardless of the $E(\vec{k})$ relation or whether such a relation even exists. Of course, in general it may not be easy to replace the summation by an integral, since the separation between energy levels may not follow a simple analytical prescription like Eq. (6.2.2). But we can still calculate a continuous DOS from Eq. (6.2.14) by broadening each spike into a Lorentzian (see Eq. (1.3.2)):

$$\delta(E - \varepsilon_\alpha) \to \frac{\gamma/2\pi}{(E - \varepsilon_\alpha)^2 + (\gamma/2)^2} \tag{6.2.15}$$

The DOS will look continuous if the individual energies ε_α are spaced closely compared to the broadening γ, which is usually true for large systems.

Separable problems: An interesting result that can be proved using Eq. (6.2.14) is that if the eigenvalue problem is separable, then the overall DOS is given by the convolution of the individual densities. For example, suppose we have a 2D problem that separates into x- and y-components (as shown in Eq. (2.3.6) for 3D problems) such that the overall energies are given by the sum of the x-energy and the y-energy:

$$\varepsilon(n, m) = \varepsilon_x(n) + \varepsilon_y(m) \tag{6.2.16}$$

We could define an x-DOS and a y-DOS:

$$D_x(E) = \sum_n \delta[E - \varepsilon_x(n)] \tag{6.2.17a}$$

$$D_y(E) = \sum_m \delta[E - \varepsilon_y(m)] \tag{6.2.17b}$$

and it is straightforward to show that the total DOS

$$D(E) = \sum_n \sum_m \delta[E - \varepsilon_x(n) - \varepsilon_y(m)] \tag{6.2.17c}$$

can be written as a convolution product of the x-DOS and the y-DOS:

$$D(E) = \int_{-\infty}^{+\infty} \mathrm{d}E' D_x(E')\, D_y(E - E') \tag{6.2.18}$$

6.3 Minimum resistance of a wire

Now that we have discussed the concept of subbands, we are ready to answer a very interesting fundamental question. Consider a wire of cross-section $L_y L_z$ with a voltage V applied across it (Fig. 6.3.1). What is the conductance of this wire if the contacts were perfect, and we reduce its length to very small dimensions? Based on Ohm's law, we might be tempted to say that the conductance will increase indefinitely as the length of the wire is reduced, since the resistance (inverse of conductance) is proportional to the length. However, as I pointed out in Section 1.3 the maximum conductance $G = I/V$ for a *one-level conductor* is a fundamental constant given by

$$G_0 \equiv q^2/h = 38.7\ \mu\text{S} = (25.8\ \text{k}\Omega)^{-1} \tag{6.3.1}$$

We are now ready to generalize this concept to a wire with a finite cross-section. It has been established experimentally that once the length of a wire has been reduced sufficiently that an electron can cross the wire without an appreciable chance of scattering (ballistic transport) the conductance will approach a constant value given by (assuming "perfect" contacts)

$$G = [M(E)]_{E=E_F}\, G_0 \tag{6.3.2}$$

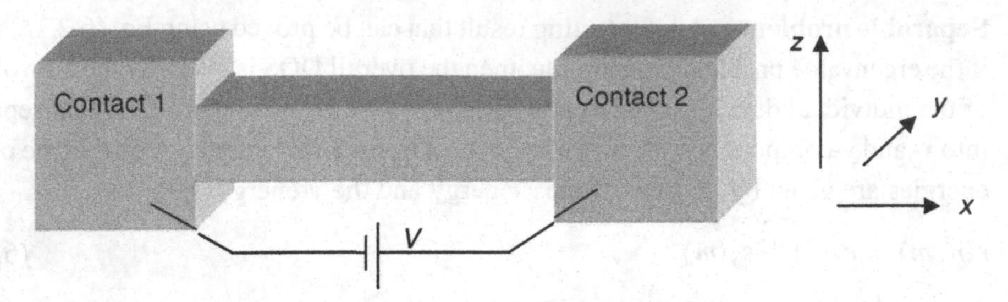

Fig. 6.3.1 A wire of cross-section $L_y L_z$ with a voltage V applied across large contact.

where $M(E)$ is the number of "modes" or subbands at an energy E. The actual number of modes $M(E)$ at a given energy depends on the details of the wire, but the maximum conductance per mode is G_0 independent of these details. This can be shown as follows.

Maximum current in a single-moded wire: Consider a mode or subband ν with a dispersion relation $E_\nu(k_x)$. The current carried by the states having a positive group velocity can be written as

$$I = \frac{-q}{L} \sum_{v_x(k_x)>0} v_x(k_x)$$

We have seen earlier that for free electrons with $E = \hbar^2 k^2/2m$, the velocity is given by $\hbar k/m$, which is equal to the momentum $\hbar k$ divided by the mass m. But what is the appropriate velocity for electrons in a periodic solid having some dispersion relation $E_\nu(\vec{k})$? The answer requires careful discussion which we will postpone for the moment (see Section 6.4) and simply state that the correct velocity is the group velocity generally defined as the gradient of the dispersion relation

$$\hbar \vec{v}(\vec{k}) = \vec{\nabla}_k\, E_\nu(\vec{k})$$

so that in our 1D example we can write $v_x(k_x) = \partial E_\nu(k_x)/\partial k_x$ and the current is given by

$$I = \frac{-q}{L} \sum_{v(k_x)>0} \frac{1}{\hbar} \frac{\partial E_\nu(k_x)}{\partial k_x}$$

$$= -q \int \frac{\mathrm{d}k_x}{2\pi} \frac{1}{\hbar} \frac{\partial E_\nu(k_x)}{\partial k_x} = \frac{-q}{h} \int \mathrm{d}E_\nu \tag{6.3.3}$$

showing that each mode of a wire carries a current of (q/h) per unit energy. At equilibrium there is no net current because states with positive and negative velocities are all equally occupied. An applied voltage V changes the occupation of the levels over an energy range $E_\mathrm{F} \pm (qV/2)$ creating a non-equilibrium situation whose details depend on the nature of the coupling to the contacts. But regardless of all these details, it is

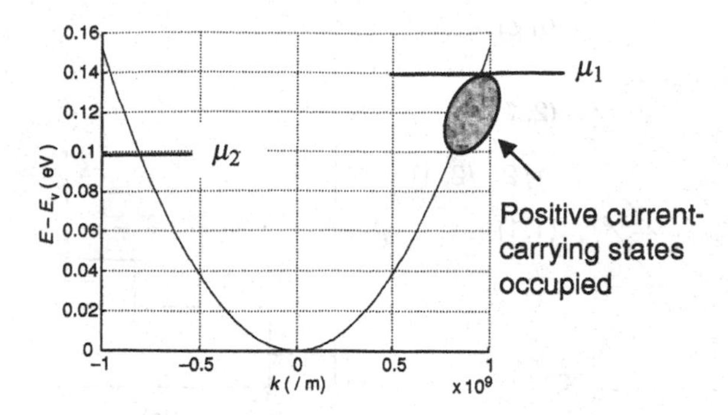

Fig. 6.3.2

apparent that the maximum net current will be established if the positive velocity states are occupied up to $\mu_1 = E_F + (qV/2)$ while the negative velocity states are occupied up to $\mu_2 = E_F - (qV/2)$, so that in the energy range

$$E_F - (qV/2) < E < E_F + (qV/2)$$

only the positive velocity states are occupied (Fig. 6.3.2). From Eq. (6.3.3) we can write the current carried by these states belonging to mode ν as

$$I = \frac{-q}{h}(\mu_1 - \mu_2) \rightarrow \frac{q^2}{h}V$$

so that the maximum conductance of mode ν is equal to q^2/h as stated earlier (see Eq. (6.3.1)). Note that this result is independent of the dispersion relation $E_\nu(k_x)$ for the mode ν.

Number of modes: How many modes $M(E)$ we actually have at a given energy, however, is very dependent on the details of the problem at hand. For example, if the relevant energy range involves the bottom of the conduction band then we may be able to approximate the band diagram with a parabolic relation (see Fig. 6.1.2, Eq. (6.1.1)). The subbands can then be catalogued with two indices (n, p) as shown in Fig. 6.3.3

$$E_{n,p}(k_x) \approx E_c + n^2\varepsilon_y + p^2\varepsilon_z + \frac{\hbar^2 k_x^2}{2m_c} \tag{6.3.4}$$

where $\varepsilon_y = \pi^2\hbar^2/2m_cL_y^2$ and $\varepsilon_z = \pi^2\hbar^2/2m_cL_z^2$, assuming that the electrons are confined in the wire by infinite potential wells of width L_y and L_z in the y- and z-directions respectively. The mode density $M(E)$ then looks as shown in Fig. 6.3.2, increasing with energy in increments of one every time a new subband becomes available.

The details of the subbands in the valence band are much more complicated, because of the multiple bands that are coupled together giving rise to complicated dispersion

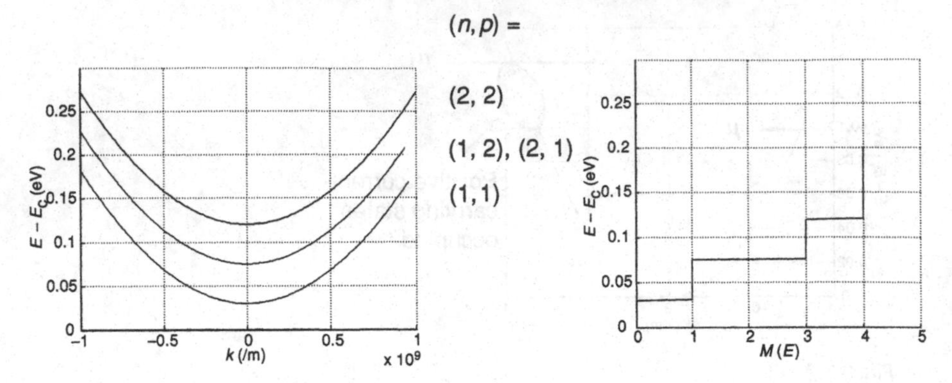

Fig. 6.3.3 Energy dispersion relation showing the four lowest conduction band subbands (see Eq. (6.3.4)) in a rectangular wire with $L_y = L_z = 10\,\text{nm}$, $m_c = 0.25m$.

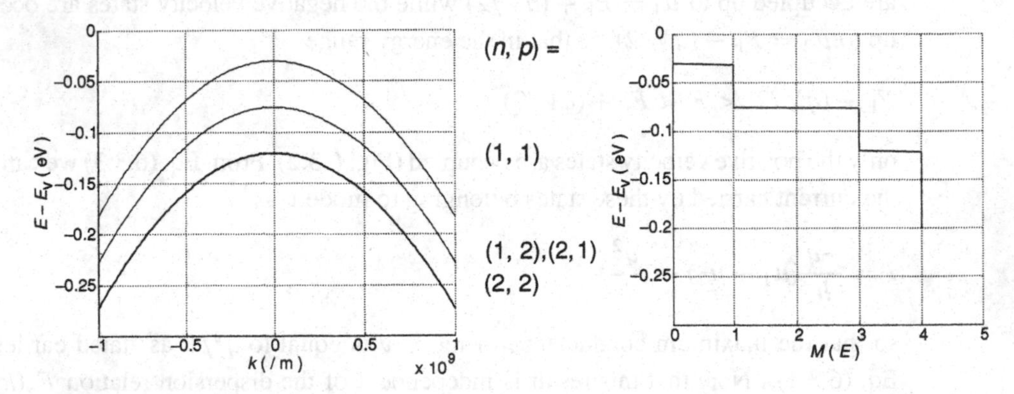

Fig. 6.3.4 Energy dispersion relation showing the four lowest valence band subbands (see Eqs. (6.3.5) and (6.3.6)) in a rectangular wire with $L_y = L_z = 10\,\text{nm}$, $m_v = 0.25m$.

relations. For simple back-of-an-envelope estimates we could approximate with a simple *inverted* parabola

$$h(\vec{k}) = E_v - \frac{\hbar^2 k^2}{2m_v} \tag{6.3.5}$$

We then get inverted dispersion relations for the different subbands

$$E_{n,p}(k_x) \approx E_v - n^2\varepsilon_y - p^2\varepsilon_z - \frac{\hbar^2 k_x^2}{2m_v} \tag{6.3.6}$$

with a mode density $M(E)$ that increases with decreasing energy as shown in Fig. 6.3.4.

Minimum resistance of a wide conductor: We have see that there is a minimum resistance (h/q^2) that a wire can have per mode. An interesting consequence of this is that there is a minimum resistance that a given conductor could have. For example, we could model a field effect transistor (FET) as a two-dimensional conductor with a width W and a length L as shown in the figure below. As the length L is reduced, it would approach a ballistic conductor whose resistance is dominated by the contact resistance. This is a well-known fact and device engineers work very hard to come up with contacting procedures that minimize the contact resistance. What is not as well-recognized is that there is a fundamental limit to how small the contact resistance can possibly be – even with the best conceivable contacts.

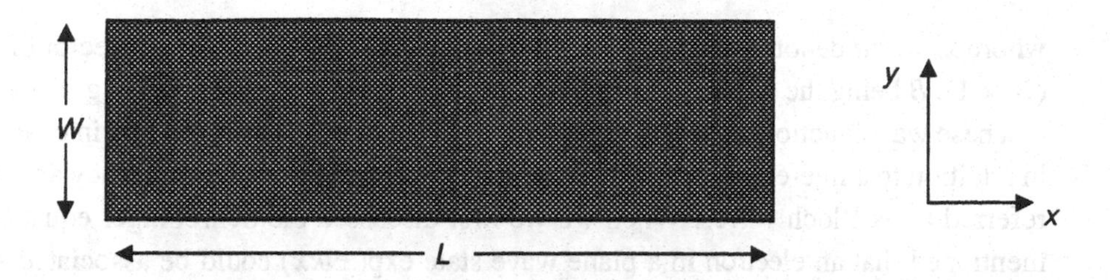

To estimate this minimum contact resistance, let us assume an n-type semiconductor where conduction takes place through the conduction band states described by a parabolic dispersion relation of the form given in Eq. (6.3.4). Assuming that it has only one subband along the z-direction we can write the electron density per unit area at $T = 0$ K as (see Table 6.2.1)

$$n_s = \frac{m_c}{\pi \hbar^2}(E_F - E_c) \tag{6.3.7}$$

The maximum conductance is given by

$$G_{max} = \frac{2q^2}{h} \text{Int} \sqrt{\frac{E_F - E_c}{\hbar^2 \pi^2 / 2m_c W^2}} \approx W \frac{2q^2}{h} \sqrt{\frac{2n_s}{\pi}}$$

where we have made use of Eq. (6.3.7) and used the symbol Int(x) to denote the largest integer smaller than x. The minimum resistance is given by the inverse of G_{max}:

$$R_{min} W = \frac{h}{2q^2} \sqrt{\frac{\pi}{2n_s}} = \frac{16.28 \text{ k}\Omega}{\sqrt{n_s}} \tag{6.3.8}$$

With a carrier density of $n_s = 10^{12}/\text{cm}^2$, this predicts $R_{min} W \approx 160\ \Omega\ \mu\text{m}$. Note that we have assumed only one subband arising from the z-confinement. For silicon, the six conduction valleys give rise to six sets of subbands, of which the lowest two are degenerate. This means that if the L_z-dimension is small enough, there will be two degenerate subbands arising from the z-confinement and the corresponding minimum contact resistance will be half our estimate which was based on one z-subband: $R_{min} W \approx 80\ \Omega\ \mu\text{m}$.

6.4 Velocity of a (sub)band electron

Electrons in periodic structures have energy levels that form (sub)bands and the corresponding wavefunctions have the form

$$\{\psi\}_n = \{\psi\}_0 \exp(\pm\, ikx_n) \tag{6.4.1}$$

where $x_n = na$ denotes points on a discrete lattice and $\{\psi\}_n$ is a column vector of size $(B \times 1)$, B being the number of basis functions describing the unit cell (Fig. 6.4.1).

These wavefunctions thus have a unit cell component that is atomic-like in character in addition to a free-electron or plane-wave-like $\sim \exp(\pm ikx)$ character. They are often referred to as Bloch waves. When we we first discussed the Schrödinger equation, I mentioned that an electron in a plane wave state $\exp(\pm ikx)$ could be associated with a velocity of $\hbar k/m$ since the probability current was equal to this quantity times the probability density (see Eq. (2.1.20)). What is the velocity of an electron in a Bloch wave state?

A plausible conjecture is that the velocity must be $\hbar k/m^*$, m^* being the effective mass. This is indeed true under certain conditions, but a more general discussion is called for since not all bands/subbands can be described in terms of an effective mass, especially in low-dimensional structures like quantum wires. Besides, it is not clear that we should be looking for a single number. In general we could have a velocity matrix of size $(B \times B)$.

Velocity matrix: To identify this velocity matrix, we start by noting that the Hamiltonian for a quantum wire can be written in a block tridiagonal form

$$H_{\text{quantum wire}} = \begin{bmatrix} \alpha & \beta & 0 & 0 & 0 & \cdots \\ \beta^+ & \alpha & \beta & 0 & 0 & \cdots \\ 0 & \beta^+ & \alpha & \beta & 0 & \cdots \\ \cdots & & \cdots & & \cdots & \end{bmatrix} \tag{6.4.2}$$

which can be visualized as a 1D array of unit cells each of which has a very large number of basis functions B equal to the number of atoms in a cross-section times the

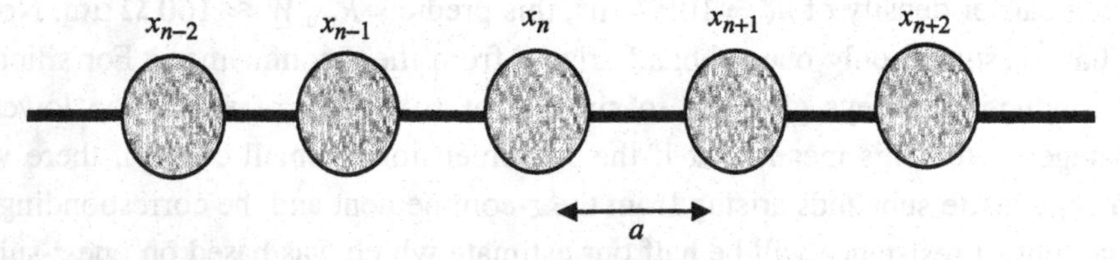

Fig. 6.4.1

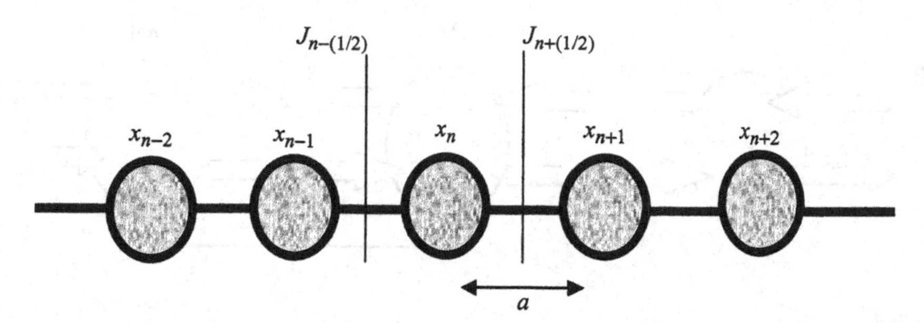

Fig. 6.4.2

number of basis functions per atom. The time-dependent Schrödinger equation for the quantum wire can be written in the form

$$i\hbar\frac{d\psi_n}{dt} = \alpha\psi_n + \beta\psi_{n+1} + \beta^+\psi_{n-1} \tag{6.4.3}$$

where ψ are $(B \times 1)$ column vectors, while α and β are $(B \times B)$ matrices.

To discover the appropriate expression for the current, we need to look at the time rate of change in the total probability density in a particular unit cell, say the nth one. This can be written as $(\psi_n^+\psi_n)$ which is a (1×1) matrix or a scalar number equal to the sum of the squared magnitudes of all B components of ψ (note that $\psi_n\psi_n^+$ is not suitable for this purpose since it is $(B \times B)$ matrix). From Eq. (6.4.3) we can write the time derivative of $(\psi_n^+\psi_n)$ as

$$i\hbar\frac{d}{dt}\psi_n^+\psi_n = \psi_n^+\beta\psi_{n+1} + \psi_n^+\beta^+\psi_{n-1} - \psi_{n-1}^+\beta\psi_n - \psi_{n+1}^+\beta^+\psi_n \tag{6.4.4}$$

Let us see if we can write the right-hand side as the difference between the current to the left of the unit and the current to the right of the unit (see Fig. 6.4.2)

$$\frac{d}{dt}\psi_n^+\psi_n = \frac{J_{n-(1/2)} - J_{n+(1/2)}}{a} \tag{6.4.5}$$

motivated by the fact that in a continuum representation, the continuity equation requires that $dn/dt = -dJ/dx$. It is straightforward to check that if we define the currents as

$$J_{n+(1/2)} \equiv \frac{\psi_{n+1}^+\beta^+\psi_n - \psi_n^+\beta\psi_{n+1}}{i\hbar}\, a \tag{6.4.6}$$

then Eq. (6.4.5) is equivalent to (6.4.4). We can make use of Eq. (6.4.1) for a (sub)band electron to express ψ_{n+1} in Eq. (6.4.4) in terms of ψ_n to obtain

$$J_{n+(1/2)} \equiv \psi_n^+\frac{a}{i\hbar}\,[\beta^+\exp(-ika) - \beta\exp(+ika)]\psi_n \tag{6.4.7}$$

suggesting that we define

$$[v(k)] \equiv \frac{a}{i\hbar}[\beta^+\exp(-ika) - \beta\exp(+ika)] \tag{6.4.8}$$

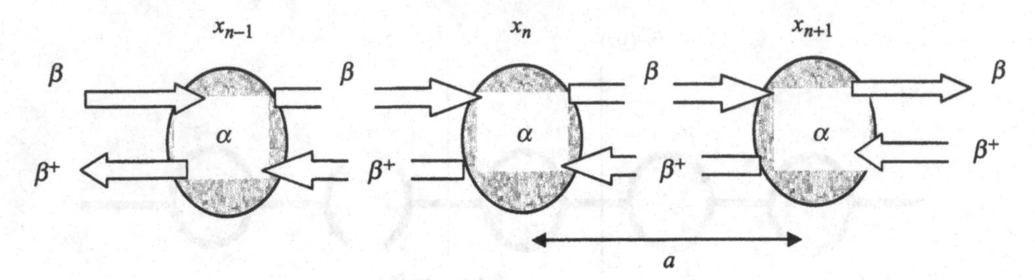

Fig. 6.4.3

as the velocity matrix for an electron (sub)band state with a given k. Now that we are finished with the mathematical argument, let us see if it makes physical sense.

Firstly we note that the energy eigenvalues are obtained by diagonalizing the matrix (this follows from the basic result, Eq. (5.2.4), underlying the calculation of bandstructure)

$$[h(k)] = \alpha + \beta \exp(+ika) + \beta^+ \exp(-ika) \tag{6.4.9}$$

It is straightforward to check that this Hamiltonian matrix is related to the velocity matrix we just obtained (Eq. (6.4.8)) by the relation

$$[v(k)] = \frac{1}{\hbar}\frac{\mathrm{d}}{\mathrm{d}k}[h(k)] \tag{6.4.10}$$

This means that if the two matrices ($[h]$ and $[v]$) were simultaneously diagonalized, then the eigenvalues would indeed obey the relation

$$\hbar\vec{v}(\vec{k}) = \vec{\nabla}_k E_\nu(\vec{k})$$

that we used to derive the maximum current carried by a subband (see Eq. (6.3.3)).

Can we diagonalize $[h]$ and $[v]$ simultaneously? It is easy to see from the definitions of these matrices (Eqs. (6.4.8) and (6.4.9)) that the answer is yes if the matrices $[\alpha]$ and $[\beta]$ can be simultaneously diagonalized. We can visualize the tridiagonal Hamiltonian matrix in Eq. (6.4.2) as a 1D array where the matrices $[\alpha]$ and $[\beta]$ are each of size $(B \times B)$, B being the number of basis functions needed to represent the cross-section of the wire (see Fig. 6.4.3). Now if we can diagonalize both $[\alpha]$ and $[\beta]$ then in this representation we can visualize the quantum wire in terms of independent modes or subbands that conduct in parallel as shown in Fig. 6.4.4.

The same representation will also diagonalize the velocity matrix and its diagonal elements will give us the velocity for each mode. One good example of a structure where this is possible is the simple square lattice shown in Fig. 6.4.5a, which we can represent as a 1D array of the form shown in Fig. 6.4.5b.

The matrices $[\beta]$, $[\beta^+]$ represent the coupling between one column of atoms with the corresponding ones on the next column and can be written as an identity matrix $[I]$.

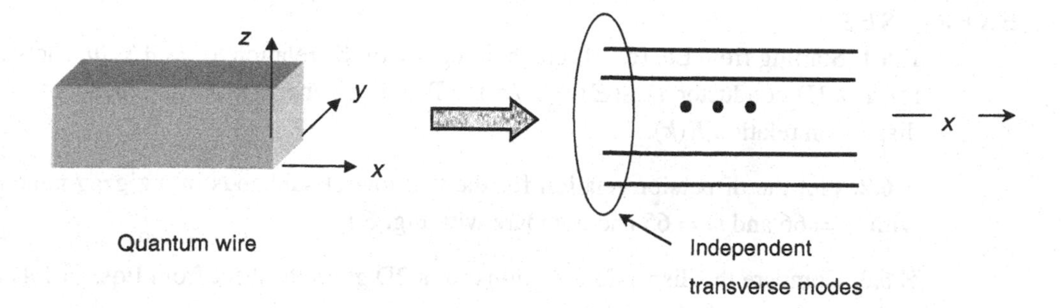

Fig. 6.4.4

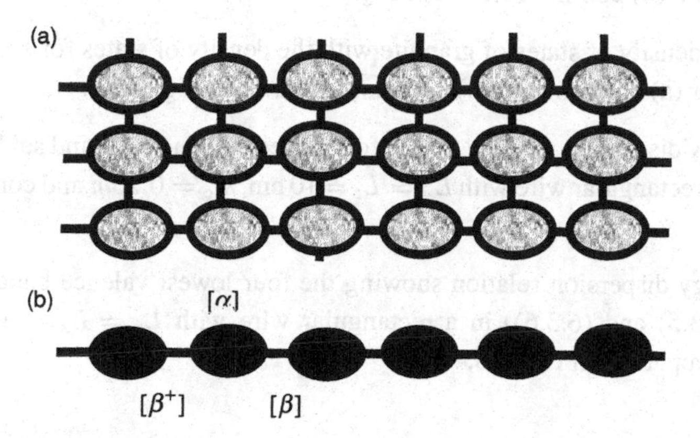

Fig. 6.4.5

This means that once we find a suitable representation to diagonalize $[\alpha]$, the matrix $[\beta]$ will also be diagonal, since the identity matrix is unchanged by a basis transformation. Hence, we can always diagonalize $[\alpha]$, $[\beta]$, $[h]$, $[v]$ simultaneously and use this representation to visualize a quantum wire as a collection of independent single-moded wires in parallel.

Mode density: We saw in the last section that the mode density $M(E)$ plays an important role in the physics of quantum wires, comparable to the role played by the density of states $D(E)$ for bulk semiconductors. If a wire of length L has eigenstates catalogued by two indices (α, k) then we could define the mode density $M(E)$ formally as

$$M(E) = \sum_{\alpha}\sum_{k} \delta(E - \varepsilon_{\alpha,k})\frac{1}{L}\frac{\partial \varepsilon_{\alpha,k}}{\partial k} \tag{6.4.11}$$

This expression can be evaluated simply if the eigenvalues are separable in the two indices: $\varepsilon_{\alpha,k} = \varepsilon'_{\alpha} + \varepsilon''_{k}$. But this is possible only if the subbands can be decoupled as discussed above.

EXERCISES

E.6.1. Starting from Eq. (6.2.4) and making use of the relation $\hbar\nu = \partial E/\partial k$ show that for any 1D conductor $D_{1D}(E) = L/\pi\hbar\nu$. This is a general result independent of the dispersion relation $E(k)$.

E.6.2. Plot the dispersion relation for the two lowest subbands of a zigzag nanotube with $m = 66$ and $m = 65$ and compare with Fig. 6.1.7.

E.6.3. Compare the dispersion relations for a 2D graphite sheet from Eqs. (6.1.4) and (6.1.11) as shown in Fig. 6.1.9.

E.6.4. Compare the 3D density of states with the density of states for a 2D box of thickness (a) 5 nm and (b) 20 nm as shown in Fig. 6.2.1.

E.6.5. Compare the density of states of graphite with the density of states for a zigzag nanotube of diameter (a) 15.4 nm and (b) 61.7 nm as shown in Fig. 6.2.2.

E.6.6. Plot the energy dispersion relation for the four lowest conduction band subbands (see Eq. (6.3.4)) in a rectangular wire with $L_y = L_z = 10\,\text{nm}$, $m_c = 0.25m$ and compare with Fig. 6.3.3.

E.6.7. Plot the energy dispersion relation showing the four lowest valence band subbands (see Eqs. (6.3.5) and (6.3.6)) in a rectangular wire with $L_y = L_z = 10\,\text{nm}$, $m_v = 0.25m$ and compare with Fig. 6.3.4.

7 Capacitance

In Chapter 1 I stated that the full quantum transport model required us to generalize each of the parameters from the one-level model into its corresponding matrix version. Foremost among these parameters is the Hamiltonian matrix $[H]$ representing the energy levels and we are almost done with this aspect. This chapter could be viewed as a transitional one where we discuss an ***equilibrium*** problem that can be handled using $[H]$ alone, without knowledge of other parameters like broadening that we will discuss here.

The problem we will discuss is the following. How does the electron density inside the device change as a function of the gate voltage V_G, assuming that the source and the drain are held at the same potential (drain voltage $V_D = 0$, see Fig. 7.1)? Strictly speaking this too is a non-equilibrium problem since the gate contact is not in equilibrium with the source and drain contacts (which are in equilibrium with each other). However, the insulator isolates the channel from the gate and lets it remain essentially in equilibrium with the source and drain contacts, which have the same electrochemical potential $\mu_1 = \mu_2 \equiv \mu$. The density matrix (whose diagonal elements in a real-space representation give us the electron density $n(\vec{r})$) is given by

$$[\rho] = f_0([H] - \mu[I]) \tag{7.1}$$

and can be evaluated simply from $[H]$ without detailed knowledge of the coupling to the source and drain. I am assuming that the channel is large enough that its energy levels are nearly continuous so that the broadening due to the source and drain coupling makes no significant difference.

The matrix $[H]$ includes two parts (see Section 1.4)

$$H = H_0 + U([\delta\rho]) \tag{7.2}$$

where H_0 represents just the isolated materials deduced from a knowledge of their bandstructure, while U represents the potential due to the applied gate voltage and any *change* in the density matrix from the reference condition described by H_0. Neglecting any corrections for correlation effects (see Section 3.2), we can calculate U from the

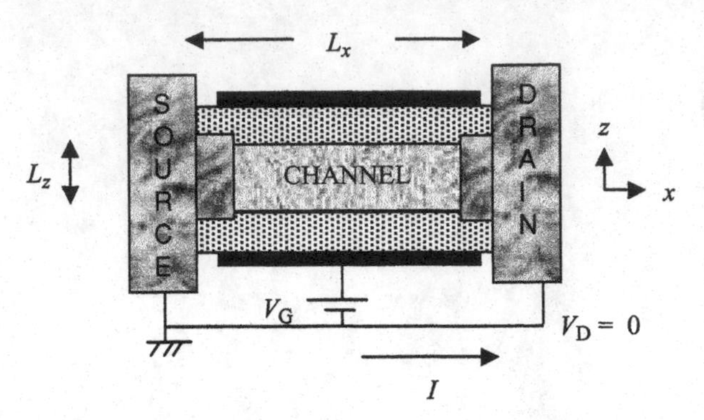

Fig. 7.1 A metal–insulator–semiconductor (MIS) capacitor. The problem is to find the charge induced in the channel in response to an applied gate voltage V_G.

Poisson equation describing Coulomb interactions (ε_r is the relative permittivity which could vary spatially):

$$\vec{\nabla} \cdot (\varepsilon_r \vec{\nabla}\, U) = -\frac{q^2}{\varepsilon_0}[n(\vec{r}) - n_0] \tag{7.3}$$

subject to the boundary conditions: $[U]_{\text{source}} = [U]_{\text{drain}} = 0$

$[U]_{\text{gate}} = -q\,V_G$

In this chapter I will use this problem to illustrate how we choose the Hamiltonian $[H_0]$ to describe an inhomogeneous structure like a transistor (Section 7.1), how we evaluate the density matrix $[\rho]$ (Section 7.2), and finally (Section 7.3) how the capacitance C obtained from a self-consistent solution of Eqs. (7.1)–(7.3) can be viewed as a series combination of an electrostatic capacitance C_E, which depends on the dielectric constant, and a quantum capacitance C_Q, which depends on the density of eigenstates in the channel. From this chapter onwards we will use the Hamiltonian matrix $[H]$ derived from a single-band effective mass equation, although the conceptual framework we describe is quite general and can be used in conjunction with other models. The supplementary Section 7.4 provides some details about another commonly used model, namely multi-band effective mass equations, though we have not used it further in this book.

7.1 Model Hamiltonian

Atomistic Hamiltonian: Let us start with the question of how we write $[H_0]$ to represent the inhomogeneous collection of isolated materials that comprise the device, from a knowledge of their individual bandstructures. For example, we could model the channel material with a $[H_0]$ that can be represented schematically as a network

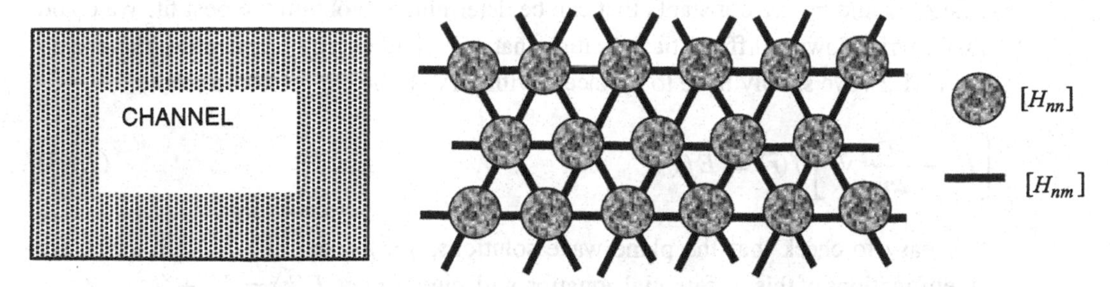

Fig. 7.1.1 Any part of the device (e.g. the channel) can be represented by an atomistic Hamiltonian matrix that can be depicted schematically as a 3D network of unit cells described by matrices $[H_{nn}]$ and bonds described by matrices $[H_{nm}]$, $n \neq m$. We have arranged the unit cells in an FCC-like network since that is the arrangement for most common semiconductors. Note that the matrices $[H_{nm}]$ for different neighbors m are in general different though we have represented them all with the same symbol.

of unit cells $[H_{nn}]$ interconnected by 'bonds' $[H_{nm}]$ of the same size $(b \times b)$. Each of these matrices is of size $b \times b$, b being the number of basis functions per unit cell (Fig. 7.1.1).

We saw in Chapter 5 that, knowing all the $[H_{nm}]$, the full bandstructure can be calculated from the eigenvalues of the $(b \times b)$ matrix

$$[h(\vec{k})] = \sum_m [H_{nm}] e^{i\vec{k}\cdot(\vec{d}_m - \vec{d}_n)} \tag{7.1.1}$$

(which is independent of n) for each value of $\vec{k}$. Conversely, we can write down the matrices $[H_{nm}]$ from a knowledge of the bandstructure and thereby write down the matrix $[H_0]$ representing a periodic solid which is of size $(Nb \times Nb)$, N being the total number of unit cells.

The insulator material would obviously be described by a different set of matrices that can be deduced from its bandstructure. The difficult part to model is the *interface*. This is partly due to our ignorance of the actual atomistic structure of the actual interface. But assuming that we know the microstructure exactly, it is still not straightforward to figure out the appropriate bond matrix $[H_{nm}]$ between two unit cells n and m belonging to different materials A and B. Clearly this information is not contained in the individual bandstructures of either A or B and it requires a more careful treatment. We will not get into this question but will simply represent an A–B bond using the average of the individual $[H_{nm}]$ matrices for A–A bonds and B–B bonds.

Effective mass Hamiltonian: We have seen before (see Fig. 5.1.2) that the energy levels around the conduction band minimum can often be described by a simple relation like

$$h(\vec{k}) = E_c + \frac{\hbar^2 k^2}{2m_c} \tag{7.1.2}$$

where E_c and m_c are constants that can be determined to obtain the best fit. We could easily write down a differential equation that will yield energy eigenvalues that match Eq. (7.1.2). We simply have to replace $\vec{k}$ with $-i\vec{\nabla}$ in the expression for $h(\vec{k})$:

$$\left[E_c - \frac{\hbar^2}{2m_c}\nabla^2\right] f(\vec{r}) = Ef(\vec{r}) \tag{7.1.3}$$

It is easy to check that the plane wave solutions, $f(\vec{r}) = \exp(i\vec{k}\cdot\vec{r})$ with any $\vec{k}$ are eigenfunctions of this differential equation with eigenvalues $E(\vec{k}) = E_c + (\hbar^2k^2/2m_c)$. We could use the finite difference method (Section 2.2) to convert Eq. (7.1.3) into a Hamiltonian matrix that is much simpler than the original atomistic Hamiltonian. For example, in one dimension we could write a tridiagonal matrix with $E_c + 2t_0$ on the diagonal and $-t_0$ on the upper and lower diagonals (see Eq. (2.3.1)) that can be represented in the form shown in Fig. 7.1.2.

We can use the basic bandstructure equation in Eq. (7.1.1) to write down the corresponding dispersion relation:

$$h(k_x) = (E_c + 2t_0) - t_0\,e^{ik_xa} - t_0\,e^{-ik_xa} = E_c + 2t_0(1 - \cos k_xa)$$

For a general 3D structure the effective mass Hamiltonian has the form shown in Fig. 7.1.3, leading to the dispersion relation

$$h(\vec{k}) = E_c + 2t_0(1 - \cos k_xa) + 2t_0(1 - \cos k_ya) + 2t_0(1 - \cos k_za) \tag{7.1.4a}$$

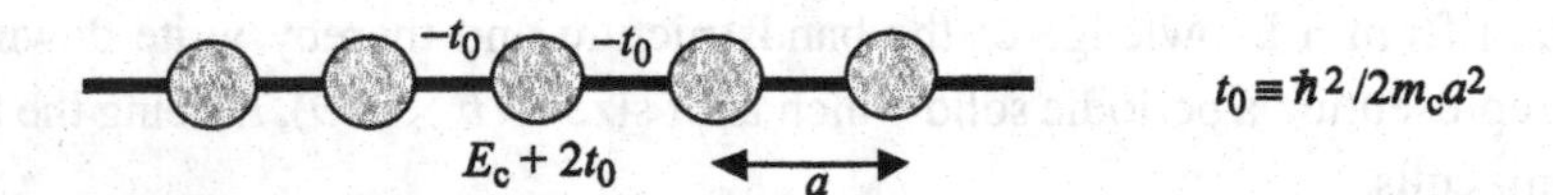

Fig. 7.1.2 The effective mass Hamiltonian matrix in 1D can be visualized as a 1D array of unit cells each with energy $E_c + 2t_0$ bonded to its nearest neighbors by $-t_0$.

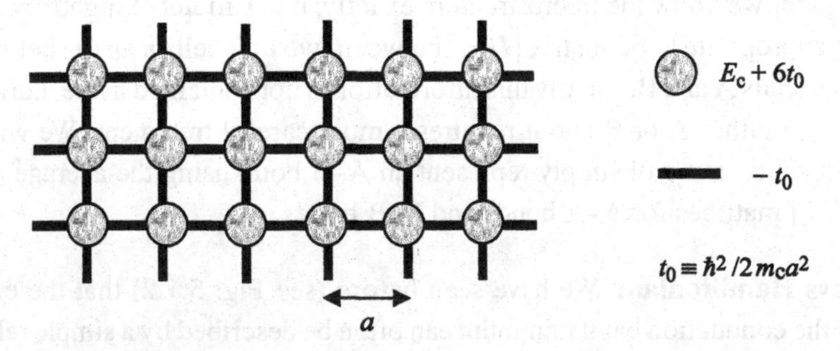

Fig. 7.1.3 The effective mass Hamiltonian matrix can be depicted schematically as a 3D network of unit cells (unrelated to the actual crystal structure) each with energy $E_c + 6t_0$ bonded to its nearest neighbors by $-t_0$.

which reduces to the parabolic relation in Eq. (7.1.2) if $k_x a$ is small enough that $(1 - \cos k_x a)$ can be approximated with $(k_x a)^2/2$ (and the same with $k_y a$ and $k_z a$):

$$h(\vec{k}) = E_c + \frac{\hbar^2}{2m_c}\left(k_x^2 + k_y^2 + k_z^2\right) \tag{7.1.4b}$$

This Hamiltonian only describes the eigenstates around the bottom of the conduction band where Eq. (7.1.2) provides an adequate approximation, unlike an atomistic Hamiltonian that describes the full bandstructure (see Eq. (7.1.1)). What we gain, however, is simplicity. The resulting Hamiltonian matrix $[H_0]$ is much smaller than the atomistic counterpart for two reasons. Firstly, the matrices $[H_{nm}]$ representing a unit cell or a bond are scalar numbers rather than $(b \times b)$ matrices. Secondly the unit cells in Fig. 7.1.3 do not have to correspond to atomic unit cells as in atomistic Hamiltonians (see Fig. 7.1.1). The lattice can be simple cubic rather than face-centered cubic and the lattice constant a can be fairly large depending on the energy range over which we want the results to be accurate. A simple rule of thumb is that a should be small enough that the corresponding t_0 is larger than the energy range (above E_c) we are interested in. Since $t_0 \equiv \hbar^2/2m_c a^2$, this means that for a given energy range, we can use a larger a if the effective mass m_c is small. But it is important to remember that the wavefunction does not provide information on an atomic scale. It only provides information on a coarse spatial scale and is sometimes referred to as an "envelope function."

Spatially varying effective mass: Effective mass equations are often used to model "heterostructures" consisting of different materials such that the conduction band edge E_c and/or the effective mass m_c appearing in Eq. (7.1.3):

$$\left[E_c - \frac{\hbar^2}{2m_c}\nabla^2\right] f(\vec{r}) = Ef(\vec{r})$$

vary spatially. The variation of E_c leads to no special problems, but the variation of m_c cannot be incorporated simply by writing

$$\left[E_c(\vec{r}) - \frac{\hbar^2}{2m_c(\vec{r})}\nabla^2\right] f(\vec{r}) = Ef(\vec{r})$$

The correct version is

$$\left[E_c(\vec{r}) - \frac{\hbar^2}{2m_c}\vec{\nabla}\cdot\left(\frac{1}{m_c(\vec{r})}\vec{\nabla}\right)\right] f(\vec{r}) = Ef(\vec{r}) \tag{7.1.5}$$

and it can be shown that if we apply the finite difference method to this version at an interface where the effective mass changes from m_1 to m_2 then we obtain a Hamiltonian matrix that can be represented as shown in Fig. 7.1.4.

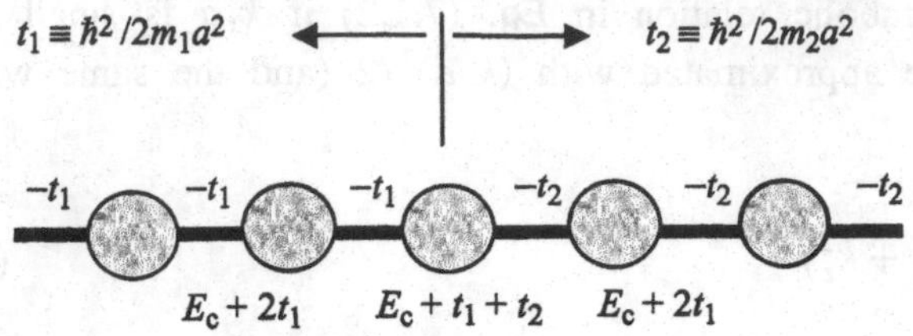

$t_1 \equiv \hbar^2/2m_1a^2$ $t_2 \equiv \hbar^2/2m_2a^2$

$-t_1$ $-t_1$ $-t_1$ $-t_2$ $-t_2$ $-t_2$

$E_c + 2t_1$ $E_c + t_1 + t_2$ $E_c + 2t_1$

Fig. 7.1.4

The point to note is that the resulting Hamiltonian matrix

$$\begin{bmatrix} E_c + 2t_1 & -t_1 & 0 \\ -t_1 & E_c + t_1 + t_2 & -t_2 \\ 0 & -t_2 & E_c + 2t_2 \end{bmatrix}$$

is hermitian as needed to ensure that the energy eigenvalues are real and current is conserved. By contrast if we start from one of the other possibilities like

$$\left[E_c - \frac{\hbar^2}{2m_c(\vec{r})} \nabla^2 \right] f(\vec{r}) = Ef(\vec{r})$$

and use the finite difference method we will end up with a Hamiltonian matrix of the form ($t_0 \equiv (t_1 + t_2)/2$)

$$\begin{bmatrix} E_c + 2t_1 & -t_1 & 0 \\ -t_1 & E_c + 2t_0 & -t_0 \\ 0 & -t_2 & E_c + 2t_2 \end{bmatrix}$$

which is clearly *non-hermitian*.

As we have mentioned before, writing down the appropriate Hamiltonian for the interface region requires knowledge of the interfacial microstructure and simple approximations are often used. But the important point to note is that whatever approximation we use, a fundamental zero-order requirement is that the Hamiltonian matrix should be *hermitian*. Otherwise we can run into serious inconsistencies due to the non-conservation of probability density and the resulting lack of continuity in electron flow.

An example: One-band effective mass models are widely used to model heterostructures of materials that are not too different. Consider, for example, a GaAs quantum well sandwiched between $Al_{0.3}Ga_{0.7}As$ barriers. For GaAs we use $E_c = 0$ eV, $m_c = 0.07m$, while for AlAs, $E_c = 1.25$ eV, $m_c = 0.15m$ and interpolate linearly to obtain E_c and m_c for the AlAs–GaAs alloy. Figure 7.1.5 shows the energies of the two lowest

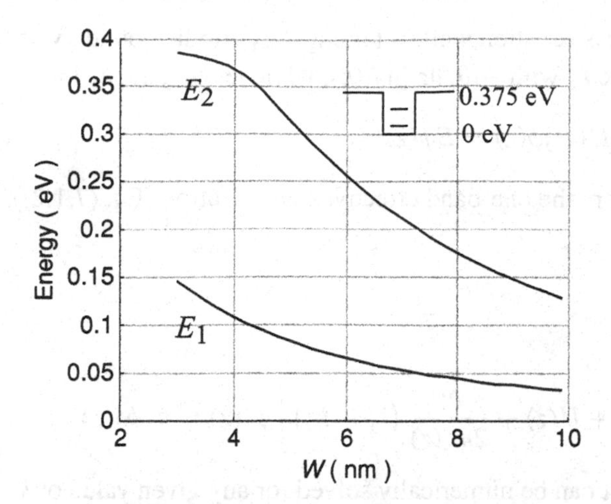

Fig. 7.1.5 Energy of the two lowest energy levels of a GaAs quantum well sandwiched between $Al_{0.3}Ga_{0.7}As$ barriers (shown in inset) as a function of the well width W calculated from a one-band effective mass model.

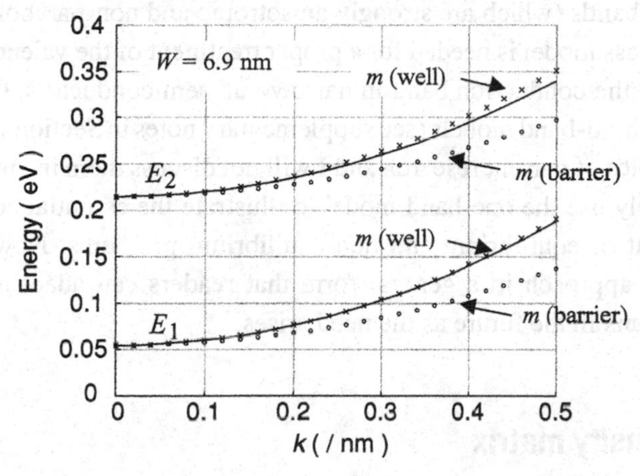

Fig. 7.1.6 Solid curves show the dispersion relation $E(\vec{k})$ as a function of the magnitude of the in-plane wavevector $\vec{k} = \{k_x \quad k_y\}$ for the two lowest subbands of a quantum well of width $W = 6.9$ nm, calculated from the one-band effective mass model. The ×'s and o's show the dispersion expected for an effective mass equal to that in the well and in the barrier respectively.

levels in the GaAs quantum well as a function of the well width, while Fig. 7.1.6 shows the dispersion relation $E(\vec{k})$ as a function of the magnitude of the in-plane wavevector $\vec{k} = \{k_x \quad k_y\}$ for the two lowest subbands of a quantum well of width $W = 6.9$ nm.

Problems like this are essentially one-dimensional and easy to solve numerically. The basic idea is that our usual prescription for obtaining the effective mass equation is to replace $\vec{k}$ with $-\mathrm{i}\vec{\nabla}$ which consists of three simultaneous replacements:

$$k_x \to -\mathrm{i}\partial/\partial x \qquad k_y \to -\mathrm{i}\partial/\partial y \qquad k_z \to -\mathrm{i}\partial/\partial z$$

To obtain a 1D effective mass equation while retaining the periodic boundary condition in the x–y plane, we replace k_z with $-\mathrm{i}\partial/\partial z$ in $h(\vec{k})$, while leaving k_x, k_y intact:

$$[h(k_x, k_y; k_z \Rightarrow -\mathrm{i}\partial/\partial z) + U(z)]\phi(z) = E\phi(z) \tag{7.1.6}$$

For example, if we start from the one-band effective mass relation (Eq. (7.1.2)),

$$h(\vec{k}) = E_\mathrm{c} + \frac{\hbar^2 k^2}{2m_\mathrm{c}}$$

we obtain

$$\left[E_\mathrm{c} - \frac{\hbar^2}{2}\frac{\partial}{\partial z}\left(\frac{1}{m_\mathrm{c}(z)}\frac{\partial}{\partial z}\right) + U(z) + \frac{\hbar^2}{2m_\mathrm{c}(z)}\left(k_x^2 + k_y^2\right)\right]\phi_\alpha(z) = \varepsilon_\alpha \phi_\alpha(z) \tag{7.1.7}$$

which is a 1D equation that can be numerically solved for any given value of k_x, k_y.

The one-band effective mass model works very well when we have an isotropic parabolic band that is well separated from the other bands. This is usually true of the conduction band in wide-bandgap semiconductors. But the valence band involves multiple closely spaced bands (which are strongly anisotropic and non-parabolic) and a multi-band effective mass model is needed for a proper treatment of the valence band (p-type devices) or even the conduction band in narrow-gap semiconductors. General device models based on multi-band models (see supplementary notes in Section 7.4) and atomistic models are topics of current research and I will not discuss them in any depth in this book. I will largely use the one-band model to illustrate the essential concepts underlying the treatment of equilibrium and non-equilibrium problems. However, I will try to describe the approach in a general form that readers can adapt to more sophisticated Hamiltonians in the future as the need arises.

7.2 Electron density/density matrix

Once we have identified the basic Hamiltonian matrix H_0 representing the isolated materials comprising a device, the next step is to evaluate the density matrix (especially the diagonal elements in a real-space representation, which give us the electron density)

$$n(\vec{r}) = 2\rho(\vec{r}, \vec{r}) = 2\sum_\alpha |\phi_\alpha(\vec{r})|^2 f_0(\varepsilon_\alpha - \mu) \tag{7.2.1}$$

where $\phi_\alpha(\vec{r})$ are the eigenfunctions of $[H]$ with eigenvalues ε_α, with $[H]$ given by

$$H = H_0 + U([\delta\,\rho]) \tag{7.2.2}$$

where U represents the potential due to the applied gate voltage and due to any *change* in the density matrix from the reference condition described by H_0. In general the matrix representation $[U]$ of the function $U(\vec{r})$ requires a knowledge of the basis functions, but

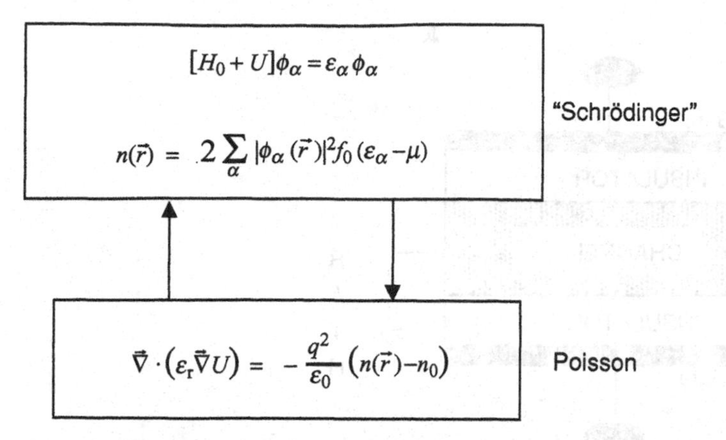

Fig. 7.2.1 Modeling a device in equilibrium generally requires a self-consistent solution of a "Schrödinger" equation with the Poisson equation.

if the potential varies slowly from one unit cell to the next, then we can simply assume the potential to have a constant value $U(\vec{r} = \vec{d}_n)$ throughout a unit cell n so that

$$[U_{nn}] = U(\vec{r} = \vec{d}_n)[I] \quad \text{and} \quad [U_{nm}] = [0] \quad \text{for} \quad m \neq n \tag{7.2.3}$$

where $[I]$ and $[0]$ are the identity matrix and the null matrix of appropriate size.

How do we calculate $U(\vec{r})$? Neglecting any corrections for correlation effects (see Section 3.2), we can use the Poisson equation describing Coulomb interactions (ε_r is the relative permittivity, which could vary spatially):

$$\vec{\nabla}\cdot(\varepsilon_r \vec{\nabla} U) = -\frac{q^2}{\varepsilon_0}(n(\vec{r}) - n_0) \tag{7.2.4}$$

subject to the boundary conditions: $[U]_{\text{source}} = [U]_{\text{drain}} = 0$

$[U]_{\text{gate}} = -qV_G$

What we need is a "Schrödinger–Poisson solver" that solves the two aspects of the problem self-consistently as shown schematically in Fig. 7.2.1. In general, 3D solutions are needed but this is numerically difficult and we will use an essentially 1D example to illustrate the "physics."

1D Schrödinger–Poisson solver: There are many problems that can be modeled with a 1D Schrödinger–Poisson solver: the metal–insulator–semiconductor (MIS) capacitor (Fig. 7.2.2) we mentioned at the beginning of this chapter represents such an example if we neglect any boundary effects in the x–y plane. What does the 1D version of the equations in Fig. 7.2.1 look like? Let us assume we are using the one-band effective mass Hamiltonian. We might guess that we should first solve a 1D version

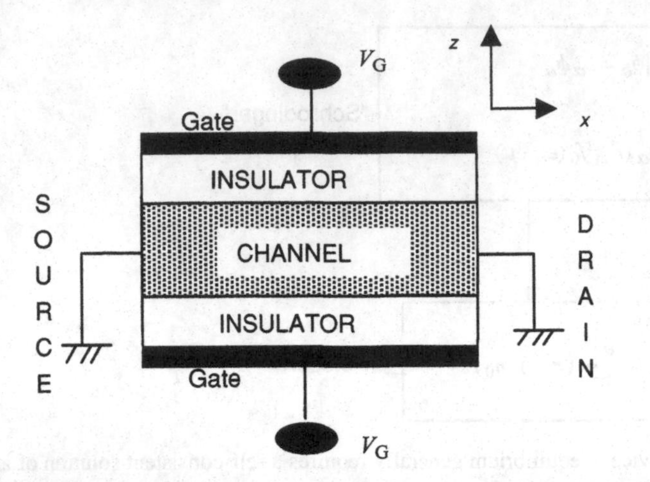

Fig. 7.2.2 The MIS capacitor can be modeled with a 1D Schrödinger equation if we neglect any boundary effects in the x–y plane.

of Eq. (7.1.7)

$$\left[E_c - \frac{\hbar^2}{2m_c}\frac{\partial^2}{\partial z^2} + U(z)\right]\phi_m(z) = \varepsilon_m\phi_m(z) \tag{7.2.5}$$

then evaluate the electron density from the 1D version of Eq. (7.2.1)

$$n(z) = 2\sum_m |\phi_m(z)|^2 f_0(\varepsilon_m - \mu) \quad \text{(WRONG!)} \tag{7.2.6}$$

and do all this self-consistently with a 1D version of Eq. (7.2.4) with the reference electron density n_0 set equal to zero.

$$-\frac{\mathrm{d}}{\mathrm{d}z}\left(\varepsilon_r\frac{\mathrm{d}U}{\mathrm{d}z}\right) = \frac{q^2}{\varepsilon_0}n(z) \tag{7.2.7}$$

All the 1D versions listed above are correct except for Eq. (7.2.6). The Fermi function f_0 appearing in this equation should be replaced by a new function f_{2D} defined as

$$f_{2D}(E) = N_0 \ln[1 + \exp(-E/k_BT)] \quad \text{with} \quad N_0 \equiv \frac{m_c k_B T}{2\pi\hbar^2} \tag{7.2.8}$$

(cf. $f_0(E) = 1/[1 + \exp(E/k_BT)]$). The correct 1D version of the Schrödinger–Poisson solver is shown in Fig. 7.2.3.

Where does this new function f_{2D} come from? As long as the structure can be assumed to be uniformly periodic in the x–y plane and we can neglect all boundary effects, the eigenfunctions can still be written in the form of plane waves in the x- and y-directions, normalized to lengths L_x and L_y, respectively:

$$\phi_\alpha(\vec{r}) = \frac{\exp[ik_x x]}{\sqrt{L_x}}\frac{\exp[ik_y y]}{\sqrt{L_y}}\phi_m(z) \tag{7.2.9}$$

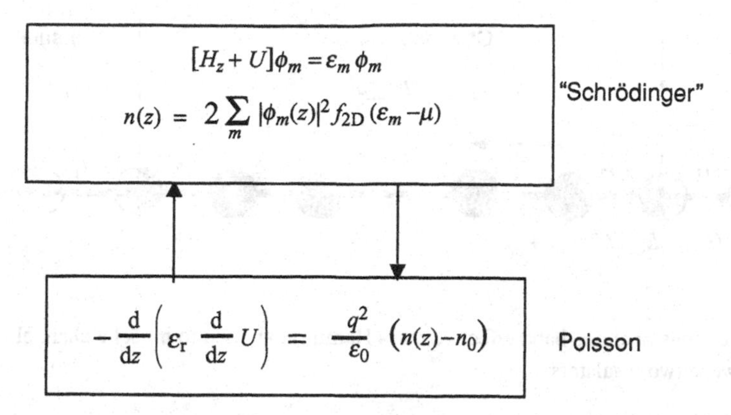

Fig. 7.2.3 The 1D Schrödinger–Poisson solver.

and the electron density is obtained from Eq. (7.2.1) after summing over all three indices $\{m, k_x, k_y\} \equiv \alpha$:

$$n(z) = 2\sum_m \sum_{k_x, k_y} |\phi_m(z)|^2 f_0(\varepsilon_\alpha - \mu) \tag{7.2.10}$$

Equation (7.2.6) is wrong because it simply ignores the summations over k_x, k_y. The correct version is obtained by noting that

$$\varepsilon_\alpha = \varepsilon_m + \frac{\hbar^2}{2m_c}\left(k_x^2 + k_y^2\right) \tag{7.2.11}$$

which follows from Eq. (7.1.7) with a constant (z-independent) effective mass. Note that things could get more complicated if the mass itself varies with z since the extra term

$$\frac{\hbar^2}{2m_c(z)}\left(k_x^2 + k_y^2\right)$$

would no longer be a constant that can just be added on to obtain the total energy. Under some conditions, this may still be effectively true since the wavefunction is largely confined to a region with a constant effective mass (see, for example, Exercise 7.1c), but it is not generally true. Also, the simple parabolic relation in Eq. (7.2.11) usually does not hold for the multi-band effective mass equation (see Fig. 7.4.2).

Using Eq. (7.2.11), the summation over k_x, k_y can now be performed analytically to show that

$$\sum_{k_x, k_y} f_0(\varepsilon_\alpha - \mu) = \sum_{k_x, k_y} f_0\left(\varepsilon_m - \mu + \frac{\hbar^2}{2m_c}\left(k_x^2 + k_y^2\right)\right)$$
$$= f_{2D}(\varepsilon_m - \mu) \tag{7.2.12}$$

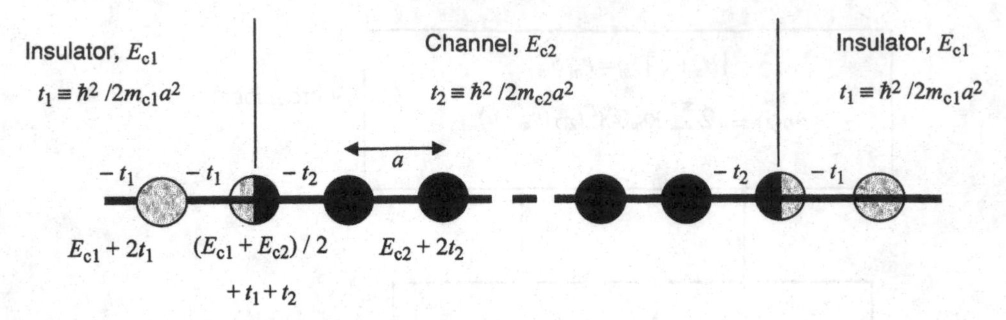

Fig. 7.2.4 One-dimensional one-band effective mass Hamiltonian used to model a channel sandwiched between two insulators.

This is shown as follows:

$$\frac{1}{S}\sum_{k_x,k_y} f_0\left(E+\frac{\hbar^2}{2m_c}(k_x^2+k_y^2)\right)$$

$$=\int_0^\infty \frac{2\pi k\,dk}{4\pi^2}\frac{1}{1+A\exp(\hbar^2k^2/2m_ck_BT)} \quad \text{where} \quad A\equiv\exp(E/k_BT)$$

$$=\frac{m_ck_BT}{2\pi\hbar^2}\int_0^\infty \frac{dy}{1+A\,e^y}$$

$$=\frac{m_ck_BT}{2\pi\hbar^2}\{\ln[A+e^{-y}]\}_\infty^0$$

$$=\frac{m_ck_BT}{2\pi\hbar^2}\ln[1+\exp(-E/k_BT)]\equiv f_{2D}(E)$$

Equation (7.2.12) allows us to simplify Eq. (7.2.10) to obtain the equation listed in Fig. 7.2.3:

$$n(z)=2\sum_m|\phi_m(z)|^2 f_{2D}(\varepsilon_m-\mu) \tag{7.2.13}$$

A numerical example: We can model the MOS capacitor shown in Fig. 7.2.2 by setting up a 1D Schrödinger–Poisson solver assuming the cross-section to be uniform in the x–y plane with periodic boundary conditions. We set up a lattice along the z-direction with a 1D Hamiltonian H_z that looks like Fig. 7.2.4, where we have assumed that both the conduction band edge E_c and the effective mass m_c could be different for the insulator and the channel. However, we will use Eq. (7.2.13) with N_0 (see Eq. (7.2.8)) given by the channel effective mass, since the wavefunctions are strongly excluded from the insulator region.

Once we have set up this Hamiltonian H_z it is straightforward to evaluate the electron density $n(z)$ which can be viewed as the diagonal elements of the 1D density matrix given by

$$\rho(z, z') = \sum_m \phi_m(z) f_{2D}(\varepsilon_m - \mu)\phi_m^*(z') \tag{7.2.14a}$$

As we have discussed before (see discussion following Eq. (4.3.8)) we could view Eq. (7.2.14a) as the real-space representation of a more general matrix relation

$$\rho = f_{2D}(H_z - \mu I) \tag{7.2.14b}$$

As before, the function of a matrix $[H_z]$ is evaluated by firstly diagonalizing $[H_z]$, secondly calculating the density matrix in the eigenrepresentation, and thirdly transforming it back to the real-space lattice. This can be achieved in MATLAB using the set of commands

```
(1) [V, D] = eig(Hz); D = diag (D)
(2) rho = log(1 + (exp((mu-D)./kT)))
(3) rho = V* diag(rho)* V'; N = diag(rho)
```

The electron density $n(z)$ is obtained from N by multiplying by $2N_0$ (see Eq. (7.2.8), with an extra factor of two for spin) and dividing by the size of a unit cell a: $n(z) = N \times 2N_0/a$.

We can use the same lattice to solve the Poisson equation

$$-\frac{d}{dz}\left(\varepsilon_r \frac{dU}{dz}\right) = \frac{q^2}{\varepsilon_0} n(z)$$

which looks just like the Schrödinger equation and can be solved by the method of finite differences in exactly the same way:

$$[D2]\{U\} = \frac{q^2}{\varepsilon_0\, a}(2N_0 a^2)\{N\} + \{U_{\text{bdy}}\} \tag{7.2.15}$$

where $[D2]$ is the matrix operator representing the second derivative. For a constant ε_r,

$$[D2] \equiv \varepsilon_r \begin{bmatrix} 2 & -1 & 0 & 0 & 0 & \cdots \\ -1 & 2 & -1 & 0 & 0 & \cdots \\ 0 & -1 & 2 & -1 & 0 & \cdots \\ \cdots & \cdots & \cdots & \cdots & \cdots & \cdots \end{bmatrix} \tag{7.2.16}$$

Spatial variations in ε_r can be handled in the same way that we handled spatially varying effective masses. The boundary term comes from the non-zero values of U at the two boundaries:

$$\{U_{\text{bdy}}\}^{\text{T}} = \varepsilon_r\{-qV_G \quad 0 \quad \cdots \quad \cdots \quad 0 \quad -qV_G\} \tag{7.2.17}$$

Knowing N, we can calculate the potential U from Eq. (7.2.15):

$$\{U\} = \frac{q^2}{\varepsilon_0 a}(2N_0 a^2)[D2]^{-1}\{N\} + [D2]^{-1}\{U_{\text{bdy}}\} \quad (7.2.18)$$

Figures 7.2.5a, b show the equilibrium band diagram and electron density for a 3 nm wide channel and a 9 nm wide channel respectively. We have assumed that the conduction band edge E_c is zero in the channel and at 3 eV in the insulator. We use a relative dielectric constant $\varepsilon_r = 4$ and an effective mass $m_c = 0.25m$ everywhere. Also, we have assumed that the electrochemical potential μ is equal to E_c. In real structures the proper location of μ is set by the work function of the metal. The thickness of the oxide is assumed to be 2.1 nm and the calculation was done using a discrete lattice with $a = 0.3$ nm. The gate voltage V_G is assumed to be 0.25 V.

For the 3 nm channel, the charge density is peaked near the middle of the channel as we might expect for the wavefunction corresponding to the lowest energy level of a "particle in a box" problem. By contrast, the semi-classical charge density piles up near the edges of the channel as we might expect from purely electrostatic considerations. This is an example of what is referred to as *size quantization*. It disappears as we make the channel wider, since the "particle in a box" levels get closer together and many of them are occupied at low temperatures. Consequently the electron distribution looks more classical for the 9 nm channel. Also shown (in dashed lines) is the electron density if the gate voltage is applied asymmetrically, i.e. 0 V on one gate and 0.25 V on the other gate. Note that for the 9 nm channel there is a significant skewing of the distribution when the bias is applied asymmetrically, as we would expect intuitively. But for the 3 nm channel the electron distribution is only slightly changed from the symmetric to the asymmetric bias. The wavefunction remains relatively unaffected by the applied bias, because the eigenstates are further separated in energy.

Semi-classical method: Figure 7.2.5 also shows a comparison of the electron density with that obtained from a semi-classical approach which works like this. In a homogeneous structure the eigenfunctions are given by

$$\phi_\alpha(\vec{r}) = \frac{\exp[ik_x x]}{\sqrt{L_x}}\frac{\exp[ik_y y]}{\sqrt{L_y}}\frac{\exp[ik_z z]}{\sqrt{L_z}} \quad (7.2.19)$$

so that the electron density obtained from Eq. (7.2.1) after summing over all three indices $\{k_x, k_y, k_z\} \equiv \alpha$ is uniform in space ($\Omega = L_x L_y L_z$):

$$n = 2\,(\text{for spin}) \times \frac{1}{\Omega}\sum_{k_x, k_y, k_z} f_0(\varepsilon_\alpha - \mu)$$

with

$$\varepsilon_\alpha = E_c + \frac{\hbar^2}{2m_c}\left(k_x^2 + k_y^2 + k_z^2\right) \quad (7.2.20)$$

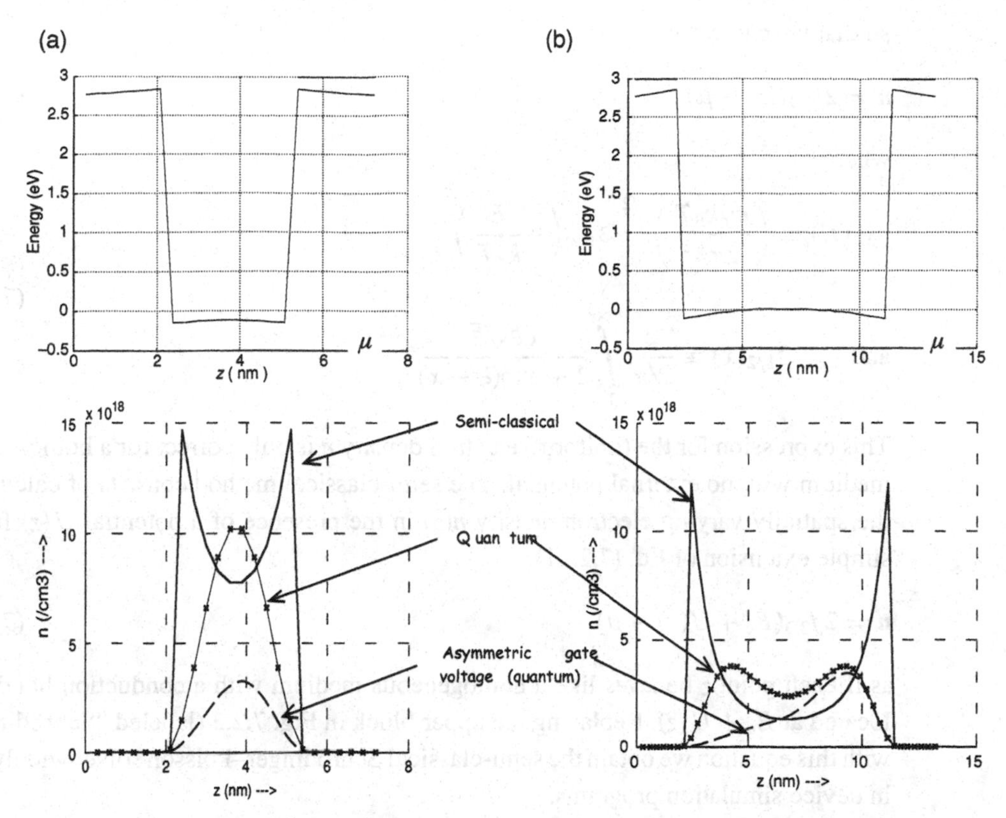

Fig. 7.2.5 An MOS capacitor (see Fig. 7.2.2) with a channel thickness of (a) 3 nm and (b) 9 nm. We assume $\mu = 0$, $E_c = 0$ in the silicon and $E_c = 3$ eV in the oxide. The top figures show the equilibrium band diagram: the solid curves include both the conduction band profile (dashed) and the self-consistent potential. The lower figures show the electron density. The thickness of the oxide is assumed to be 2.1 nm and the calculation was done using a discrete lattice with $a = 0.3$ nm. The gate voltage V_G is assumed to be 0.25 V. The dashed lines show the electron density when the voltage is applied asymmetrically: 0.25 V on one gate, 0 V on the other.

This summation can be performed following the same procedure as described in connection with the 2D version in Eq. (7.2.12):

$$\frac{1}{\Omega}\sum_{k_x,\,k_y,\,k_z} f_0\left(E_c - \mu + \frac{\hbar^2}{2m_c}\left(k_x^2 + k_y^2 + k_z^2\right)\right)$$

$$= \int_0^\infty \frac{4\pi k^2\,\mathrm{d}k}{8\pi^3}\,\frac{1}{1 + A\exp(\hbar^2k^2/2m_c k_B T)} \quad \text{where } A \equiv \exp[(E_c - \mu)/k_B T]$$

$$= \left(\frac{m_c k_B T}{2\pi\hbar^2}\right)^{3/2}\frac{2}{\sqrt{\pi}}\int_0^\infty \frac{\mathrm{d}y\sqrt{y}}{1 + A\,\mathrm{e}^y}$$

so that we can write

$$n = 2f_{3D}(E_c - \mu) \tag{7.2.21}$$

where

$$f_{3D}(E) \equiv \left(\frac{m_c k_B T}{2\pi\hbar^2}\right)^{3/2} \Im_{1/2}\left(-\frac{E}{k_B T}\right)$$

$$\text{and} \quad \Im_{1/2}(x) \equiv \frac{2}{\sqrt{\pi}} \int_0^\infty \frac{d\xi\sqrt{\xi}}{1 + \exp(\xi - x)} \tag{7.2.22}$$

This expression for the (uniform) electron density n is only correct for a homogeneous medium with no external potential. The semi-classical method consists of calculating the spatially varying electron density $n(z)$ in the presence of a potential $U(z)$ from a simple extension of Eq. (7.2.21):

$$n = 2f_{3D}(E_c + U(z) - \mu) \tag{7.2.23}$$

as if each point z behaves like a homogeneous medium with a conduction band edge located at $E_c + U(z)$. Replacing the upper block in Fig. 7.2.3 (labeled "Schrödinger") with this equation we obtain the semi-classical Schrödinger–Poisson solver widely used in device simulation programs.

7.3 Quantum vs. electrostatic capacitance

The electron density in the channel per unit area n_S is obtained by integrating $n(z)$ in Eq. (7.2.13) and noting that the wavefunctions are normalized, that is $\int dz|\phi_m(z)|^2 = 1$:

$$n_S = \int dz\, n(z) = 2\sum_m f_{2D}(\varepsilon_m - \mu) \tag{7.3.1}$$

Figure 7.3.1 shows the electron density n_S as a function of the gate voltage V_G applied symmetrically to both gates.

The basic physics is illustrated in Fig. 7.3.2. A positive gate voltage lowers the overall density of states (DOS) and increases the electron density n_S. As long as the electrochemical potential μ is located below the lowest energy level, the device is in the OFF state. Once μ moves into the energy range with a non-zero DOS the device is in the ON state. Figure 7.3.1 shows that it takes a higher threshold voltage to turn on the device with the 3 nm channel relative to the one with the 9 nm channel. This is because of the increase in the lowest energy level due to size quantization.

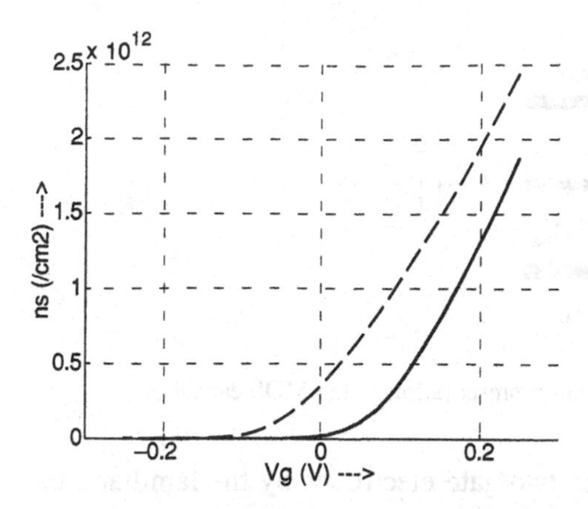

Fig. 7.3.1 Electron density per unit area, n_S in a 3 nm (solid curve) and a 9 nm (dashed curve) channel as a function of the gate voltage V_G applied symmetrically to both gates calculated numerically using the model described in Section 7.2.

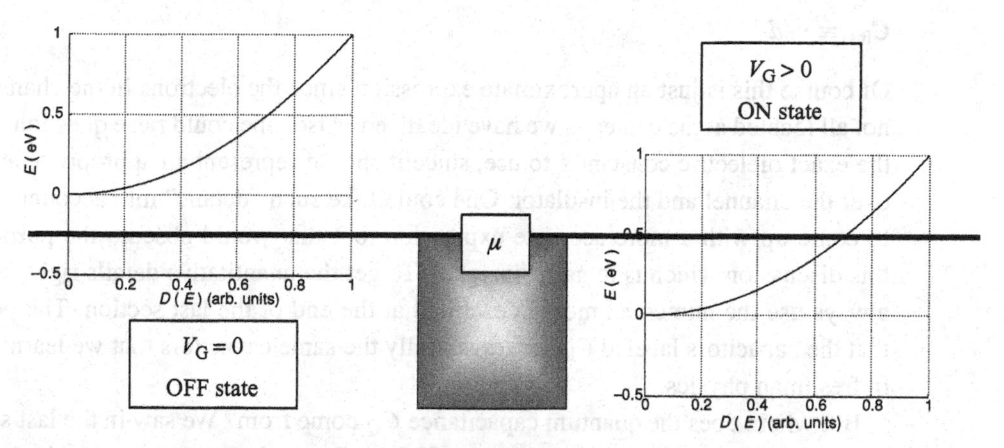

Fig. 7.3.2 A positive voltage V_G applied to the gate moves the density of states $D(E)$ downwards. Since the electrochemical potential μ remains fixed, this increases the number of occupied states and hence the number of electrons N.

Equivalent circuit: An interesting question to ask is the following: how does the potential V_C in the channel change as the gate voltage V_G is changed? It is easy to answer this question in two extreme situations. If the channel is in the OFF state, then it behaves basically like an insulator, and the channel potential V_C is equal to V_G. But if the channel is in the ON state then it behaves like the negative plate of a parallel plate capacitor, so that the channel potential V_C is equal to the source (or drain) potential which we have assumed to be the ground. What is not obvious is the answer in intermediate situations when the channel is neither an insulator nor a conductor. The approximate equivalent circuit shown in Fig. 7.3.3 can be used to answer this question. Let me explain where it comes from.

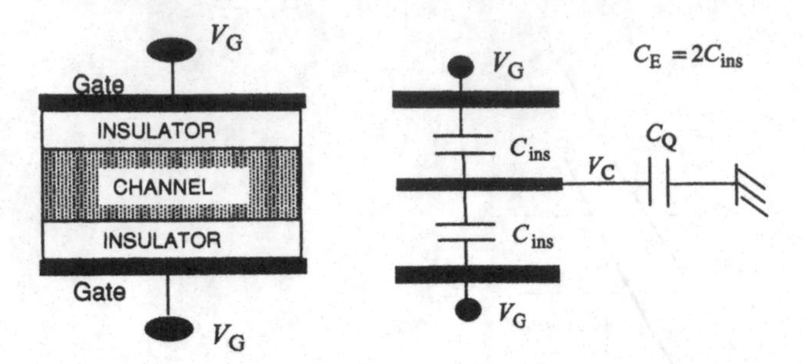

Fig. 7.3.3 Approximate equivalent circuit representation of the MOS capacitor.

The channel is connected to the two gate electrodes by the familiar parallel plate capacitors; the capacitance (per unit area) is proportional to the effective dielectric constant ε and inversely proportional to the distance d between the center of the channel and the gate electrode:

$$C_{ins} \equiv \varepsilon/d \tag{7.3.2}$$

Of course this is just an approximate expression since the electrons in the channel are not all located at the center as we have idealized. Also, one could raise questions about the exact dielectric constant ε to use, since it should represent an appropriate average over the channel and the insulator. One could take such "details" into account and try to come up with a more accurate expression, but that would obscure the purpose of this discussion which is to gain "insight." To get the quantitative details right, we can always use the numerical model described at the end of the last section. The point is that the capacitors labeled C_{ins} are essentially the same capacitors that we learnt about in freshman physics.

But where does the quantum capacitance C_Q come from? We saw in the last section that we have to perform a simultaneous solution of two relations connecting the electron density to the potential (see Fig. 7.2.3): an electrostatic relation (Poisson) and a quantum relation (Schrödinger). The electrostatic relation can be written as

$$U = U_L + [q^2(N - N_0)/C_E] \tag{7.3.3}$$

where U_L is the channel potential obtained from a solution to the *Laplace* equation assuming zero charge, while $(N - N_0)$ tells us the extra electron density relative to the number N_0 required to keep it neutral. The quantum relation can be written as

$$N = \int_{-\infty}^{+\infty} dE\, D(E - U) f_0(E - \mu) \tag{7.3.4}$$

where $D(E - U)$ is the density of states (per unit area) shifted by the potential U. This is a non-linear relation and we could get some insight by linearizing it around an appropriate point. For example, we could define a "*neutral potential*" $U = U_N$, for which $N = N_0$ and keeps the channel exactly neutral:

$$N_0 = \int_{-\infty}^{+\infty} dE\, D(E - U_N) f_0(E - \mu)$$

Any increase in U will raise the energy levels and reduce N, while a decrease in U will lower the levels and increase N. So, for small deviations from the neutral condition, we could write

$$N - N_0 \approx C_Q[U_N - U]/q^2 \tag{7.3.5}$$

where

$$C_Q \equiv -q^2[dN/dU]_{U=U_N} \tag{7.3.6}$$

is called the quantum capacitance and depends on the density of states. We can substitute this linearized relation into Eq. (7.3.3) to obtain

$$U = U_L + \frac{C_Q}{C_E}(U_N - U)$$

$$\rightarrow U = \frac{C_E U_L + C_Q U_N}{C_E + C_Q} \tag{7.3.7}$$

Equation (7.3.7) is easily visualized in terms of the capacitive network shown in Fig. 7.3.3. The actual channel potential U is intermediate between the Laplace potential U_L and the neutral potential U_N. How close it is to one or the other depends on the relative magnitudes of the electrostatic capacitance C_E and the quantum capacitance C_Q.

From Eqs. (7.3.4) and (7.3.6) it is straightforward to show that the quantum capacitance C_Q is proportional to the DOS averaged over a few k_BT around μ:

$$C_Q \equiv q^2 D_0 \tag{7.3.8}$$

$$D_0 \equiv \int_{-\infty}^{+\infty} dE\, D(E - U_N) F_T(E - \mu) \tag{7.3.9}$$

where $F_T(E)$ is the thermal broadening function defined as

$$F_T(E) \equiv -\frac{df_0}{dE} = \frac{1}{4k_BT} \operatorname{sech}^2\left(\frac{E}{2k_BT}\right) \tag{7.3.10}$$

Figure 7.3.4 shows a sketch of the thermal broadening function; its maximum value is $(1/4k_BT)$ while its width is proportional to k_BT; it is straightforward to show that the area obtained by integrating this function is equal to one, independent of k_BT: $\int_{-\infty}^{+\infty} dE\, F_T(E) = 1$. This means that at low temperatures $F_T(E)$ becomes very large but

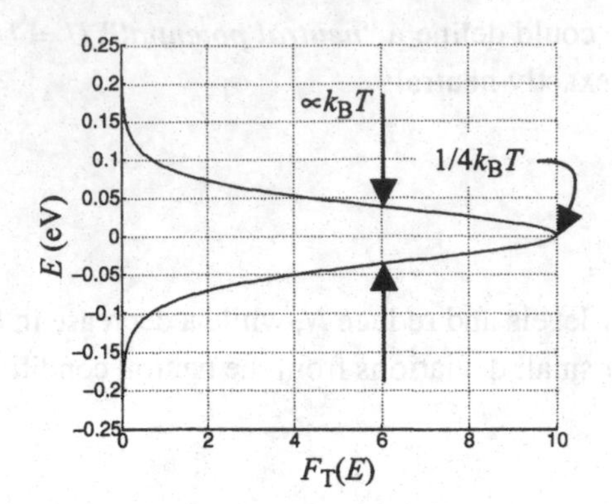

Fig. 7.3.4 Plot of the thermal broadening function $F_T(E)$ (Eq. (7.3.10)) with $k_BT = 0.025$ eV.

very narrow while maintaining a constant area of one. Such a function can be idealized as a delta function: $F_T(E) \to \delta(E)$, which allows us to simplify the expression for the quantum capacitance at low temperatures

$$C_Q \approx q^2 D(E = \mu + U_N) \tag{7.3.11}$$

showing that it is proportional to the density of states around the electrochemical potential μ after shifting by the neutral potential U_N.

It is easy to see from the equivalent circuit in Fig. 7.3.3 that

$$V_C = V_G \frac{C_E}{C_E + C_Q} \quad \text{where} \quad C_E = 2C_{ins} \tag{7.3.12}$$

Devices in the OFF state have zero C_Q, so that $V_C = V_G$. But in the ON state, C_Q is non-zero and V_C is thus smaller than V_G. The measured capacitance C is the series combination of the electrostatic and quantum capacitances

$$C = \frac{C_E C_Q}{C_E + C_Q} \tag{7.3.13}$$

and is dominated by the smaller of the two.

We can get an approximate feeling for the magnitude of the quantum capacitance C_Q in the ON state by noting that a 2D conductor described by a parabolic dispersion relation has a constant DOS (Table 6.2.1): $D(E) = Sm_c/\pi\hbar^2$, so that we can write the quantum capacitance approximately as

$$C_Q = \frac{q^2 m_c S}{\pi \hbar^2} = \frac{\varepsilon S}{a_0^*/4} \tag{7.3.14}$$

where a_0^* is given by an expression similar to that for the Bohr radius that we defined in Chapter 2 (see Eq. (2.1.5)). But it is larger than the Bohr radius (= 0.053 nm) because the conduction band effective mass (m_c) is smaller than the free electron mass (m) and the dielectric constant (ε) is larger than that for free space (ε_0):

$$a_0^* \equiv \frac{4\pi\varepsilon\hbar^2}{m_c q^2} = (0.053\,\text{nm})\frac{m}{m_c}\frac{\varepsilon}{\varepsilon_0} \tag{7.3.15}$$

The quantum capacitance is thus equal to that of a parallel plate capacitor whose plates are separated by $a_0^*/4$. Since this is usually a very small number, the quantum capacitance in the ON state is typically well in excess of the electrostatic capacitance C_E and the measured capacitance is dominated by the latter (see Eq. (7.3.13)). But in materials with a small effective mass, the quantum capacitance can be small enough to have a significant effect, especially if the insulator is very thin making C_E large.

OFF regime: In the OFF regime, the density of states close to $E = \mu$ is negligible, and so is the quantum capacitance C_Q. Consequently the channel potential V_C is essentially equal to the gate voltage V_G, so that we can write

$$N = \int_{-\infty}^{+\infty} dE\, f_0(E - \mu - qV_G)D(E)$$

$$\approx \int_{-\infty}^{+\infty} dE\, \exp\left(-\frac{E - \mu - qV_G}{k_B T}\right) D(E)$$

since $E - \mu - qV_G \gg k_B T$ in the energy range where $D(E)$ is non-zero. In this regime, the number of electrons changes exponentially with gate voltage:

$$N \sim N_0 \exp\left(\frac{qV_G}{k_B T}\right)$$

so that

$$\log_{10}\left(\frac{N}{N_0}\right) \approx \left(\frac{qV_G}{2.3\,k_B T}\right) \tag{7.3.16}$$

This is basically the well-known result that in the OFF regime, the number of electrons increases by a decade (i.e. a factor of ten) for every $2.3k_BT$ ($\sim 60\,\text{meV}$ at room temperature) increase in the gate voltage. This relation can be verified by replotting Fig. 7.3.1 on a logarithmic scale and looking at the slope in the OFF regime.

ON regime: In the ON regime, the electrochemical potential μ lies well inside the band of states where $D(E)$ (and hence the quantum capacitance C_Q) is significant. The actual capacitance is a series combination of the quantum and electrostatic capacitances as explained above. From the slope of the n_S vs. V_G curve in the ON region (see Fig. 7.3.1), we deduce a capacitance of approximately $1.8 \times 10^{-6}\,\text{F/cm}^2$ for the 3 nm channel. If we equate this to $2\varepsilon/d$, we obtain $d = 3.9\,\text{nm}$, which compares well with the number obtained by adding half the channel width (1.5 nm) to the oxide thickness (2.1 nm), showing that the effective capacitance is largely electrostatic (rather than quantum) in origin.

7.4 Supplementary notes: multi-band effective mass Hamiltonian

The one-band effective mass model works very well when we have an isotropic parabolic band that is well separated from the other bands. This is usually true of the conduction band in wide-bandgap semiconductors. But the valence band involves multiple closely spaced bands that are strongly anisotropic and non-parabolic. Close to the Γ point the energy dispersion can usually be expressed in the form (*A*, *B*, and *C* are constants)

$$E(\vec{k}) = E_v - Ak^2 \pm \sqrt{B^2k^4 + C^2(k_x^2k_y^2 + k_y^2k_z^2 + k_z^2k_x^2)} \tag{7.4.1}$$

This dispersion relation can be described by a 4 × 4 matrix of the form (*I* is a 4 × 4 identity matrix)

$$[h(\vec{k})] = -PI - T \tag{7.4.2}$$

where

$$[T] \equiv \begin{bmatrix} Q & 0 & -S & R \\ 0 & Q & R^+ & S^+ \\ -S^+ & R & -Q & 0 \\ R^+ & S & 0 & -Q \end{bmatrix}$$

$$P \equiv E_v + \frac{\hbar^2\gamma_1}{2m}(k_x^2 + k_y^2 + k_z^2)$$

$$Q \equiv \frac{\hbar^2\gamma_2}{2m}(k_x^2 + k_y^2 - 2k_z^2)$$

$$R \equiv \frac{\hbar^2}{2m}\left[-\sqrt{3}\gamma_2(k_x^2 - k_y^2) + i\,2\sqrt{3}\gamma_3 k_x k_y\right]$$

$$S \equiv \frac{\hbar^2\gamma_3}{2m}2\sqrt{3}(k_x - i k_y)k_z$$

The Luttinger parameters γ_1, γ_2, and γ_3 are available in the literature for all common semiconductors (see, for example, Lawaetz (1971)). One can argue that Eqs. (7.4.1) and (7.4.2) are equivalent since it can be shown using straightforward algebra that

$$[T]^2 = (Q^2 + R^2 + S^2)I = \left[B^2k^4 + C^2(k_x^2k_y^2 + k_y^2k_z^2 + k_z^2k_x^2)\right]I$$

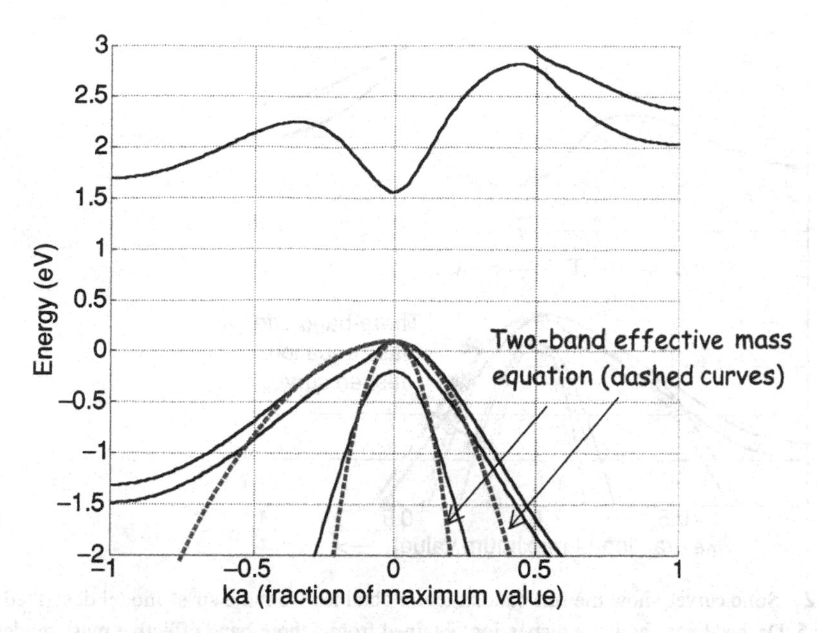

Fig. 7.4.1 Solid curves show the full bandstructure obtained from the sp^{3}s* model described in Chapter 5. Dashed curves show the dispersion obtained from a two-band effective mass model (Eq. (7.4.2)) with parameters adjusted for best fit.

It can be seen from Fig. 7.4.1 that the eigenvalues of $[h(\vec{k})]$ describe well the two highest valence bands (light hole and heavy hole) very close to the Γ point. To get better agreement over a wider range of k-values and to include the split-off band (see Fig. 7.4.2), one often uses a three-band $[h(\vec{k})]$ of the form

$$[h(\vec{k})] = -\begin{bmatrix} P+Q & 0 & -S & R & -S/\sqrt{2} & \sqrt{2}R \\ 0 & P+Q & R^+ & S^+ & -\sqrt{2}R^+ & -S^+/\sqrt{2} \\ R^+ & 0 & P-Q & 0 & -\sqrt{2}Q & \sqrt{3/2}\,S \\ 0 & R^+ & 0 & P-Q & \sqrt{3/2}S^+ & \sqrt{2}Q^+ \\ -S^+/\sqrt{2} & -\sqrt{2}R & -\sqrt{2}Q^+ & \sqrt{3/2}S & P+\Delta & 0 \\ \sqrt{2}R^+ & -S/\sqrt{2} & \sqrt{3/2}S^+ & \sqrt{2}Q & 0 & P+\Delta \end{bmatrix} \tag{7.4.3}$$

We can use either the two-band $[h(\vec{k})]$ in Eq. (7.4.2) or the three-band $[h(\vec{k})]$ in Eq. (7.4.3) to construct an effective mass equation for the valence band using the same principle that we used for the conduction band (cf. Eqs. (7.1.2), (7.1.3)). We simply replace $\vec{k}$ with $-\mathrm{i}\vec{\nabla}$ in the expression for $h(\vec{k})$ to obtain a coupled differential equation of the form

$$[h(\vec{k} \to -\mathrm{i}\vec{\nabla})]\{f(\vec{r})\} = E\{f(\vec{r})\} \tag{7.4.4}$$

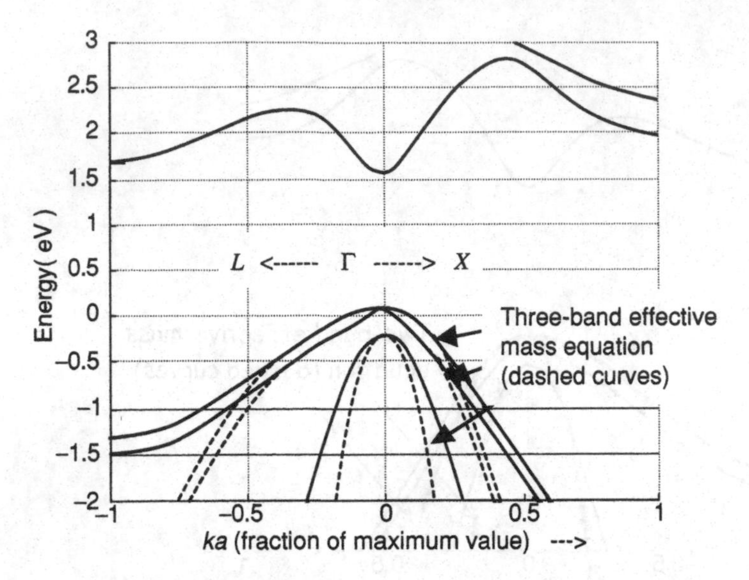

Fig. 7.4.2 Solid curves show the full bandstructure obtained from the sp^3s^* model described in Chapter 5. Dashed lines show the dispersion obtained from a three-band effective mass model (Eq. (7.4.3)) with parameters adjusted for best fit.

where the "wavefunction" $\{f(\vec{r})\}$ now has four (or six) components. It is easy to check that plane wave solutions of the form $\{f(\vec{r})\} = \{f_0\}\exp(i\vec{k}\cdot\vec{r})$ with any $\vec{k}$ will satisfy Eq. (7.4.4) provided $\{f_0\}$ is an eigenfunction of $[h(\vec{k})]$

$$[h(\vec{k})]\{f_0\} = E\{f_0\} \tag{7.4.5}$$

This means that the effective mass equation in Eq. (7.4.4) will generate a bandstructure identical to that obtained from the original $[h(\vec{k})]$.

We could use the finite difference method to convert Eq. (7.4.4) into a Hamiltonian matrix, the same way we went from Eq. (7.1.3) to the matrix depicted in Fig. 7.4.2. In Fig. 7.4.3, the unit cell matrices $[H_{nn}]$ and the bond matrices $[H_{nm}, n \neq m]$ will all be (4×4) or (6×6) matrices depending on whether we start from the two-band (Eq. (7.4.2)) or the three-band (Eq. (7.4.3)) model. For example, the two-band model leads to matrices $[H_{nm}]$ of the form

$$[H_{nm}] \equiv \begin{bmatrix} -P_{nm} - Q_{nm} & 0 & S_{nm} & -R_{nm} \\ 0 & -P_{nm} - Q_{nm} & -R^*_{mn} & -S^*_{mn} \\ S^*_{mn} & -R_{nm} & -P_{nm} + Q_{nm} & 0 \\ -R^*_{mn} & -S_{nm} & 0 & -P_{nm} + Q_{nm} \end{bmatrix}$$

where the individual terms P_{nm}, Q_{nm}, etc. are obtained from the corresponding functions $P(\vec{k})$, $Q(\vec{k})$ using the same procedure that we used to obtain a one-band effective mass Hamiltonian (see Fig. 7.1.3) from the one-band $h(\vec{k})$ (see Eq. (7.1.2)).

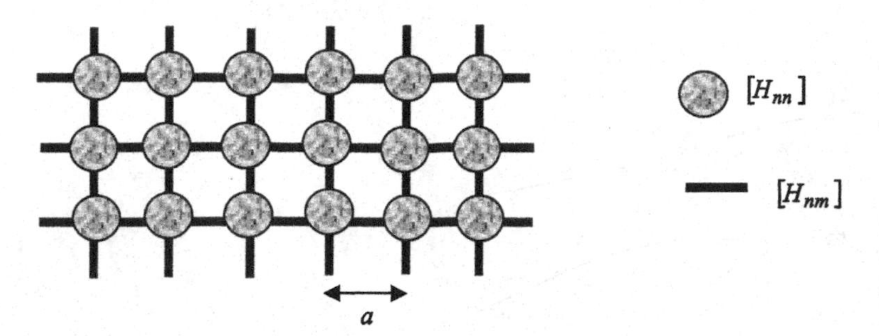

Fig. 7.4.3 The multi-band effective mass Hamiltonian matrix can be depicted schematically as a 3D network of unit cells (unrelated to the actual crystal structure) described by $[H_{nn}]$ bonded to its nearest neighbors by $[H_{nm}]$. These matrices will be (4×4) or (6×6) depending on whether we start from the two-band (Eq. (7.4.2)) or the three-band (Eq. (7.4.3)) model.

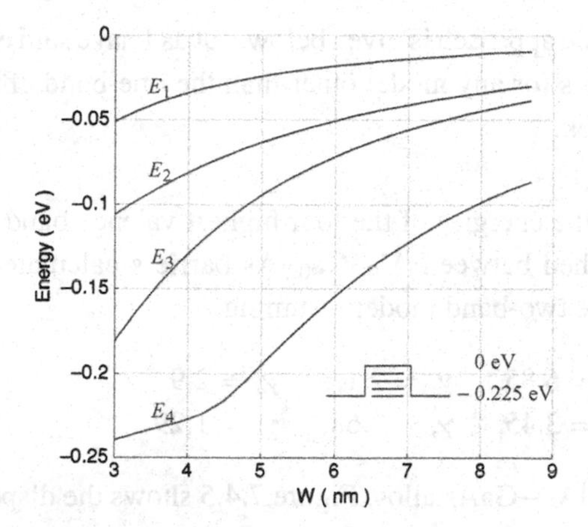

Fig. 7.4.4 Energy of the four highest energy levels of a GaAs quantum well sandwiched between $Al_{0.3}Ga_{0.7}As$ barriers (shown in inset) as a function of the well width W calculated from a two-band effective mass model.

The same approach is used to write down the $[H_{nm}]$ matrices for the three-band model (Eq. (7.4.3)). For narrow-gap semiconductors, it is common to use a four-band model where the matrices $[H_{nm}]$ are (8×8) in size. Multi-band effective mass models may not appear to represent much of a simplification relative to an atomistic model like the sp^3s^* model. However, the simplification (numerical and even conceptual) can be considerable for two reasons:

(1) the matrices $[H_{nm}]$ are somewhat smaller (cf. (20×20) for the sp^3s^* model)
(2) the lattice can be much coarser and can have a simpler structure (simple cubic rather than FCC) than the real atomic lattice, resulting in a smaller overall Hamiltonian that is also easier to visualize.

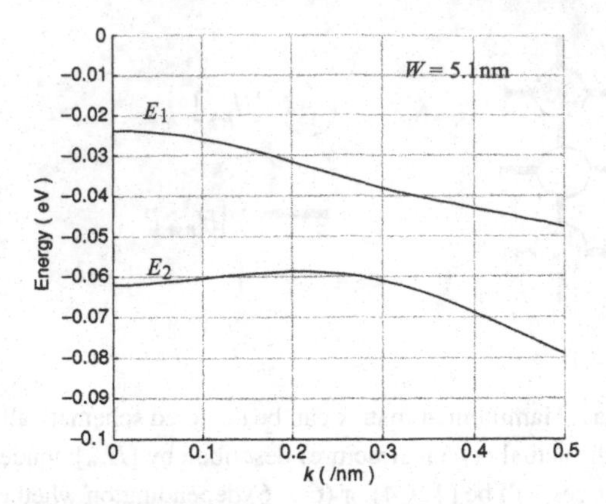

Fig. 7.4.5 The dispersion relation $E(\vec{k})$ as a function of k_y with $k_x = 0$ for the two highest subbands of a quantum well with $W = 5.1$ nm, calculated from the two-band effective mass model.

A short example to illustrate the basic approach is given below, but as I have said earlier, I will not discuss multi-band models (or any model other than the one-band effective mass model) any further in this book.

An example: Figure 7.4.4 shows the energies of the four highest valence band levels of a GaAs quantum well sandwiched between $Al_{0.3}Ga_{0.7}As$ barriers calculated as a function of the well width using the two-band model assuming

for GaAs:	$E_v = 0\,\text{eV}$,	$\gamma_1 = 6.85$,	$\gamma_2 = 2.1$,	$\gamma_3 = 2.9$
for AlAs:	$E_v = 0.75\,\text{eV}$,	$\gamma_1 = 3.45$,	$\gamma_2 = 0.68$,	$\gamma_3 = 1.29$

and interpolating linearly for the AlAs–GaAs alloy. Figure 7.4.5 shows the dispersion relation $E(\vec{k})$ as a function of k_y with $k_x = 0$ for the four highest valence subbands of a quantum well of width $W = 5.1$ nm.

EXERCISES

E.7.1.

(a) Plot $E(k)$ along Γ–X and Γ–L from

$$h(\vec{k}) = E_c + \frac{\hbar^2 k^2}{2m_c}$$

and compare with the plot from the sp^3s^* model (see Exercise E.5.2) over the appropriate energy and wavevector range (cf. Fig. 6.1.2). What values of E_c and m_c give the best fit?

(b) Use a one-band effective mass model to calculate the energies of the two lowest levels of a GaAs quantum well sandwiched between $Al_{0.3}Ga_{0.7}As$ barriers as a function of the well width. Assume that for GaAs, $E_c = 0$ eV, $m_c = 0.07m$; and for

AlAs, $E_c = 1.25$ eV, $m_c = 0.15m$ and interpolate linearly to obtain E_c and m_c for the AlAs–GaAs alloy (cf. Fig. 7.1.5).

(c) Use a one-band model to calculate the dispersion relation $E(\vec{k})$ as a function of the magnitude of the in-plane wavevector $\vec{k} = \{k_x \quad k_y\}$ for the two lowest subbands of a quantum well with $W = 6.9$ nm, using the same parameters as in part (b) above (cf. Fig. 7.1.6).

E.7.2.

(a) Consider the MOS capacitor shown in Fig. 7.2.2 and calculate the self-consistent conduction band profile and the electron density using a discrete lattice with $a = 0.3$ nm. Assume that: (1) the thickness of the oxide is 2.1 nm and the channel thickness is 3 nm; (2) $\mu = 0$, $E_c = 0$ in the silicon and $E_c = 3$ eV in the oxide; (3) dielectric constant $\varepsilon = 4\varepsilon_0$ and $m_c = 0.25m$ everywhere; and (4) the gate voltage $V_G = 0.25$ V. Repeat with a channel thickness of 9 nm and also with the gate voltage applied asymmetrically with 0 V on one gate and 0.25 V on the other gate. Compare with Fig. 7.2.5.

(b) Calculate the electron density per unit area as a function of the gate voltage (applied symmetrically to both gates) for the structure with a 3 nm channel and with a 9 nm channel. Compare with Fig. 7.3.1. Calculate the effective capacitance from the slope of the curve in the ON state and deduce an effective plate separation d by equating the capacitance to $2\varepsilon/d$.

E.7.3.

(a) Plot $E(k)$ along Γ–X and Γ–L from the two-band model (see Eq. (7.4.2)) and compare with the plot from the sp^{3}s* model (see Exercise E.5.3) over the appropriate energy and wavevector range (cf. Fig. 7.4.1). Use $\gamma_1 = 6.85$, $\gamma_2 = 2.1$, $\gamma_3 = 2.9$.

(b) Use the two-band model to calculate the energies of the four highest valence band levels of a GaAs quantum well sandwiched between $Al_{0.3}Ga_{0.7}As$ barriers as a function of the well width. Assume that for GaAs: $E_v = 0$ eV, $\gamma_1 = 6.85$, $\gamma_2 = 2.1$, $\gamma_3 = 2.9$ and for AlAs: $E_v = -0.75$ eV, $\gamma_1 = 3.45$, $\gamma_2 = 0.68$, $\gamma_3 = 1.29$ and interpolate linearly for the AlAs–GaAs alloy. Compare with Fig. 7.4.4.

(c) Use the two-band model to calculate the dispersion relation $E(\vec{k})$ as a function of k_y with $k_x = 0$ for the four highest valence subbands of a quantum well with $W = 5.1$ nm, using the same parameters as in part (b). Compare with Fig. 7.4.5.

E.7.4. A small device described by a (2 × 2) Hamiltonian matrix

$$H = \begin{bmatrix} 0 & t_0 \\ t_0 & 0 \end{bmatrix}$$

is in equilibrium with a reservoir having an electrochemical potential μ with temperature $T = 0$. Write down the density matrix if (i) $\mu = 0$, (ii) $\mu = 2$ assuming (a) $t_0 = +1$ and (b) $t_0 = -1$.

E.7.5. A metallic nanotube of radius a has only one subband in the energy range of interest, whose dispersion relation is given by $\varepsilon(k) = \hbar v k$, v being a constant and k being the wavenumber along the length of the nanotube. Assume that (i) the electrochemical potential $\mu = 0$ and (ii) a coaxial gate of radius b is wrapped around the nanotube such that the electrostatic capacitance $C_E = 2\pi\varepsilon_r\varepsilon_0/\ln(b/a)$.

(a) How is the change (from the equilibrium value) in the electron density per unit length related to the gate voltage V_G?

(b) Would this make a good transistor?

E.7.6. Write down the appropriate effective mass equation if the dispersion relation (in 1D) is given by

$$E = E_c + \frac{\hbar^2 k^2}{2m_c} + \alpha k^4$$

8 Level broadening

In Chapter 1, we saw that current flow typically involves a channel connected to two contacts that are out of equilibrium with each other, having two distinct electrochemical potentials. One contact keeps filling up the channel while the other keeps emptying it causing a net current to flow from one contact to the other. In the next chapter we will take up a quantum treatment of this problem. My purpose in this chapter is to set the stage by introducing a few key concepts using a simpler example: *a channel connected to just one contact* as shown in Fig. 8.1.

Since there is only one contact, the channel simply comes to equilibrium with it and there is no current flow under steady-state conditions. As such this problem does not involve the additional complexities associated with multiple contacts and non-equilibrium conditions. This allows us to concentrate on a different physics that arises simply from connecting the channel to a large contact: the set of discrete levels broadens into a continuous density of states as shown on the right-hand side of Fig. 8.1.

In Chapter 1 I introduced this broadening without any formal justification, pointing out the need to include it in order to get the correct value for the conductance. My objective in this chapter is to provide a quantum mechanical treatment whereby the broadening will arise naturally along with the "uncertainty" relation $\gamma = \hbar/\tau$ connecting it to the escape rate $1/\tau$ for an electron from the channel into the contact. Moreover, we will see that in general the broadening is not just a number γ as we assumed, but a matrix $[\Gamma]$ of the same size as the Hamiltonian matrix $[H]$, which can be energy dependent (unlike $[H]$).

I will start in Section 8.1 from a Schrödinger equation describing the entire system, channel + contact (see Fig. 8.2):

$$E\begin{Bmatrix}\psi\\ \Phi\end{Bmatrix} = \begin{bmatrix}H & \tau\\ \tau^+ & H_R\end{bmatrix}\begin{Bmatrix}\psi\\ \Phi\end{Bmatrix} \tag{8.1}$$

and show that the contact (or reservoir, of size R) can be eliminated to obtain an equation for only the channel that has the form

$$E\{\psi\} = [H + \Sigma]\{\psi\} + \{S\} \tag{8.2}$$

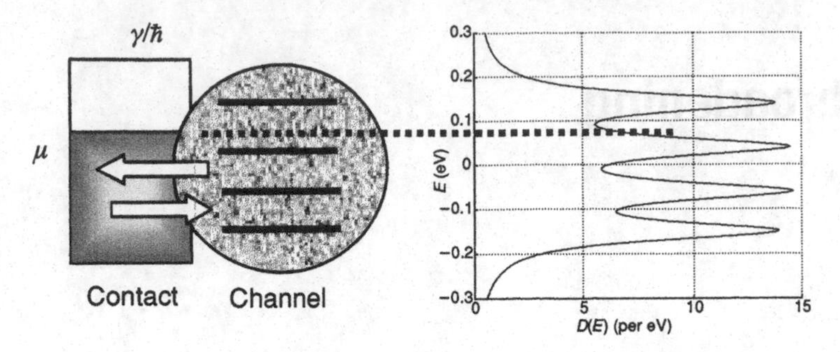

Fig. 8.1 A channel connected to one contact. The set of discrete levels broaden into a continuous density of states as shown.

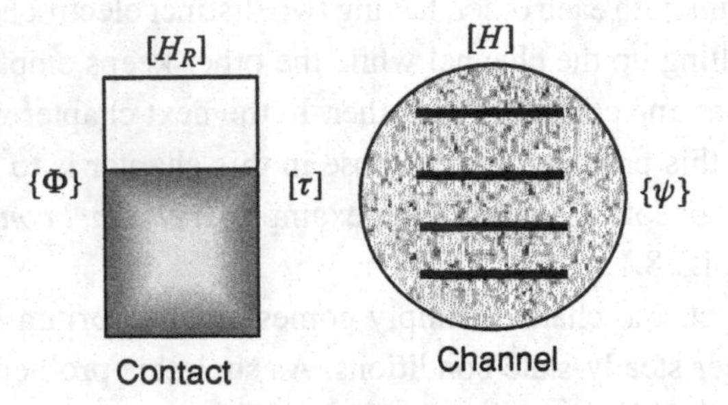

Fig. 8.2 A channel described by $[H]$ is connected through $[\tau]$ to a contact described by $[H_R]$. We can write an equation for the channel alone that has the form shown in Eq. (8.2).

This is the central result that we will use as a starting point in Section 8.1 when we discuss current flow between two contacts. Here $\{S\}$ is a "source" term representing the excitation of the channel by electron waves from the contact, while the *self-energy matrix* $[\Sigma]$ could be viewed as a modification of the Hamiltonian $[H]$ so as to incorporate the "boundary conditions," in somewhat the same way that we added a couple of terms to $[H]$ to account for the periodic boundary conditions (see Eq. (1.3.3)).

However, there are two factors that make $[\Sigma]$ much more than a minor modification to $[H]$. Firstly, $[\Sigma]$ is energy dependent, which requires a change in our viewpoint from previous chapters where we viewed the system as having resonant energies given by the eigenvalues of $[H]$. Since $[\Sigma]$ is energy dependent, we would need to find each of the eigenvalues ε_n iteratively, so that it is an eigenvalue of $[H + \Sigma(E = \varepsilon_n)]$. It is more convenient to think of the energy E as an *independent variable* and look for the response of the device to incident electrons with different energies and that is the viewpoint we will adopt from hereon.

Note that this represents a significant departure from the viewpoint we have held so far, regardless of whether the $[\Sigma]$ is energy-dependent or not. In the past, we have been asking for the eigenenergies of an isolated channel, which is analogous to finding the resonant frequencies of a guitar string (see Fig. 2.1.3). But now we wish to know how the "string" responds when it is excited by a tuning fork with any given frequency, that is, how our channel responds when excited by an electron of any given energy from the contact.

The second distinguishing feature of the self-energy Σ is that, unlike $[H]$, it is NOT hermitian and so the eigenvalues of $H + \Sigma$ are complex. Indeed the anti-hermitian part of Σ

$$\Gamma = \mathrm{i}[\Sigma - \Sigma^+] \tag{8.3}$$

can be viewed as the matrix version of the broadening γ introduced earlier for a one-level device and in Section 8.2 we will relate it to the broadened density of states in the channel. In Section 8.3, we will relate the broadening to the finite lifetime of the electronic states, reflecting the fact that an electron introduced into a state does not stay there forever, but leaks away into the contact.

You might wonder how we managed to obtain a non-hermitian matrix $[H + \Sigma]$ out of the hermitian matrix in Eq. (8.1). Actually we do not really start from a hermitian matrix: we add an infinitesimal quantity $\mathrm{i}0^+$ to the reservoir Hamiltonian H_R making it a "tiny bit" non-hermitian. This little infinitesimal for the reservoir gives rise to a finite broadening Γ for the channel whose magnitude is independent of the precise value of 0^+. But this seemingly innocuous step merits a more careful discussion, for it essentially converts a reversible system into an irreversible one. As we will explain in Section 8.4, it also raises deeper questions about how large a system needs to be in order to function as a reservoir that leads to irreversible behavior.

In this chapter we are using the concept of self-energy to account for the contacts (like the source and drain) to the channel. However, the concept of self-energy is far more general and can be used to describe all kinds of interactions (reversible and irreversible) with the surroundings and not just the contacts. Indeed this is one of the seminal concepts of many-body physics that is commonly used to describe complicated interactions, compared to which our problem of contacts is a relatively trivial one that could be treated with more elementary methods, though not quite so "elegantly." My objective, however, is not so much to provide an elegant treatment of a simple problem, as to introduce a deep and profound concept in a simple context. In Chapter 10 we will extend this concept to describe less trivial "contacts," like the interaction with photons and phonons.

8.1 Open systems

Our objective in this section is to obtain an equation of the form (see Eq. (8.2))

$$E\{\psi\} = [H + \Sigma]\{\psi\} + \{S\} \tag{8.1.1}$$

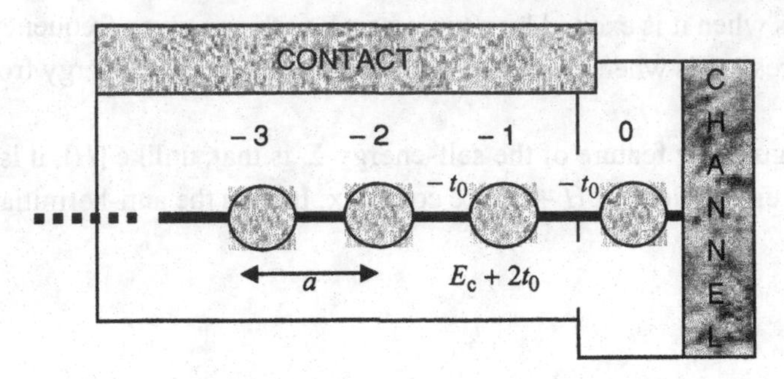

Fig. 8.1.1 Toy example: a semi-infinite wire described by a one-band effective mass Hamiltonian. The first point "0" is treated as the channel and the rest as the contact.

describing an open system, unlike the equation $E\{\psi\} = [H]\{\psi\}$ that we have been using for closed systems so far. The basic idea is easy to see using a simple toy example.

Toy example: Consider a semi-infinite 1D wire described by a one-band effective mass Hamiltonian of the form shown in Fig. 8.1.1. Let us treat the first point of the wire labeled "0" as our channel and the rest of the wire labeled n, $n < 0$, as the contact.

If the "channel" were decoupled from the "contact" it would be described by the equation:

$$E\psi = \underbrace{(E_c + 2t_0)\psi}_{H\psi}$$

Once we couple it to the "contact" this equation is modified to

$$E\psi = (E_c + 2t_0)\psi - t_0\Phi_{-1} \tag{8.1.2}$$

where the contact wavefunctions Φ_n satisfy an infinite series of equations ($n < 0$)

$$E\Phi_n = -t_0\Phi_{n-1} + (E_c + 2t_0)\Phi_n - t_0\Phi_{n+1} \tag{8.1.3}$$

Now, since the equations in this infinite set all have the same structure we can use the basic principle of bandstructure calculation (see Eq. (5.2.4)) to write the solutions in the form of plane waves, labeled by k. Assuming the solution to consist of an incident wave from the contact and a reflected wave back from the channel, we can write

$$\Phi_n = B\exp(+ikna) + C\exp(-ikna) \tag{8.1.4}$$

where

$$E = E_c + 2t_0(1 - \cos ka) \tag{8.1.5}$$

Using Eq. (8.1.4) we can write

$$\psi \equiv \Phi_0 = B + C$$

and

$$\Phi_{-1} = B\exp(-ika) + C\exp(+ika)$$

so that

$$\Phi_{-1} = \psi\exp(+ika) + B[\exp(-ika) - \exp(+ika)]$$

Substituting back into Eq. (8.1.2) we obtain

$$E\psi = \underbrace{(E_c + 2t_0)\psi}_{H\psi} - \underbrace{t_0\exp(+ika)\psi}_{\Sigma\psi} + \underbrace{t_0 B[\exp(+ika) - \exp(-ika)]}_{S} \tag{8.1.6}$$

which has exactly the form we are looking for with

$$\Sigma = -t_0\exp(+ika) \tag{8.1.7a}$$

and

$$S = \mathrm{i}t_0\,2B\sin ka \tag{8.1.7b}$$

Note that the self-energy Σ is non-hermitian and is independent of the amplitudes B, C of the contact wavefunction. It represents the fact that the channel wavefunction can leak out into the contact. The source term S, on the other hand, represents the excitation of the channel by the contact and is proportional to B. Let us now go on to a general treatment with an arbitrary channel connected to an arbitrary contact.

General formulation: Consider first a channel with no electrons which is disconnected from the contacts as shown in Fig. 8.1.2a. The electrons in the contact have wavefunctions $\{\Phi_R\}$ that obey the Schrödinger equation for the isolated contact

$$[EI_R - H_R]\{\Phi_R\} = \{0\}$$

where $[H_R]$ is the Hamiltonian for the contact and $[I_R]$ is an identity matrix of the same size as $[H_R]$. Let me modify this equation to write

$$[EI_R - H_R + \mathrm{i}\eta]\{\Phi_R\} = \{S_R\} \tag{8.1.8a}$$

where $[\eta] = 0^+[I_R]$ is a small positive infinitesimal times the identity matrix, whose significance we will discuss further in Section 8.4. At this point, let me simply note that the term $\mathrm{i}[\eta]\{\Phi_R\}$ on the left of Eq. (8.1.8a) represents the extraction of electrons from the contact while the term $\{S_R\}$ on the right of Eq. (8.1.8a) represents the reinjection of electrons from external sources: such extraction and reinjection are essential to maintain the contact at a constant electrochemical potential.

The results we will derive next are independent of the term $\{S_R\}$ that I have introduced in Eq. (8.1.8a). Indeed, it might appear that we could set it equal to the term $\mathrm{i}[\eta]\{\Phi_R\}$, thereby reducing Eq. (8.1.8a) to the Schrödinger equation $[EI_R - H_R]\{\Phi_R\} = \{0\}$.

(a)

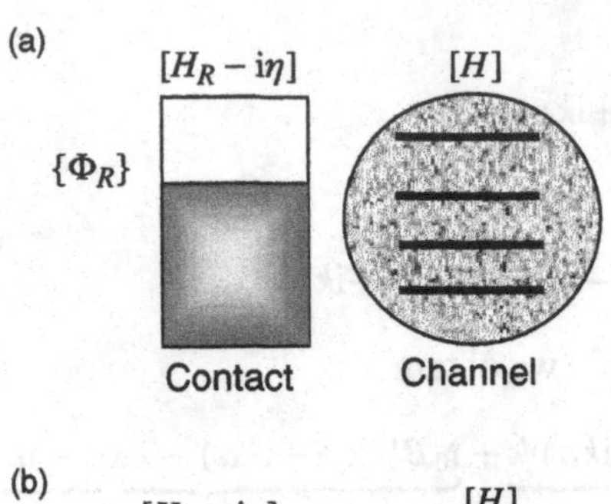

(b)

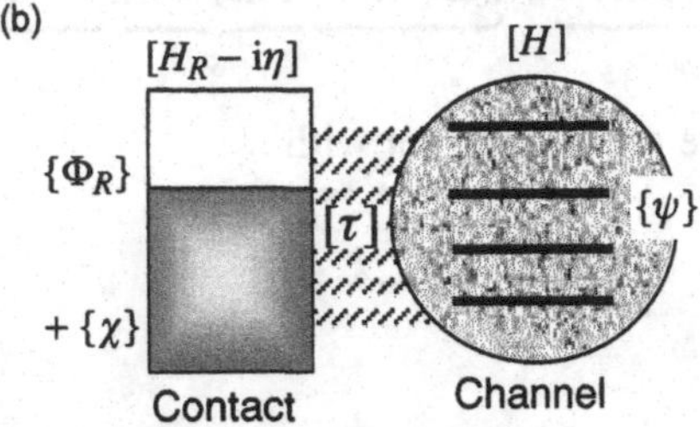

Fig. 8.1.2 (a) Channel contains no electrons and is disconnected from the contact where the electrons occupy the states described by $\{\Phi_R\}$. (b) On connecting to the contact, the contact wavefunctions $\{\Phi_R\}$ "spill over" into the device giving rise to a wavefunction $\{\psi\}$ in the channel which in turn generates a scattered wave $\{\chi\}$ in the contact.

However, as I mentioned in the introduction, the transition from the Schrödinger equation to Eq. (8.1.8a) represents a fundamental change in viewpoint: E is no longer an eigenenergy, but an independent variable representing the energy of excitation from external sources. With the Schrödinger equation, the $\{\Phi_R\}$ are essentially the eigenfunctions of $[H_R]$ that are non-zero only when the energy E matches one of the eigenenergies of $[H_R]$. On the other hand, the $\{\Phi_R\}$ in Eq. (8.1.8a) are non-zero for all energies E with peaks around the eigenenergies of $[H_R]$, whose sharpness depends on the infinitesimal 0^+.

When we couple the channel to the contact as shown in Fig. 8.1.2b, the contact wavefunctions will "spill over" giving rise to a wavefunction $\{\psi\}$ inside the device which in turn will excite scattered waves $\{\chi\}$. The overall wavefunction will satisfy the composite Schrödinger equation for the composite contact–device system, which we can write in two blocks:

$$\begin{array}{l} \\ \text{contact} \\ \text{device} \end{array} \begin{array}{c} \begin{array}{cc} \text{contact} & \text{device} \end{array} \\ \begin{pmatrix} EI_R - H_R + i\eta & -\tau^+ \\ -\tau & EI - H \end{pmatrix} \end{array} \begin{Bmatrix} \Phi_R + \chi \\ \psi \end{Bmatrix} = \begin{Bmatrix} S_R \\ 0 \end{Bmatrix} \tag{8.1.8b}$$

where $[H]$ is the device Hamiltonian. The different quantities appearing in Eq. (8.1.8b) are not numbers (except for the energy E). They are matrices of different sizes:

Contact Hamiltonian $[H_R]$, identity matrix $[I_R]$, damping $[\eta]$: $(R \times R)$
Channel Hamiltonian $[H]$, identity matrix $[I]$: $(d \times d)$
Coupling Hamiltonian $[\tau]$: $(d \times R)$, $[\tau^+]$: $(R \times d)$

or column vectors:

Contact wavefunction $\{\Phi_R\}$, $\{\chi\}$, source $\{S_R\}$: $(R \times 1)$
Device wavefunction $\{\psi\}$: $(d \times 1)$

Note that in going from Eq. (8.1.8a) to Eq. (8.1.8b), the term $\{S_R\}$ on the right-hand side, representing the reinjection of electrons from external sources, is assumed to remain unchanged. This allows us to make use of Eq. (8.1.8a) to eliminate $\{S_R\}$ from Eq. (8.1.8b) to write:

$$[EI_R - H_R + i\eta]\{\chi\} - [\tau^+]\{\psi\} = \{0\}$$

$$[EI - H]\{\psi\} - [\tau]\{\chi\} = [\tau]\{\Phi_R\}$$

We can use straightforward matrix algebra to express $\{\chi\}$ in terms of $\{\psi\}$ from the first equation

$$\{\chi\} = G_R\,\tau^+\{\psi\}$$

where

$$G_R \equiv [EI_R - H_R + i\eta]^{-1} \tag{8.1.9}$$

and

$$[\eta] = 0^+[I_R]$$

and substitute into the second equation to obtain

$$[EI - H - \Sigma]\{\psi\} = \{S\} \tag{8.1.10}$$

where

$$\Sigma \equiv \tau G_R \tau^+ \qquad S \equiv \tau\Phi_R \tag{8.1.11}$$

Equation (8.1.10) has exactly the form of the result (see Eq. (8.1.1)) that we are trying to prove, while Eq. (8.1.11) gives us a formal expression that we can use to evaluate Σ and S. It is apparent from Eq. (8.1.9) that the quantity G_R represents a property of the isolated contact since it only involves the contact Hamiltonian H_R. It is called the *Green's function* for the isolated contact, the physical significance of which we will discuss in Section 8.2.

Evaluation of Σ and S: Looking at Eq. (8.1.11), it is not clear how we could evaluate it for specific examples, since the matrix G_R is of size $(R \times R)$, which is typically huge since the size of the reservoir R is often infinite. However, we note that although the matrix $[\tau]$ is formally of size $(d \times R)$, in "real space" it only couples the r surface elements of the reservoir next to the channel. So we could truncate it to a $(d \times r)$ matrix and write

$$\underset{(d\times d)}{\Sigma} \equiv \underset{(d\times r)}{\tau} \; \underset{(r\times r)}{g_R} \; \underset{(r\times d)}{\tau^+} \tag{8.1.12a}$$

and

$$\underset{(d\times 1)}{S} \equiv \underset{(d\times r)}{\tau} \; \underset{(r\times 1)}{\phi_R} \tag{8.1.12b}$$

where the surface Green's function g_R represents an $(r \times r)$ subset of the full Green's function G_R involving just the r points at the surface, and $\{\phi_R\}$ represents an $(r \times 1)$ subset of the contact wavefunction Φ_R. For the toy example that we discussed at the beginning of this section, we can show that

$$\tau = -t_0 \qquad \phi_R = -\text{i}2B \sin ka \tag{8.1.13a}$$

and

$$g_R = -(1/t_0)\exp(\text{i}ka) \tag{8.1.13b}$$

which, when substituted into Eqs. (8.1.12a, b), yields the same results that we obtained earlier (cf. Eqs. (8.1.7a, b)). The expression for ϕ_R is obtained by noting that it is equal to the wavefunction Φ_{-1} that we would have in the contact (at the point that is connected to the channel) if it were decoupled from the channel. This decoupling would impose the boundary condition that $\Phi_0 = 0$, making $C = -B$, and the corresponding Φ_{-1} is equal to $(-\text{i}2B \sin ka)$ as stated above. The expression for g_R takes a little more algebra to work out and we will delegate this to exercise E.8.1 at the end of the chapter.

Another way to evaluate Σ and S is to work in the *eigenstate representation* of the contact, so that the contact Hamiltonian H_R is diagonal and the Green's function G_R is easily written down in terms of the eigenvalues ε_n of H_R:

$$[G_R(E)] = \begin{bmatrix} \dfrac{1}{E - \varepsilon_1 + \text{i}0^+} & 0 & \cdots \\ 0 & \dfrac{1}{E - \varepsilon_2 + \text{i}0^+} & \cdots \\ \cdots & \cdots\;\cdots & \cdots \end{bmatrix} \tag{8.1.14}$$

In this representation, the coupling matrix $[\tau]$ cannot be truncated to a smaller size and we have to evaluate an infinite summation over the eigenstates of the reservoir:

$$\Sigma_{ij}(E) = \sum_n \frac{[\tau]_{in}\,[\tau^+]_{nj}}{E - \varepsilon_n + \text{i}0^+} \tag{8.1.15a}$$

$$S_i(E) \equiv \sum_n [\tau]_{in}\,\{\Phi_R\}_n \tag{8.1.15b}$$

However, this summation can often be carried out analytically after converting to a summation. As an example, Exercise E.8.2 shows how we can obtain our old results (Eqs. (8.1.7a)) for the toy problem starting from Eq. (8.1.15a).

Before moving on let me briefly summarize what we have accomplished. A channel described by a Hamiltonian $[H]$ of size $(d \times d)$ is coupled to a contact described by a $(R \times R)$ matrix $[H_R]$, where the reservoir size R is typically much larger than d $(R \gg d)$. We have shown that the effect of the reservoir on the device can be described through a self-energy matrix $\Sigma(E)$ of size $(d \times d)$ and a source term $S(E)$ of size $(d \times 1)$.

8.2 Local density of states

We have just seen that a channel coupled to a contact can be described by a modified Schrödinger equation of the form $E\{\psi\} = [H + \Sigma]\{\psi\} + \{S\}$ where $\{S\}$ represents the excitation from the contact and the self-energy Σ represents the modification of the channel by the coupling. Unlike $[H]$, $[H + \Sigma]$ has complex eigenvalues and the imaginary part of the eigenvalues both broadens the density of states *and* gives the eigenstates a finite lifetime. In this section we will talk about the first effect and explain how we can calculate the density of states in an open system. In the next section we will talk about the second effect.

Consider the composite system consisting of the channel and the contact. Earlier, in Eq. (6.2.14), we agreed that a system with a set of eigenvalues ε_α has a density of states given by

$$D(E) = \sum_\alpha \delta(E - \varepsilon_\alpha) \tag{8.2.1}$$

How can different energy levels have different weights as implied in the broadened lineshape on the right-hand side of Fig. 8.1? Doesn't Eq. (8.2.1) tell us that each energy level gives rise to a delta function whose weight is one? The problem is that the density of states in Eq. (8.2.1) does not take into account the spatial distribution of the states. If we want to know the local density of states in the channel we need to weight each state by the fraction of its squared wavefunction that resides in the channel denoted by d:

$$D(d; E) = \sum_\alpha |\phi_\alpha(d)|^2 \delta(E - \varepsilon_\alpha)$$

For example, suppose the device with one energy level ε were decoupled from the reservoir with a dense set of energy levels $\{\varepsilon_R\}$. The total density of states would then be given by

$$D(E) = \sum_\alpha \delta(E - \varepsilon_\alpha) = \delta(E - \varepsilon) + \sum_n \delta(E - \varepsilon_n)$$

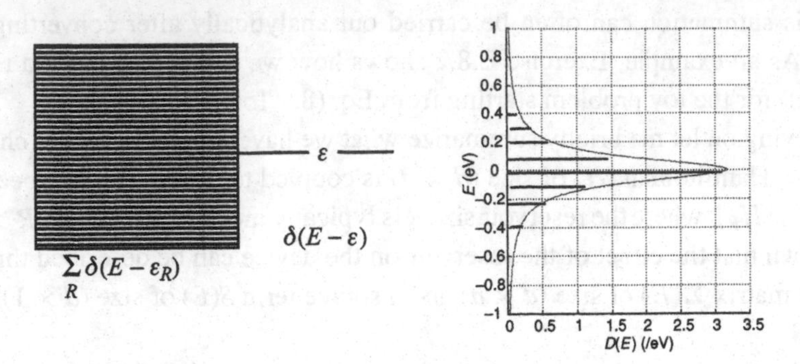

Fig. 8.2.1 A channel with a single energy level ε coupled to a reservoir with a dense set of energy levels $\{\varepsilon_n\}$. The local density of states on the channel shows a single sharp level before being coupled to the reservoir. But on being coupled, it shows a series of levels of varying heights reflecting the fraction of their squared wavefunction that reside in the channel.

while the local density of states on the channel would simply be given by

$$D(d; E) = \sum_{\alpha} |\phi_\alpha(d)|^2 \delta(E - \varepsilon_\alpha) = \delta(E - \varepsilon)$$

since the reservoir states have wavefunctions that have no amplitude in the channel at all. Once we couple the channel to the reservoir, things will not be so clear cut any more. There will be one level with its wavefunction largely on the channel, but there will be many other neighboring states with their wavefunctions residing partially on the channel. If we look at the local density of states in the channel we see a series of energy levels with varying heights, reflecting the fraction of the squared wavefunction residing in the channel (Fig. 8.2.1).

In general we can define a local density of states (LDOS) $D(\vec{r}; E)$ that weights each level by the square of its wavefunction at the location $\vec{r}$:

$$D(\vec{r}; E) = \sum_{\alpha} |\Phi_\alpha(\vec{r})|^2 \delta(E - \varepsilon_\alpha) \tag{8.2.2}$$

which can be viewed as the diagonal element (divided by 2π) of a more general concept called the spectral function [$A(E)$]:

$$A(\vec{r}, \vec{r}\,'; E) = 2\pi \sum_{\alpha} \phi_\alpha(\vec{r}) \delta(E - \varepsilon_\alpha) \phi_\alpha^*(\vec{r}\,') \tag{8.2.3}$$

just as the electron density

$$n(\vec{r}) = \sum_{\alpha} |\phi_\alpha(\vec{r})|^2 f_0(\varepsilon_\alpha - \mu) \tag{8.2.4}$$

can be viewed as the diagonal element of the density matrix:

$$\rho(\vec{r}, \vec{r}\,') = \sum_{\alpha} \phi_\alpha(\vec{r}) f_0(\varepsilon_\alpha - \mu) \phi_\alpha^*(\vec{r}\,') \tag{8.2.5}$$

We argued in Section 4.3 that Eq. (8.2.5) is just the real-space representation of the matrix relation:

$$[\rho] = f_0([H] - \mu[I]) \tag{8.2.6}$$

Using the same argument we could write the spectral function as

$$[A(E)] = 2\pi\delta(E[I] - [H]) \tag{8.2.7}$$

and view Eq. (8.2.3) as its real-space representation. If we use the eigenstates of H as our basis then $[H]$ is diagonal:

$$[H] = \begin{bmatrix} \varepsilon_1 & 0 & 0 & \cdots \\ 0 & \varepsilon_2 & 0 & \cdots \\ 0 & 0 & \varepsilon_3 & \cdots \\ & \cdots & & \cdots \end{bmatrix}$$

and so is $[A(E)]$:

$$[A(E)] = 2\pi \begin{bmatrix} \delta(E-\varepsilon_1) & 0 & 0 & \cdots \\ 0 & \delta(E-\varepsilon_2) & 0 & \cdots \\ 0 & 0 & \delta(E-\varepsilon_3) & \cdots \\ \cdots & \cdots & \cdots & \cdots \end{bmatrix} \tag{8.2.8}$$

Equation (8.2.3) transforms this matrix into a real-space representation. In principle we could write the spectral function in any representation and its diagonal elements will tell us the LDOS (times 2π) at energy E in that representation, just as the diagonal elements of the density matrix tell us the local electron density in that representation. The total number of electrons N is given by the sum of all the diagonal elements of $[\rho]$ or the trace of $[\rho]$, which is independent of representation:

$$N = \text{Trace}[\rho] = \sum_\alpha f_0(\varepsilon_\alpha - \mu)$$

Similarly, the total density of states given by the trace of the spectral function $[A]$ divided by 2π is independent of representation and is readily written down from the eigenstate representation:

$$D(E) = \frac{1}{2\pi}\text{Trace}[A(E)] = \sum_\alpha \delta(E - \varepsilon_\alpha) \tag{8.2.9}$$

Sum rule: An important point to note is that if we look at the total number of states at any point integrated over all energy, the answer is one. If we start with a device having one level and couple it to a reservoir, it will broaden into a series of levels (Fig. 8.2.1) of varying strengths representing the fact that the wavefunction for each level contributes to different extents to the device. But if we add up the strengths of all the levels the answer is the same as that of the original level. What the device loses from its one level due to hybridization, it gains back from the other levels so that the broadened level in the device can accommodate exactly the same number of electrons that the one discrete level could accommodate before it got coupled to the reservoir. This sum rule could be stated as follows:

$$\int_{-\infty}^{+\infty} \mathrm{d}E D(\vec{r}; E) = 1$$

and can be proved by noting that $\int_{-\infty}^{+\infty} \mathrm{d}E D(\vec{r}; E)$ is basically the diagonal element of the matrix $\int_{-\infty}^{+\infty} \mathrm{d}E[A(E)]/2\pi$ evaluated in the real-space representation. It is easy to see from Eq. (8.2.8) that in the eigenstate representation

$$\int_{-\infty}^{+\infty} \frac{\mathrm{d}E}{2\pi}[A(E)] = \begin{bmatrix} 1 & 0 & 0 & \cdots \\ 0 & 1 & 0 & \cdots \\ 0 & 0 & 1 & \cdots \\ \cdots & \cdots & \cdots & \cdots \end{bmatrix} \quad \text{since} \quad \int_{-\infty}^{+\infty} \mathrm{d}E\delta(E-\varepsilon) = 1$$

The point is that this quantity will look the same in *any representation* since the identity matrix remains unchanged by a change in basis.

Green's function: In evaluating the spectral function it is convenient to make use of the identity

$$2\pi\,\delta(E-\varepsilon_\alpha) = \left[\frac{2\eta}{(E-\varepsilon_\alpha)^2+\eta^2}\right]_{\eta\to 0^+}$$

$$= \mathrm{i}\left[\frac{1}{E-\varepsilon_\alpha+\mathrm{i}0^+} - \frac{1}{E-\varepsilon_\alpha-\mathrm{i}0^+}\right] \tag{8.2.10a}$$

to write

$$2\pi\,\delta(EI-H) = \mathrm{i}\{[(E+\mathrm{i}0^+)I-H]^{-1} - [(E-\mathrm{i}0^+)I-H]^{-1}\} \tag{8.2.10b}$$

where 0^+ denotes a positive infinitesimal (whose physical significance we will discuss at length in Section 8.4). Equation (8.2.10b) would be a simple extension of (8.2.10a) if the argument $(EI-H)$ were an ordinary number. But since $(EI-H)$ is a matrix, Eq. (8.2.10b) may seem like a big jump from Eq. (8.2.10a). However, we can justify it by going to a representation that diagonalizes $[H]$, so that both sides of Eq. (8.2.10b) are diagonal matrices and the equality of each diagonal element is ensured by Eq. (8.2.10a). We can thus establish the matrix equality, Eq. (8.2.10b) in the eigenstate representation, which should ensure its validity in any other representation.

Using Eqs. (8.2.7) and (8.2.10b) we can write

$$A(E) = \mathrm{i}[G(E) - G^+(E)] \tag{8.2.11}$$

where the retarded Green's function is defined as

$$G(E) = [(E + i0^+)I - H]^{-1} \tag{8.2.12a}$$

and the advanced Green's function is defined as

$$G^+(E) = [(E - i0^+)I - H]^{-1} \tag{8.2.12b}$$

In the next section we will see how the Green's function (and hence the spectral function) can be evaluated for open systems.

Density matrix: Starting from $[\rho] = f_0([H] - \mu[I])$ (Eq. (8.2.6)) we can write

$$[\rho] = \int_{-\infty}^{+\infty} dE f_0(E - \mu)\delta([EI - H])$$
$$= \int_{-\infty}^{+\infty} \frac{dE}{2\pi} f_0(E - \mu)[A(E)] \tag{8.2.13}$$

which makes good sense if we note that $[A(E)]/2\pi$ is the matrix version of the density of states $D(E)$, in the same way that the density matrix $[\rho]$ is the matrix version of the total number of electrons N. We could view Eq. (8.2.10) as the matrix version of the common sense relation

$$N = \int_{-\infty}^{+\infty} dE f_0(E - \mu)D(E)$$

which simply states that the number of electrons is obtained by multiplying the number of states $D(E)$ dE by the probability $f_0(E)$ that they are occupied and adding up the contributions from all energies.

Why should we want to use Eq. (8.2.13) rather than Eq. (8.2.6)? In previous chapters we have evaluated the density matrix using Eq. (8.2.6) and it may not be clear why we might want to use Eq. (8.2.13) since it involves an extra integration over energy. Indeed if we are dealing with the entire system described by a matrix $[H]$ then there is no reason to do so. But if we are dealing with an open system described by a matrix of the form (see Fig. 8.2)

$$\overline{H} = \begin{bmatrix} H & \tau \\ \tau^+ & H_R \end{bmatrix}$$

then Eq. (8.2.6) requires us to deal with the entire matrix which could be huge compared with $[H]$ since the reservoir matrix $[H_R]$ is typically huge – that is why we call it a reservoir! The spectral function appearing in Eq. (8.2.13) and the Green's function are technically just as large

$$\bar{A} = \begin{bmatrix} A & A_{dR} \\ A_{Rd} & A_{RR} \end{bmatrix} \qquad \bar{G} = \begin{bmatrix} G & G_{dR} \\ G_{Rd} & G_{RR} \end{bmatrix}$$

but we only care about the top $(d \times d)$ subsection of this matrix and the great advantage of the Green's function approach is that this subsection of $[G(E)]$, and hence $[A(E)]$, can be calculated without the need to deal with the full matrix. This is what we will show next, where we will encounter the same self-energy matrix Σ that we encountered in the last section.

Self-energy matrix – all over again: The overall Green's function can be written from Eq. (8.2.12a) as

$$\bar{G} \equiv \begin{bmatrix} G & G_{dR} \\ G_{Rd} & G_{RR} \end{bmatrix} = \begin{bmatrix} (E + i0^+)I - H & -\tau \\ -\tau^+ & (E + i0^+)I - H_R \end{bmatrix}^{-1} \tag{8.2.14}$$

The power of the Green's function method comes from the fact that we can evaluate the $(d \times d)$ subsection $[G]$ that we care about *exactly* from the relation

$$G = [(E + i0^+)I - H - \Sigma(E)]^{-1} \tag{8.2.15}$$

where $\Sigma(E)$ is the self-energy matrix given in Eq. (8.1.11).

Equation (8.2.15) follows from Eq. (8.2.14) using straightforward matrix algebra. The basic result we make use of is the following. If

$$\begin{bmatrix} a & b \\ c & d \end{bmatrix} = \begin{bmatrix} A & B \\ C & D \end{bmatrix}^{-1} \tag{8.2.16a}$$

then

$$\begin{bmatrix} A & B \\ C & D \end{bmatrix} \begin{bmatrix} a & b \\ c & d \end{bmatrix} = \begin{bmatrix} I & 0 \\ 0 & I \end{bmatrix}$$

so that $Aa + Bc = I$ and $Ca + Dc = 0 \rightarrow c = -D^{-1}Ca$. Hence

$$a = (A - BD^{-1}C)^{-1} \tag{8.2.16b}$$

Comparing Eq. (8.2.16a) with Eq. (8.2.14) and making the obvious replacements we obtain from Eq. (8.2.16b):

$$G = \left[(E + i0^+)I - H - \tau G_R \tau^+\right]^{-1} \quad \text{where} \quad G_R = \left[(E + i0^+)I - H_R\right]^{-1}$$

which yields the result stated above in Eq. (8.2.15).

Equation (8.2.16b) is a well-known result that is often used to find the inverse of large matrices by *partitioning* them into smaller ones. Typically in such cases we are interested in finding all the component matrices a, b, c, and d and they are all approximately equal in size. In our problem, however, the matrices a, A are much smaller than the matrices d, D and we only want to find a. Equation (8.2.15) allows us to evaluate $[G]$ by inverting a matrix of size $(d \times d)$ rather than the full $(d + R) \times (d + R)$ matrix in Eq. (8.2.14). This can be a major practical advantage since R is typically much larger than d. But the idea of describing the effect of the surroundings on a device through a self-energy function $[\Sigma]$ is not just a convenient numerical tool. It represents a major conceptual step and we will try to convey some of the implications in the next section. For the moment, let us look at a couple of examples, one analytical and one numerical.

Analytical example: Consider a uniform infinite 1D wire modeled with a one-band effective mass Hamiltonian of the form shown in Fig. 8.2.2. Since this is a uniform wire the eigenstates can be catalogued in terms of k obeying a dispersion relation and we can use our previous results to write the DOS per unit cell as

$$D(E) = a/\pi\hbar v(E) \quad \text{where} \quad v = (1/\hbar)\,\mathrm{d}E/\mathrm{d}k$$

Now let us obtain this same result using the Green's function method developed in this section. We replace the infinite 1D wire with a *single* unit cell and add self-energy terms to account for the two semi-infinite wires on either side (Fig. 8.2.3).

The Green's function for this single cell is a (1×1) matrix or a number

$$G(E) = \frac{1}{E - (E_c + 2t_0 - t_0 \exp[ika] - t_0 \exp[ika])}$$

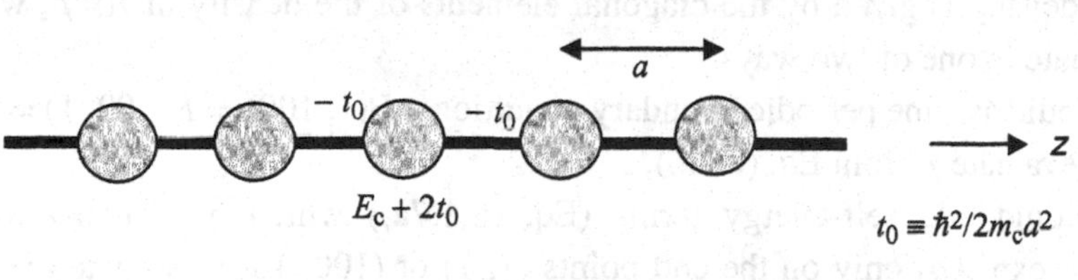

Fig. 8.2.2

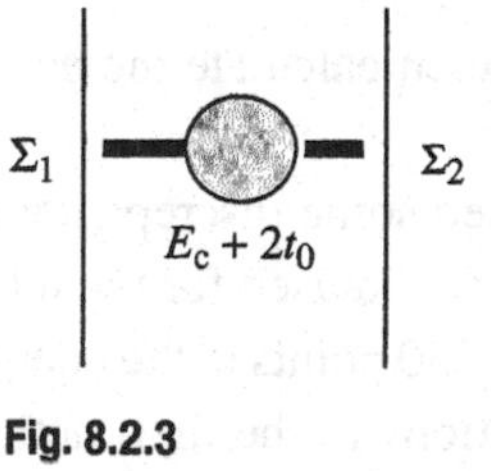

Fig. 8.2.3

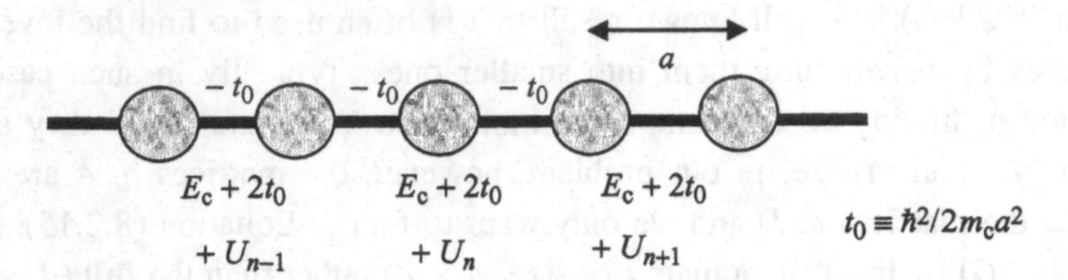

Fig. 8.2.4

which is simplified making use of the dispersion relation $E = E_c + 2t_0(1 - \cos ka)$ to obtain

$$G(E) = \frac{1}{t_0 \exp[ika] - t_0 \exp[-ika]} = \frac{1}{2it_0 \sin ka}$$

from which the DOS is obtained:

$$D(E) = i[G - G^+]/2\pi$$
$$= 1/2\pi t_0 \sin ka = a/\pi\hbar v$$

which is the same as the previous result since $\hbar v = dE/dk = 2at_0 \sin ka$

Numerical example: To get a feeling for the self-energy method, it is instructive to redo the problem of finding the equilibrium electron density in a 1D box that we discussed in Chapter 4 (see Fig. 4.3.1). We consider a similar problem, namely, a 1D box with a potential $U(x)$ that changes linearly from -0.1 eV at one end to $+0.1$ eV at the other. We model it using a one-band Hamiltonian with a lattice of 50 points spaced by $a = 2$ Å and with the effective mass m_c equal to 0.25 times the free electron mass m (see Fig. 8.2.4).

We wish to evaluate the electron density $n(z)$ in the box assuming that it is in equilibrium with an electrochemical potential $\mu = E_c + 0.25$ eV and $k_B T = 0.025$ eV. The electron density is given by the diagonal elements of the density marix ρ, which we can evaluate in one of two ways.

(1) We could assume periodic boundary conditions: $H(1, 100) = H(100, 1) = -t_0$ and then evaluate ρ from Eq. (8.2.6).

(2) We could add self-energy terms (Eq. (8.1.7a)) which have non-zero values of $-t_0 \exp[ika]$ only on the end points (1, 1) or (100, 100), evaluate the Green's function from Eq. (8.2.15)

$$G = [(E + i0^+)I - H - \Sigma_1 - \Sigma_2]^{-1}$$

obtain the spectral function from Eq. (8.2.11), and then calculate the equilibrium density matrix from Eq. (8.2.13).

Figure 8.2.5 shows that the two results agree well. Indeed some discrepancy is likely due to errors introduced by the discreteness of the energy grid used for the integration in the last step of the method. We have used a grid having 250 points in the energy range $E_c - 0.1\text{ eV} < E < E_c + 0.4\text{ eV}$. However, the oscillations in the first method arise

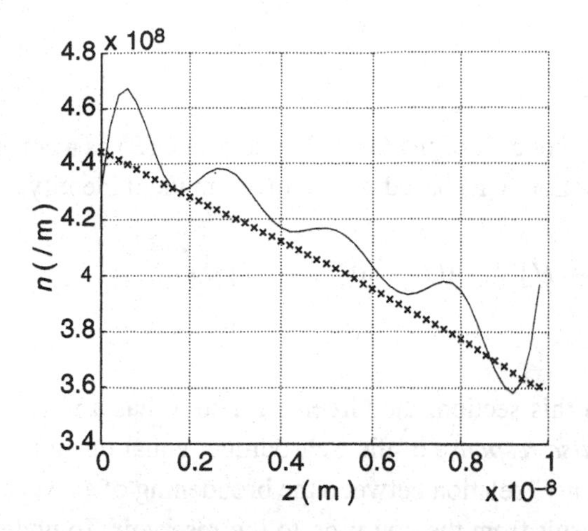

Fig. 8.2.5 Plot of electron density $n(z)$ calculated for a 1D wire with a linear potential $U(z)$ using periodic boundary conditions (solid curve) and using the self-energy method to enforce open boundary conditions (crosses).

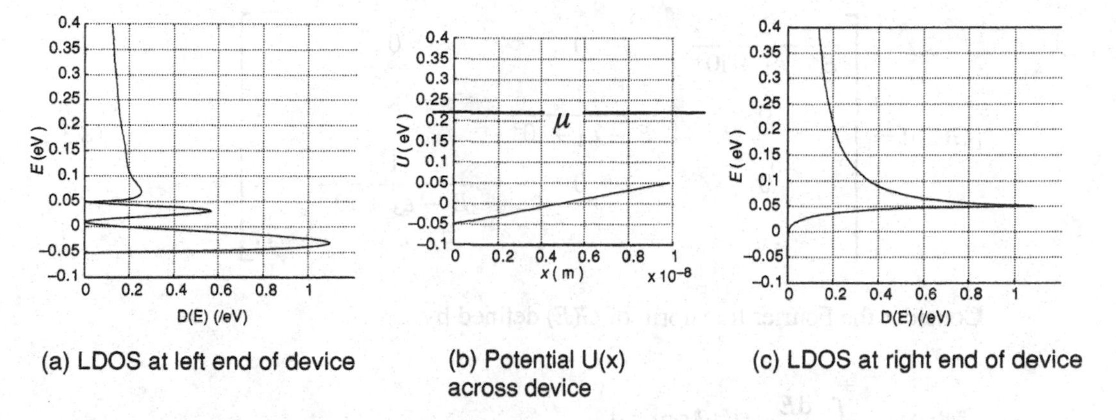

Fig. 8.2.6 Local density of states (LDOS) at two ends of a 1D wire with a linear potential $U(x)$ impressed across it. The corresponding electron density is shown in Fig. 8.2.5. $\mu = E_c + 0.25$ eV and $k_B T = 0.025$ eV.

from the standing waves in a closed ring (resulting from the use of periodic boundary conditions) which are absent in the open system modeled by the self-energy method. The oscillations in method 1 will be less pronounced for longer devices (or for a larger effective mass) because the energy levels will become closer together.

As we know, the effect of the self-energy is to broaden the energy level, but its integrated strength is unchanged because of the sum rule mentioned earlier. Consequently the distinction between the two methods is somewhat obscured when we look at the electron density since it involves an integration over energy. The self-energy method allows us to investigate in detail the local density of states in different parts of the device (see Fig. 8.2.6).

8.3 Lifetime

In Section 8.2 we introduced the concept of the Green's function $G(E)$ as a convenient way to evaluate the spectral function $A(E)$ based on the mathematical identity:

$$2\pi\delta(EI - H) = \mathrm{i}\{[(E + \mathrm{i}0^+)I - H]^{-1} - [(E - \mathrm{i}0^+)I - H]^{-1}\}$$
$$A(E) = \mathrm{i}[G(E) - G^+(E)]$$

However, as we will explain in this section, the Green's function has a physical significance of its own as the *impulse response* of the Schrödinger equation and this will help us understand the "uncertainty" relation between the broadening of a level and the finite lifetime, both of which result from the coupling to the reservoir. To understand the meaning of the Green's function let us use the eigenstates of H as our basis so that the Green's function is diagonal:

$$[G(E)] = \begin{bmatrix} \dfrac{1}{E - \varepsilon_1 + \mathrm{i}0^+} & 0 & 0 & \cdots \\ 0 & \dfrac{1}{E - \varepsilon_2 + \mathrm{i}0^+} & 0 & \cdots \\ 0 & 0 & \dfrac{1}{E - \varepsilon_3 + \mathrm{i}0^+} & \cdots \\ \cdots & \cdots & \cdots & \cdots \end{bmatrix} \tag{8.3.1}$$

Consider the Fourier transform of $G(E)$ defined by

$$[\tilde{G}^{\mathrm{R}}(t)] = \int_{-\infty}^{+\infty} \frac{\mathrm{d}E}{2\pi\hbar} \mathrm{e}^{+\mathrm{i}Et/\hbar}[G(E)]$$

which is also diagonal and looks like this:

$$[\tilde{G}^{\mathrm{R}}(t)] = \frac{-\mathrm{i}}{\hbar}\vartheta(t)\,\mathrm{e}^{-0^+t} \begin{bmatrix} \exp(-\mathrm{i}\varepsilon_1 t/\hbar) & 0 & 0 & \cdots \\ 0 & \exp(-\mathrm{i}\varepsilon_2 t/\hbar) & 0 & \cdots \\ 0 & 0 & \exp(-\mathrm{i}\varepsilon_3 t/\hbar) & \cdots \\ \cdots & \cdots & \cdots & \cdots \end{bmatrix} \tag{8.3.2}$$

It takes a little work (involving contour integration on a complex plane) to get from Eq. (8.3.1) to Eq. (8.3.2). But it is quite straightforward to go the other way from

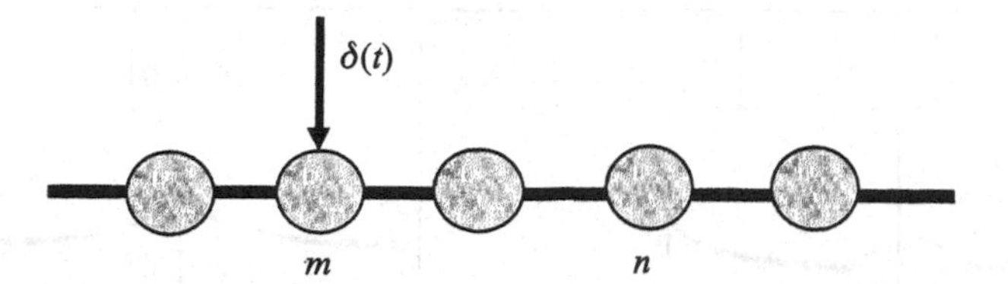

Fig. 8.3.1

Eq. (8.3.2) to Eq. (8.3.1) using the inverse transform relation:

$$[G(E)] = \int_{-\infty}^{+\infty} \mathrm{d}t\, \mathrm{e}^{+\mathrm{i}Et/\hbar}[\tilde{G}^{\mathrm{R}}(t)]$$

$$= \frac{-\mathrm{i}}{\hbar}\int_{-\infty}^{+\infty} \mathrm{d}t\, \mathrm{e}^{\mathrm{i}Et/\hbar}\vartheta(t)\mathrm{e}^{-\mathrm{i}\varepsilon t/\hbar}\mathrm{e}^{-0^{+}t}$$

$$= \frac{-\mathrm{i}}{\hbar}\int_{-\infty}^{+\infty} \mathrm{d}t\, \mathrm{e}^{\mathrm{i}(E-\varepsilon)t/\hbar}\mathrm{e}^{-0^{+}t} = \frac{1}{E-\varepsilon+\mathrm{i}0^{+}}$$

I should mention that here I am not using the superscript "R" to denote reservoir. I am using it to denote "retarded" which refers to the fact that the function $\tilde{G}^{\mathrm{R}}(t)$ is zero at all times $t < 0$. It is easy to see that the diagonal elements of this function satisfy the differential equation

$$\left(\mathrm{i}\hbar\frac{\partial}{\partial t} - \varepsilon_{\alpha}\right)\tilde{G}^{\mathrm{R}}_{\alpha\alpha}(t) = \delta(t)$$

so that we can write

$$\left(\mathrm{i}\hbar\frac{\partial}{\partial t} - [H]\right)[\tilde{G}^{\mathrm{R}}(t)] = [I]\delta(t) \tag{8.3.3}$$

suggesting the interpretation of $\tilde{G}^{\mathrm{R}}(t)$ as the impulse response of the Schrödinger equation

$$\left(\mathrm{i}\hbar\frac{\partial}{\partial t} - [H]\right)\{\Psi(t)\} = 0 \tag{8.3.4}$$

The (n, m) element of this matrix $\tilde{G}^{\mathrm{R}}_{nm}(t)$ tells us the nth component of the wavefunction if the system is given an impulse excitation at its mth component (see Fig. 8.3.1).

From this point of view it seems natural to expect that the Green's function should be "retarded," since we cannot have a response before the impulse is applied (which is at $t = 0$). Mathematically, however, this is not the only solution to Eq. (8.3.3). It is straightforward to show that the "advanced" Green's function

$$[\tilde{G}^{\mathrm{A}}(t)] = [\tilde{G}^{\mathrm{R}}(-t)]^{*} \tag{8.3.5}$$

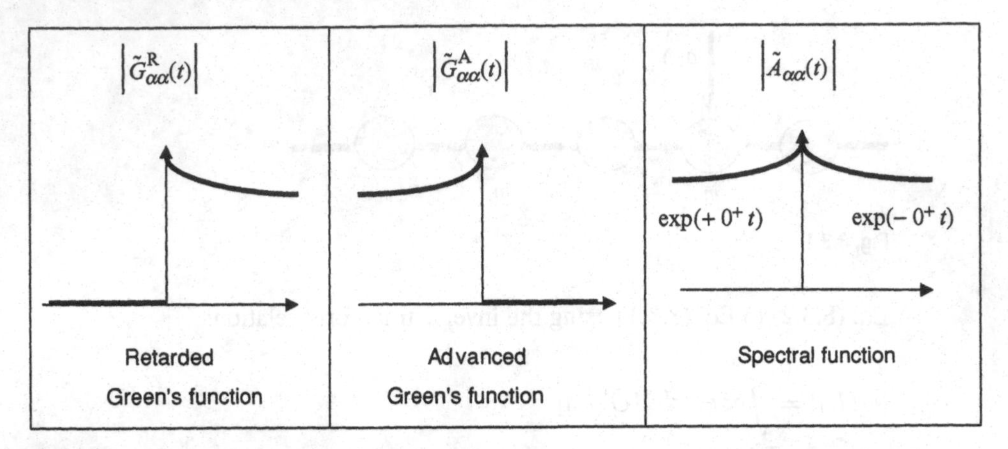

Fig. 8.3.2 Sketch of the magnitude of any diagonal element (in the eigenstate representation) of the retarded and advanced Green's functions and the spectral function in the time domain.

satisfies the same equation

$$\left(i\hbar\frac{\partial}{\partial t} - [H]\right)[G^{A}(t)] = [I]\,\delta(t) \tag{8.3.6}$$

but it is zero at all times *after* $t = 0$. In the eigenstate representation we can write from Eqs. (8.3.2) and (8.3.5):

$$[\tilde{G}^{A}(t)] = \frac{i}{\hbar}\vartheta(-t)e^{+0^{+}t}\begin{bmatrix} \exp(-i\varepsilon_1 t/\hbar) & 0 & 0 & \cdots \\ 0 & \exp(-i\varepsilon_2 t/\hbar) & 0 & \cdots \\ 0 & 0 & \exp(-i\varepsilon_3 t/\hbar) & \cdots \\ \cdots & \cdots & \cdots & \cdots \end{bmatrix} \tag{8.3.7}$$

This is actually the Fourier transform of $G^{+}(E)$ (once again it is easier to do the inverse transform). The difference between the retarded and advanced Green's function in the energy domain:

Retarded	*Advanced*
$G(E) = [(E + i0^{+})I - H]^{-1}$	$G^{+}(E) = [(E - i0^{+})I - H]^{-1}$

looks very minor: the two only differ in the sign of an infinitesimally small quantity 0^{+}; one is tempted to conclude wrongly that they differ only in some insignificant sense. In the time domain, however, their difference is hard to miss. One is zero for $t < 0$ (causal) and the other is zero for $t > 0$ (non-causal). One is interpreted as the response to an impulse excitation at $t = 0$; the other has no physical interpretation but is a mathematically valid solution of the same equation with a different unphysical initial condition. Figure 8.3.2 shows the magnitude of one of the diagonal elements of $\tilde{G}^{R}_{\alpha\alpha}(t)$ and $\tilde{G}^{A}_{\alpha\alpha}(t)$. Note that the spectral function is proportional to the difference between the retarded and advanced Green's functions (see Eq. (8.2.11)):

$$\tilde{A}^{\mathrm{R}}_{\alpha\alpha}(t) = \mathrm{i}\left[\tilde{G}^{\mathrm{R}}_{\alpha\alpha}(t) - \tilde{G}^{\mathrm{A}}_{\alpha\alpha}(t)\right]$$

Since both Green's functions satisfy the same differential equation, the spectral function $[A(t)]$ satisfies the homogeneous differential equation without the impulse excitation:

$$\left(\mathrm{i}\hbar\frac{\partial}{\partial t} - [H]\right)[\tilde{A}(t)] = [0]$$

and hence has no discontinuity at $t = 0$ unlike $\tilde{G}^{\mathrm{R}}(t)$ and $\tilde{G}^{\mathrm{A}}(t)$ as shown in Fig. 8.3.2.

Physical meaning of the self-energy: We saw in Section 8.2 that we can calculate the device subsection of the full Green's function (Eq. (8.2.14)):

$$\overline{G} \equiv \begin{bmatrix} G & G_{dR} \\ G_{Rd} & G_{RR} \end{bmatrix} = \begin{bmatrix} (E + \mathrm{i}0^+)I - H & -\tau \\ -\tau^+ & (E + \mathrm{i}0^+)I - H_R \end{bmatrix}^{-1}$$

exactly from the relation (Eq. (8.2.15)).

$$G = [(E + \mathrm{i}0^+)I - H - \Sigma(E)]^{-1}$$

where the self-energy

$$\Sigma(E) = \tau g_R(E)\tau^+$$

can be calculated from a knowledge of the surface property of the reservoir (g_R) and the device–reservoir coupling (τ).

Now that we have interpreted the time domain Green's function as the impulse response of the Schrödinger equation (see Eq. (8.3.4)), we could write a similar equation for the device subset of the Green's function by Fourier transforming Eq. (8.2.15). This would be straightforward if the self-energy Σ were independent of the energy E:

$$\left(\mathrm{i}\hbar\frac{\partial}{\partial t} - [H] - [\Sigma]\right)[\tilde{G}^{\mathrm{R}}(t)] = [I]\,\delta(t) \tag{8.3.8a}$$

If we take the energy dependence into account then the Fourier transform looks more complicated. The product of $\Sigma(E)$ and $G(E)$ when transformed becomes a convolution in the time domain:

$$\left(\mathrm{i}\hbar\frac{\partial}{\partial t} - [H]\right)[\tilde{G}^{\mathrm{R}}(t)] - \int \mathrm{d}t'\,[\tilde{\Sigma}(t - t')][\tilde{G}^{\mathrm{R}}(t')] = [I]\,\delta(t) \tag{8.3.8b}$$

To get some insight into the physical meaning of Σ let us ignore this "detail." In fact, let us make the problem very simple by considering a small device with just a single energy level ε (Fig. 8.3.3) so that $[H]$ and $[\Sigma]$ are both simple numbers rather than matrices:

$$\left(\mathrm{i}\hbar\frac{\partial}{\partial t} - \varepsilon - \Sigma\right)\tilde{G}^{\mathrm{R}}(t) = \delta(t)$$

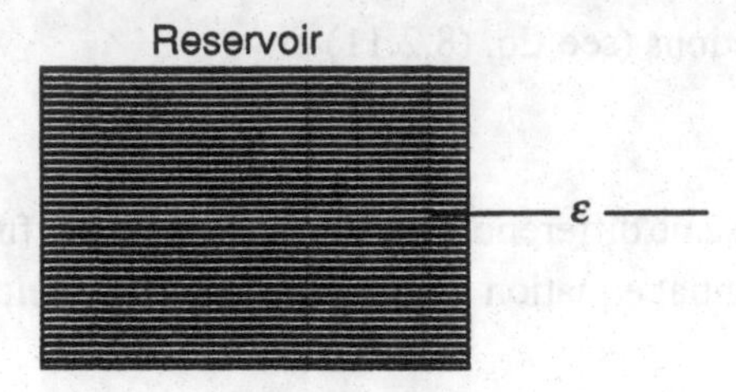

Fig. 8.3.3

The solution to this equation

$$\tilde{G}^{R}(t) = -\frac{i}{\hbar}e^{-i(\varepsilon+\Sigma)t/\hbar}\vartheta(t)$$

tells us the wavefunction in response to an impulse excitation of the device at $t = 0$. We can write

$$\tilde{G}^{R}(t) = -\frac{i}{\hbar}e^{-i\varepsilon' t/\hbar}e^{-\gamma t/2\hbar}\vartheta(t) \tag{8.3.9}$$

where

$$\varepsilon' = \varepsilon + \text{Re}\Sigma \quad \text{and} \quad \gamma = -2\text{Im}\Sigma \tag{8.3.10}$$

The real part of the self-energy causes a shift in the device energy level from ε to ε', while the imaginary part has the effect of giving the eigenstates a finite lifetime. This is evident from the squared magnitude of this wavefunction which tells us how the probability decays with time after the initial excitation:

$$|\tilde{G}^{R}(t)|^2 = \frac{1}{\hbar^2}\vartheta(t)\exp(-\gamma t/\hbar)$$

Clearly we can relate the lifetime of the state to the imaginary part of the self-energy:

$$\frac{1}{\tau} = -\frac{\gamma}{\hbar} = -\frac{2\text{Im}\Sigma}{\hbar} \tag{8.3.11}$$

We can identify this as the "uncertainty" relation between lifetime and broadening if we note that the imaginary part of the self-energy is equal to the broadening of the density of states. To see this we note that the Fourier transform of the simple version of the Green's function in Eq. (8.3.9) is given by

$$G(E) = \frac{1}{E - \varepsilon' + i\gamma/2}$$

so that

$$\frac{A(E)}{2\pi} \equiv D(E) = i\left(\frac{1}{E - \varepsilon' + i\gamma/2} - \frac{1}{E - \varepsilon' - i\gamma/2}\right)$$
$$= \frac{\gamma}{(E - \varepsilon')^2 + (\gamma/2)^2}$$

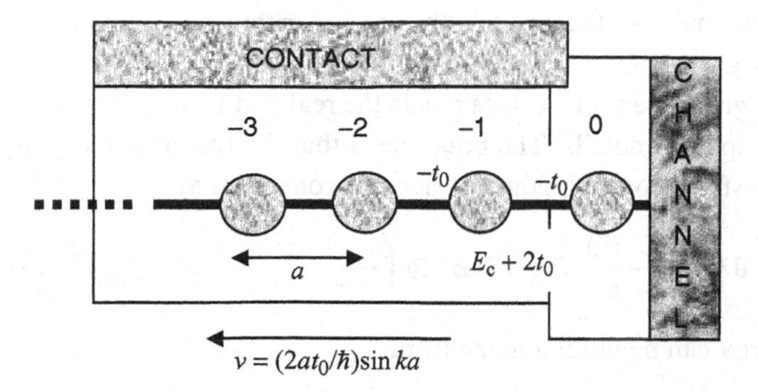

Fig. 8.3.4

showing that the LDOS on the device is broadened into a Lorentzian of width γ equal to twice the imaginary part of the self-energy. Of course, the lineshape in general need not be Lorentzian. We have obtained this result because in this discussion we ignored the energy dependence of the self-energy (in the time domain one would call it the memory effect of the reservoir) and used Eq. (8.3.8a) instead of (8.3.8b) for the purpose of clarity in this physical discussion.

We saw in Section 8.1 that the self-energy for a one-dimensional contact is diagonal with two non-zero entries:

$$\Sigma(1,1) = -t_0 \exp(ika) = -t_0 \cos(ka) - it_0 \sin(ka)$$

From Eq. (8.3.11) we could write the corresponding lifetime for site 1 as

$$\frac{1}{\tau} = \frac{\gamma}{\hbar} = \frac{2t_0 \sin ka}{\hbar}$$

It is interesting to note that the velocity associated with a particular k in the wire is given by

$$v = \frac{1}{\hbar}\frac{\partial E}{\partial k} = \frac{1}{\hbar}\frac{\partial}{\partial k}[2t_0(1 - \cos ka)] = \frac{2at_0}{\hbar}\sin ka$$

so that we can write

$$\frac{1}{\tau} = \frac{\gamma}{\hbar} = \frac{v}{a}$$

which is intuitively satisfying since we expect the escape rate from a given cell to equal the escape velocity divided by the size of a cell (see Fig. 8.3.4). Indeed one could use this principle to write down the imaginary part of the self-energy approximately for more complicated geometries where an exact calculation of the surface Green's function may not be easy:

$$\text{Im}\Sigma(E) \approx \hbar v(E)/R \tag{8.3.12}$$

Here R is a linear dimension of the unit cell, the precise arithmetic factor depending on the specific geometry.

Knowing the imaginary part, one can calculate the real part from a general principle independent of the specific details. The principle is that the real and imaginary parts must be Hilbert transforms of each other (⊗ denotes convolution)

$$\text{Re}\,\Sigma(E) = -\frac{1}{\pi}\int \text{d}E' \frac{\text{Im}\Sigma(E')}{E - E'} = \text{Im}\Sigma(E) \otimes \left(-\frac{1}{\pi E}\right) \tag{8.3.13}$$

so that the self-energy can be written in the form

$$\Sigma(E) = [\text{Re}\Sigma(E)] + \text{i}\,[\text{Im}\Sigma(E)] = \text{iIm}\Sigma(E) \otimes \left(\delta(E) + \frac{\text{i}}{\pi E}\right) \tag{8.3.14}$$

This principle is obeyed by any function whose Fourier transform is causal (that is, the Fourier transform is zero for $t < 0$). The self-energy function is causal because it is proportional to the surface Green's function of the reservoir (see Eq. (8.1.11)), which is causal as we discussed earlier. To see why causal functions obey this principle, we note that $\delta(E) + (\text{i}/\pi E)$ is the Fourier transform of the unit step function: $\vartheta(t)$. This means that any time domain function of the form $\vartheta(t)f(t)$ has a Fourier transform that can be written as (product in the time domain becomes a convolution in the transform domain)

$$F(E) \otimes \left(\delta(E) + \frac{\text{i}}{\pi E}\right)$$

where the transform of $f(t)$ is $F(E)$ which can be identified with i Im $\Sigma(E)$ in Eq. (8.3.14).

Broadening matrix: In the simple case of a one-level device we have seen that the imaginary part of the self-energy gives us the broadening or inverse lifetime of the level (see Eq. (8.3.11)). More generally, the self-energy is a matrix and one can define a broadening matrix $\Gamma(E)$ equal to its anti-hermitian component:

$$\Gamma(E) = \text{i}[\Sigma(E) - \Sigma^{+}(E)] \tag{8.3.15}$$

This component of the self-energy is responsible for the broadening of the level, while the hermitian component

$$\Sigma_{\text{H}}(E) = \frac{1}{2}[\Sigma(E) + \Sigma^{+}(E)]$$

can conceptually be viewed as a correction to the Hamiltonian [H]. Overall, we could write

$$H + \Sigma(E) = [H + \Sigma_{\text{H}}(E)] - \frac{\text{i}\Gamma(E)}{2}$$

We have often made use of the fact that we can simplify our description of a problem by using the eigenstates of the Hamiltonian [H] as our basis. For open systems we would want to use a representation that diagonalizes [$H + \Sigma_H$] in our energy range of interest. If the same representation also diagonalizes [Γ], then the problem could be viewed simply in terms of many one-level devices in parallel. But in general this may not be the case. The representation that diagonalizes [$H + \Sigma_H$] may not diagonalize [Γ] and vice versa. We can then diagonalize one or the other but not both and interesting new physics beyond the one-level example can result.

8.4 What constitutes a contact (reservoir)?

Let me quickly summarize what we did in this chapter before we discuss an important conceptual issue. We started in Section 8.1 from the "Schrödinger" equation for the channel + the contact

$$\begin{pmatrix} EI_R - H_R + i\eta & -\tau^+ \\ -\tau & EI - H \end{pmatrix} \begin{Bmatrix} \Phi_R + \chi \\ \psi \end{Bmatrix} = \begin{Bmatrix} S_R \\ 0 \end{Bmatrix} \quad (8.4.1)$$

and obtained an equation for the channel alone

$$[EI - H - \Sigma]\{\psi\} = \{S\} \quad (8.4.2)$$

whose solution can be written in the form

$$\{\psi\} = [G]\{S\}, \quad \text{where } [G] = [EI - H - \Sigma]^{-1} \quad (8.4.3)$$

We then discussed the significance of the Green's function G and the self-energy Σ in both the energy and time domains.

Now if we look back at the steps in Section 8.1 that lead from Eq. (8.4.1) to (8.4.2), it should be apparent that someone could just as well treat the contact as primary and represent the effect of the channel on it through a self-energy. Mathematically, [H] and [H_R] are on an equal footing. The only "asymmetry" comes from the terms $i\eta$ and S_R which were added "by hand" to account for the extraction and reinjection of electrons from the contact by an external source (usually $i\eta$ is justified as a mathematical artifice designed to ensure the convergence of a Fourier transform). But these infinitesimal terms could just as well have been added "symmetrically" and so it is natural to ask what distinguishes the channel from the contact. A related question is how small is the infinitesimal η: neV? meV? Does it matter?

To answer these questions let us look at a simple example. Consider a "molecule" described by a (2 × 2) Hamiltonian matrix

$$H = \begin{bmatrix} \varepsilon & \tau \\ \tau^* & \varepsilon_1 \end{bmatrix} \quad (8.4.4)$$

2π . (LDOS at site 1)

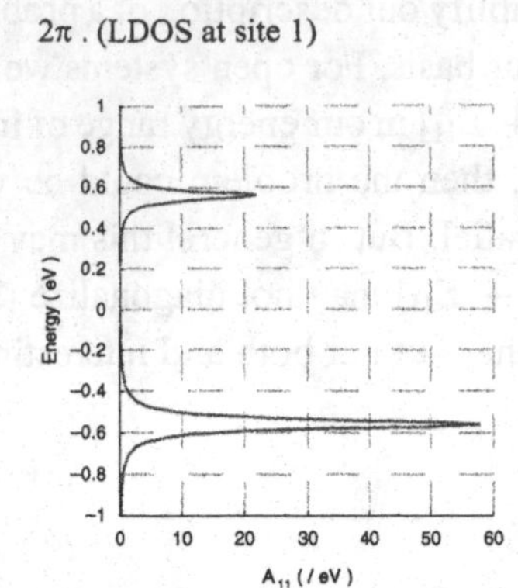

2π . (LDOS at site 2)

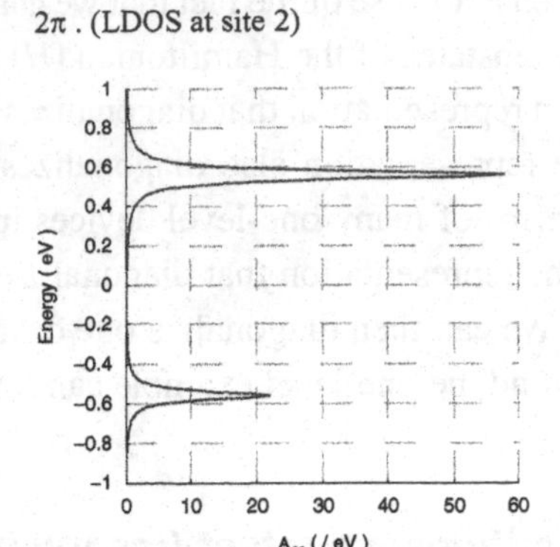

Fig. 8.4.1

We could calculate the (2×2) spectral function $[A(E)]$ from the relation ($[\eta] = 0^+[I]$, I being the (2×2) identity matrix)

$$[A] = \mathrm{i}[G - G^+], \quad \text{where } [G] = [EI - H + \mathrm{i}\eta]^{-1} \tag{8.4.5}$$

and look at its diagonal elements to obtain the local density of states (LDOS) on each of the two sites. For example with $\varepsilon = -0.25$ eV, $\varepsilon_1 = +0.25$ eV, $\tau = 0.5$ eV and $0^+ = 0.025$ eV we obtain the plots shown in Fig. 8.4.1. Note that there are two peaks located around each of the two eigenenergies obtained by setting the determinant of ($EI - H$) equal to zero:

$$\det \begin{bmatrix} E - \varepsilon & -\tau \\ -\tau^* & E - \varepsilon_1 \end{bmatrix} = 0 \tag{8.4.6}$$

The two peaks are of unequal height reflecting the fact that the eigenfunction corresponding to the lower eigenenergy is skewed towards site "1" while the eigenfunction corresponding to the higher energy is skewed towards site "2".

Now if we are interested primarily in site 1 we could represent the effect of site 2 through a self-energy

$$\Sigma(E) = \frac{|\tau|^2}{E - \varepsilon_1 + \mathrm{i}0^+} \tag{8.4.7}$$

and calculate the LDOS from a (1×1) spectral function:

$$[a] = \mathrm{i}[g - g^+], \quad \text{where } [g] = [E - \varepsilon - \Sigma]^{-1} \tag{8.4.8}$$

The result would be exactly what we obtained earlier from the (2×2) spectral function. On the other hand if we are interested primarily in site "2" we could represent the effect of site 1 through a self-energy

$$\Sigma_1(E) = \frac{|\tau|^2}{E - \varepsilon + \mathrm{i}0^+} \tag{8.4.9}$$

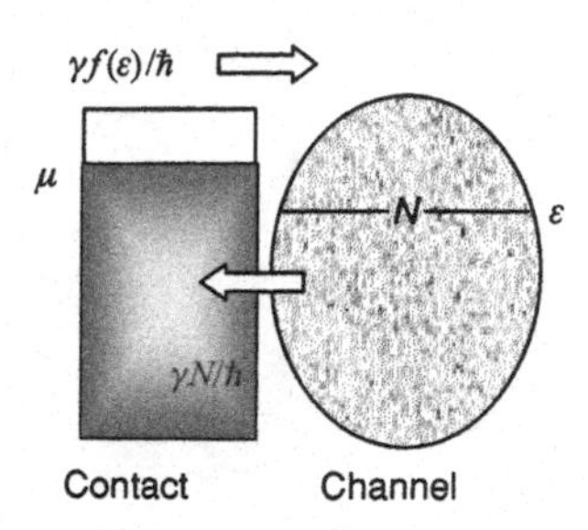

Fig. 8.4.2

and calculate the LDOS from a (1 × 1) spectral function:

$$[a_1] = \mathrm{i}[g_1 - g_1^+], \quad \text{where } [g_1] = [E - \varepsilon_1 - \Sigma_1]^{-1} \tag{8.4.10}$$

Again the answer would match the result obtained from the (2 × 2) spectral function. In short, we could treat site 1 as the channel and site 2 as the contact or vice versa. Either choice is mathematically acceptable.

Neither of these atomic contacts, however, can be called a "reservoir" in the sense it is commonly understood. In Chapter 1 I wrote the current flow between a contact and a channel in the form (see Eq. (1.2.2a) and Fig. 8.4.2)

$$I = \text{Inflow} - \text{Outflow} = (1/h)\,(\gamma f - \gamma N)$$

The essential characteristic of a "reservoir" is that the rate constant γ for outflow and the rate of inflow γf remain unaffected by the filling and emptying of states and other dynamical details.

The atomic "contacts" we have just described do not satisfy this criterion. For example, if we consider site 2 as the contact (Eq. (8.4.7)) the broadening (or escape rate) is given by

$$\gamma = \mathrm{i}[\Sigma - \Sigma^+] = \frac{|\tau|^2\, 0^+}{(E - \varepsilon_1)^2 + (0^+)^2} \tag{8.4.11}$$

It is not only energy-dependent (sharply peaked around $E = \varepsilon_1$) but also affected strongly by the precise value of the infinitesimal 0^+. This means that the escape rate (γ/h) for electrons from the "channel" is affected by the rate at which they are extracted from the "contact." A "proper" reservoir should provide a constant escape rate independent of the value of 0^+. This is possible if it consists of numerous closely spaced energy levels, as we will now show.

Consider a channel with a single state with $\varepsilon = 0$ (Fig. 8.4.3) coupled to a contact with numerous closely spaced levels such that the overall system is described by a large

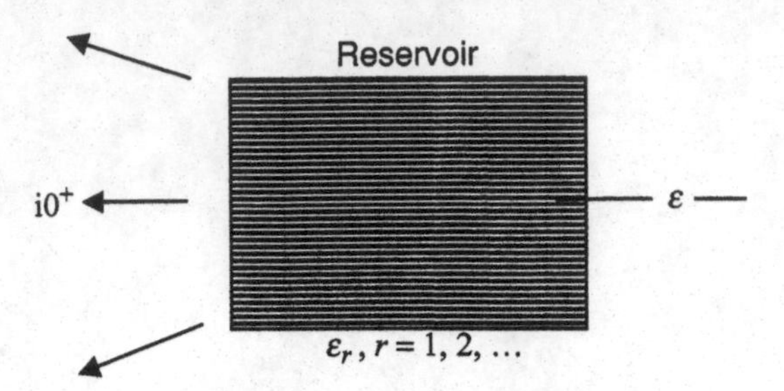

Fig. 8.4.3 A device with a single state with $\varepsilon = 0$ is coupled to a reservoir that consists of numerous closely spaced energy levels $\{\varepsilon_r, r = 1, 2, \ldots\}$.

Hamiltonian matrix of the form

$$[H] = \begin{bmatrix} \varepsilon & \tau_1 & \tau_2 & \tau_3 & \cdots \\ \tau_1^* & \varepsilon_1 & 0 & 0 & \cdots \\ \tau_2^* & 0 & \varepsilon_2 & 0 & \cdots \\ \tau_3^* & 0 & 0 & \varepsilon_3 & \cdots \\ & \cdots & & & \end{bmatrix}$$

We can describe the effect of the contact through a (1×1) self-energy "matrix" (see Eq. (8.1.15a)) given by

$$\Sigma = \sum_r \frac{|\tau_r|^2}{E - \varepsilon_r + \mathrm{i}0^+} \tag{8.4.12}$$

so that the broadening is given by

$$\gamma = \mathrm{i}[\Sigma - \Sigma^+] = \sum_r \frac{|\tau_r|^2\, 0^+}{(E - \varepsilon_r)^2 + (0^+)^2} \tag{8.4.13}$$

If the levels are very closely spaced, we can replace the summation by an integral

$$\gamma = \int \mathrm{d}\varepsilon_r\, D_R(\varepsilon_r)\, |\tau_r|^2 \frac{0^+}{(E - \varepsilon_r)^2 + (0^+)^2} \tag{8.4.14}$$

where D_R is the number of states per unit energy. The last function in the integrand is sharply peaked over a small range of energies around $\varepsilon_r = E$ and if the other terms are nearly constant for all states within this range, we can pull them out of the integral to obtain

$$\begin{aligned} \gamma &= D_R(E)\, |\tau(E)|^2 \int \mathrm{d}\varepsilon_r \frac{0^+}{(E - \varepsilon_r)^2 + (0^+)^2} \\ &= 2\pi\, D_R(E)\, |\tau(E)|^2 \end{aligned} \tag{8.4.15}$$

This result (Eq. (8.4.15)), often referred to as *Fermi's golden rule*, is widely used for many different problems. Indeed we could also write our earlier result (Eq. (8.4.11))

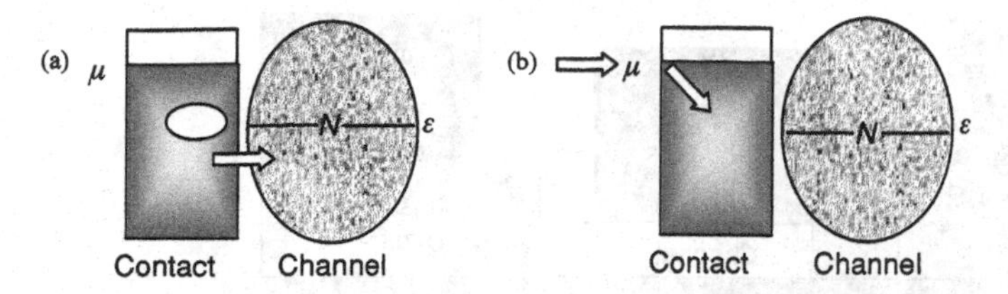

Fig. 8.4.4

for an atomic contact in the same form by noting that the density of states of the atomic "reservoir" is given by

$$D_R(E) \rightarrow \frac{0^+/2\pi}{(E-\varepsilon_1)^2+(0^+)^2}$$

As we mentioned earlier, this makes a very poor reservoir because the DOS is sharply varying in energy and strongly dependent on 0^+. By contrast, the reservoir with many closely spaced energy levels has a constant DOS independent of 0^+ if our level of interest is coupled approximately equally to all of them, with each state broadened by an amount 0^+ that is well in excess of the level spacing. Typical contacts in the real world are huge macroscopic objects with level spacings of picoelectron-volts or less and even the slightest external perturbation can provide the required broadening. In calculating the self-energy $\Sigma(E)$ analytically, it is common to take the limit as the level spacing and the level broadening 0^+ both tend to zero (the former is inforced by letting the volume tend to infinity). But it is important to take this limit in the proper order with the broadening always exceeding the spacing. In numerical calculations, it is common to use finite-sized contacts to save computation time. Since finite-sized contacts also have finite level spacings $\Delta\varepsilon$, it is important to choose a value of the "infinitesimal" 0^+ that is in excess of $\Delta\varepsilon$ in order to turn them into well-behaved reservoirs.

There is actually a more important property that is required of a good reservoir: it must not only provide a constant escape rate γ but also a constant inflow rate $\gamma f, f$ being the Fermi function in the contact. This requires that external sources constantly maintain the contact in local equilibrium, which is possible only if there is good communication among the states in the contact. For example, if an electron enters the channel from the contact it will leave a hole in the contact that is way below the electrochemical potential μ as sketched in Fig. 8.4.4a. This creates a highly non-equilibrium situation and the incoming electron at μ quickly loses energy to the solid and fills up the hole through screening and "inelastic scattering processes" of the type to be discussed in Chapter 10. This is why much of the Joule heating (I^2R) occurs in the contact rather than in the channel.

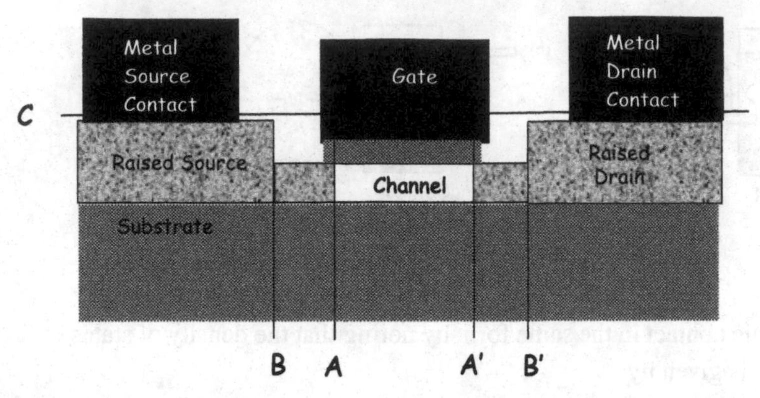

Fig. 8.4.5 An ultrathin body silicon-on-insulator field effect transistor (FET). Possible planes for drawing the line between H and Σ are shown (courtesy of M. S. Lundstrom).

For a contact to function like a good reservoir, it is important that this restoration of equilibrium occurs fast enough that the contact can always be assumed to remain in local equilibrium. This assumption may need to be revisited as we work more on "nanocontacts," especially if spin relaxation is involved.

Where do we draw the line between the Hamiltonian [*H*] and the self-energy [Σ]? For practical reasons, we would like to make the region described by $[H]$ as small as possible, but we have to ensure that the region described by $[\Sigma]$ can be assumed to be in local equilibrium without significant loss of accuracy. The self-energy method described in this chapter is accurate and useful even if a particular contact does not have a smooth LDOS on a constant escape rate (see, for example, Klimeck *et al.* (1995)). What is more important is that the scattering processes be strong enough to maintain local equilibrium. For example, Fig. 8.4.5 shows an ultrathin nanotransistor with raised source and drain regions designed to lower the resistance. If we draw the line between H and Σ at A, A′ we incur significant error since the region to the left of A (or right of A′) has the same cross-section as the channel and there is no reason to expect it to be any more in equilibrium than the rest of the channel. It is thus necessary to move the lines to B, B′, or perhaps even to C and beyond. For a discussion of some of these issues, see Venugopal *et al.* (2004).

Let me end by noting that the role of contacts is really quite ubiquitous and goes far beyond what these specific examples might suggest. When we calculate the optical absorption by a semiconductor, we implicitly assume the valence and conduction bands to be the "source" and "drain" contacts that are maintained in local equilibrium (see Exercise E.10.4) and there are examples of nanodevices where these "contacts" get driven significantly off equilibrium. In modeling the interaction of electrons with lattice vibrations or phonons, the latter are usually assumed to be maintained in equilibrium, but this "contact" too can get driven off equilibrium. A more subtle example is the Overhauser effect where electronic spins drive nuclear spins significantly off equilibrium because the latter cannot relax easily due to their isolation from the surroundings (see, for example, Salis *et al.* (2001)). The point is that quantum processes inevitably involve all kinds of "contacts" which need not always be metallurgical like the ones in Fig. 8.4.5 and it is not safe to assume that they will function like well-behaved reservoirs without scrutinizing their structure and dynamics.

EXERCISES

E.8.1. Assume a 1D one-band effective mass model for a 1D lead as shown in Fig. E.8.1.

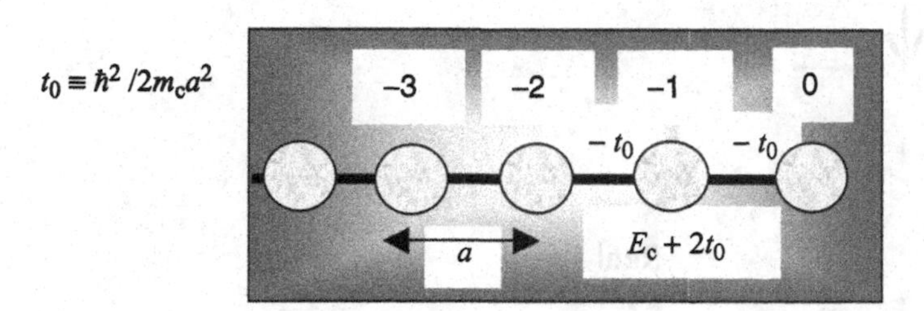

Fig. E.8.1

Starting from Eq. (8.1.9) $G_R = [(E + i0^+)I - H_R]^{-1}$
Show that $G_R(0, 0) \equiv g_R = -\exp(ika)/t_0$.

E.8.2.
(a) In section 8.1 we have derived the self-energy for a one-dimensional lead using elementary techniques:

$$\Sigma(E) = -t_0 \exp(ika) \tag{8.1.7a}$$

where

$$E = E_c + 2t_0(1 - \cos ka) \tag{8.1.5}$$

Eliminate ka from Eq. (8.1.7a) to obtain an expression for the self-energy explicitly in terms of E:

$$\Sigma(E)/t_0 = (x - 1) - i\sqrt{2x - x^2}$$

where

$$x \equiv (E - E_c)/2t_0, \quad 0 \le x \le 2$$

The sign of the radical has been chosen so that the imaginary part of Σ is negative. For values of x where the quantity under the radical sign is negative (this corresponds to energy values outside the allowed band) we should choose the sign such that Σ goes to zero for large values of x:

$$\Sigma(E)/t_0 = (x-1) + \sqrt{x^2 - 2x}, \quad -\infty \leq x \leq 0$$
$$\Sigma(E)/t_0 = (x-1) - \sqrt{x^2 - 2x}, \quad 2 \leq x \leq +\infty$$

Plot the self-energy over the range $-1 \leq x \leq 3$.

You should obtain a plot like this:

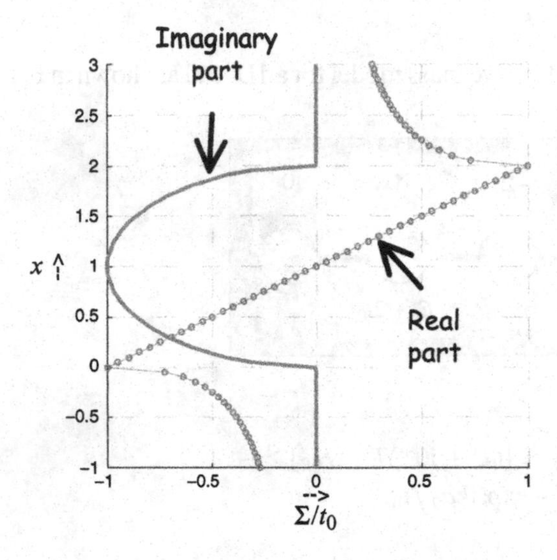

(b) It is instructive to re-derive Eq. (8.1.7a) using the more general formula given in Eq. (8.1.15a). For a semi-infinite one-dimensional lead with N points spaced by a, the eigenenergies are given by

$$\varepsilon_n = 2t_0(1 - \cos k_n a)$$

The corresponding normalized wavefunction is written as

$$\psi_n(x_\alpha) = \sqrt{2/N}\sin(k_n x_\alpha) \quad \text{where } k_n = n\pi/a \quad \text{and} \quad x_\alpha = \alpha a$$

so that $[\tau]_{in} = [\tau]_{i\alpha}[\tau]_{\alpha n} = -t_0\sqrt{2/N}\sin k_n a$

From Eq. (8.1.15a),

$$\Sigma(E) = \frac{2}{N}\sum_n \frac{t_0^2 \sin^2 k_n a}{E - \varepsilon_n + i0^+}$$

Convert the summation to an integral to show that

$$\Sigma(E) = \frac{2}{N}\int_0^{\pi/a} \frac{dk_n L}{\pi}\frac{t_0^2\sin^2 k_n a}{E - \varepsilon_n + i0^+} = \int_0^{4t_0} \frac{d\varepsilon_n}{\pi}\frac{t_0 \sin k_n a}{E - \varepsilon_n + i0^+}$$

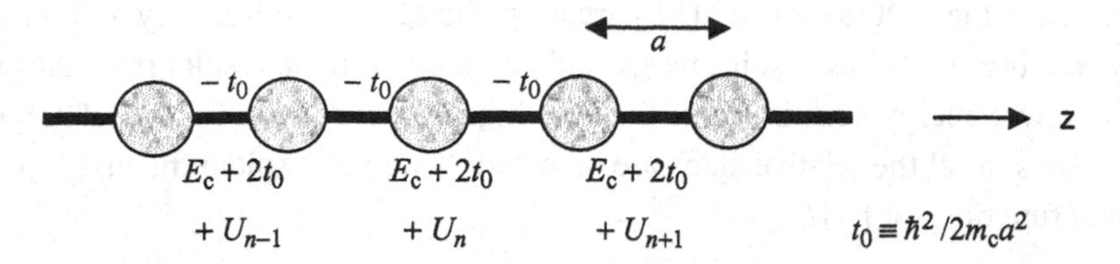

Fig. E.8.3

(c) It is possible to evaluate the self-energy analytically using contour integration techniques, but somewhat difficult. What is relatively straightforward is to evaluate the broadening function $\Gamma(E)$ which is equal to twice the imaginary part of the self-energy. Show that the broadening function is given by

$$\Gamma(E) = \mathrm{i}(\Sigma - \Sigma^{+}) = 2t_0 \sin ka, \quad E_c \le E \le E_c + 4t_0$$

where $E = E_c + 2t_0(1 - \cos ka)$

and is zero outside the range $E_c \le E \le E_c + 4t_0$. Hint: You may find Eq. (8.2.10a) useful.

(d) Show that $\Sigma(E) = \int_{-\infty}^{\infty} \mathrm{d}E' \frac{\Gamma(E')/2\pi}{E-E'+\mathrm{i}0^{+}}$

and hence

$$\mathrm{Re}\Sigma(E) = P\left(\int_{-\infty}^{\infty} \mathrm{d}E' \frac{\Gamma(E')/2\pi}{E - E'}\right), \quad \mathrm{Im}\Sigma(E) = -\Gamma(E)/2$$

where P stands for principal part.

E.8.3. Consider a 1D wire with a potential $U(x)$ that changes linearly from −0.1 eV at one end to +0.1 eV at the other end (Fig. E.8.3) and model it using a one-band Hamiltonian with a lattice of 50 points spaced by $a = 2$ Å and with the effective mass m_c equal to 0.25 times the free electron mass m.

Calculate the electron density $n(z)$ in the wire assuming that it is in equilibrium with an electrochemical potential $\mu = E_c + 0.25$ eV and $k_B T = 0.025$ eV, using (a) periodic boundary conditions and (b) the self-energy method. Compare with Fig. 8.2.5. Calculate the LDOS at the two ends of the box from the self-energy method and compare with Fig. 8.2.6.

E.8.4. Consider the problem we discussed at the beginning of Section 8.4 with two coupled atoms described by a (2 × 2) Hamiltonian, $H = \begin{bmatrix} \varepsilon & \tau \\ \tau^{*} & \varepsilon_1 \end{bmatrix}$ and use the same parameters as in the text: $\varepsilon = -0.25$ eV, $\varepsilon_1 = +0.25$ eV, $\tau = 0.5$ eV and $0^{+} = 0.01$ eV.

(a) Calculate the (2 × 2) spectral function and plot the LDOS on sites '1' and '2' as a function of energy.

(b) Calculate the LDOS on site '1' by treating site '2' as a self-energy and on site '2' by treating site '1' as a self-energy and compare with the results from part (a).

(c) The lower energy peak is larger than the higher energy peak for the LDOS on site '1' (on site '2' the relative sizes are reversed). Can you explain this in terms of the eigenfunctions of $[H]$?

E.8.5. Consider a linear conductor with a repulsive potential $U(x) = U_0\ \delta(x)$. Calculate the local density of states (LDOS) as a function of x at $E = 0.005$ eV and at $E = 0.05$ eV. Use a one-band tight-binding model with 101 lattice sites spaced by $a = 0.25$ nm assuming $U_0 = 5$ eV nm and an effective mass of $0.25m$. You should obtain plots like this

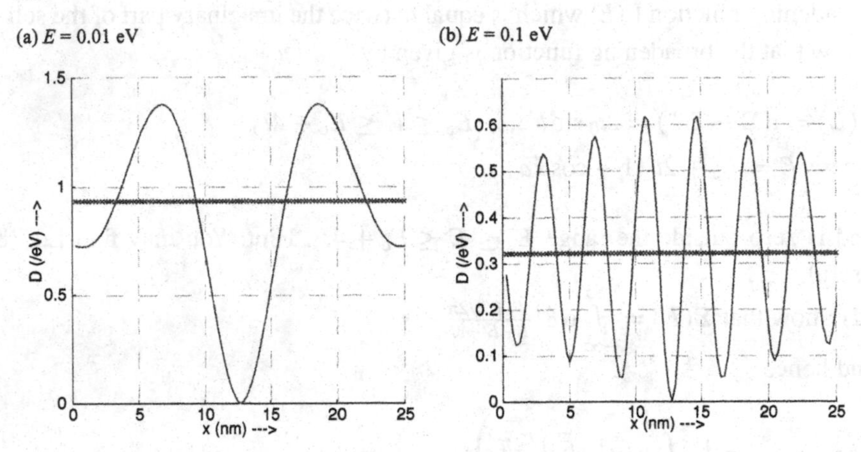

The LDOS goes to zero right at the impurity, but oscillates around it with a period determined by the electron wavelength corresponding to the energy E. Such oscillations in the LDOS have been observed by scanning tunneling microscopy. The reader may find it instructive to plot $D(x, E)$ using a gray scale plot on the x–E plane similar to that shown for example, in Fig. 9 of Lake and Datta (1992a).

9 Coherent transport

The reader may wish to review Section 1.6 before reading this chapter.

9.1 Overview

Since this chapter is rather long, let me start with a detailed overview, that can also serve as a "summary." In Chapter 1, I described a very simple model for current flow, namely a single level ε which communicates with two contacts, labeled the source and the drain. The strength of the coupling to the source (or the drain) was characterized by the rate $\gamma_1/\hbar$ (or $\gamma_2/\hbar$) at which an electron initially occupying the level would escape into the source (or the drain).

I pointed out that the flow of current is due to the difference in "agenda" between the source and the drain, each of which is in a state of local equilibrium, but maintained at two different electrochemical potentials and hence with two distinct Fermi functions:

$$f_1(E) \equiv f_0(E - \mu_1) = \frac{1}{\exp[(E - \mu_1)/k_B T] + 1} \tag{9.1.1a}$$

$$f_2(E) \equiv f_0(E - \mu_2) = \frac{1}{\exp[(E - \mu_2)/k_B T] + 1} \tag{9.1.1b}$$

by the applied bias V: $\mu_2 - \mu_1 = -qV$. The source would like the number of electrons occupying the level to be equal to $f_1(\varepsilon)$ while the drain would like to see this number be $f_2(\varepsilon)$. The actual steady-state number of electrons N lies somewhere between the two and the source keeps pumping in electrons while the drain keeps pulling them out, each hoping to establish equilibrium with itself. In the process, a current flows in the external circuit (Fig. 9.1.1).

My purpose in this chapter is essentially to carry out a generalized version of this treatment applicable to an arbitrary multi-level device (Fig. 9.1.2) whose energy levels and coupling are described by matrices rather than ordinary numbers:

$\varepsilon \rightarrow [H]$ *Hamiltonian* matrix

$\gamma_{1,2} \rightarrow [\Gamma_{1,2}]$ *Broadening* matrices $\Gamma_{1,2} = \mathrm{i}[\Sigma_{1,2} - \Sigma_{1,2}^{+}]$

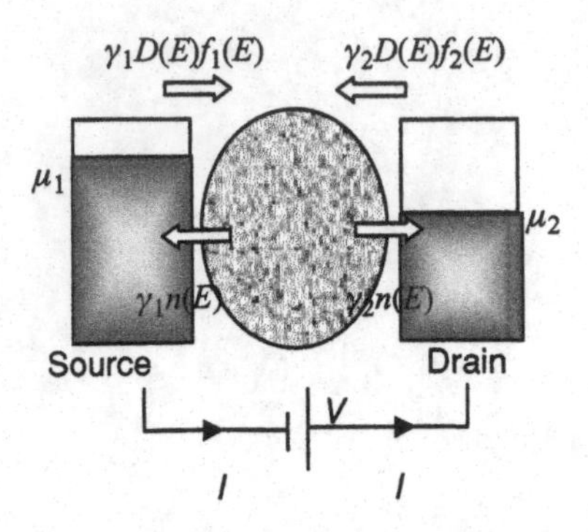

Fig. 9.1.1 Flux of electrons into and out of a channel: independent level model, see Eqs. (1.6.4)–(1.6.6).

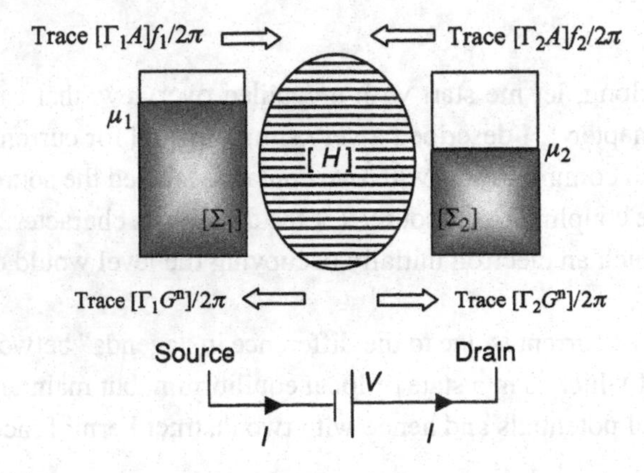

Fig. 9.1.2 Inflow and outflow for an arbitrary multi-level device whose energy levels are described by a Hamiltonian matrix *[H]* and whose coupling to the source and drain contacts is described by self-energy matrices $[\Sigma_1(E)]$ and $[\Sigma_2(E)]$ respectively.

In Chapter 8 we saw that connecting a device to a reservoir broadens its energy levels and it is convenient to talk in terms of a continuous independent energy variable E, rather than a discrete set of eigenstates. The density matrix can be written in the form (see Eq. (8.2.13))

$$[\rho] = \int_{-\infty}^{+\infty} (\mathrm{d}E/2\pi)[G^{\mathrm{n}}(E)] \tag{9.1.2a}$$

where, in equilibrium,

$$[G^{\mathrm{n}}(E)]_{\mathrm{eq}} = [A(E)] f_0(E-\mu) \tag{9.1.2b}$$

Just as the spectral function $[A]$ represents the matrix version of the density of states per unit energy, the correlation function $[G^n]$ is the matrix version of the electron density per unit energy.

Non-equilibrium density matrix: In Section 9.2 the first result we will prove is that when the device is connected to two contacts with two distinct Fermi functions $f_1(E)$ and $f_2(E)$, the density matrix is given by Eq. (9.1.2a) with (dropping the argument E for clarity)

$$[G^n] = [A_1]f_1 + [A_2]f_2 \tag{9.1.3}$$

where

$$A_1 = G\Gamma_1 G^+ \quad \text{and} \quad A_2 = G\Gamma_2 G^+ \tag{9.1.4}$$

$$G = [EI - H - \Sigma_1 - \Sigma_2]^{-1} \tag{9.1.5}$$

suggesting that a fraction $[A_1]$ of the spectral function remains in equilibrium with the source Fermi function f_1, while another fraction $[A_2]$ remains in equilibrium with the drain Fermi function f_2. We will show that these two partial spectral functions indeed add up to give the total spectral function $[A]$ that we discussed in Chapter 8:

$$[A] \equiv \mathrm{i}[G - G^+] = [A_1] + [A_2] \tag{9.1.6}$$

Current: Next (in Section 9.3) we will show that the current I_i at terminal i can be written in the form

$$I_i = (-q/h) \int_{-\infty}^{+\infty} \mathrm{d}E\, \tilde{I}_i(E) \tag{9.1.7a}$$

with

$$\tilde{I}_i = \mathrm{Trace}[\Gamma_i A] f_i - \mathrm{Trace}[\Gamma_i G^n] \tag{9.1.7b}$$

representing a dimensionless current per unit energy. This leads to the picture shown in Fig. 9.1.2 which can be viewed as the quantum version of our elementary picture from Chapter 1 (Fig. 9.1.1).

One-level model: In Chapter 1, we went through an example with just one level so that the electron density and current could all be calculated from a rate equation with a simple model for broadening. I then indicated that in general we need a matrix version of this "scalar model" and that is what the rest of the book is about (see Fig. 1.6.5).

It is instructive to check that the full "matrix model" we have stated above (and will derive in this chapter) reduces to our old results (Eqs. (1.6.4)–(1.6.6)) when we specialize to a one-level system so that all the matrices reduce to pure numbers.

From Eq. (9.1.5), $G(E) = [E - \varepsilon + (i\Gamma/2)]^{-1}$

From Eq. (9.1.4), $A_1(E) = \dfrac{\Gamma_1}{(E-\varepsilon)^2 + (\Gamma/2)^2}$, $A_2(E) = \dfrac{\Gamma_2}{(E-\varepsilon)^2 + (\Gamma/2)^2}$

From Eq. (9.1.6), $A(E) = \dfrac{\Gamma}{(E-\varepsilon)^2 + (\Gamma/2)^2}$

From Eq. (9.1.3), $G^n(E) = A(E)\left(\dfrac{\Gamma_1}{\Gamma} f_1(E) + \dfrac{\Gamma_2}{\Gamma} f_2(E)\right)$

which can be compared with Eq. (1.6.4). Similarly, from Eqs. (9.1.7) the current at the two terminals is given by (cf. Eqs. (1.6.5a, b)):

$$I_1 = \frac{q}{h}\int_{-\infty}^{+\infty} dE\,\Gamma_1[A(E)f_1(E) - G^n(E)]$$

$$I_2 = \frac{q}{h}\int_{-\infty}^{+\infty} dE\,\Gamma_2[A(E)f_2(E) - G^n(E)]$$

Transmission: Equation (9.1.7) can be combined with (9.1.3) and (9.1.6) to write

$$\overline{I}_1 = -\overline{I}_2 = \overline{T}(E)(f_1(E) - f_2(E))$$

where

$$\overline{T}(E) \equiv \text{Trace}[\Gamma_1 A_2] = \text{Trace}[\Gamma_2 A_1] \tag{9.1.8}$$

The current I in the external circuit is given by

$$I = (q/h)\int_{-\infty}^{+\infty} dE\,\overline{T}(E)(f_1(E) - f_2(E)) \tag{9.1.9}$$

The quantity $\overline{T}(E)$ appearing in the current equation (Eq. (9.1.9)) is called the *transmission* function, which tells us the rate at which electrons transmit from the source to the drain contacts by propagating through the device. Knowing the device Hamiltonian [H] and its coupling to the contacts described by the self-energy matrices $\Sigma_{1,2}$, we can calculate the current either from Eqs. (9.1.7) or from Eq. (9.1.9). This procedure can be used to analyze any device as long as the evolution of electrons through the device is *coherent*. Let me explain what that means.

The propagation of electrons is said to be coherent if it does not suffer phase-breaking scattering processes that cause a change in the state of an external object. For example, if an electron were to be deflected from a rigid (that is unchangeable) defect in the lattice, the propagation would still be considered coherent. The effect could be incorporated through an appropriate defect potential in the Hamiltonian [H] and we could still calculate the current from Fig. 9.1.2. But, if the electron transferred some energy

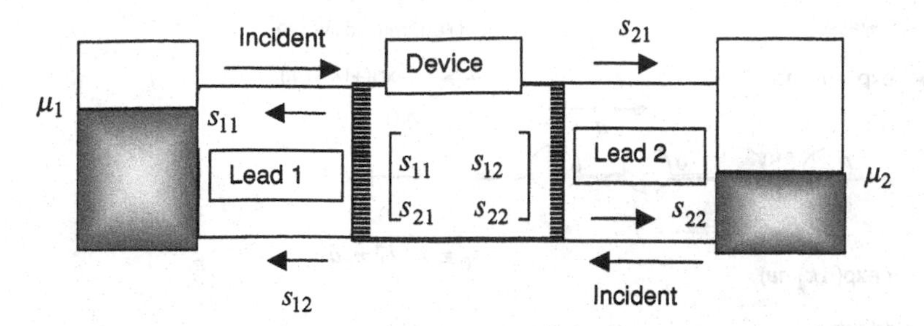

Fig. 9.1.3 The transmission formalism assumes the device to be connected via ideal multi-moded quantum wires to the contacts and the transmission function is related to the *S*-matrix between these leads.

to the atomic lattice causing it to start vibrating that would constitute a phase-breaking process and the effect cannot be included in [*H*]. How it can be included is the subject of Chapter 10.

I should mention here that coherent transport is commonly treated using the transmission formalism which starts with the assumption that the device is connected to the contacts by two ideal leads which can be viewed as multi-moded quantum wires so that one can calculate an *S*-matrix for the device (Fig. 9.1.3), somewhat like a microwave waveguide. The transmission matrix s_{21} (or s_{12}) is of size $M \times N$ (or $N \times M$) if lead 1 has *N* modes and lead 2 has *M* modes and the transmission function is obtained from its trace: $\overline{T}(E) = \text{Trace}[s_{12}s_{12}^+] = \text{Trace}[s_{21}s_{21}^+]$. This approach is widely used and seems quite appealing especially to those familiar with the concept of *S*-matrices in microwave waveguides.

Transmission from Green's function: For coherent transport, one can calculate the transmission from the Green's function method, using the relation

$$\overline{T}(E) \equiv \text{Trace}[\Gamma_1 G \Gamma_2 G^+] = \text{Trace}[\Gamma_2 G \Gamma_1 G^+] \tag{9.1.10}$$

obtained by combining Eq. (9.1.8) with (9.1.4). In Sections 9.2 and 9.3 we will derive all the equations (Eq. (9.1.2a)–(9.1.7a)) given in this section. But for the moment let me just try to justify the expression for the transmission (Eq. (9.1.10)) using a simple example. Consider now a simple 1D wire modeled with a discrete lattice (Fig. 9.1.4). We wish to calculate the transmission coefficient

$$\overline{T}(E) = (v_2/v_1)|t|^2 \tag{9.1.11}$$

where the ratio of velocities (v_2/v_1) is included because the transmission is equal to the ratio of the transmitted to the incident current, and the current is proportional to the velocity times the probability $|\psi|^2$.

To calculate the transmission from the Green's function approach, we start from the Schrödinger equation $[EI - H]\{\psi\} = \{0\}$, describing the entire infinite system and use

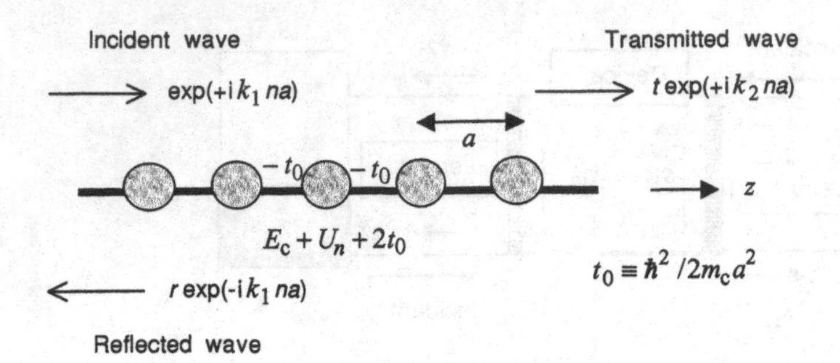

Fig. 9.1.4

the same approach described in Section 8.1 to eliminate the semi-infinite leads

$$[EI - H - \Sigma_1 - \Sigma_2]\{\psi\} = \{S\} \rightarrow \{\psi\} = [G]\{S\} \tag{9.1.12}$$

where $[G]$ is given by Eq. (9.1.5). Σ_1 and Σ_2 are matrices that represent the effects of the two leads: each has only one non-zero element (see Eq. (8.1.7a)):

$$\Sigma_1(1,1) = -t_0 \exp(ik_1 a) \qquad \Sigma_2(N,N) = -t_0 \exp(ik_2 a)$$

corresponding to the end point of the channel (1 or N) where the lead is connected. The source term $\{S\}$ is a column vector with just one non-zero element corresponding to the end point (1) on which the electron wave is incident (see Eq. (8.1.7b)):

$$S(1) = \mathrm{i}2t_0 \sin k_1 a = \mathrm{i}(\hbar v_1/a)$$

Note that in general for the same energy E, the k-values (and hence the velocities) can be different at the two ends of the lattice since the potential energy U is different:

$$E = E_c + U_1 + 2t_0 \cos k_1 a = E_c + U_N + 2t_0 \cos k_2 a$$

From Eq. (9.1.12) we can write $t = \psi(N) = G(N,1)\,S(1)$ so that from Eq. (9.1.11)

$$\overline{T}(E) = (\hbar v_1/a)(\hbar v_2/a)\,|G(1,N)|^2$$

which is exactly what we get from the general expression in Eq. (9.1.10).

This simple example is designed to illustrate the relation between the Green's function and transmission points of view. I believe the advantages of the Green's function formulation are threefold.

(1) The generality of the derivation shows that the basic results apply to arbitrarily shaped channels described by $[H]$ with arbitrarily shaped contacts described by $[\Sigma_1]$, $[\Sigma_2]$. This partitioning of the channels from the contacts is very useful when dealing with more complicated structures.
(2) The Green's function approach allows us to calculate the density matrix (hence the electron density) as well. This can be done within the transmission formalism, but less straightforwardly (Cahay *et al.*, 1987).

(3) The Green's function approach can handle incoherent transport with phase-breaking scattering, as we will see in Chapter 10. Phase-breaking processes can only be included phenomenologically within the transmission formalism (Büttiker, 1988).

We will derive the expressions for the density matrix, Eqs. (9.1.3)–(9.1.6), in Section 9.2, the expression for the current, Eq. (9.1.7), in Section 9.3, discuss the relation with the transmission formalism in Section 9.4, and finally present a few illustrative examples in Section 9.5. In short, Sections 9.2–9.4 derive the expressions stated in this section, while Section 9.5 applies them. The reader is encouraged to take a look ahead at Section 9.5, as it might help motivate him/her to suffer through the intervening sections.

9.2 Density matrix

In this section we will derive the results stated in Section 9.1 for the non-equilibrium density matrix (Eqs. (9.1.3)–(9.1.6)) for a channel connected to two contacts. In the next section we will derive the current expressions (Eqs. (9.1.7)–(9.1.9)).

Channel with one contact: I would like to start by revisiting the problem of a channel connected to one contact and clearing up a conceptual issue, before we take on the real problem with two contacts. In Section 8.1 we started from a Schrödinger equation for the composite contact–channel system

$$\begin{pmatrix} EI_R - H_R + i\eta & -\tau^+ \\ -\tau & EI - H \end{pmatrix} \begin{Bmatrix} \Phi_R + \chi \\ \psi \end{Bmatrix} = \begin{Bmatrix} S_R \\ 0 \end{Bmatrix} \tag{9.2.1}$$

and showed that the scattered waves $\{\psi\}$ and $\{\chi\}$ can be viewed as arising from the "spilling over" of the wavefunction $\{\Phi_R\}$ in the isolated contact (Fig. 9.2.1). Using straightforward matrix algebra we obtained

$$\{\chi\} = G_R \tau^+ \{\psi\} \tag{9.2.2}$$

where

$$G_R \equiv [EI_R - H_R + i\eta]^{-1} \tag{9.2.3}$$

$$\{\psi\} = [G]\{S\} \tag{9.2.4}$$

$$G \equiv [EI - H - \Sigma]^{-1} \tag{9.2.5}$$

$$\Sigma \equiv \tau G_R \tau^+ \tag{9.2.6}$$

$$\{S\} = \tau\{\Phi_R\} \tag{9.2.7}$$

Since there is only one contact this is really an equilibrium problem and the density matrix is obtained simply by filling up the spectral function

$$A(E) = i[G - G^+] \tag{9.2.8}$$

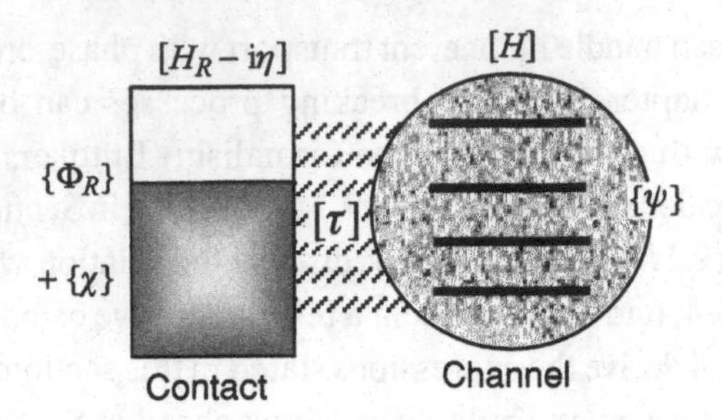

Fig. 9.2.1

according to the Fermi function as stated in Eqs. (9.1.2a, b). What I would like to do now is to obtain this result in a completely different way. I will assume that the source waves $\{\Phi_R\}$ from the contact are filled according to the Fermi function and the channel itself is filled simply by the spilling over of these wavefunctions. I will show that the resulting density matrix in the channel is identical to what we obtained earlier. Once we are clear about the approach we will extend it to the real problem with two contacts.

Before we connect the contact to the device, the electrons will occupy the contact eigenstates α according to its Fermi function, so that we can write down the density matrix for the contact as

$$\rho_R(\vec{r}, \vec{r}') = \sum_{\alpha} \phi_\alpha(\vec{r}) f_0(\varepsilon_\alpha - \mu) \phi_\alpha^*(\vec{r}')$$

or in matrix notation as

$$[\rho_R] = \sum_{\alpha} f_0(\varepsilon_\alpha - \mu)\{\phi_\alpha\}\{\phi_\alpha\}^+ \tag{9.2.9}$$

Now we wish to calculate the device density matrix by calculating the response of the device to the excitation $\tau\{\Phi\}$ from the contact. We can write the source term due to each contact eigenstate α as $\{S_\alpha\} = \tau\{\phi_\alpha\}$, find the resulting device wavefunction from Eqs. (9.2.4) and (9.2.7) $\{\psi_\alpha\} = G\tau\{\phi_\alpha\}$, and then obtain the device density matrix by adding up the individual components weighted by the appropriate Fermi factors for the original contact eigenstate α:

$$\begin{aligned}
[\rho] &= \sum_{\alpha} f_0(\varepsilon_\alpha - \mu)\{\psi_\alpha\}\{\psi_\alpha\}^+ \\
&= \int \mathrm{d}E\, f_0(E - \mu) \sum_{\alpha} \delta(E - \varepsilon_\alpha)\{\psi_\alpha\}\{\psi_\alpha\}^+ \\
&= \int \mathrm{d}E\, f_0(E - \mu) G\tau \left[\sum_{\alpha} \delta(E - \varepsilon_\alpha)\{\phi_\alpha\}\{\phi_\alpha\}^+\right] \tau^+ G^+ \\
&= \int \frac{\mathrm{d}E}{2\pi} f_0(E - \mu) G\tau A_R \tau^+ G^+
\end{aligned} \tag{9.2.10}$$

making use of the expression for the spectral function in the contact Eq. (8.2.3):

$$A_R(E) = \sum_{\alpha} \delta(E - \varepsilon_\alpha)\{\phi_\alpha\}\{\phi_\alpha\}^+ \tag{9.2.11}$$

From Eq. (9.2.6),

$$\Gamma = \mathrm{i}[\Sigma - \Sigma^+] = \tau A_R \tau^+ \tag{9.2.12}$$

so that from Eq. (9.2.10) we can write

$$[G^n] = [G\Gamma G^+] f_0(E - \mu) \tag{9.2.13}$$

where we have made use of Eq. (9.1.2a). To show that this is the same as our earlier result (Eqs. (9.1.2a, b)), we need the following *important* identity. If

$$[G] = [EI - H - \Sigma]^{-1} \quad \text{and} \quad \Gamma = \mathrm{i}[\Sigma - \Sigma^+]$$

then

$$A \equiv \mathrm{i}[G - G^+] = G\Gamma G^+ = G^+\Gamma G \tag{9.2.14}$$

This is shown by writing $(G^+)^{-1} - G^{-1} = \Sigma - \Sigma^+ = -\mathrm{i}\Gamma$, then pre-multiplying with G and post-multiplying with G^+ to obtain

$$G - G^+ = -\mathrm{i}G\Gamma G^+ \rightarrow A = G\Gamma G^+$$

Alternatively if we pre-multiply with G^+ and post-multiply with G we obtain

$$G - G^+ = -\mathrm{i}G^+\Gamma G \rightarrow A = G^+\Gamma G$$

I have used this simple one-contact problem to illustrate the important physical principle that the different eigenstates are *uncorrelated* and so we should calculate their contributions to the density matrix independently and then add them up.

This is a little bit like Young's two-slit experiment shown below. If the two slits are illuminated coherently then the intensity on the screen will show an interference pattern.

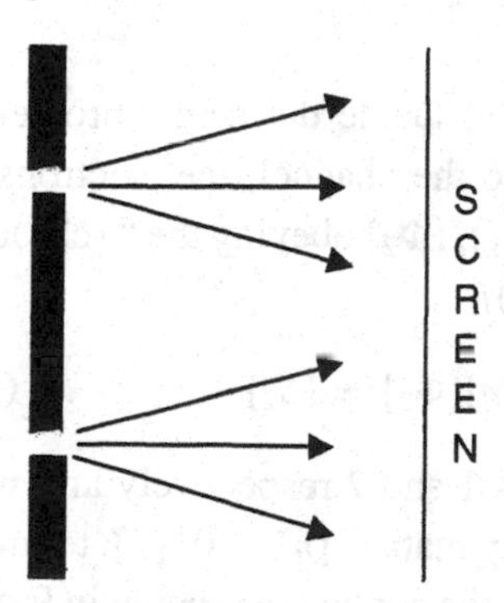

But if the slits are illuminated incoherently then the intensity on the screen is simply the sum of the intensities we would get from *each slit independently*. Each

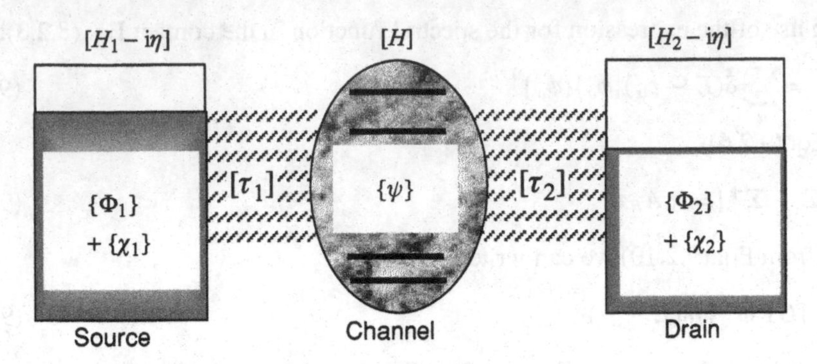

Fig. 9.2.2 A channel connected to two contacts.

eigenstate α is like a "slit" that "illuminates" the device and the important point is that the "slits" have no phase coherence. That is why we calculate the device density matrix for each "slit" α independently and add them up.

Now that we have been through this exercise once, it is convenient to devise the following rule for dealing with the contact and channel wavefunctions

$$\{\Phi_R\}\{\Phi_R^+\} \Rightarrow \int (\mathrm{d}E/2\pi) f_0(E-\mu)[A_R(E)] \tag{9.2.15a}$$

$$\{\psi\}\{\psi^+\} \Rightarrow \int (\mathrm{d}E/2\pi)[G^{\mathrm{n}}(E)] \tag{9.2.15b}$$

reflecting the fact that the electrons in the contact are distributed according to the Fermi function $f_0(E-\mu)$ in a continuous distribution of eigenstates described by the spectral function $A_R(E)$. This rule can be used to shorten the algebra considerably. For example, to evaluate the density matrix we first write down the result for a single eigenstate

$$\{\psi\} = G\tau\{\Phi_R\} \rightarrow \{\psi\}\{\psi\}^+ = G\tau\{\Phi_R\}\{\Phi_R\}^+\tau^+G^+$$

and then apply Eq. (9.2.11) to obtain $[G^{\mathrm{n}}] = [G\tau A_R\tau^+G^+]f_0(E-\mu)$, which reduces to Eq. (9.2.13) making use of Eq. (9.2.12).

Channel with two contacts: Now we are ready to tackle the actual problem with two contacts. We assume that before connecting to the channel, the electrons in the source and the drain contact have wavefunctions $\{\Phi_1\}$, $\{\Phi_2\}$ obeying the "Schrödinger" equations for the isolated contacts (see Eq. (8.1.8b)):

$$[EI - H_1 + \mathrm{i}\eta]\{\Phi_1\} = \{S_1\} \quad \text{and} \quad [EI - H_2 + \mathrm{i}\eta]\{\Phi_2\} = \{S_2\} \tag{9.2.16}$$

where $[H_1]$, $[H_2]$ are the Hamiltonians for contacts 1 and 2 respectively and we have added a small positive infinitesimal times an identity matrix, $[\eta] = 0^+[I]$, to introduce dissipation as before. When we couple the device to the contacts as shown in Fig. 9.2.2, these electronic states from the contacts "spill over" giving rise to a wavefunction $\{\psi\}$

inside the device which in turn excites scattered waves $\{\chi_1\}$ and $\{\chi_2\}$ in the source and drain respectively.

The overall wavefunction will satisfy the composite Schrödinger equation for the composite contact-1–device–contact-2 system which we can write in three blocks (cf. Eq. (8.1.8b)):

$$\begin{pmatrix} EI - H_1 + i\eta & -\tau_1^+ & 0 \\ -\tau_1 & EI - H & -\tau_2 \\ 0 & -\tau_2^+ & EI - H_2 + i\eta \end{pmatrix} \begin{Bmatrix} \Phi_1 + \chi_1 \\ \psi \\ \Phi_2 + \chi_2 \end{Bmatrix} = \begin{Bmatrix} S_1 \\ 0 \\ S_2 \end{Bmatrix} \qquad (9.2.17)$$

where $[H]$ is the channel Hamiltonian. Using straightforward matrix algebra we obtain from the first and last equations

$$\{\chi_1\} = G_1 \tau_1^+ \{\psi\} \quad \text{and} \quad \{\chi_2\} = G_2 \tau_2^+ \{\psi\} \qquad (9.2.18)$$

where

$$G_1 = [EI - H_1 + i\eta]^{-1} \quad \text{and} \quad G_2 = [EI - H_2 + i\eta]^{-1} \qquad (9.2.19)$$

are the Green's functions for the isolated reservoirs. Using Eqs. (9.2.18) to eliminate $\{\chi_1\}, \{\chi_2\}$ from the middle equation in Eq. (9.2.17) we obtain

$$[EI - H - \Sigma_1 - \Sigma_2]\{\psi\} = \{S\} \qquad (9.2.20)$$

where

$$\Sigma_1 = \tau_1 G_1 \tau_1^+ \quad \text{and} \quad \Sigma_2 = \tau_2 G_2 \tau_2^+ \qquad (9.2.21)$$

are the self-energy matrices that we discussed in Chapter 8. The corresponding broadening matrices (Eq. (9.2.12)) are given by

$$\Gamma_1 = \tau_1 A_1 \tau_1^+ \quad \text{and} \quad \Gamma_2 = \tau_2 A_2 \tau_2^+ \qquad (9.2.22)$$

where $A_1 = i[G_1 - G_1^+]$ and $A_2 = i[G_2 - G_2^+]$ are the spectral functions for the isolated contacts 1 and 2 respectively. Also,

$$\{S\} \equiv \tau_1\{\Phi_1\} + \tau_2\{\Phi_2\} \qquad (9.2.23)$$

is the sum of the source terms $\tau_1 \Phi_1$ (from the source) and $\tau_2 \Phi_2$ (from the drain) as shown in Fig. 9.2.3.

To evaluate the density matrix, we define the channel Green's function

$$G \equiv [EI - H - \Sigma_1 - \Sigma_2]^{-1} \qquad (9.2.24)$$

and use it to express the channel wavefunction in terms of the source terms from Eq. (9.2.20):

$$\{\psi\} = G\{S\} \rightarrow \{\psi\}\{\psi\}^+ = G\{S\}\{S\}^+ G^+ \qquad (9.2.25)$$

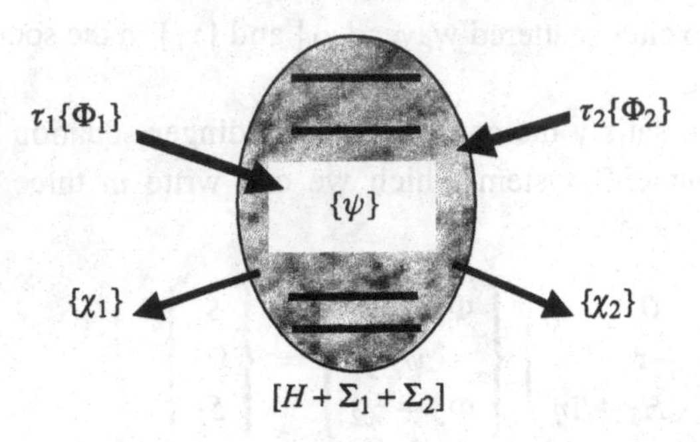

Fig. 9.2.3 Channel excited by $\tau_1\Phi_1$ (from the source) and $\tau_2\Phi_2$ (from the drain). The channel response is decribed by Eq. (9.2.20) and it in turn generates $\{\chi_1\}$, $\{\chi_2\}$ in the contacts (see Eq. (9.2.18)).

Note that the cross-terms in the source

$$SS^+ = \tau_1\Phi_1\Phi_1^+\tau_1^+ + \tau_2\Phi_2\Phi_2^+\tau_2^+$$
$$\underbrace{+\ \tau_1\Phi_1\Phi_2^+\tau_2^+ + \tau_2\Phi_2\Phi_1^+\tau_1^+}_{\text{Cross-terms} = 0}$$

are zero since Φ_1 and Φ_2 are the wavefunctions (before connecting to the channel) in the source and drain contacts which are physically disjoint and unconnected. The direct terms are evaluated using the basic principle (see Eqs. (9.2.15)) that we formulated earlier with the one-contact problem:

$$\{\Phi_1\}\{\Phi_1^+\} \Rightarrow \int (\mathrm{d}E/2\pi) f_1(E)[A_1(E)] \tag{9.2.26a}$$

$$\{\Phi_2\}\{\Phi_2^+\} \Rightarrow \int (\mathrm{d}E/2\pi) f_2(E)[A_2(E)] \tag{9.2.26b}$$

to write down the density matrix from $\{\psi\}\{\psi\}^+ = G\{S\}\{S\}^+G^+$

$$\rho = \int (\mathrm{d}E/2\pi)\{[G\tau_1 A_1 \tau_1^+ G^+] f_1 + [G\tau_2 A_2 \tau_2^+ G^+] f_2\}$$

Making use of Eq. (9.2.22) we can simplify this expression to write

$$G^{\mathrm{n}} = G\Sigma^{\mathrm{in}}G^+ \tag{9.2.27}$$

and

$$[\Sigma^{\mathrm{in}}] = [\Gamma_1]f_1 + [\Gamma_2]f_2 \tag{9.2.28}$$

noting that $[\rho] = \int (\mathrm{d}E/2\pi)[G^{\mathrm{n}}]$ as defined in Eq. (9.1.2a). Just as G^{n} is obtained from $\{\psi\}\{\psi\}^+$, Σ^{in} is obtained from $\{S\}\{S\}^+$. One could thus view Eq. (9.2.27) as a relation between the "electron density" in the device created by the source term $\{S\}$ representing the spill-over of electrons from the contacts.

Partial spectral function: Substituting Eq. (9.2.28) into Eq. (2.9.27) we can write

$$[G^n] = [A_1]f_1 + [A_2]f_2 \tag{9.2.29}$$

where

$$A_1 = G\Gamma_1 G^+ \quad \text{and} \quad A_2 = G\Gamma_2 G^+$$

Comparing this with the equilibrium result (see Eq. (9.1.2b)), $[G^n] = [A]f_0$, it seems natural to think of the total spectral function $[A(E)]$ as consisting of two parts: $[A_1(E)]$ arising from the spill-over (or propagation) of states in the left contact and $[A_2(E)]$ arising from the spill-over of states in the right contact. The former is filled according to the left Fermi function $f_1(E)$ while the latter is filled according to the right Fermi function $f_2(E)$. To show that the two partial spectra indeed add up to give the correct total spectral function, $A = A_1 + A_2$, we note from Eq. (9.2.14) that since the self-energy Σ has two parts Σ_1 and Σ_2 coming from two contacts, $A = G[\Gamma_1 + \Gamma_2]G^+ = A_1 + A_2$ as stated in Eq. (9.1.6).

Exclusion principle? An important conceptual point before we move on. Our approach is to use the Schrödinger equation to calculate the evolution of a specific eigenstate Φ_α from one of the contacts and then superpose the results from distinct eigenstates to obtain the basic rule stated in Eqs. (9.2.15) or Eq. (9.2.26). It may appear that by superposing all these individual fluxes we are ignoring the Pauli exclusion principle. Wouldn't the presence of electrons evolving out of one eigenstate *block* the flux evolving out of another eigenstate? The answer is no, as long as the evolution of the electrons is coherent. This is easiest to prove in the time domain, by considering two electrons that originate in distinct eigenstates $\{\Phi_1\}$ and $\{\Phi_2\}$. Initially there is no question of one blocking the other since they are orthogonal: $\{\Phi_1\}^+\{\Phi_2\} = 0$. At later times their wavefunctions can be written as

$$\{\psi_1(t)\} = \exp[-\mathrm{i}Ht/\hbar]\{\Phi_1\}$$
$$\{\psi_2(t)\} = \exp[-\mathrm{i}Ht/\hbar]\{\Phi_2\}$$

if both states evolve coherently according to the Schrödinger equation:

$$\mathrm{i}\hbar\mathrm{d}\{\psi\}/\mathrm{d}t = [H]\{\psi\}$$

It is straightforward to show that the overlap between any two states does not change as a result of this evolution: $\{\psi_1(t)\}^+\{\psi_2(t)\} = \{\Phi_1\}^+\{\Phi_2\}$. Hence, wavefunctions originating from orthogonal states remain orthogonal at all times and never "Pauli block" each other. Note, however, that this argument cannot be used when phase-breaking processes (briefly explained in the introduction to this chapter) are involved since the evolution of electrons cannot be described by a one-particle Schrödinger equation.

9.3 Inflow/outflow

Now that we have derived the results for the non-equilibrium density matrix (see Eqs. (9.1.3)–(9.1.6)), let us discuss the current flow at the terminals (Eqs. (9.1.7)–(9.1.9)). As before let us start with the "one-contact" problem shown in Fig. 9.2.1.

Channel with one contact: Consider again the problem of a channel connected to one contact described by:

$$E\begin{Bmatrix}\psi\\ \Phi\end{Bmatrix}=\begin{bmatrix}H & \tau\\ \tau^+ & H_R-\mathrm{i}\eta\end{bmatrix}\begin{Bmatrix}\psi\\ \Phi\end{Bmatrix}$$

which is the same as Eq. (9.2.1) with $\{\Phi\}\equiv\{\Phi_R+\chi\}$ and $\{S_R\}$ dropped for clarity. How can we evaluate the current flowing between the channel and the contact? Just as we did in Section 6.4 (when discussing the velocity of a band electron), we need to look at the time-dependent version of this equation

$$\mathrm{i}\hbar\frac{\mathrm{d}}{\mathrm{d}t}\begin{Bmatrix}\psi\\ \Phi\end{Bmatrix}=\begin{bmatrix}H & \tau\\ \tau^+ & H_R-\mathrm{i}\eta\end{bmatrix}\begin{Bmatrix}\psi\\ \Phi\end{Bmatrix}$$

and obtain an expression for the time rate of change in the probability density inside the channel, which is given by

$$\mathrm{Trace}[\psi\psi^+]=\mathrm{Trace}[\psi^+\psi]=\psi^+\psi$$

(note that $\psi^+\psi$ is just a number and so it does not matter if we take the trace or not):

$$I\equiv\frac{\mathrm{d}}{\mathrm{d}t}\psi^+\psi=\frac{\mathrm{Trace}[\psi^+\tau\Phi-\Phi^+\tau^+\psi]}{\mathrm{i}\hbar}\tag{9.3.1}$$

Noting that $\{\Phi\}\equiv\{\Phi_R+\chi\}$, we can divide this net current I conceptually into an inflow, proportional to the "incident" wave $\{\Phi_R\}$, and an outflow proportional to the "scattered" wave $\{\chi\}$:

$$I=\underbrace{\frac{\mathrm{Trace}[\psi^+\tau\Phi_R-\Phi_R^+\tau^+\psi]}{\mathrm{i}\hbar}}_{\text{Inflow}}-\underbrace{\frac{\mathrm{Trace}[\chi^+\tau^+\psi-\psi^+\tau\chi]}{\mathrm{i}\hbar}}_{\text{Outflow}}\tag{9.3.2}$$

Making use of Eqs. (9.2.4) and (9.2.7) we can write the inflow as

$$\text{Inflow}=\mathrm{Trace}[S^+G^+S-S^+GS]/\mathrm{i}\hbar=\mathrm{Trace}[SS^+A]/\hbar$$

since $\mathrm{i}[G-G^+]=[A]$. To obtain the total inflow we need to sum the inflows due to each contact eigenstate α, all of which as we have seen are taken care of by the replacement

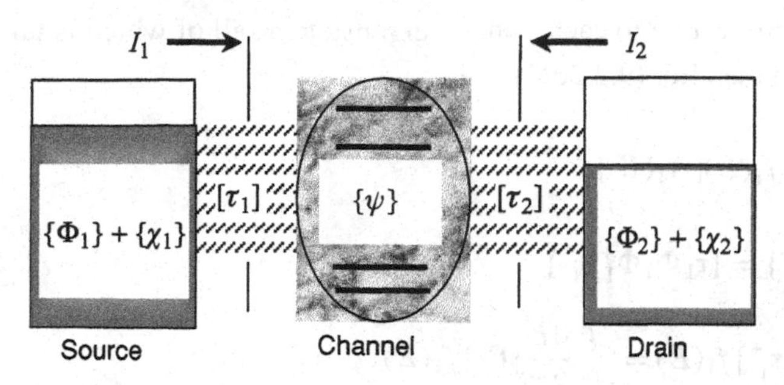

Fig. 9.3.1

(see Eq. (9.2.15a))

$$\{\Phi_R\}\{\Phi_R^+\} \Rightarrow \int \frac{\mathrm{d}E}{2\pi} f_0(E-\mu)[A_R(E)]$$

Since $S = \tau\Phi_R$, this leads to

$$SS^+ \Rightarrow \int \frac{\mathrm{d}E}{2\pi} f_0(E-\mu)\tau A_R \tau^+ = \int \frac{\mathrm{d}E}{2\pi} f_0(E-\mu)[\Gamma]$$

so that the inflow term becomes

$$\text{Inflow} = \frac{1}{\hbar}\int \frac{\mathrm{d}E}{2\pi} f_0(E-\mu)\,\text{Trace}[\Gamma A] \tag{9.3.3a}$$

Similarly, we make use of Eqs. (9.2.2) and (9.2.12) to write the outflow term as

$$\text{Outflow} = \text{Trace}[\psi^+\tau G_R^+\tau^+\psi - \psi^+\tau G_R\tau^+\psi]/i\hbar = \text{Trace}[\psi\psi^+\Gamma]/\hbar$$

On summing over all the eigenstates, $\psi\psi^+ \Rightarrow \int \mathrm{d}E\, G^{\mathrm{n}}/2\pi$, so that

$$\text{Outflow} = \frac{1}{\hbar}\int \frac{\mathrm{d}E}{2\pi}\,\text{Trace}[\Gamma G^{\mathrm{n}}] \tag{9.3.3b}$$

It is easy to see that the inflow and outflow are equal at equilibrium, since $G^{\mathrm{n}} = Af_0$ (see Eq. (9.1.2b)).

Channel with two contacts: Now we are ready to calculate the inflow and outflow for the channel with two contacts (see Fig. 9.3.1). We consider one of the interfaces, say the one with the source contact, and write the inflow as (cf. Eq. (9.3.2))

$$I_1 = \underbrace{\frac{\text{Trace}[\psi^+\tau_1\Phi_1 - \Phi_1^+\tau_1^+\psi]}{i\hbar}}_{\text{Inflow}} - \underbrace{\frac{\text{Trace}[\chi_1^+\tau_1^+\psi - \psi^+\tau_1\chi_1]}{i\hbar}}_{\text{Outflow}}$$

Making use of the relations $\psi = GS$ (with S and G defined in Eqs. (9.2.23) and (9.2.24)) and $\{S_1\} \equiv \tau_1\{\Phi_1\}$, we can write

$$\text{Inflow} = \text{Trace}[S^+G^+S_1 - S_1^+GS]/i\hbar = \text{Trace}[S_1S_1^+A]/\hbar$$

since $S = S_1 + S_2$ and $S_1^+S_2 = S_2^+S_1 = 0$.

Next we sum the inflow due to each contact eigenstate α, all of which is taken care of by the replacement (see Eq. (9.2.26a))

$$\{\Phi_1\}\{\Phi_1^+\} \Rightarrow \int \frac{\mathrm{d}E}{2\pi} f_1(E)[A_1(E)]$$

leading to $\{S_1\}\{S_1^+\} = [\tau_1 \Phi_1 \Phi_1^+ \tau_1^+]$

$$\Rightarrow \int \frac{\mathrm{d}E}{2\pi}[\tau_1 A_1 \tau_1^+] f_1(E) = \int \frac{\mathrm{d}E}{2\pi}[\Gamma_1] f_1(E)$$

so that

$$\text{Inflow} = \frac{1}{\hbar} \int \frac{\mathrm{d}E}{2\pi} f_1(E)\, \text{Trace}[\Gamma_1 A] \tag{9.3.4a}$$

Similarly we make use of Eqs. (9.2.18) and (9.2.22) to write the outflow term as

$$\text{Outflow} = \text{Trace}[\psi^+ \tau_1 G_1^+ \tau_1^+ \psi - \psi^+ \tau_1 G_1 \tau_1^+ \psi]/i\hbar = \text{Trace}[\psi \psi^+ \Gamma_1]/\hbar$$

On summing over all the eigenstates, $\{\psi\}\{\psi^+\} \Rightarrow \int (\mathrm{d}E/2\pi)[G^{\mathrm{n}}]$, so that

$$\text{Outflow} = (1/\hbar) \int \frac{\mathrm{d}E}{2\pi} \text{Trace}[\Gamma_1 G^{\mathrm{n}}] \tag{9.3.4b}$$

The net current I_i at terminal i is given by the difference between the inflow and the outflow (multiplied by the charge $-q$ of an electron) as stated in Eq. (9.1.7a)

$$I_i = (-q/\hbar) \int_{-\infty}^{+\infty} \frac{\mathrm{d}E}{2\pi} \tilde{I}_i(E)$$

with

$$\tilde{I}_i = \text{Trace}[\Gamma_i A] f_i - \text{Trace}[\Gamma_i G^{\mathrm{n}}] \tag{9.3.5}$$

and illustrated in Fig. 9.1.2.

9.4 Transmission

In the last section we obtained expressions for the current at each of the contacts, which can be expressed as the difference between an inflow and an outflow. In this section we will express the current in a slightly different form that gives a different perspective to the problem of current flow and helps establish a connection with the transmission formalism widely used in the literature. We start by combining Eq. (9.1.7b)

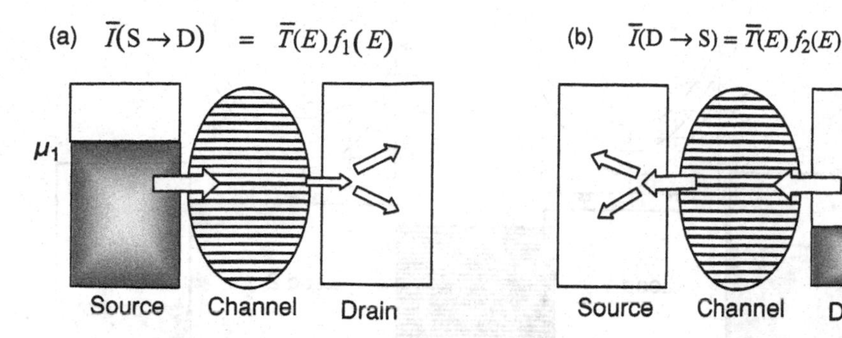

Fig. 9.4.1 The net current through the device (Eq. (9.4.1)) can be viewed as the difference between two counterpropagating fluxes from electrons: one from the source to the drain $f_2 = 0$ (a) and the other from the drain to the source $f_1 = 0$ (b).

with Eqs. (9.1.5) and (9.1.6) to write

$$\bar{I}_1 = \bar{T}_{12}(E)[f_1(E) - f_2(E)] \qquad \text{where } \bar{T}_{12}(E) \equiv \text{Trace}[\Gamma_1 A_2]$$
$$\bar{I}_2 = \bar{T}_{21}(E)[f_2(E) - f_1(E)] \qquad \text{where } \bar{T}_{21}(E) \equiv \text{Trace}[\Gamma_2 A_1]$$

We expect the currents at the two terminals to be equal and opposite and this is ensured if $\text{Trace}[\Gamma_1 A_2] = \text{Trace}[\Gamma_2 A_1]$. To show that they are indeed equal, we make use of Eq. (9.2.14) to show that

$$\text{Trace}[\Gamma_1 A] = \text{Trace}[\Gamma_1 G \Gamma G^+] = \text{Trace}[\Gamma G^+ \Gamma_1 G] = \text{Trace}[\Gamma A_1]$$

Subtracting $\text{Trace}[\Gamma_1 A_1]$ from both sides gives the desired result that $\text{Trace}[\Gamma_1 A_2] = \text{Trace}[\Gamma_2 A_1]$ (noting that $\Gamma = \Gamma_1 + \Gamma_2$ and $A = A_1 + A_2$). This allows us to write the current as (noting that $2\pi\hbar = h$)

$$I = (q/h) \int_{-\infty}^{+\infty} dE\, \bar{T}(E)[f_1(E) - f_2(E)] \tag{9.4.1}$$

where

$$\begin{aligned} \bar{T}(E) &\equiv \text{Trace}[\Gamma_1 A_2] = \text{Trace}[\Gamma_2 A_1] \\ &= \text{Trace}[\Gamma_1 G \Gamma_2 G^+] = \text{Trace}[\Gamma_2 G \Gamma_1 G^+] \end{aligned} \tag{9.4.2}$$

is called the transmission function. Physically we can view the current in Eq. (9.4.1) as the difference between two counterpropagating fluxes, one from the source to the drain and the other from the drain to the source as sketched in Fig. 9.4.1. One could view the device as a "semi-permeable membrane" that separates two reservoirs of electrons (source and drain) and the transmission function $\bar{T}(E)$ as a measure of the permeability of this membrane to electrons with energy E. We will show that the same function $\bar{T}(E)$ will govern both fluxes as long as transport is coherent.

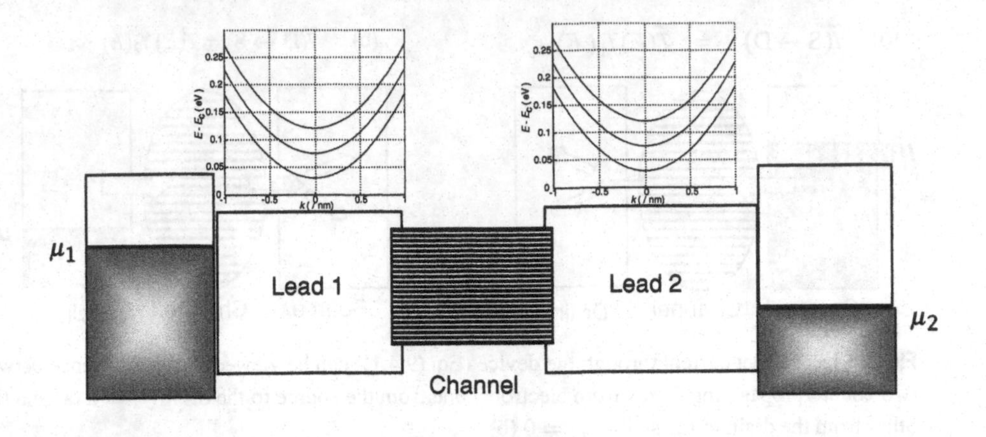

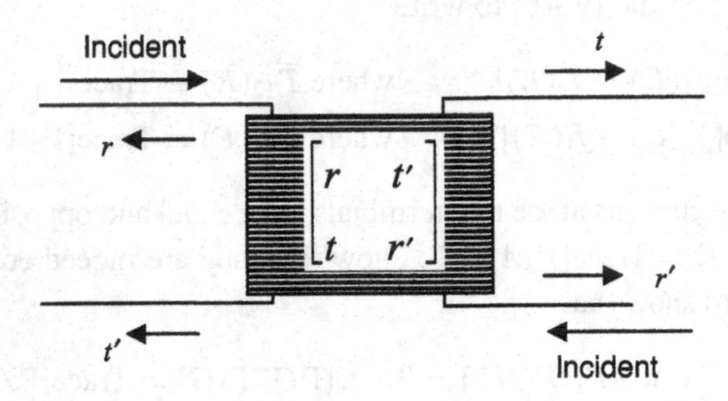

Fig. 9.4.2 In the transmission formalism, the channel is assumed to be connected to the contacts by two uniform leads that can be viewed as quantum wires with multiple subbands (see Chapter 6) having well-defined E–k relationships as shown. This allows us to define an S-matrix for the device analogous to a microwave waveguide.

Transmission formalism: In the transmission formalism (sometimes referred to as the Landauer approach) the channel is assumed to be connected to the contacts by two uniform leads that can be viewed as quantum wires with multiple modes or subbands (see Chapter 6) having well-defined E–k relationships as sketched in Fig. 9.4.2. This allows us to define an S-matrix for the device analogous to a microwave waveguide where the element t_{nm} of the t-matrix tells us the amplitude for an electron incident in mode m in lead 1 to transmit to a mode n in lead 2. It can then be shown that the current is given by Eq. (9.4.1) with the transmission function given by

$$\overline{T}(E) = \sum_m \sum_n |t_{nm}|^2 = \text{Trace}[tt^+] \tag{9.4.3}$$

This viewpoint, which is very popular, has the advantage of being based on relatively elementary concepts and also allows one to calculate the transmission function by solving a scattering problem. In the next section we will show with simple examples that this approach yields the same result as that obtained from $\overline{T} = \text{Trace}[\Gamma_2 G \Gamma_1 G^+]$ applied to devices with uniform leads.

Landauer formula: Landauer pioneered the use of the scattering theory of transport as a conceptual framework for clarifying the meaning of electrical conductance and stressed its fundamental connection to the transmission function: "Conductance *is* transmission." This basic relation can be seen starting from Eq. (9.4.1) (making use of Eqs. (9.1.1))

$$I = (q/h) \int_{-\infty}^{+\infty} \mathrm{d}E\, \overline{T}(E)[f_0(E-\mu_1) - f_0(E-\mu_2)]$$

and noting that the current is zero at equilibrium since $\mu_1 = \mu_2$. A small bias voltage V changes each of the functions $\overline{T}$, μ_1, and μ_2, and the resulting current can be written to first order as (δ denotes a small change)

$$I \approx (q/h) \int_{-\infty}^{+\infty} \mathrm{d}E\, \delta\overline{T}(E)[f_0(E-\mu_1) - f_0(E-\mu_2)]$$

$$+ (q/h) \int_{-\infty}^{+\infty} \mathrm{d}E\, \overline{T}(E)\, \delta[f_0(E-\mu_1) - f_0(E-\mu_2)]$$

The first term is zero and the second can be written as

$$I \approx (q^2 V/h) \int_{-\infty}^{+\infty} \mathrm{d}E\, \overline{T}(E)\, (-\partial f_0(E)/\partial E)_{E=\mu}$$

so that the conductance is given by

$$G = (q^2/h) T_0 \qquad \text{where } T_0 \equiv \int_{-\infty}^{+\infty} \mathrm{d}E\, \overline{T}(E)\, F_T(E-\mu) \tag{9.4.4}$$

and F_T is the thermal broadening function discussed in Chapter 7, which is peaked sharply around $E = \mu$ with a width proportional to $k_B T$ (see Fig. 7.3.4). The conductance is thus proportional to the transmission function averaged over an energy range of a few $k_B T$ around the equilibrium electrochemical potential μ, just as the quantum capacitance is proportional to the averaged density of states (see Eq. (7.3.8)).

The maximum value of the transmission function (and hence the conductance) is obtained if each of the M subbands or modes in one lead transmits perfectly to the other lead (see Fig. 9.4.2). The matrix $[tt^+]$ is then a diagonal matrix of size ($M \times M$) with ones along the diagonal, so that the transmission is equal to M. This suggests that the maximum transmission is equal to the number of modes M in the leads. But

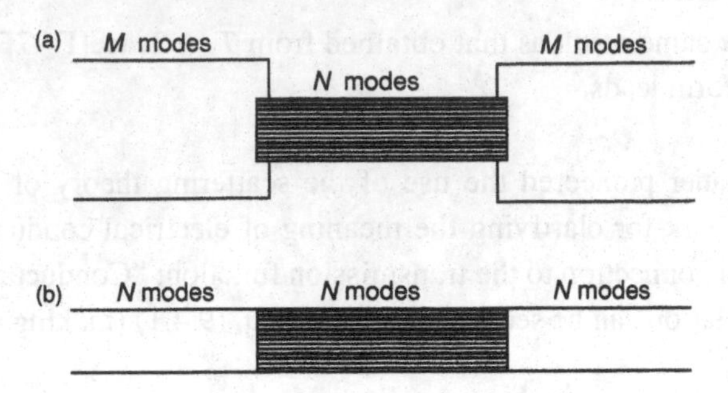

Fig. 9.4.3

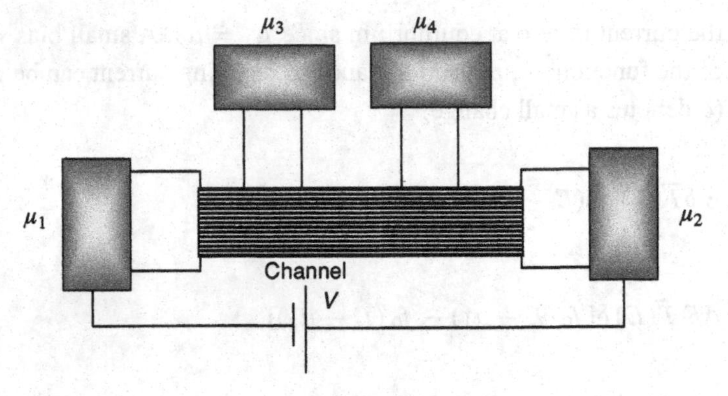

Fig. 9.4.4 Conductance measurements are commonly carried out in a four-probe configuration that can be analyzed using the Büttiker equations.

what happens if the device is narrower than the lead and has only N modes, $N < M$ (Fig. 9.4.3a)?

It can be argued that such a structure could not have a transmission any greater than a structure with the leads the same size as the channel (Fig. 9.4.3b) since in either case the electrons have to transmit through the narrow device region (assuming that the device is not so short as to allow direct tunneling). Since this latter structure has a maximum transmission of N that must be true of the first structure as well and detailed calculations do indeed show this to be the case. In general we can expect that the maximum transmission is equal to the number of modes in the narrowest segment. Earlier, in Chapter 6, we argued that the maximum conductance of a wire with N modes is equal to $(q^2/h)N$ based on the maximum current it could possibly carry.

Büttiker equations: Conductance measurements are often performed using a four-probe structure (Fig. 9.4.4) and their interpretation in small structures was initially unclear till Büttiker came up with an elegant idea (Büttiker, 1988). He suggested that the Landauer formula

$$G = (q^2/h)\tilde{T} \rightarrow I = (q/h)\tilde{T}[\mu_1 - \mu_2]$$

be extended to structures with multiple terminals by writing the current I_i at the ith terminal as

$$I_i = (q/h)\sum_j \tilde{T}_{ij}[\mu_i - \mu_j] \qquad (9.4.5)$$

where $\tilde{T}_{ij}$ is the average transmission from terminal j to i. We know the electrochemical potentials μ at the current terminals (1 and 2) but we do not know them at the voltage terminals, which float to a suitable potential so as to make the current zero. How do we calculate the currents from Eq. (9.4.5) since we do not know all the potentials? The point is that of the eight variables (four potentials and four currents), if we know any four, we can calculate the other four with simple matrix algebra. Actually, there are six independent variables. We can always set one of the potentials to zero, since only potential differences give rise to currents. Also, Kirchhoff's law requires all the currents to add up to zero, so that knowing any three currents we can figure out the fourth. So, it is convenient to set the potential at one terminal (say terminal 2) equal to zero and write Eq. (9.4.5) in the form of a (3 × 3) matrix equation:

$$\begin{Bmatrix} I_1 \\ I_3 \\ I_4 \end{Bmatrix} = \frac{q}{h}\begin{bmatrix} \tilde{T}_{12}+\tilde{T}_{13}+\tilde{T}_{14} & -\tilde{T}_{13} & -\tilde{T}_{14} \\ -\tilde{T}_{31} & \tilde{T}_{31}+\tilde{T}_{32}+\tilde{T}_{34} & -\tilde{T}_{34} \\ -\tilde{T}_{41} & -\tilde{T}_{43} & \tilde{T}_{41}+\tilde{T}_{42}+\tilde{T}_{43} \end{bmatrix}\begin{Bmatrix} \mu_1 \\ \mu_3 \\ \mu_4 \end{Bmatrix}$$

Knowing $\mu_1, I_3 = 0, I_4 = 0$, we can calculate I_1, μ_3, μ_4 and hence the four-probe conductance:

$$G_{\text{four-probe}} = (\mu_3 - \mu_4)/qI_1$$

We can visualize the Büttiker equations with a simple circuit model if the transmission coefficients are reciprocal, that is, if $\tilde{T}_{ij} = \tilde{T}_{ji}$. These equations are then identical to Kirchhoff's law applied to a network of conductors $G_{ij} \propto \tilde{T}_{ij} = \tilde{T}_{ji}$ connecting each pair of contacts i and j (see Fig. 9.4.5). However, this picture cannot be used if the transmission coefficients are non-reciprocal: $\tilde{T}_{ij} \neq \tilde{T}_{ij}$, as they are in Hall effect measurements where a magnetic field is present, some of the most notable applications of the Büttiker equations, Eq. (9.4.5), are to the interpretation of such measurements.

Büttiker probes: We mentioned earlier that the scattering theory of transport can only be used if the electrons transmit coherently through the device so that an S-matrix can be defined. But floating probes effectively extract electrons from the device and reinject them after phase randomization, thus effectively acting as phase-breaking scatterers. This is a seminal observation due to Büttiker that provides a simple phenomenological

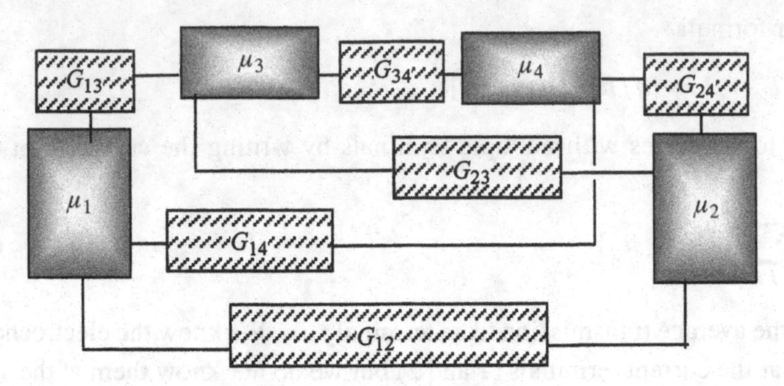

Fig. 9.4.5 The Büttiker equations can be visualized in terms of a conductor network if the transmission between terminals is reciprocal.

technique for including the effects of phase-breaking processes in the calculation of current. We simply connect one or more purely conceptual floating probes to the device and then calculate the net current using the Büttiker equations, which can be applied to any number of terminals.

We could even use the general current equation (see Eq. (9.4.1)), rather than the low-bias conductance relation, extended to include multiple floating probes:

$$I_i = (q/h) \int_{-\infty}^{+\infty} \mathrm{d}E\, \overline{I}_i(E) \tag{9.4.6}$$

where

$$\overline{I}_i(E) = \sum_j \overline{T}_{ij}(E)[f_i(E) - f_j(E)] \tag{9.4.7}$$

One could then adjust the potential μ_j to make the current at each energy equal to zero: $\overline{I}_j(E) = 0$. In principle this could result in different values for μ_j at different energies. Alternatively, we could require a single value for μ_j at all energies, adjusted to make the total current at all energies equal to zero, $\int \mathrm{d}E\, \overline{I}_j(E) = 0$. One could then have positive values of $\overline{I}_j(E)$ at certain energies balanced by negative values at other energies making the total come out to zero, indicating a flow of electrons from one energy to another due to the scattering processes that the "probe" is expected to simulate. This makes the detailed implementation more complicated since different energy channels get coupled together.

The transmission coefficients at a given energy are usually calculated from the S-matrix for the composite device including the conceptual probes:

$$\overline{T}_{ij} = \mathrm{Trace}[s_{ij}(E)s_{ij}^{+}(E)] \tag{9.4.8}$$

But we could just as well combine this phenomenological approach with our Green's function method using separate self-energy matrices $[\Sigma_i]$ to represent different floating probes and then use the expression

$$\overline{T}_{ij}(E) = \text{Trace}[\Gamma_i G \Gamma_j G^+] \tag{9.4.9}$$

to evaluate the transmission. This expression can be derived using the same procedure described earlier for two-terminal structures. The current at terminal i is given by the difference between the inflow and outflow:

$$I_i(E) = (1/\hbar)\ \text{Trace}\{[\Gamma_i(E)][[A(E)]f_i - [G^n(E)]]\}$$

Making use of the relations (see Eqs. (9.2.14), (9.2.27), and (9.2.28)) $A = \sum_j G\Gamma_j G^+$ and $G^n = \sum_j G\Gamma_j G^+ f_j$ we can write

$$I_i(E) = (1/\hbar) \sum_q \text{Trace}[\Gamma_i G \Gamma_j G^+](f_i - f_j)$$

so that the current can be written as

$$I_i(E) = (1/\hbar) \sum_j \overline{T}_{ij}(f_i - f_j) \tag{9.4.10}$$

in terms of the transmission function defined above in Eq. (9.4.9).

Sum rule: A very useful result in the scattering theory of transport is the requirement that the sum of the rows or columns of the transmission matrix equals the number of modes:

$$\sum_j \overline{T}_{ij} = \sum_j \overline{T}_{ji} = M_i \tag{9.4.11}$$

where M_i is the number of modes in lead i. One important consequence of this sum rule is that for a *two-terminal* structure $\overline{T}_{12} = \overline{T}_{21}$, even in a magnetic field, since with a (2×2) $\overline{T}$ matrix:

$$\begin{bmatrix} \overline{T}_{11} & \overline{T}_{12} \\ \overline{T}_{21} & \overline{T}_{22} \end{bmatrix} \quad \text{we have } \overline{T}_{11} + \overline{T}_{12} = M_1 = \overline{T}_{11} + \overline{T}_{21} \rightarrow \overline{T}_{12} = \overline{T}_{21}$$

Note that a similar argument would not work with more than two terminals. For example, with a three-terminal structure we could show that $\overline{T}_{12} + \overline{T}_{13} = \overline{T}_{21} + \overline{T}_{31}$, but we could not prove that $\overline{T}_{12} = \overline{T}_{21}$ or that $\overline{T}_{13} = \overline{T}_{31}$.

The Green's function-based expression for the transmission (see Eq. (9.4.9)) also yields a similar sum rule:

$$\sum_j \overline{T}_{ij} = \sum_j \overline{T}_{ji} = \text{Trace}[\Gamma_i A] \tag{9.4.12}$$

This is shown by noting that

$$\sum_j \bar{T}_{ij} = \sum_j \text{Trace}[\Gamma_i G \Gamma_j G^+] = \text{Trace}[\Gamma_i G \Gamma G^+] = \text{Trace}[\Gamma_i A]$$

where we have made use of Eq. (9.2.14) in the last step. Similarly,

$$\sum_j \bar{T}_{ji} = \sum_j \text{Trace}[\Gamma_j G \Gamma_i G^+] = \text{Trace}[\Gamma_i G^+ \Gamma G] = \text{Trace}[\Gamma_i A]$$

The quantity Trace$[\Gamma_i A]$ thus plays the same role that the number of modes M_i plays in the scattering theory of transport. Interestingly, while M_i is an integer, Trace $[\Gamma_i \text{A}]$ can take on any non-integer value. For example, if the device were a really small one having just one level with $E = \varepsilon$, communicating with multiple reservoirs, then

$$\Gamma_i A = \frac{\Gamma_i}{(E - \varepsilon)^2 + (\Gamma/2)^2} \qquad \text{with } \Gamma = \sum_i \Gamma_i$$

which has the shape of a Lorentzian if the broadening is energy independent. Clearly this can have any fractional value.

9.5 Examples

9.5.1 An analytical example

Scattering theory: To see that the Green's function formalism gives the same answer as the scattering theory of transport it is instructive to go through a simple example where the results are easily worked out on paper. Consider, for example, a linear conductor with a repulsive potential $U(z) = U_0\, \delta(z)$ at $z = 0$ (Fig. 9.5.1). The coefficients r and t (in Fig. 9.5.1) are obtained by requiring that the wavefunction be continuous at $z = 0$:

$$[\psi]_{z=0^+} - [\psi]_{z=0^-} = 0 \rightarrow t - (1 + r) = 0 \tag{9.5.1a}$$

and that the derivative be discontinuous by

$$\left[\frac{\mathrm{d}\psi}{\mathrm{d}z}\right]_{z=0^+} - \left[\frac{\mathrm{d}\psi}{\mathrm{d}z}\right]_{z=0^-} = \frac{2mU_0}{\hbar^2}[\psi]_{z=0} \rightarrow \mathrm{i}k[t - (1 - r)] = \frac{2mU_0 t}{\hbar^2} \tag{9.5.1b}$$

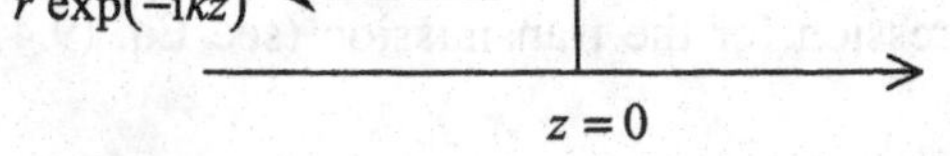

Fig. 9.5.1

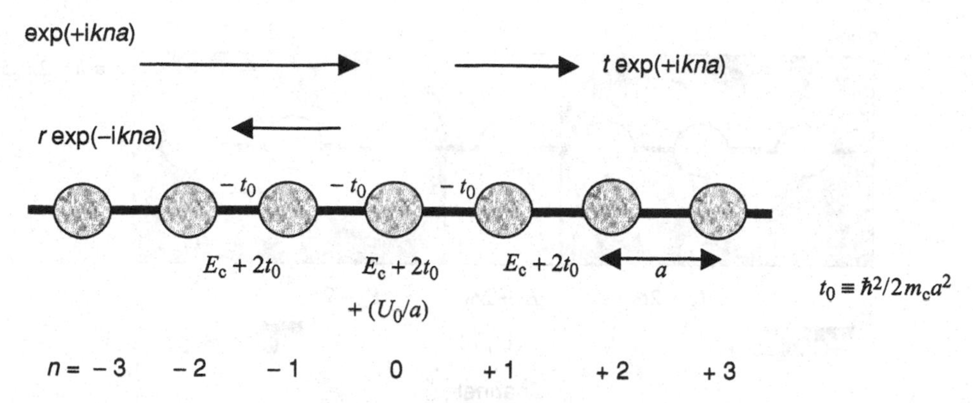

Fig. 9.5.2

Equations (9.5.1a, b) are solved to yield

$$t = \frac{i\hbar v}{i\hbar v - U_0} \rightarrow \overline{T} = |t|^2 = \frac{\hbar^2 v^2}{\hbar^2 v^2 + U_0^2}$$

Scattering theory on a discrete lattice: Let us now redo this problem using a discrete lattice with points spaced by a, the central cell having an extra potential (U_0/a) for the delta function (Fig. 9.5.2). We can carry out a discrete lattice version of the calculation described above, starting from

$$E\psi_0 = [E_c + 2t_0 + (U_0/a)]\psi_0 - t_0\psi_{-1} - t_0\psi_{+1} \tag{9.5.2}$$

and then writing

$$\psi_0 = 1 + r = t$$
$$\psi_{+1} = t\exp(+ika)$$
$$\psi_{-1} = \exp(-ika) + r\exp(+ika)$$

so that

$$\psi_{+1} = \psi_0 \exp(+ika)$$
$$\psi_{-1} = -2\mathrm{i}\sin ka + \psi_0 \exp(+ika) \tag{9.5.3}$$

Substituting back into Eq. (9.5.2) yields

$$[E - E_c - 2t_0 - (U_0/a) + 2t_0\exp(+ika)]\psi_0 = 2\mathrm{i}t_0 \sin ka$$

Making use of the dispersion relation

$$E = E_c + 2t_0(1 - \cos ka) \rightarrow \hbar v(E) = 2at_0 \sin ka \tag{9.5.4}$$

this is simplified to $[-(U_0/a) + 2\mathrm{i}t_0 \sin ka]\psi_0 = 2\mathrm{i}t_0 \sin ka$

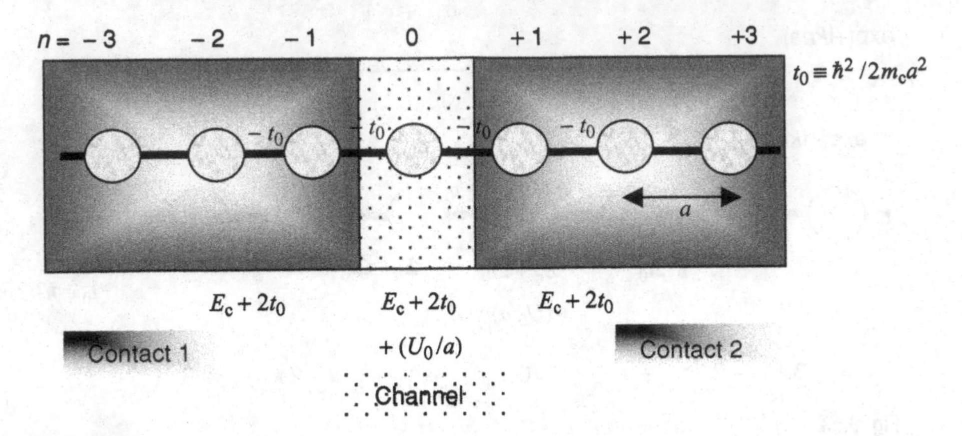

Fig. 9.5.3

that is,

$$\psi_0 = \frac{i\hbar v}{i\hbar v - U_0} \tag{9.5.5}$$

Hence the transmission is given by

$$\overline{T}(E) = |t|^2 = |\psi_0|^2 = \frac{\hbar^2 v(E)^2}{\hbar^2 v(E)^2 + U_0^2} \tag{9.5.6}$$

Green's function method: Finally, let us do this problem using the Green's function formulation presented in this chapter. We treat just one point as the "device" with a (1×1) Hamiltonian given by (see Fig. 9.5.3)

$$[H] = E_c + 2t_0 + (U_0/a)$$

while the effects of the two semi-infinite leads (one on each side) are represented by (1×1) self-energy matrices as discussed in Chapter 8:

$$[\Sigma_1(E)] = -t_0 \exp(ika) \quad \text{and} \quad [\Sigma_2(E)] = -t_0 \exp(ika)$$

where ka is related to the energy E by the dispersion relation (see Eq. (9.5.4)), so that

$$[\Gamma_{1,2}(E)] = i[\Sigma_{1,2} - \Sigma_{1,2}^+] = 2t_0 \sin ka = \hbar v/a$$

Since all matrices are (1×1) in size, it is easy to write down the Green's function:

$$G = [EI - H - \Sigma_1 - \Sigma_2]^{-1}$$
$$= [E - E_c - 2t_0 + 2t_0 \exp(ika) - (U_0/a)]^{-1}$$

Using the dispersion relation to simplify as before

$$G = [i2t_0 \sin ka - (U_0/a)]^{-1} = a/(i\hbar v - U_0)$$

so that the transmission is given by

$$\overline{T}(E) = \text{Trace}[\Gamma_1 G \Gamma_2 G^+] = \frac{\hbar^2 v(E)^2}{\hbar^2 v(E)^2 + U_0^2}$$

in agreement with the earlier result (Eq. (9.5.6)).

9.5.2 Numerical example

The real power of the Green's function method, of course lies not in simple problems like this, but in its ability to handle complex problems without the need for any additional formulation or setting up. Given a Hamiltonian [*H*] and self-energy matrices $\Sigma_1(E)$ and $\Sigma_2(E)$, the procedure is mechanical: Eqs. (9.1.5) and (9.1.10) can be applied blindly to evaluate the transmission. Of course, complicated contacts can require some extra effort to evaluate the appropriate self-energy matrices, but it is a one-time effort. Besides, as we have mentioned earlier, one can make a reasonable guess based on Eqs. (8.3.12) and (8.3.14) without a detailed calculation – a procedure that can be justified physically by arguing that one never knows the precise shape of the contacts anyway. The examples we discuss below are all based on one-dimensional leads for which the self-energy is written down easily.

We use a *one-dimensional* discrete lattice with $a = 0.3$ nm to model each of the following devices which are assumed to be single-moded in the transverse (x- and y-) directions (Fig. 9.5.4). The barrier regions indicated by the "brick wall" pattern have a conduction band that is 0.4 eV higher than the rest. We assume that the effective mass ($m_c = 0.25m$) is the same everywhere. Figure 9.5.5 shows the (non-self-consistent)

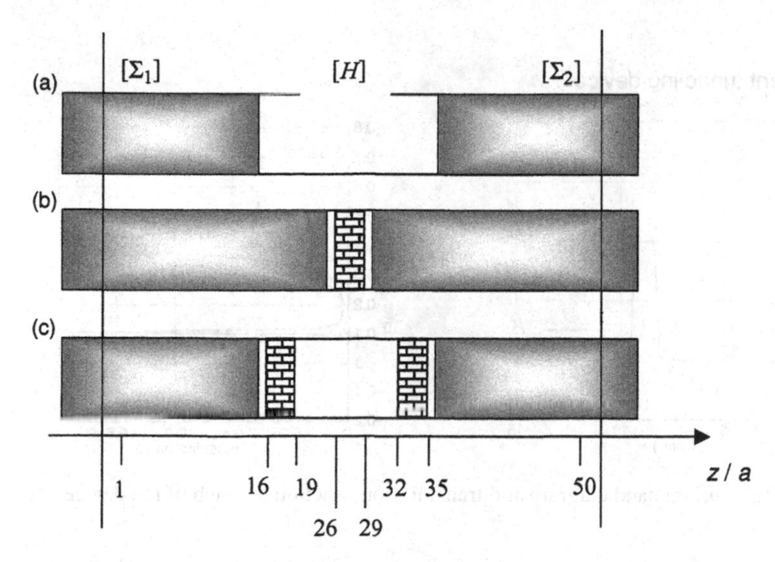

Fig. 9.5.4 Three device examples: (a) ballistic device; (b) tunneling device; (c) resonant tunneling device.

(a) Ballistic device

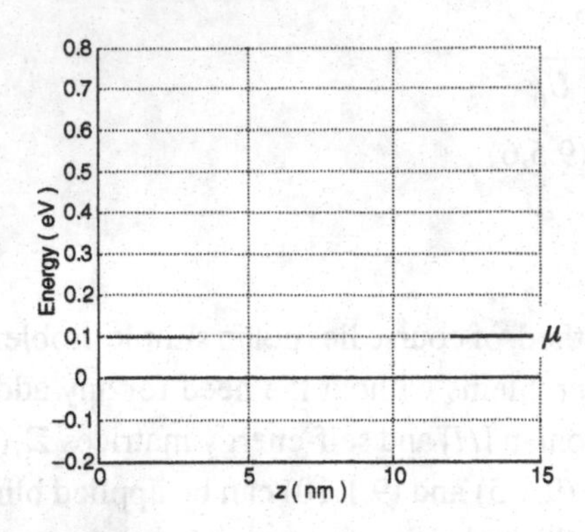

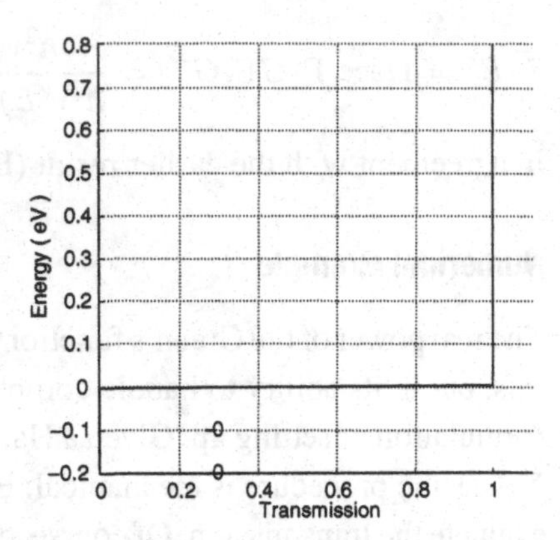

(b) Tunneling device

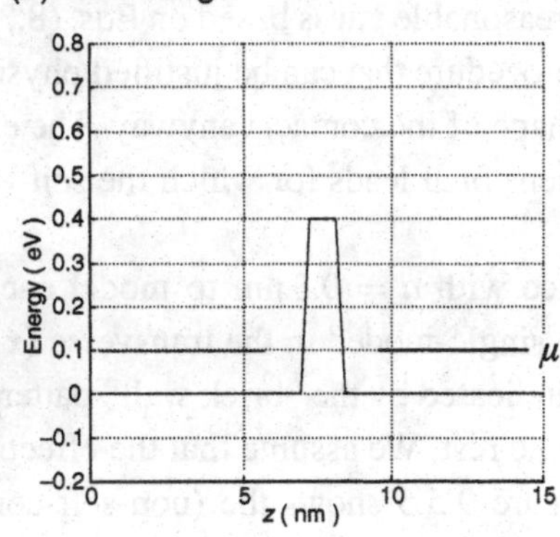

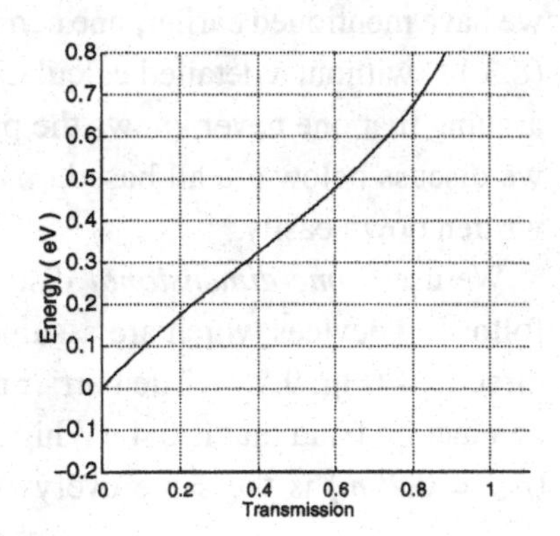

(c) Resonant tunneling device

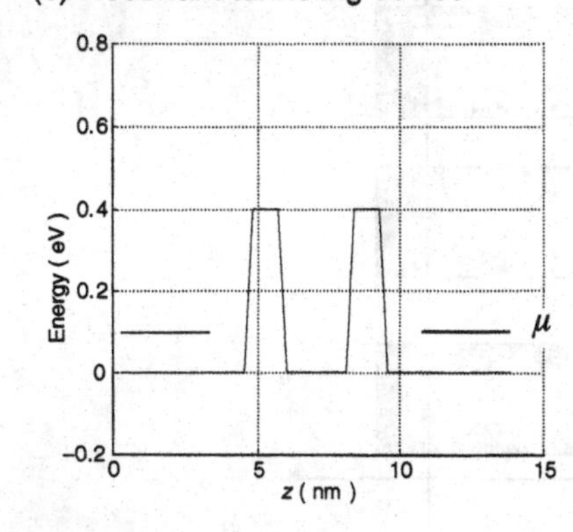

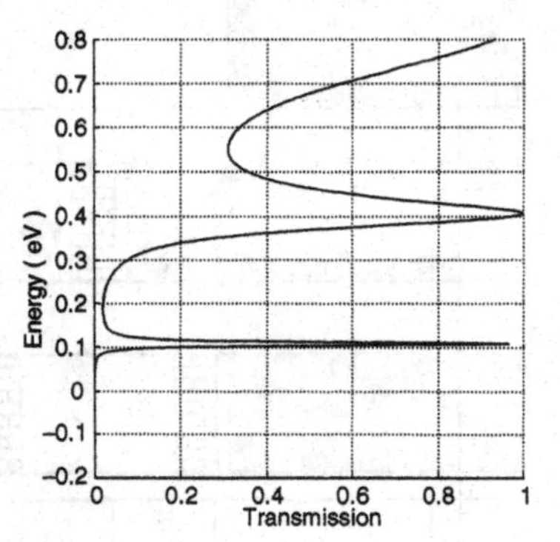

Fig. 9.5.5 Equilibrium band diagram and transmission function for each of the devices in Fig. 9.5.4.

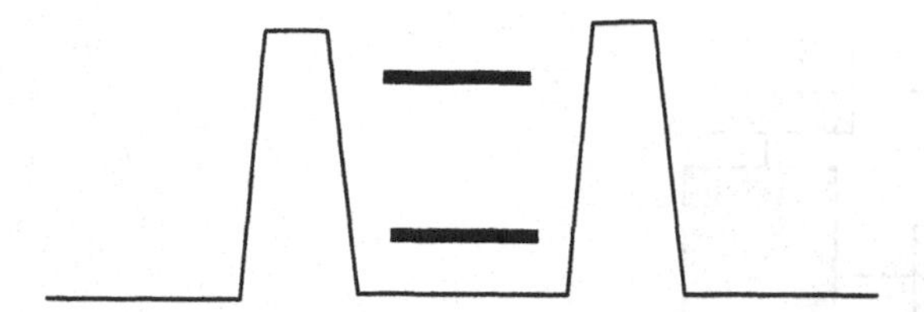

Fig. 9.5.6

equilibrium band diagram and transmission functions $\overline{T}(E)$ calculated numerically for each of these devices from the Hamiltonian matrix [H] and the self-energy matrices $\Sigma_{1,2}(E)$.

For the ballistic device the transmission is zero for energies below the band edge E_c and increases to one above the band edge. For the tunneling device, the transmission increases from zero to one, though more slowly. The transmission for a resonant tunneling device, on the other hand, shows a very different behavior with two sharp resonances that can be understood by noting that the two barriers create a "box" with discrete energy levels (Fig. 9.5.6, see Section 2.1). The transmission from left to right peaks whenever the energy matches one of these levels. It is possible to obtain the same results by matching wavefunctions and derivatives across different sections, but the process quickly gets cumbersome. Arbitrary potential profiles, however, are easily included in the Hamiltonian [H] and the transmission is then calculated readily from the Green's function formalism: $\overline{T}(E) = \text{Trace}[\Gamma_1 G \Gamma_2 G^+]$.

In calculating the transmission through devices with sharp resonances (like the resonant tunneling device) it is often convenient to include a Büttiker probe (see Section 9.4). The reason is that it is easy to miss very sharp resonances in a numerical calculation if the energy grid is not fine enough. A Büttiker probe simulates the role of phase-breaking processes thereby broadening the resonance. The effective transmission is calculated by solving the Büttiker equations (see Eq. (9.4.5)) as explained in the last section. In this case the transmission between different terminals is reciprocal so that we can calculate the effective transmission from a simple resistor network (see Fig. 9.5.7) adapted to three terminals.

Noting that the conductance is proportional to the transmission we can write the effective transmission using the elementary law of addition for conductors in series and in parallel:

$$\overline{T}_{\text{eff}}(E) = \overline{T}_{12}(E) + \frac{\overline{T}_{13}(E)\overline{T}_{23}(E)}{\overline{T}_{13}(E) + \overline{T}_{23}(E)} \tag{9.5.7}$$

Figure 9.5.8 shows the effective transmission for a resonant tunneling device with one Büttiker probe attached to the center of the device. Compared to the earlier result without a probe, the resonances are broadened somewhat, especially the sharpest one.

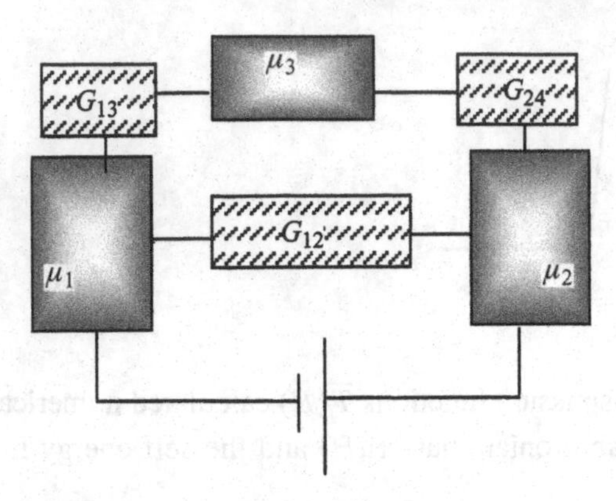

Fig. 9.5.7

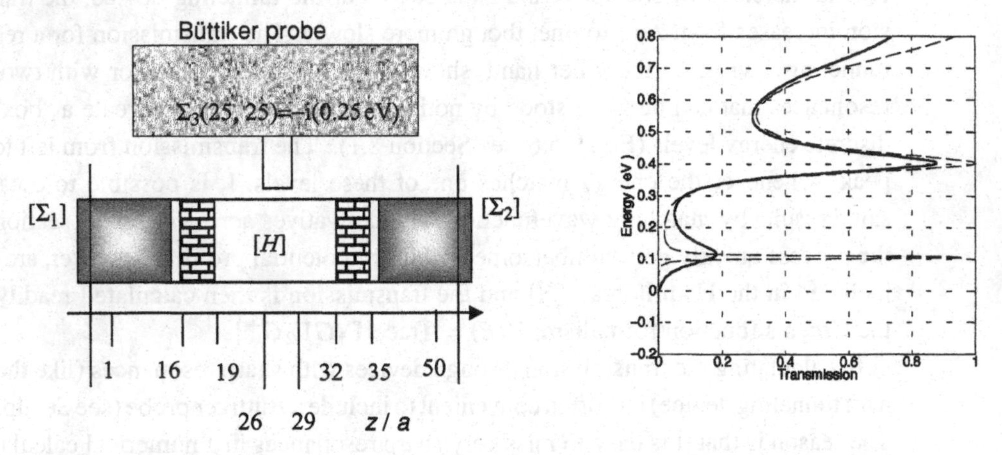

Fig. 9.5.8 Effective transmission function for a resonant tunneling device including a Büttiker probe located at lattice site number 25 at the center of the device to simulate the effect of phase-breaking processes phenomenologically. The dashed curve shows result from Fig. 9.5.5c without a Büttiker probe.

Current (*I*)–voltage (*V*) characteristics: Equation (9.1.9) can be used to calculate the *I–V* characteristics of any coherent device, provided we know how the applied voltage drops across the device. This is not important if we are only interested in the low-bias conductance (or "linear response"), but can be of paramount importance in determining the shape of the full current–voltage characteristics as discussed in Section 1.4.

In general, for quantitatively correct results, it is important to solve for the potential profile *self-consistently*. Just like the equilibrium problem (see Fig. 7.2.1), we should include a self-consistently determined potential U in the total Hamiltonian $H = H_0 + U([\delta\rho])$.

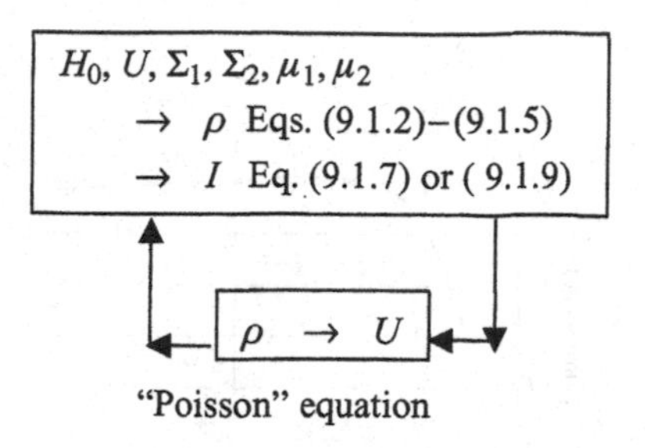

Fig. 9.5.9

This potential U represents the average potential that an electron feels due to the change $\delta\rho$ in the electron density, or more generally the density matrix. The first step in this process is to calculate the electron density from the diagonal elements of the density matrix. This electron density can then be used in the Poisson equation to calculate the potential which is then included in the Hamiltonian to recalculate the electron density and so on till the process converges as sketched in Fig. 9.5.9. A full self-consistent calculation like this can be time-consuming (we will describe a simple one in Section 11.4) and so it is common to assume a "reasonable" potential profile. What is a reasonable profile?

The basic principle is straightforward. If the channel were insulating (low quantum capacitance, see Eq. (7.3.8)), the potential profile would be given by the Laplace potential $U_L(\vec{r})$, obtained by solving the Laplace equation. But if it were metallic (large quantum capacitance), the profile would be given by the "neutral potential" $U_N(\vec{r})$ obtained from the transport equation assuming perfect space charge neutrality everywhere. The correct potential profile is intermediate between these extremes. In regions of low density of states the quantum capacitance is small and the potential profile will tend to follow $U_L(\vec{r})$ while in regions with high density of states the quantum capacitance is large and the potential profile will tend to follow $U_N(\vec{r})$. The common practice for choosing a "reasonable profile" is to assume that the potential follows $U_N(\vec{r})$ (that needed to maintain charge neutrality) at the ends which should be regions of high density of states, while in the central channel region the profile is assumed to follow the Laplace potential $U_L(\vec{r})$.

Figure 9.5.10 shows the I–V characteristics for (a) the ballistic device, (b) the tunneling device, and (c) the resonant tunneling device calculated assuming that the potential drops linearly across the central unshaded regions in Fig. 9.5.4. This assumed potential profile gives reasonable qualitative features, but it is easy to check that the results can change quantitatively if we choose different profiles. We will talk about this further in Section 11.4 when we discuss the factors that influence the ON current of a nanotransistor.

(a) Ballistic device

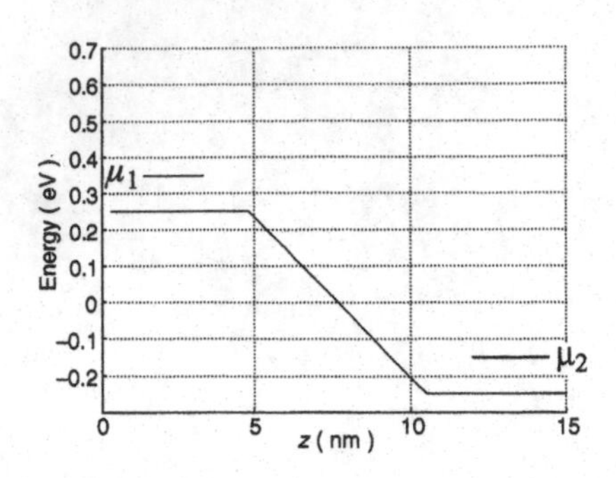

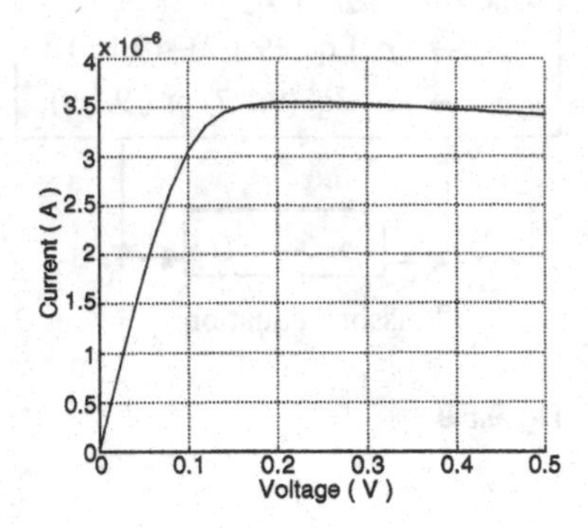

(b) Tunneling device

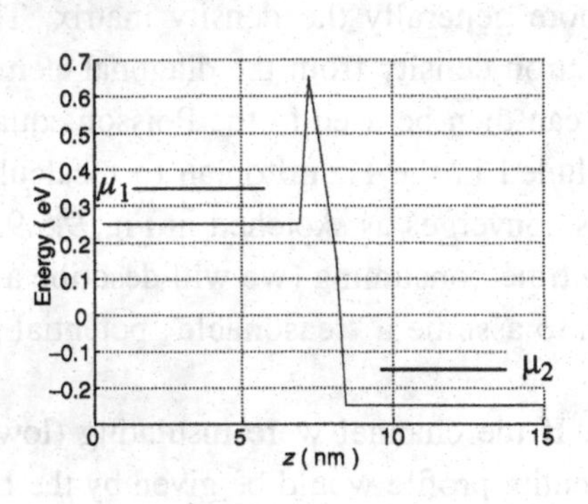

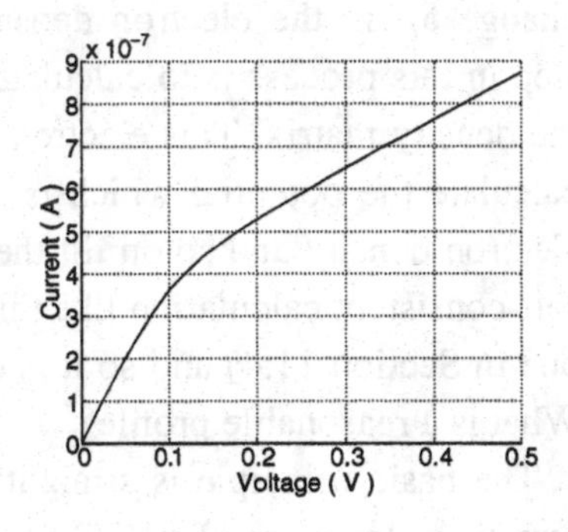

(c) Resonant tunneling device

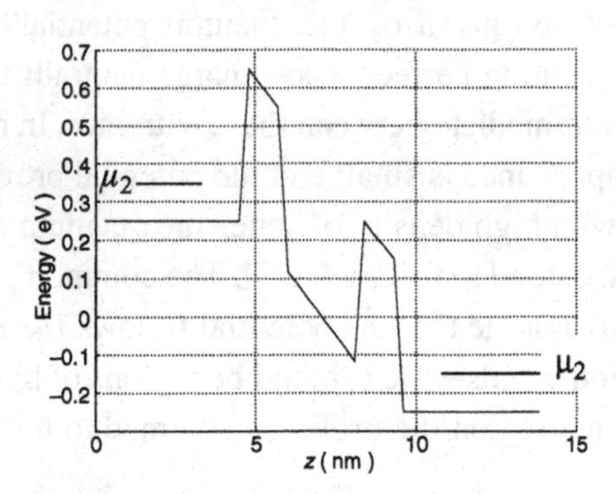

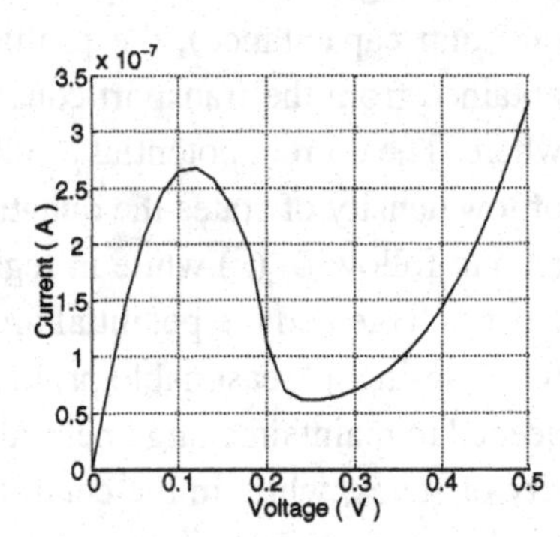

Fig. 9.5.10 Current (I) versus voltage (V) characteristics of the three devices shown in Fig. 9.5.4 calculated assuming the linear potential profile shown. The left-hand plots show the assumed band diagrams at a bias of 0.5 V.

EXERCISES

E.9.1. Use a *one-dimensional* discrete lattice with $a = 0.3$ nm to model each of the devices shown in Fig. 9.5.4 which are assumed to be single-moded in the transverse (x- and y-) directions. Assume that the effective mass ($m_c = 0.25m$) is the same everywhere. The barrier regions indicated by the "brick wall" pattern have a conduction band that is 0.4 eV higher than the rest. (a) Set up an energy grid over the range $-0.2\,\text{eV} < E < 0.8\,\text{eV}$ and plot the transmission probability as a function of energy. Compare with Fig. 9.5.5. (b) Plot the transmission probability as a function of energy for the *resonant tunneling device* using a Büttiker probe as indicated in Fig. 9.5.8.

E.9.2. (a) Calculate the current (I)–voltage (V) characteristics in the bias range of $0 < V < 1$ V assuming that the applied bias drops across the device following the profile shown in Fig. 9.5.10. Assume the equilibrium Fermi energy to be $E_F = 0.1$ eV and the chemical potentials in the two contacts under bias to be $\mu_1 = E_F + qV/2$ and $\mu_2 = E_F - qV/2$. The energy integration needs to be carried out only over the range $\mu_1 + 4k_BT < E < \mu_2 - 4k_BT$. Use an energy grid with $\Delta E \approx 0.2k_BT$. (b) Calculate the electron density $n(x)$ per unit length assuming that the applied bias of 0.5 V drops across the *tunneling device* following the profile shown in Fig. 9.5.10.

E.9.3. Transfer Hamiltonian: See Fig. E.9.3. Starting from the expression for the transmission in Eq. (9.4.2), $\overline{T}(E) = \text{Trace}[\Gamma_1 G \Gamma_2 G^+]$, and making use of the expressions for the broadening matrices in Eq. (9.2.22) show that

$$\overline{T}(E) = \text{Trace}[A_1 M A_2 M^+]$$

where A_1 and A_2 are the spectral functions in the two contacts and the matrix element M is given by

$$M = \tau_1^+ G \tau_2$$

This form is similar to the version often seen in connection with the transfer Hamiltonian formalism (see for example, Eq. (2.3.5) on p. 69 of Chen (1993)). In the transfer Hamiltonian formalism the matrix element M is assumed to be unaffected by the coupling to the contacts which is assumed to be small, but in the present formulation G and hence M are affected by the contacts through the self-energy due to the contacts.

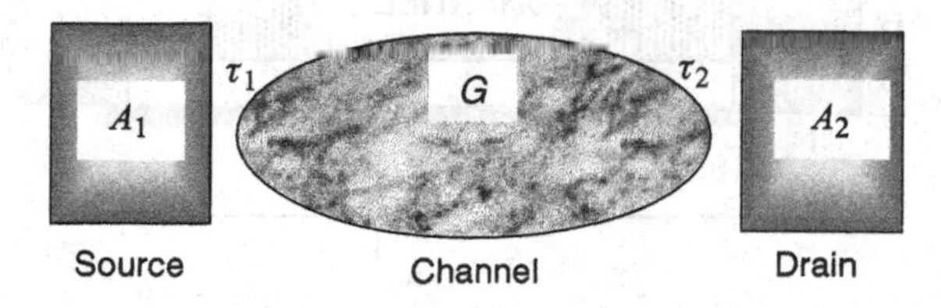

Fig. E.9.3

E.9.4. 2D cross-section: In the examples of Section 9.5 we have assumed that the device is one-dimensional. The 2D cross-section can be included in a simple way if we assume periodic boundary conditions and assume that all the transverse modes are decoupled as we did when calculating the capacitance in Chapter 7. We could then simply sum our 1D result over all the transverse modes represented by the two-dimensional vector $\vec{k}$ to write ($\varepsilon_{\vec{k}} = \hbar^2 k^2/2m_c$):

$$I = \frac{q}{2\pi\hbar} \sum_{\vec{k}} \int_{-\infty}^{+\infty} \mathrm{d}E\, \overline{T}(E)[f_0(E + \varepsilon_{\vec{k}} - \mu_1) - f_0(E + \varepsilon_{\vec{k}} - \mu_2)]$$

The transmission function depends only on the longitudinal energy E while the Fermi functions are determined by the total energy $E + \varepsilon_{\vec{k}}$. The summation over $\vec{k}$ can be carried out analytically to write (where S is cross-sectional area)

$$\frac{I}{S} = \frac{q}{\pi\hbar} \int_{-\infty}^{+\infty} \mathrm{d}E\, \overline{T}(E)[f_{2D}(E - \mu_1) - f_{2D}(E - \mu_2)]$$

This means that the current in a device with a 2D cross-section is obtained using the same procedure that we used for a 1D device, provided we use the k-summed Fermi function f_{2D} (see Eq. (7.2.12)) in place of the usual Fermi function. Repeat Exercise E.9.2 using f_{2D} (see Eq. (7.2.12)) instead of the Fermi function f_0 to account for a device with a 2D cross-section. The current should now be expressed in A/m^2 and the electron density should be expressed in /m^3.

E.9.5. 1D cross-section: In analyzing field effect transistors, we often have a 1D cross-section (y-direction) to sum over, while the transmission has to be calculated from a 2D problem in the z–x plane (Fig. E.9.5). Assuming periodic boundary conditions in the y-direction show that the 1D k-sum can be done analytically to obtain

$$\frac{I}{W} = \frac{q}{\pi\hbar} \int_{-\infty}^{+\infty} \mathrm{d}E\, \overline{T}(E)[f_{1D}(E - \mu_1) - f_{1D}(E - \mu_2)]$$

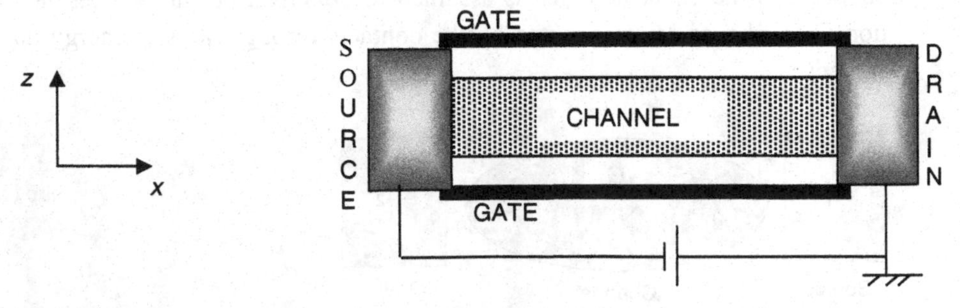

Fig. E.9.5

where the 1D k-summed Fermi function is given by

$$f_{1D}(E) \equiv \left(\frac{m_c k_B T}{2\pi\hbar^2}\right) \Im_{-1/2}\left(-\frac{E}{k_B T}\right)$$

with

$$\Im_{-1/2}(x) \equiv \frac{1}{\sqrt{\pi}} \int_0^{+\infty} \frac{\mathrm{d}y}{1+\exp(y-x)} \frac{1}{\sqrt{x}} = \frac{\mathrm{d}}{\mathrm{d}x} \Im_{1/2}(x)$$

where $\Im_{1/2}(x)$ was defined in Eq. (7.2.22).

E.9.6. In Section 9.3 we obtained expressions (Eqs. (9.1.7)–(9.1.9)) for the terminal currents by considering the overall inflow and outflow from the channel. However, this does not give us the detailed current flow pattern inside the channel. Once the correlation function G^n has been obtained from (Eq. (9.1.3), the current flowing from nodel 'a' to node 'b' within the channel can be evaluated from the expression:

$$I_{a\to b} = (-q/h) \int_{-\infty}^{+\infty} dE \tilde{I}_{a\to b}(E)$$

with $\tilde{I}_{a\to b}(E) = i[H_{ab} G^n_{ba} - G^n_{ab} H_{ba}]$

Obtain this expression by generalizing the 1-D current operator in Eq. (6.4.6) to write

$$J_{a\to b} = [\Psi_b^+ H_{ba} \Psi_a - \Psi_a^+ H_{ab} \Psi_b]/i\hbar$$

and then making use of Eq. (9.2.15b). Incidentally, this expression also provides an alternative way to evaluate the current at a particular termial: Draw a plane that separates the terminal from the rest of the device and sum $I_{a\to b}$ over all points 'a' and 'b' lying on opposite sides of this plane.

E.9.7. For more examples on coherent transport the reader can look at Datta (2000): MATLAB codes used to generate the figures are available on request.

10 Non-coherent transport

In Chapter 9, we discussed a quantum mechanical model that describes the flow of electrons *coherently* through a channel. All dissipative/phase-breaking processes were assumed to be limited to the contacts where they act to keep the electrons in local equilibrium. In practice, such processes are present in the channel as well and their role becomes increasingly significant as the channel length is increased. Indeed, prior to the advent of mesoscopic physics, the role of contacts was assumed to be minor and quantum transport theory was essentially focused on the effect of such processes. By contrast, we have taken a "bottom-up" view of the subject and now that we understand how to model a small coherent device, we are ready to discuss dissipative/phase-breaking processes.

Phase-breaking processes arise from the interaction of one electron with the surrounding bath of photons, phonons, and other electrons. Compared to the coherent processes that we have discussed so far, the essential difference is that phase-breaking processes involve a change in the "surroundings." In coherent interactions, the background is rigid and the electron interacts elastically with it, somewhat like a ping pong ball bouncing off a truck. The motion of the truck is insignificant. In reality, the background is not quite as rigid as a truck and is set in "motion" by the passage of an electron and this excitation of the background is described in terms of phonons, photons, etc. This is in general a difficult problem with no exact solutions and what we will be describing here is the lowest order approximation, sometimes called the self-consistent Born approximation, which usually provides an adequate description. Within this approximation, these interactions can essentially be viewed as a coupling of the channel from the "standard" configuration with $\{N_\omega\}$ phonons/photons (in different modes with different frequencies $\{\omega\}$) to a neighboring configuration with one less (absorption) or one more (emission) phonon/photon as depicted in Fig. 10.1.

This coupling to neighboring configurations results in an outflow of electrons from our particular subspace and a subsequent return or inflow back into this subspace. A general model for quantum transport needs to include this inflow and outflow into the coherent transport model from Chapter 9, through an additional terminal "s" described

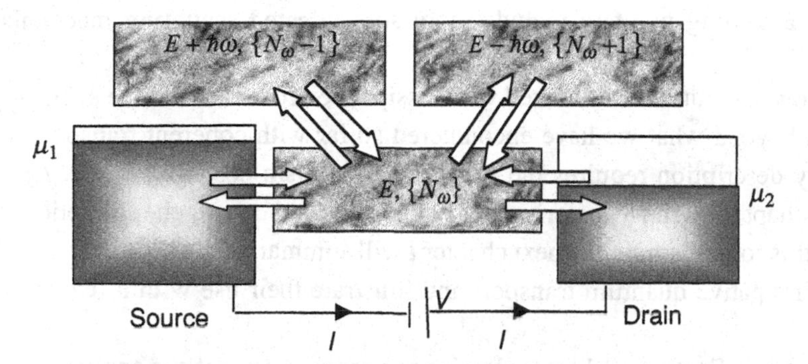

Fig. 10.1 Phase-breaking introduces a coupling to neighboring configurations having one more or one less number of excitations $\{N_\omega\}$ than the original.

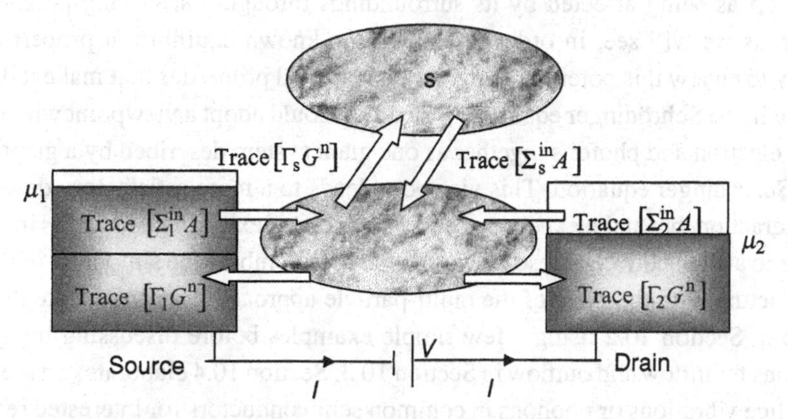

Fig. 10.2 Non-coherent quantum transport: inflow and outflow.

by the additional terms Σ_s^{in} and Σ_s (see Fig. 10.2). My objective in this chapter is to explain how these additional terms are calculated. We have seen that for the regular contacts, the inscattering is related to the broadening:

$$\Sigma_1^{in} = \Gamma_1 f_1 \quad \text{and} \quad \Sigma_2^{in} = \Gamma_2 f_2$$

However, for the scattering terminal both Σ_s^{in} and Σ_s have to be determined separately since there is no Fermi function f_s describing the scattering "terminal" and hence no simple connection between Σ_s^{in} and Σ_s (or Γ_s), unlike the contacts. Of course, one could adopt a phenomenological point of view and treat the third terminal like another contact whose chemical potential μ_s is adjusted to ensure zero current at this terminal. That would be in the spirit of the "Büttiker probe" discussed in Chapter 9 and could well be adequate for many applications. However, I will describe microscopic (rather than phenomenological) models for Σ_s^{in} and Σ_s that can be used to benchmark any phenomenological models that the reader may choose to use. They can also use these models as a starting point to include more sophisticated scattering mechanisms as needed.

The inflow and outflow associated with dissipative processes involve subtle conceptual issues beyond what we have encountered so far with coherent transport. A fully satisfactory description requires the advanced formalism described in the Appendix, but in this chapter I will try to derive the basic results and convey the subtleties without the use of this formalism. In the next chapter I will summarize the complete set of equations for dissipative quantum transport and illustrate their use with a few interesting examples.

We will start in Section 10.1 by explaining two viewpoints that one could use to model the interaction of an electron with its surroundings, say the electromagnetic vibrations or photons. One viewpoint is based on the one-particle picture where we visualize the electron as being affected by its surroundings through a scattering potential U_s. However, as we will see, in order to explain the known equilibrium properties it is necessary to endow this potential U_s with rather special properties that make it difficult to include in the Schrödinger equation. Instead we could adopt a viewpoint whereby we view the electron and photons together as one giant system described by a giant multi-particle Schrödinger equation. This viewpoint leads to a more satisfactory description of the interaction, but at the expense of conceptual complexity. In general it is important to be able to switch between these viewpoints so as to combine the simplicity of the one-particle picture with the rigor of the multi-particle approach. I will illustrate the basic principle in Section 10.2 using a few simple examples before discussing the general expressions for inflow and outflow in Section 10.3. Section 10.4 elaborates on the nature of the lattice vibrations or phonons in common semiconductors for interested readers.

10.1 Why does an atom emit light?

We started this book by noting that the first great success of the Schrödinger equation was to explain the observed optical spectrum of the hydrogen atom. It was found that the light emitted by a hot vapor of hydrogen atoms consisted of discrete frequencies $\omega = 2\pi\nu$ that were related to the energy eigenvalues from the Schrödinger equation: $\hbar\omega = \varepsilon_n - \varepsilon_m$. This is explained by saying that if an electron is placed in an excited state $|2\rangle$, it relaxes to the ground state $|1\rangle$, and the difference in energy is radiated in the form of light or photons (Fig. 10.1.1). Interestingly, however, this behavior does not really follow from the Schrödinger equation, unless we add something to it.

To see this let us write the time-dependent Schrödinger equation (Eq. (2.1.8)) in the form of a matrix equation

$$i\hbar\frac{\mathrm{d}}{\mathrm{d}t}\{\psi\} = [H]\{\psi\} \tag{10.1.1}$$

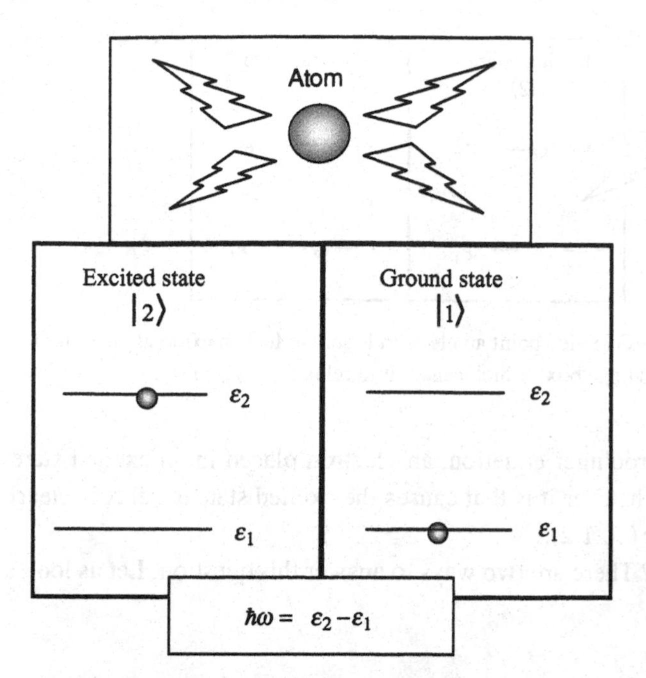

Fig. 10.1.1 If an electron is placed in an excited state $|2\rangle$, it will lose energy by radiating light and relax to the ground state $|1\rangle$. However, this behavior does not follow from the Schrödinger equation, unless we modify it appropriately.

using a suitable set of basis functions. If we use the eigenfunctions of $[H]$ as our basis then this equation has the form:

$$i\hbar \frac{d}{dt}\begin{Bmatrix} \psi_1 \\ \psi_2 \\ \cdots \\ \cdots \end{Bmatrix} = \begin{bmatrix} \varepsilon_1 & 0 & 0 & \cdots \\ 0 & \varepsilon_2 & 0 & \cdots \\ 0 & 0 & \varepsilon_3 & \cdots \\ & \cdots & \cdots & \cdots \end{bmatrix} \begin{Bmatrix} \psi_1 \\ \psi_2 \\ \cdots \\ \cdots \end{Bmatrix}$$

which decouples neatly into a set of independent equations:

$$i\hbar \frac{d}{dt}\{\psi_n\} = [\varepsilon_n]\{\psi_n\} \tag{10.1.2}$$

one for each energy eigenvalue ε_n. It is easy to write down the solution to Eq. (10.1.2) for a given set of initial conditions at $t = 0$:

$$\psi_n(t) = \psi_n(0)\exp(-i\varepsilon_n t/\hbar) \tag{10.1.3a}$$

This means that the probability P_n for finding an electron in state n does not change with time:

$$P_n(t) = |\psi_n(t)|^2 = |\psi_n(0)|^2 = P_n(0) \tag{10.1.3b}$$

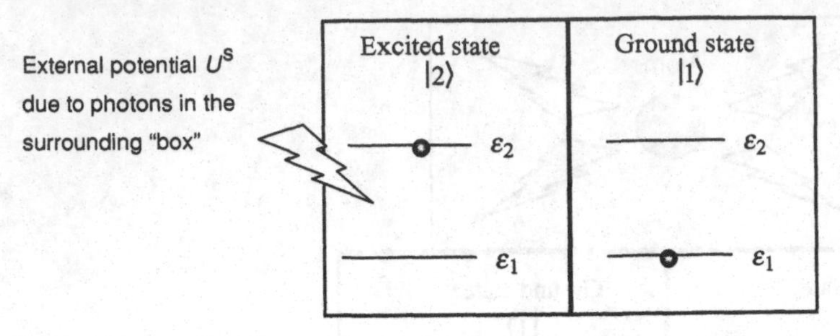

Fig. 10.1.2 In the one-particle viewpoint an electron is said to feel an external potential U^S due to the photons in the surrounding "box" which causes it to relax from $|2\rangle$ to $|1\rangle$.

According to the Schrödinger equation, an electron placed in an excited state would stay there for ever! Whatever it is that causes the excited state to relax is clearly not a part of Eq. (10.1.1) or (10.1.2).

So what is missing? There are two ways to answer this question. Let us look at these one by one.

One-particle viewpoint: This viewpoint says that an electron feels a random external potential U^S due to the photons in the surrounding "box" which causes it to relax to the ground state (Fig. 10.1.2). This potential gives rise to off-diagonal terms in the Hamiltonian that couple the different states together. With just two states we could write

$$i\hbar\frac{d}{dt}\begin{Bmatrix}\psi_1\\ \psi_2\end{Bmatrix}=\begin{bmatrix}\varepsilon_1 & U_{12}^S\\ U_{21}^S & \varepsilon_2\end{bmatrix}\begin{Bmatrix}\psi_1\\ \psi_2\end{Bmatrix}\tag{10.1.4}$$

Without getting into any details it is clear that if the electron is initially in state $|2\rangle$, the term U_{12}^S will tend to drive it to state $|1\rangle$. But this viewpoint is not really satisfactory. Firstly, one could ask why there should be any external potential U^S at zero temperature when all thermal excitations are frozen out. The answer usually is that even at zero temperature there is some noise present in the environment, and these so-called zero-point fluctuations tickle the electron into relaxing from $|2\rangle$ to $|1\rangle$. But that begs the second question: why do these zero-point fluctuations not provide any transitions from $|1\rangle$ to $|2\rangle$? Somehow we need to postulate a scattering potential for which (note that $\varepsilon_2 > \varepsilon_1$)

$$U_{21}^S = 0 \quad \text{but} \quad U_{12}^S \neq 0$$

at zero temperature.

For non-zero temperatures, U_{21}^S need not be zero, but it will still have to be much smaller than U_{12}^S, so as to stimulate a greater rate $S(2 \to 1)$ of transitions from 2 to 1 than from 1 to 2. For example, we could write

$$S(2 \to 1) = K_{2\to 1} f_2(1 - f_1) \quad \text{and} \quad S(1 \to 2) = K_{1\to 2} f_1(1 - f_2)$$

where $f_1(1 - f_2)$ is the probability for the system to be in state $|1\rangle$ (level 1 occupied with level 2 empty) and $f_2(1 - f_1)$ is the probability for it to be in state $|2\rangle$ (level 2 occupied with level 1 empty). At equilibrium the two rates must be equal, which requires that

$$\frac{K_{1\to 2}}{K_{2\to 1}} = \frac{f_2(1 - f_1)}{f_1(1 - f_2)} = \frac{(1 - f_1)/f_1}{(1 - f_2)/f_2} \tag{10.1.5}$$

But at equilibrium, the occupation factors f_1 and f_2 are given by the Fermi function:

$$f_n = \frac{1}{1 + \exp[(\varepsilon_n - \mu)/k_B T]} \to \frac{1 - f_n}{f_n} = \exp\left(\frac{\varepsilon_n - \mu}{k_B T}\right)$$

Hence from Eq. (10.1.5),

$$\left(\frac{K_{1\to 2}}{K_{2\to 1}}\right)_{\text{equilibrium}} = \exp\left(-\frac{\varepsilon_2 - \varepsilon_1}{k_B T}\right) \tag{10.1.6}$$

Clearly at equilibrium, $K_{2\to 1} \gg K_{1\to 2}$, as long as the energy difference $(\varepsilon_2 - \varepsilon_1) \gg k_B T$.

Early in the twentieth century, Einstein argued that if the number of photons with energy $\hbar\omega$ present in the box is N, then the rate of downward transitions is proportional to $(N + 1)$ while the rate of upward transitions is proportional to N:

$$\begin{aligned} K(1 \to 2) &= \alpha N && \textit{photon absorption} \\ K(2 \to 1) &= \alpha(N + 1) && \textit{photon emission} \end{aligned} \tag{10.1.7}$$

This ensures that at equilibrium Eq. (10.1.6) is satisfied since the number of photons is given by the Bose–Einstein factor

$$[N]_{\text{equilibrium}} = \frac{1}{\exp(\hbar\omega/k_B T) - 1} \tag{10.1.8}$$

and it is easy to check that

$$\left(\frac{K_{1\to 2}}{K_{2\to 1}}\right)_{\text{equilibrium}} = \left(\frac{N}{N + 1}\right)_{\text{equilibrium}} = \exp\left(-\frac{\hbar\omega}{k_B T}\right) \tag{10.1.9}$$

Since $\hbar\omega = \varepsilon_2 - \varepsilon_1$, Eqs. (10.1.9) and (10.1.6) are consistent.

What is not clear is why the external potential should stimulate a greater rate of downward transitions $(2 \to 1)$ than upward transitions $(1 \to 2)$, but clearly this must be the case if we are to rationalize the fact that, at equilibrium, lower energy states are more likely to be occupied than higher energy states as predicted by the Fermi function. But there is really no straightforward procedure for incorporating this effect into the

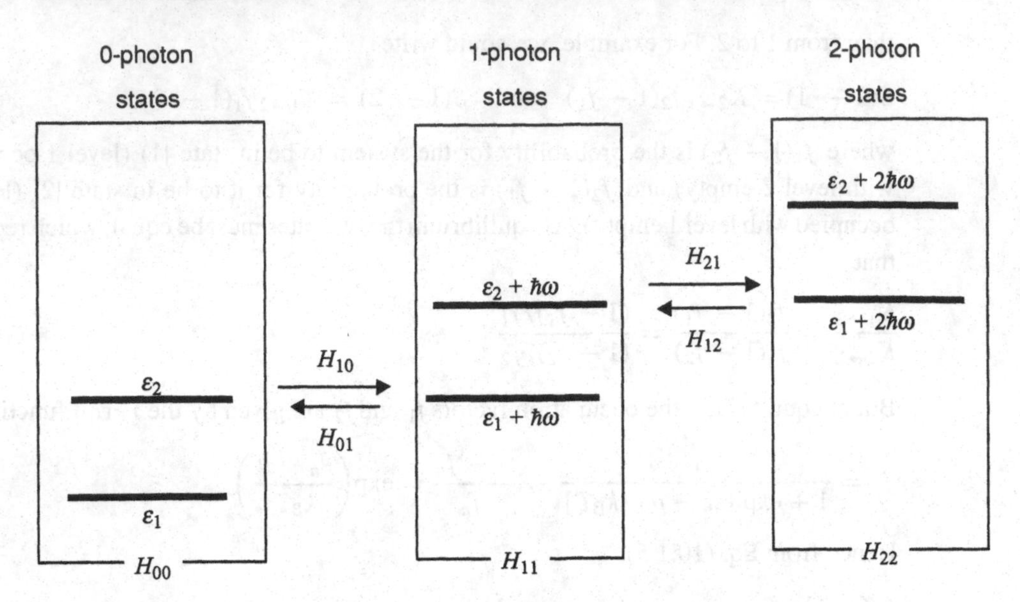

Fig. 10.1.3 In the multi-particle viewpoint, the electron–photon coupling causes transitions between $|2, N\rangle$ and $|1, N-1\rangle$, which are degenerate states of the composite system.

Schrödinger equation with an appropriate choice of the scattering potential U^{S}. Any Hermitian operator U^{S} will have $U^{\mathrm{S}}_{12} = U^{\mathrm{S}}_{21}$ and thus provide equal rates of upward and downward transitions.

Many-particle viewpoint: This brings us to the other viewpoint, which provides a natural explanation for the difference between upward and downward transition rates, but is conceptually more complicated. In this viewpoint we picture the electron + photon as one big many-particle system whose dynamics are described by an equation that formally looks just like the Schrödinger equation (Eq. (10.1.1))

$$i\hbar \frac{\mathrm{d}}{\mathrm{d}t}\{\Psi\} = [H]\{\Psi\} \tag{10.1.10}$$

However, $\{\Psi\}$ now represents a state vector in a multi-particle Hilbert space, which includes both the electron and the photon systems. The basis functions in this multi-particle space can be written as a product of the electronic and photonic subspaces (see Fig. 10.1.3):

$$|n, N\rangle = \underset{\text{electron}}{|n\rangle} \otimes \underset{\text{photon}}{|N\rangle}$$

just as the basis functions in a two-dimensional problem can be written as the product of the basis states of two one-dimensional problems:

$$|k_x, k_y\rangle = |k_x\rangle \otimes |k_y\rangle \sim \exp(\mathrm{i}k_x x)\exp(\mathrm{i}k_y y)$$

We can write Eq. (10.1.10) in the form of a matrix equation:

$$i\hbar\frac{d}{dt}\begin{Bmatrix}\Psi_0\\ \Psi_1\\ \Psi_2\end{Bmatrix}=\begin{bmatrix}H_{00} & H_{01} & 0 & \cdots\\ H_{10} & H_{11} & H_{12} & \cdots\\ 0 & H_{21} & H_{22} & \cdots\\ \cdots & \cdots & \cdots & \end{bmatrix}\begin{Bmatrix}\Psi_0\\ \Psi_1\\ \Psi_2\end{Bmatrix}$$

where $\{\Psi_N\}$ represents the N-photon component of the wavefunction. If the electronic subspace is spanned by two states $|1\rangle$ and $|2\rangle$ as shown in Fig. 10.1.3 then $\{\Psi_N\}$ is a (2×1) column vector

$$\{\Psi_N\}=\begin{Bmatrix}\psi_{1,N}\\ \psi_{2,N}\end{Bmatrix}$$

and the matrices H_{NM} are each (2×2) matrices given by

$$H_{NN}=\begin{bmatrix}\varepsilon_1+N\hbar\omega & 0\\ 0 & \varepsilon_2+N\hbar\omega\end{bmatrix}$$

$$H_{N,N+1}=\begin{bmatrix}0 & K\sqrt{N+1}\\ K^*\sqrt{N+1} & 0\end{bmatrix} \tag{10.1.11}$$

with $H_{N+1,N}=H^+_{N,N+1}$

Broadening: The point is that if we consider the N-photon subspace, it is like an open system that is connected to the $(N+1)$- and $(N-1)$-photon subspaces, just as a device is connected to the source and drain contacts. In Chapter 8, we saw that the effect of the source or drain contact could be represented by a self-energy matrix (see Eq. (8.1.11))

$$\Sigma=\tau g\tau^+$$

whose imaginary (more precisely anti-Hermitian) part represents the broadening

$$\Gamma\equiv i[\Sigma-\Sigma^+]=\tau a\tau^+$$

$a\equiv i[g-g^+]$ being the spectral function of the isolated reservoir. We could use the same relation to calculate the self-energy function that describes the effect of the rest of the photon reservoir on the N-photon subspace, which we view as the "channel." Actually the details are somewhat more complicated because (unlike coherent interactions) we have to account for the exclusion principle. For the moment, however, let us calculate the broadening (or the outflow) assuming all other states to be "empty" so that there is no exclusion principle to worry about. Also, to keep things simple, let us focus just on the diagonal element of the broadening:

$$\begin{aligned}[\Gamma_{nn}(E)]_{N,N}&=[H_{nm}]_{N,N+1}\,[a_{mm}]_{N+1,N+1}\,[H_{mn}]_{N+1,N}\\ &\quad+[H_{nm}]_{N,N-1}\,[a_{mm}]_{N-1,N-1}\,[H_{mn}]_{N-1,N}\end{aligned}$$

Assuming that the coupling from one photon subspace to the next is weak, we can approximate the spectral functions a with their unperturbed values:

$$[\Gamma_{nn}(E)]_{N,N} = \left|K_{mn}^{\mathrm{em}}\right|^2 (N+1) 2\pi\delta\left[E - \varepsilon_m - (N+1)\hbar\omega\right] + \left|K_{mn}^{\mathrm{ab}}\right|^2 N 2\pi\delta\left[E - \varepsilon_m - (N-1)\hbar\omega\right]$$

where

$$K_{mn}^{\mathrm{em}} \equiv [H_{mn}]_{N+1,N} \tag{10.1.12a}$$

and

$$K_{mn}^{\mathrm{ab}} \equiv [H_{mn}]_{N-1,N} \tag{10.1.12b}$$

Again with weak coupling between photon subspaces we can assume that the state $|n, N\rangle$ remains an approximate eigenstate with an energy $\varepsilon_n + N\hbar\omega$, so that we can evaluate the broadening at $E = \varepsilon_n + N\hbar\omega$:

$$\Gamma_{nn} = 2\pi\left|K_{mn}^{\mathrm{em}}\right|^2 (N+1)\delta\left(\varepsilon_n - \varepsilon_m - \hbar\omega\right) \quad (emission) + 2\pi\left|K_{mn}^{\mathrm{ab}}\right|^2 N\delta\left(\varepsilon_n - \varepsilon_m + \hbar\omega\right) \quad (absorption) \tag{10.1.13}$$

The first term arises from the coupling of the N-photon subspace to the $(N + 1)$-photon subspace, indicating that it represents a *photon emission* process. Indeed it is peaked for photon energies $\hbar\omega$ for which $\varepsilon_n - \hbar\omega = \varepsilon_m$, suggesting that we view it as a process in which an electron in state n transits to state m and emits the balance of the energy as a photon. The second term in Eq. (10.1.13) arises from the coupling of the N-photon subspace to the $(N - 1)$-photon subspace, indicating that it represents a *photon absorption* process. Indeed it is peaked for photon energies $\hbar\omega$ for which $\varepsilon_n + \hbar\omega = \varepsilon_m$, suggesting that we view it as a process in which an electron in state n transits to state m and absorbs the balance of the energy from a photon.

Coupling constants: How do we write down the coupling constants K appearing in Eq. (10.1.13)? This is where it helps to invoke the one-electron viewpoint (Fig. 10.1.2). The entire problem then amounts to writing down the "potential" U_S that an electron feels due to one photon or phonon occupying a particular mode with a frequency ω in the form:

$$U_S(\vec{r}, t) = U^{\mathrm{ab}}(\vec{r})\exp(-i\omega t) + U^{\mathrm{em}}(\vec{r})\exp(+i\omega t) \tag{10.1.14}$$

where $U^{\mathrm{ab}}(\vec{r}) = U^{\mathrm{em}}(\vec{r})^*$.

Once we have identified this "interaction potential," the coupling constants for emission and absorption can be evaluated simply from the matrix elements of U^{em} and U^{ab}:

$$K_{mn}^{\mathrm{em}} = \int \mathrm{d}r\,\phi_m^*(r) U^{\mathrm{em}} \phi_n(r) \equiv \langle m|U^{\mathrm{em}}|n\rangle$$

and

$$K_{mn}^{\mathrm{ab}} = \int \mathrm{d}r\phi_m^*(r)U^{\mathrm{ab}}\phi_n(r) \equiv \langle m|U^{\mathrm{ab}}|n\rangle \tag{10.1.15}$$

where ϕ_m and ϕ_n are the wavefunctions for levels m and n respectively.

Electron–phonon coupling: In Section 10.4, I will try to elaborate on the meaning of phonons. But for the moment we can simply view them as representing the vibrations of the lattice of atoms, just as photons represent electromagnetic vibrations. To write down the interaction potential for phonons, we need to write down the atomic displacement or the strain due to the presence of a single phonon in a mode with frequency ω and then multiply it by the change D in the electronic energy per unit displacement or strain. The quantity D, called the deformation potential, is known experimentally for most bulk materials of interest and one could possibly use the same parameter unless dealing with very small structures. Indeed relatively little work has been done on phonon modes in nanostructures and it is common to assume plane wave modes labeled by a wave vector $\vec{\beta}$, which is appropriate for bulk materials. The presence of a (longitudinal) phonon in such a plane wave mode gives rise to a strain (ρ is the mass density and Ω is the normalization volume)

$$S = \beta\sqrt{2\hbar/\rho\omega\Omega}\cos(\vec{\beta}\cdot\vec{r} - \omega(\beta)t) \tag{10.1.16}$$

so that the interaction potentials in Eq. (10.1.14) are given by

$$U_{\vec{\beta}}^{\mathrm{ab}}(\vec{r}) \equiv (U_{\vec{\beta}}/2)\exp(\mathrm{i}\vec{\beta}\cdot\vec{r}) \quad \text{and} \quad U_{\vec{\beta}}^{\mathrm{em}}(\vec{r}) = U_{\vec{\beta}}^{\mathrm{ab}}(\vec{r})^* \tag{10.1.17}$$

where $U_{\vec{\beta}} = D\beta\sqrt{2\hbar/\rho\omega\Omega}$.

Electron–photon coupling: The basic principle for writing down the electron–photon coupling coefficient is similar: we need to write down the interaction potential that an electron feels due to the presence of a single photon in a particular mode. However, the details are complicated by the fact that the effect of an electromagnetic field enters the Schrödinger equation through the vector potential (which we discussed very briefly in the supplementary notes in Chapter 5) rather than a scalar potential.

First, we write down the electric field due to a single photon in mode $(\vec{\beta}, \hat{\nu})$ in the form

$$\vec{E} = \hat{\nu}E_0\sin(\vec{\beta}\cdot\vec{r} - \omega(\beta)t)$$

whose amplitude E_0 is evaluated by equating the associated energy to $\hbar\omega$ (Ω is the volume of the "box"):

$$\varepsilon E_0^2\Omega/2 = \hbar\omega \rightarrow |E_0| = \sqrt{2\hbar\omega/\varepsilon\Omega}$$

The corresponding vector potential $\vec{A}$ is written as (noting that for electromagnetic waves $\vec{E} = -\partial\vec{A}/\partial t$):

$$\vec{A} = \hat{\nu} A_0 \cos(\vec{\beta}\cdot\vec{r} - \omega(\beta)t) \quad \text{with} \quad |A_0| = \sqrt{2\hbar/\omega\varepsilon\Omega} \tag{10.1.18}$$

Next we separate the vector potential due to one photon into two parts (cf. Eq. (10.1.14)):

$$\vec{A}(\vec{r}, t) = \vec{A}^{\mathrm{ab}}(\vec{r})\exp(-\mathrm{i}\omega(\beta)t) + \vec{A}^{\mathrm{em}}(\vec{r})\exp(+\mathrm{i}\omega(\beta)t) \tag{10.1.19}$$

where $\vec{A}^{\mathrm{ab}}(\vec{r}) \equiv \hat{\nu}(A_0/2)\exp(\mathrm{i}\vec{\beta}\cdot\vec{r})$ and $\vec{A}^{\mathrm{em}}(\vec{r}) = \vec{A}^{\mathrm{ab}}(\vec{r})^*$

The coupling coefficient for absorption processes is given by the matrix element for $(q/m)\vec{A}^{\mathrm{ab}}(\vec{r})\cdot\vec{p}$, while the coupling coefficient for emission processes is given by the matrix element for $(q/m)\vec{A}^{\mathrm{em}}(\vec{r})\cdot\vec{p}$ so that $(\vec{p} \equiv -\mathrm{i}\hbar\vec{\nabla})$

$$K_{mn}(\vec{\beta}, \hat{\nu}) = (qA_0/2m)\langle m|\exp(\mathrm{i}\vec{\beta}\cdot\vec{r})\vec{p}\cdot\hat{\nu}|n\rangle \quad \textit{Absorption} \tag{10.1.20a}$$

$$K_{mn}(\vec{\beta}, \hat{\nu}) = (qA_0/2m)\langle m|\exp(-\mathrm{i}\vec{\beta}\cdot\vec{r})\vec{p}\cdot\hat{\nu}|n\rangle \quad \textit{Emission} \tag{10.1.20b}$$

Note that the m appearing here stands for mass and is different from the index m we are using to catalog basis functions.

Equations (10.1.20a, b) require a slightly extended justification since we have not had much occasion to deal with the vector potential. We know that the scalar potential $\phi(\vec{r})$ enters the Schrödinger equation additively:

$$p^2/2m \rightarrow (p^2/2m) - q\phi(\vec{r})$$

and if the photon could be represented by scalar potentials the coupling coefficients would simply be given by the matrix elements of $(-q)\phi^{\mathrm{ab}}(\vec{r})$ and $(-q)\phi^{\mathrm{em}}(\vec{r})$ for absorption and emission respectively as we did in writing down the electron–phonon coupling. But photons require a vector potential which enters the Schrödinger equation as

$$p^2/2m \rightarrow (\vec{p} + q\vec{A})\cdot(\vec{p} + q\vec{A})/2m$$

so that the change due to the photon is given by

$$(q/2m)(\vec{A}\cdot\vec{p} + \vec{p}\cdot\vec{A}) + (q^2/2m)\vec{A}\cdot\vec{A} \approx (q/2m)(\vec{A}\cdot\vec{p} + \vec{p}\cdot\vec{A})$$

assuming that the vector potential is small enough that the quadratic term is negligible. Finally we note that for any scalar function $\phi(\vec{r})$

$$\vec{p}\cdot(\vec{A}\phi) = \vec{A}\cdot(\vec{p}\phi) + \phi(\vec{p}\cdot\vec{A})$$

so that we can write

$$(q/2m)(\vec{A}\cdot\vec{p} + \vec{p}\cdot\vec{A}) = (q/m)\vec{A}\cdot\vec{p}$$

as long as $\vec{p}\cdot\vec{A} = 0$. It can be checked that this is indeed true of the photon vector potential given in Eq. (10.1.19) because of the transverse nature of electromagnetic waves which requires that the wavevector $\vec{\beta}$ and the polarization $\hat{\nu}$ be orthogonal to each other: $\vec{\beta} \cdot \hat{\nu} = 0$. This allows us to obtain the coupling coefficient from the matrix element for $(q/m)\vec{A}(\vec{r}) \cdot \vec{p}$ using $\vec{A} \to \vec{A}^{\text{ab}}(\vec{r})$ for absorption processes and $\vec{A} \to \vec{A}^{\text{em}}(\vec{r})$ for emission processes.

10.2 Examples

In this section I will go through a few examples to illustrate the basic approach for describing incoherent interactions. I will take up the more general case of inflow and outflow in Section 10.3, but in this section I will assume all other states to be "empty" so that there is no exclusion principle to worry about and I will calculate the broadening (or the outflow), which can be identified with $\hbar/\tau$, τ being the lifetime of the state. This will include: (1) the photon-induced (radiative) lifetime due to atomic transitions; (2) the radiative lifetime due to interband transitions in semiconductors; and (3) the phonon-induced (non-radiative) lifetime due to intraband transitions in semiconductors (Fig. 10.2.1). The basic approach is to write down the interaction potential (see Eq. (10.1.14)), evaluate the coupling constants (see Eq. (10.1.15)), and obtain the broadening and hence the lifetime from Eq. (10.1.13).

10.2.1 Atomic transitions

From Eq. (10.1.13) it is apparent that the broadening is large when the argument of the delta function vanishes. How large it is at that point depends on the value of 0^+ (see Section 8.4) that we choose to broaden each reservoir state. As we have seen in Chapter 8, the precise value of 0^+ usually does not matter as long as the system

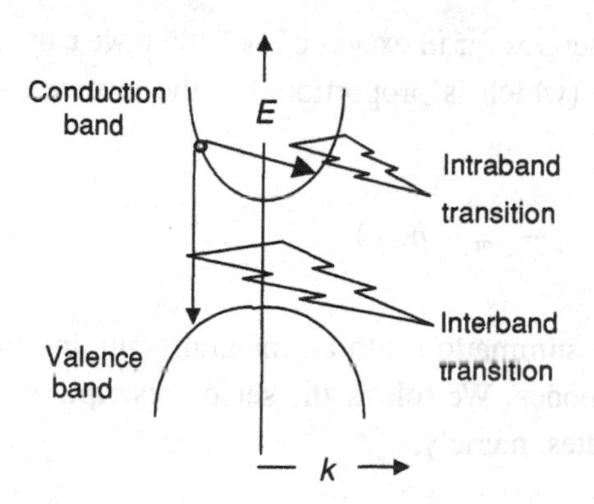

Fig. 10.2.1 Electronic transitions in semiconductors can be classified as interband and intraband. The former are associated primarily with electron–photon interactions while the latter are associated primarily with electron–phonon interactions.

Fig. 10.2.2

is coupled to a continuous distribution of reservoir states. This is true in this case, because we usually do not have a single photon mode with energy $\hbar\omega$. Instead, we have a continuous distribution of photons with different wavevectors $\vec{\beta}$ with energies given by

$$\hbar\omega(\vec{\beta}) = \hbar\bar{c}\beta \tag{10.2.1}$$

where $\bar{c}$ is the velocity of light in the solid. Consequently the states in a particular subspace do not look discrete as shown in Fig. 10.1.3, but look more like as shown in Fig. 10.2.2.

The broadening is obtained from Eq. (10.1.13) after summing over all modes $\vec{\beta}$ and the two allowed polarizations $\hat{\nu}$ for each $\vec{\beta}$:

$$\begin{aligned}\Gamma_{nn} &= \sum_{\vec{\beta},\hat{\nu}} 2\pi|K_{mn}(\vec{\beta},\hat{\nu})|^2(N_{\vec{\beta},\hat{\nu}}+1)\delta(\varepsilon_n-\varepsilon_m-\hbar\omega(\beta)) \\ &+\sum_{\vec{\beta},\hat{\nu}} 2\pi|K_{mn}(\vec{\beta},\hat{\nu})|^2 N_{\vec{\beta},\hat{\nu}}\delta(\varepsilon_n-\varepsilon_m+\hbar\omega(\beta))\end{aligned} \tag{10.2.2}$$

If the photon reservoir is in equilibrium then the number of photons in mode $\vec{\beta}$ and polarization $\hat{\nu}$ is given by the Bose–Einstein factor (Eq. (10.1.8)):

$$N_{\vec{\beta},\hat{\nu}} = \frac{1}{\exp(\hbar\omega(\beta)/k_{\mathrm{B}}T)-1} \tag{10.2.3}$$

If we consider transitions involving energies far in excess of $k_{\mathrm{B}}T$, then we can set $N_{\vec{\beta},\hat{\nu}}$ equal to zero, so that the broadening (which is proportional to the inverse radiative lifetime τ_{r}) is given by

$$\Gamma_{nn} = \left(\frac{\hbar}{\tau_{\mathrm{r}}}\right)_n = \sum_{\vec{\beta},\hat{\nu}} 2\pi|K_{mn}(\vec{\beta},\hat{\nu})|^2\delta(\varepsilon_n-\varepsilon_m-\hbar\omega_\beta) \tag{10.2.4a}$$

which is evaluated by converting the summation into an integral assuming periodic boundary conditions for the photon modes. We follow the same prescription that we have used in the past for electronic states, namely,

$$\sum_{\vec{\beta}} \rightarrow \frac{\Omega}{8\pi^3}\int_0^{\infty} d\beta\,\beta^2 \int_0^{+\pi} d\theta\,\sin\theta \int_0^{2\pi} d\phi$$

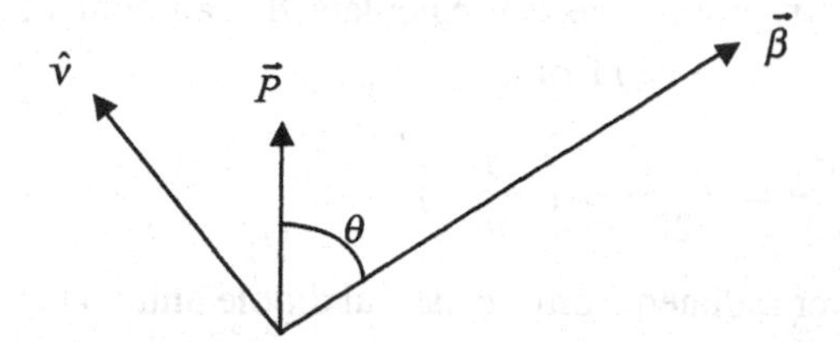

Fig. 10.2.3 Emission of photon with wavevector $\vec{\beta}$ and polarization $\hat{\nu}$ by an atomic transition with an equivalent dipole moment $\vec{P}$.

(where Ω is again the volume of the "box") to obtain

$$\Gamma_{nn} = \frac{\Omega}{8\pi^3}\sum_{\hat{\nu}}\int_0^{\infty} \mathrm{d}\beta\,\beta^2 \int_{-\pi}^{+\pi} \mathrm{d}\theta\,\sin\theta \int_0^{2\pi} \mathrm{d}\phi\, 2\pi\,|K_{mn}(\vec{\beta},\hat{\nu})|^2\delta(\varepsilon_n - \varepsilon_m - \hbar\omega_\beta) \tag{10.2.4b}$$

To proceed further, we need to insert the electron–photon coupling coefficients K from Eqs. (10.1.20a, b).

For atomic wavefunctions that are localized to extremely short dimensions (much shorter than an optical wavelength) we can neglect the factor $\exp(\mathrm{i}\vec{\beta}\cdot\vec{r})$ and write from Eq. (10.1.20a, b)

$$K_{mn}(\vec{\beta},\hat{\nu}) = \frac{q}{m}\sqrt{\frac{\hbar}{2\varepsilon\omega\Omega}}P\sin\theta \tag{10.2.5}$$

where $\vec{P} \equiv \langle m|\vec{p}\,|n\rangle$ and θ is the complement of the angle between the dipole moment of the transition and the polarization of the photon (Fig. 10.2.3).

Using Eqs. (10.2.5) and (10.2.1) we can find the radiative lifetime from Eq. (10.2.4b):

$$\Gamma = \frac{\Omega}{8\pi^3}\int_0^{\infty}\frac{\mathrm{d}\omega\,\omega^2}{\bar{c}^3}\int_{-\pi}^{+\pi}\mathrm{d}\theta\,\sin^3\theta\int_0^{2\pi}\mathrm{d}\phi\frac{q^2\hbar}{2m^2\varepsilon\omega\Omega}P^2 2\pi\,\delta(\varepsilon_n - \varepsilon_m - \hbar\omega)$$

so that

$$\frac{1}{\tau_r} = \frac{\Gamma}{\hbar} = \frac{q^2}{4\pi\varepsilon\hbar\bar{c}}\,\frac{2(\varepsilon_n - \varepsilon_m)}{3\hbar m\bar{c}^2}\,\frac{2P^2}{m} \tag{10.2.6}$$

Note that the answer is obtained without having to worry about the precise height of the delta function (which is determined by the value of 0^+ as we discussed in Section 8.4). But if the photon modes do not form a quasi-continuous spectrum (as in small nanostructures) then it is conceivable that there will be reversible effects that are affected by the precise values of 0^+.

Analogy with a classical dipole antenna: We can calculate the amount of power radiated per electron (note that $\hbar\omega = \varepsilon_n - \varepsilon_m$) from:

$$W = \frac{\hbar\omega}{\tau} = \frac{q^2}{4\pi\varepsilon\hbar\bar{c}} \frac{2(\varepsilon_n - \varepsilon_m)^2}{3\hbar m\bar{c}^2} \frac{2P^2}{m} = \frac{\omega^2}{12\pi\varepsilon\bar{c}^3}\left(\frac{2qP}{m}\right)^2 \tag{10.2.7}$$

It is interesting to note that the power radiated from a classical dipole antenna of length d carrying a current $I\cos\omega t$ is given by

$$W = \frac{\omega^2}{12\pi\varepsilon\bar{c}^3}(Id)^2$$

suggesting that an atomic radiator behaves like a classical dipole with

$$\underset{\textit{antenna}}{\vec{I}d} = \underset{\textit{atomic radiator}}{2q\vec{P}/m} = (2q/m)\langle m|\vec{p}|n\rangle \tag{10.2.8}$$

Indeed it is not just the total power, even the polarization and angular distribution of the radiation are the same for a classical antenna and an atomic radiator. The light is polarized in the plane containing the direction of observation and $\vec{P}$, and its strength is proportional to $\sim\sin^2\theta$, θ being the angle between the direction of observation and the dipole as shown in Fig. 10.2.3.

10.2.2 Interband transitions in semiconductors

The basic rule stated in Eqs. (10.1.20a, b) for the coupling coefficients can be applied to delocalized electronic states too, but we can no longer neglect the factor $\exp(i\vec{\beta}\cdot\vec{r})$ as we did when going to Eq. (10.2.5). For example, in semiconductors (see Fig. 10.2.1), the conduction (c) and valence (v) band electronic states are typically spread out over the entire solid consisting of many unit cells as shown in Fig. 10.2.4 where $|c\rangle_n$ and $|v\rangle_n$ are the atomic parts of the conduction and valence band wavefunctions in unit cell n. These functions depend on the wavevector $\vec{k}_c$ or $\vec{k}_v$, but are the same in each unit cell, except for the spatial shift. This allows us to write the coupling elements for absorption and emission from Eq. (10.1.20a, b) in the form

$$\langle v|\vec{p}\cdot\hat{\nu}|c\rangle \sum_n \frac{1}{N}\exp[i(\vec{k}_c \pm \vec{\beta} - \vec{k}_v)\cdot\vec{r}_n]$$

where we take the upper sign (+) for absorption and the lower sign (−) for emission, $\langle\ldots\rangle$ denotes an integral over a unit cell, and we have neglected the variation of the factor $\exp(i\vec{\beta}\cdot\vec{r})$ across a unit cell. This leads to non-zero values only if

$$\vec{k}_v = \vec{k}_c \pm \vec{\beta} \tag{10.2.9}$$

Eq. (10.2.9) can be viewed as a rule for momentum conservation, if we identify $\hbar\vec{k}$ as the electron momentum and $\hbar\vec{\beta}$ as the photon momentum. The final electronic momentum

Conduction band $\frac{1}{\sqrt{N}}\sum_n |c\rangle_n \exp\left(i\vec{k}_c \cdot \vec{r}_n\right)$

Valence band $\frac{1}{\sqrt{N}}\sum_n |v\rangle_n \exp\left(i\vec{k}_v \cdot \vec{r}_n\right)$

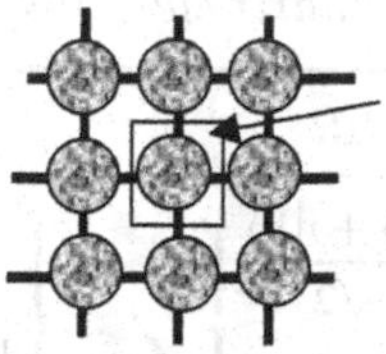

Unit cell n

Fig. 10.2.4

$\hbar\vec{k}_v$ is equal to the initial electronic momentum $\hbar\vec{k}_c$ plus or minus the photon momentum $\hbar\vec{\beta}$ depending on whether the photon is absorbed or emitted. The photon wavevector is typically very small compared to the electronic wavevector, so that radiative transitions are nearly "vertical" with $\vec{k}_v = \vec{k}_c \pm \vec{\beta} \approx \vec{k}_c$. This is easy to see if we note that the range of k extends over a Brillouin zone which is $\sim 2\pi$ divided by an atomic distance, while the photon wavevector is equal to 2π divided by the optical wavelength which is thousands of atomic distances.

Assuming that the momentum conservation rule in Eq. (10.2.9) is satisfied, we can write the coupling coefficients from Eq. (10.1.20a, b) as

$$K_{mn}(\vec{\beta}, \hat{\nu}) = \frac{q}{m}\sqrt{\frac{\hbar}{2\varepsilon\omega\Omega}}\, P\sin\theta \qquad \text{where } \vec{P} \equiv \langle v|\,\vec{p}\,|c\rangle \tag{10.2.10}$$

showing that "vertical" radiative transitions in semiconductors can be understood in much the same way as atomic transitions (see Eq. (10.2.5)) using the atomic parts of the conduction and valence band wavefunctions. For example, if we put the numbers characteristic of conduction–valence band transitions in a typical semiconductor like GaAs into Eq. (10.2.6), $\varepsilon_n - \varepsilon_m = 1.5\,\text{eV}$, $2P^2/m = 20\,\text{eV}$, $\varepsilon = 10\varepsilon_0$, we obtain $\tau_r = 0.7$ ns for the radiative lifetime.

Polarization and angular distribution: We know that the electronic states at the bottom of the conduction band near the Γ-point are isotropic or s-type denoted by $|S\rangle$. If the states at the top of the valence band were purely p_x, p_y, and p_z types denoted by $|X\rangle, |Y\rangle, |Z\rangle$, then we could view the system as being composed of three independent antennas with their dipoles pointing along x-, y-, and z-directions since

$$\langle s|\,\vec{p}\,|X\rangle = \hat{x}P \qquad \langle s|\,\vec{p}\,|Y\rangle = \hat{y}P \qquad \langle s|\,\vec{p}\,|Z\rangle = \hat{z}P$$

The resulting radiation can then be shown to be unpolarized and isotropic. In reality, however, the top of the valence band is composed of light hole and heavy hole bands which are mixtures of up- and down-spin states and the equivalent dipole moments for each of the conduction–valence band pairs can be written as shown in Table 10.2.1.

We thus have eight independent antennas, one corresponding to each conduction band–valence band pair. If the C state is occupied, then the first row of four antennas is active. If we look at the radiation coming out in the z-direction then we will see the

Table 10.2.1 *Optical matrix elements for conduction–valence band transitions*

	HH	$\overline{HH}$	LH	$\overline{LH}$
	$\begin{Bmatrix} \frac{\lvert X\rangle + i\lvert Y\rangle}{\sqrt{2}} \\ 0 \end{Bmatrix}$	$\begin{Bmatrix} 0 \\ \frac{\lvert X\rangle - i\lvert Y\rangle}{\sqrt{2}} \end{Bmatrix}$	$\begin{Bmatrix} -\sqrt{\frac{2}{3}}\lvert Z\rangle \\ \frac{\lvert X\rangle + i\lvert Y\rangle}{\sqrt{6}} \end{Bmatrix}$	$\begin{Bmatrix} \frac{\lvert X\rangle - i\lvert Y\rangle}{\sqrt{6}} \\ \sqrt{\frac{2}{3}}\lvert Z\rangle \end{Bmatrix}$
$C\begin{Bmatrix} \lvert S\rangle \\ 0 \end{Bmatrix}$	$\frac{\hat{x} + i\hat{y}}{\sqrt{2}}P$	0	$-\sqrt{\frac{2}{3}}\hat{z}P$	$\frac{\hat{x} - i\hat{y}}{\sqrt{6}}P$
$\overline{C}\begin{Bmatrix} 0 \\ \lvert S\rangle \end{Bmatrix}$	0	$\frac{\hat{x} - i\hat{y}}{\sqrt{2}}P$	$\frac{\hat{x} + i\hat{y}}{\sqrt{6}}P$	$\sqrt{\frac{2}{3}}\hat{z}P$

radiation from the C–HH and C–$\overline{LH}$ transitions. The C–HH transition will emit right circularly polarized (RCP) light, which will be three times as strong as the left circularly polarized (LCP) light from the C–$\overline{LH}$ transition. If the $\overline{C}$ state is also occupied then the $\overline{C}$–$\overline{HH}$ transition would yield three times as much LCP light as the RCP light from the $\overline{C}$–LH transition. Overall there would be just as much LCP as RCP light. But if only the C state is occupied then there would be thrice as much RCP light as LCP light. Indeed the degree of circular polarization of the emission is often used as a measure of the degree of spin polarization that has been achieved in a given experiment.

10.2.3 Intraband transitions in semiconductors

We have discussed the radiation of light due to interband transitions in semiconductors (see Fig. 10.2.1). But what about intraband transitions? Can they lead to the emission of light? We will show that the simultaneous momentum and energy conservation requirements prevent any radiation of light unless the electron velocity exceeds the velocity of light: $\hbar k/m > \bar{c}$. This is impossible in vacuum, but could happen in a solid and such Cerenkov radiation of light by fast-moving electrons has indeed been observed. However, this is usually not very relevant to the operation of solid-state devices because typical electronic velocities are about a thousandth of the speed of light. What is more relevant is the Cerenkov emission of acoustic waves or phonons that are five orders of magnitude slower than light. Electron velocities are typically well in excess of the phonon velocity leading to extensive Cerenkov emission (and absorption) of phonons, somewhat like the sonic booms generated by supersonic jets.

Criteria for Cerenkov emission: For intraband transitions, both the final and initial states have the same atomic wavefunctions and, for clarity, I will not write them down explicitly. Instead I will write the initial and final states in the form of plane waves as if we are dealing with electrons in vacuum:

$$|\vec{k}\rangle \equiv (1/\sqrt{\Omega})\exp(i\vec{k}\cdot\vec{r}) \quad \text{and} \quad |\vec{k}_f\rangle \equiv (1/\sqrt{\Omega})\exp(i\vec{k}_f\cdot\vec{r}) \qquad (10.2.11)$$

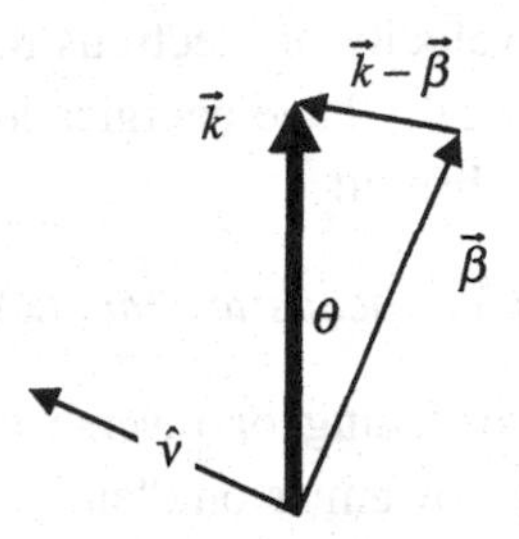

Fig. 10.2.5 Cerenkov emission: initial state $\vec{k}$, photon wavevector $\vec{\beta}$, and final state $\vec{k}-\vec{\beta}$. The photon polarization is along $\hat{\nu}$.

From Eq. (10.1.20a, b) we find that the radiative coupling constant is equal to

$$K(\vec{k}_f, \vec{k}, \vec{\beta}, \hat{\nu}) = (qA_0/2m)\hbar\vec{k}\cdot\hat{\nu} \tag{10.2.12}$$

$$\text{if} \quad \underset{\textit{Emission}}{\vec{k}_f = \vec{k}-\vec{\beta}} \qquad \underset{\textit{Absorption}}{\vec{k}_f = \vec{k}+\vec{\beta}} \tag{10.2.13}$$

and is zero otherwise. Like Eq. (10.2.9), Eq. (10.2.13) too can be interpreted as a condition for momentum conservation. Energy conservation, on the other hand, is enforced by the delta function in Eq. (10.1.13):

$$\underset{\textit{Emission}}{\varepsilon(\vec{k}_f) = \varepsilon(\vec{k}) - \hbar\omega(\vec{\beta})} \qquad \underset{\textit{Absorption}}{\varepsilon(\vec{k}_f) = \varepsilon(\vec{k}) + \hbar\omega(\vec{\beta})} \tag{10.2.14}$$

From Eqs. (10.2.13) and (10.2.14) we obtain for *emission* processes

$$\varepsilon(\vec{k}_f) - \varepsilon(\vec{k}) + \hbar\omega(\vec{\beta}) = 0 = \frac{\hbar^2}{2m}([\vec{k}-\vec{\beta}]\cdot[\vec{k}-\vec{\beta}] - k^2) + \hbar\bar{c}\beta$$

so that

$$\frac{\hbar^2}{2m}(-2k\beta\cos\theta + \beta^2) + \hbar\bar{c}\beta = 0$$

which yields

$$\cos\theta = \frac{\bar{c}}{\hbar k/m} + \frac{\beta}{2k} \tag{10.2.15}$$

The point is that, since $\cos\theta$ must be smaller than one, Cerenkov emission cannot take place unless the electron velocity $\hbar k/m$ exceeds the velocity of light $\bar{c}$. As mentioned above, this is impossible in vacuum, but possible in a solid and Cerenkov radiation has indeed been observed. The emitted light forms a cone around the electronic wavevector $\vec{k}$ with a maximum angle $\theta_{max} = \cos^{-1}(m\bar{c}/\hbar k)$ (Fig. 10.2.5).

Cerenkov emission of acoustic phonons: As stated before, Cerenkov emission of light is not very relevant to the operation of solid-state devices. But the emission (and absorption) of sound waves or acoustic phonons is quite relevant. Acoustic waves are five orders of magnitude slower than light and the velocity of electrons routinely exceeds the sound velocity. Since acoustic phonons typically have energies less than $k_B T$ they are usually present in copious numbers at equilibrium:

$$N_{\vec{\beta}} + 1 - N_{\vec{\beta}} = [\exp(\hbar\omega(\beta)/k_B T) - 1]^{-1} \cong k_B T/\hbar\omega(\beta) \quad (\textit{acoustic phonons})$$

Both terms in Eq. (10.1.13) now contribute to the broadening or inverse lifetime: Cerenkov absorption ("ab") is just as important as Cerenkov emission ("em"):

$$(\Gamma)_{\vec{k},\mathrm{em}} = \sum_{\vec{\beta}} 2\pi \frac{k_B T}{\hbar\omega(\beta)} |K(\vec{\beta})|^2 \delta[\varepsilon(\vec{k}) - \varepsilon(\vec{k} - \vec{\beta}) - \hbar\omega(\vec{\beta})] \tag{10.2.16a}$$

$$(\Gamma)_{\vec{k},\mathrm{ab}} = \sum_{\vec{\beta}} 2\pi \frac{k_B T}{\hbar\omega(\beta)} |K(\vec{\beta})|^2 \delta[\varepsilon(\vec{k}) - \varepsilon(\vec{k} + \vec{\beta}) + \hbar\omega(\vec{\beta})] \tag{10.2.16b}$$

The coupling element $K(\vec{\beta})$ is proportional to the potential that an electron feels due to the presence of a single phonon as discussed earlier (see Eq. (10.1.18)). Without getting into a detailed evaluation of Eqs. (10.2.16a, b), it is easy to relate the angle of emission θ to the magnitude of the phonon wavevector β by setting the arguments of the delta functions to zero and proceeding as we did in deriving Eq. (10.2.15):

$$\cos\theta = \frac{c_s}{\hbar k/m} + \frac{\beta}{2k} \quad (\textit{emission})$$
$$\cos\theta = \frac{c_s}{\hbar k/m} - \frac{\beta}{2k} \quad (\textit{absorption}) \tag{10.2.17}$$

where c_s is the velocity of sound waves: $\omega = c_s\beta$. The detailed evaluation of the electron lifetime due to acoustic phonon emission and absorption from Eqs. (10.2.16a, b) is described in Exercise E.10.1.

Cerenkov emission of optical phonons: In addition to acoustic phonons, there are optical phonons (see supplementary notes in Section 10.4) whose frequency is nearly constant $\omega = \omega_0$, where the phonon energy $\hbar\omega_0$ is typically a few tens of millielectron-volts, so that the number of such phonons present at equilibrium at room temperature is of order one:

$$N_{\vec{\beta}} = [\exp(\hbar\omega_0/k_B T) - 1]^{-1} \equiv N_0 \quad (\textit{optical phonons})$$

From Eq. (10.1.13) we now obtain for the emission and absorption rates

$$(\Gamma)_{\vec{k},\mathrm{em}} = (N_0 + 1) \sum_{\vec{\beta}} 2\pi |K(\vec{\beta})|^2 \delta[\varepsilon(\vec{k}) - \varepsilon(\vec{k} - \vec{\beta}) - \hbar\omega_0] \tag{10.2.18a}$$

$$(\Gamma)_{\vec{k},\mathrm{ab}} = N_0 \sum_{\vec{\beta}} 2\pi |K(\vec{\beta})|^2 \delta[\varepsilon(\vec{k}) - \varepsilon(\vec{k} + \vec{\beta}) + \hbar\omega_0] \tag{10.2.18b}$$

which can be evaluated as described in Exercise E.10.2.

10.3 Inflow and outflow

Let me now explain how we can use the concepts developed in this chapter to write down the new terms Σ_s^{in} and Γ_s appearing in Fig. 10.2 (p. 252).

Discrete levels: To start with, consider the inflow and outflow into a specific energy level ε due to transitions from levels a and b – one above it and one below it separated by an energy of $\hbar\omega$: $\varepsilon_b - \varepsilon = \varepsilon - \varepsilon_a = \hbar\omega$ (Fig 10.3.1). We saw in Section 10.1 that the rate constant (K^{ab}) for absorption processes is proportional to the number of phonons present (N_ω) while the rate constant (K^{em}) for emission processes is proportional to ($N_\omega + 1$). Let's assume the temperature is very low compared to $\hbar\omega$, so that $N_\omega \ll 1$ and we need only worry about emission. Using N, N_a, and N_b to denote the number of electrons in each of these levels we can write for the level ε:

$$\text{Inflow} = K^{em}(1 - N)N_b \quad \text{and} \quad \text{Outflow} = K^{em}N(1 - N_a)$$

We can now write the inflow term as a difference between two terms

$$\text{Inflow} = K^{em}N_b - K^{em}NN_b$$

where the second term represents the part of the inflow that is blocked by the exclusion principle. A non-obvious result that comes out of the advanced formalism (see Appendix, Section A.4) is that this part of the inflow is not really blocked. Rather the outflow is increased by this amount ($K^{em}NN_b$):

$$\text{Inflow} = K^{em}N_b \quad \text{and} \quad \text{Outflow} = K^{em}N(1 - N_a + N_b) \tag{10.3.1a}$$

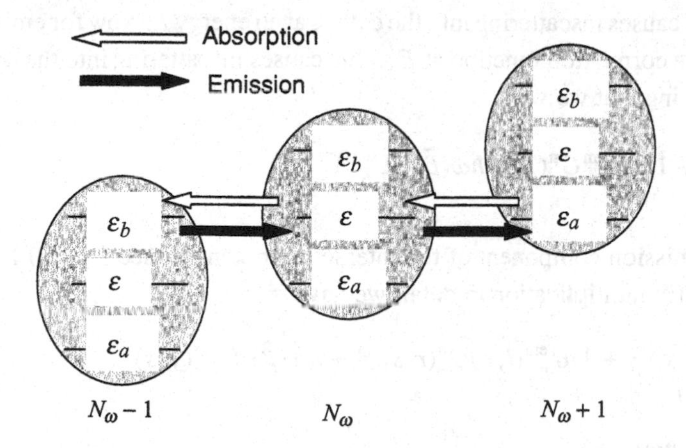

Fig. 10.3.1 An idealized device with three levels used to write down the inflow and outflow terms due to the absorption and emission of photons and phonons.

The difference between inflow and outflow is not changed from what we had guessed earlier. But the distinction is not academic, since the outflow (not the difference) determines the broadening of the level. We can compare this with the inflow and outflow between a contact and a discrete level that we discussed earlier (see Fig. 8.1):

$$\text{Inflow} = \gamma f \quad \text{and} \quad \text{Outflow} = \gamma N \tag{10.3.1b}$$

The inflow looks similar with K^{em} playing the role of γ, but the outflow involves an extra factor $(1 - N_a + N_b)$ that reduces to one if $N_a = N_b$, but not in general. Equation (10.3.1b) was earlier generalized to the form (see Fig. 9.1.2)

$$\text{Inflow} = \text{Trace}[\Gamma A] f = \text{Trace}[\Sigma^{\text{in}} A] \quad \text{and} \quad \text{Outflow} = \text{Trace}[\Gamma G^{\text{n}}]$$

Similarly in the present case we can generalize Eq. (10.3.1a) to

$$\text{Inflow} = \text{Trace}\left[\Sigma_{\text{s}}^{\text{in}} A\right] \quad \text{and} \quad \text{Outflow} = \text{Trace}[\Gamma_{\text{s}} G^{\text{n}}] \tag{10.3.2}$$

where the expressions for $\Sigma_{\text{s}}^{\text{in}}$ and Γ_{s} can be discovered heuristically by analogy. Let us do this one by one.

Inflow for continuous distribution of states: Scattering processes are very similar to ordinary contacts. The basic difference is that regular contacts are maintained in equilibrium by external sources, while for scattering processes the "contact" is the device itself (see Fig. 10.3.1) and as such can deviate significantly from equilibrium. For regular contacts we saw in Chapter 8 that the inscattering function is given by

$$\Sigma^{\text{in}}(E) = [\tau A \tau^+] f = [\tau G^{\text{n}} \tau^+]$$

where $[G^{\text{n}}] = [A] f$ is the correlation function in the contact: the contact correlation function at energy E causes inscattering into the device at an energy E. Now for emission processes, the device correlation function at $E + \hbar\omega$ causes inscattering into the device at energy E, suggesting that we write

$$\Sigma_{\text{s}}^{\text{in}}(E) = \sum_{\vec{\beta}} (N_{\vec{\beta}} + 1) \left[U_{\vec{\beta}}^{\text{em}} G^{\text{n}}(E + \hbar\omega(\vec{\beta}))(U_{\vec{\beta}}^{\text{em}})^+\right]$$

where U^{em} is the emission component of the interaction potential (see Eq. (10.1.17)). Writing out the matrix multiplication in detail we have

$$\begin{aligned}\Sigma_{\text{s}}^{\text{in}}(i, j; E) &= \sum_{p', q', \vec{\beta}} (N_{\vec{\beta}} + 1) U_{\vec{\beta}}^{\text{em}}(i, r) G^{\text{n}}(r, s; E + \hbar\omega(\vec{\beta})) U_{\vec{\beta}}^{\text{em}}(j, s)^* \\ &= \int_0^\infty \frac{\text{d}(\hbar\omega)}{2\pi} \sum_{p', q'} D^{\text{em}}(i, r; j, s; \hbar\omega) G^{\text{n}}(r, s; E + \hbar\omega)\end{aligned} \tag{10.3.3}$$

where

$$D^{em}(i,r;j,s;\hbar\omega) \equiv 2\pi \sum_{\vec{\beta}} (N_\omega + 1)\delta(\hbar\omega - \hbar\omega(\vec{\beta}))U^{em}_{\vec{\beta}}(i,r)U^{em}_{\vec{\beta}}(j,s)^* \qquad (10.3.4)$$

If we include absorption processes we obtain

$$\Sigma^{in}_s(i,j;E) = \int_0^\infty \frac{d(\hbar\omega)}{2\pi} \begin{pmatrix} D^{em}(i,r;j,s;\hbar\omega)G^n(r,s;E+\hbar\omega) \\ +D^{ab}(i,r;j,s;\hbar\omega)G^n(r,s;E-\hbar\omega) \end{pmatrix} \qquad (10.3.5)$$

where the absorption term is given by an expression similar to the emission term with N instead of $(N+1)$:

$$D^{ab}(i,r;j,s;\hbar\omega) = \sum_{\vec{\beta}} N_\omega \delta(\hbar\omega - \hbar\omega(\vec{\beta}))U^{ab}_{\vec{\beta}}(i,r)\,U^{ab}_{\vec{\beta}}(j,s)^* \qquad (10.3.6)$$

In general, the emission and absorption functions D^{em} and D^{ab} are fourth-rank tensors, but from hereon let me simplify by treating these as scalar quantities. The full tensorial results along with a detailed derivation are given in Section A.4 of the Appendix. We write the simplified emission and absorption functions as

$$D^{em}(\hbar\omega) \equiv (N_\omega + 1)D_0(\hbar\omega) \qquad (10.3.7a)$$

$$D^{ab}(\hbar\omega) \equiv N_\omega D_0(\hbar\omega) \qquad (10.3.7b)$$

So that

$$\Sigma^{in}_s(E) = \int_0^\infty \frac{d(\hbar\omega)}{2\pi} D_0(\hbar\omega)\left((N_\omega+1)G^n(E+\hbar\omega) + N_\omega G^n(E-\hbar\omega)\right) \qquad (10.3.8)$$

Outflow for continuous distribution of states: For the outflow term, in extrapolating from the discrete version in Eq. (10.3.1a) to a continuous version we replace N_b with $G^n(E+\hbar\omega)$ and $1-N_a$ with $G^p(E-\hbar\omega)$ to obtain

$$\Gamma_s(E) = \int \frac{d(\hbar\omega)}{2\pi} D^{em}(\hbar\omega)\left[G^p(E-\hbar\omega) + G^n(E+\hbar\omega)\right]$$

where I have defined a new quantity $G^p \equiv A - G^n$ that tells us the density (/eV) of empty states or holes, just as G^n tells us the density of filled states or electrons. The sum of the two is equal to the spectral function A which represents the density of states. We can extend this result as before (cf. Eqs. (10.3.5)–(10.3.10)) to include absorption terms as before to yield

$$\Gamma_s(E) = \int \frac{d(\hbar\omega)}{2\pi} D_0(\hbar\omega) \begin{pmatrix} (N_\omega+1)\left[G^p(E-\hbar\omega) + G^n(E+\hbar\omega)\right] \\ +N_\omega\left[G^n(E-\hbar\omega) + G^p(E+\hbar\omega)\right] \end{pmatrix} \qquad (10.3.9)$$

Equations (10.3.8)–(10.3.9) are the expressions for the scattering functions that we are looking for. The self-energy function Σ_s can be written as $\mathrm{Re}(\Sigma_s) + \mathrm{i}\Gamma_s/2$ where the real part can be obtained from the Hilbert transform of the imaginary part given in Eq. (10.3.9) as explained in Section 8.3.

Migdal's "theorem": If the electronic distribution can be described by an equilibrium Fermi function

$$G^n(E') \equiv f(E')A(E') \quad \text{and} \quad G^p(E') \equiv (1 - f(E'))A(E')$$

then

$$\Gamma_s(E) = \int d(\hbar\omega) \int dE' D_0(\hbar\omega) \cdot A(E') \times (N_\omega + 1 - f(E'))\delta(E - E' - \hbar\omega) + (N_\omega + f(E'))\delta(E - E' + \hbar\omega)$$

which is the expression commonly found in the literature on electron–phonon scattering in metals where it is referred to as Migdal's "theorem" (see, for example, Allen and Mitrovic, 1982). Close to equilibrium, a separate equation for the inscattering function $\Sigma_s^{in}(E)$ is not needed, since it is simply equal to $f(E)\Gamma_s(E)$, just like an ordinary contact.

Nearly elastic processes: Note that the expressions simplify considerably for low-energy scattering processes ($\hbar\omega \to 0$) for which we can set $E + \hbar\omega \approx E \approx E - \hbar\omega$:

$$\Sigma_s^{in}(E) = \int_0^\infty d(\hbar\omega)(D^{em}(\hbar\omega) + D^{ab}(\hbar\omega))\, G^n(E) \tag{10.3.10}$$

$$\Gamma_s(E) = \int_0^\infty d(\hbar\omega)(D^{em}(\hbar\omega) + D^{ab}(\hbar\omega))\, A(E) \tag{10.3.11}$$

Indeed, in this case the real part of the self-energy Σ_s does not require a separate Hilbert transform. We can simply write

$$\Sigma_s(E) = \int_0^\infty d(\hbar\omega)(D^{em}(\hbar\omega) + D^{ab}(\hbar\omega))\, G(E) \tag{10.3.12}$$

since $\mathrm{Re}(\Sigma_s)$ is related to Γ_s in exactly the same way as $\mathrm{Re}(G)$ is related to A, namely through a Hilbert transform.

The complete equations of dissipative quantum transport including the inflow/outflow terms discussed in this section are summarized in Chapter 11 along with illustrative examples.

10.4 Supplementary notes: phonons

As I have mentioned before, phonons represent lattice vibrations just as photons represent electromagnetic vibrations. In this section I will try to elaborate on this statement and clarify what I mean. At low temperatures, the atoms that make up a molecule or a solid are frozen in their positions on the lattice and this frozen atomic potential is used in calculating the energy levels for the electrons, starting from the Schrödinger equation. As the temperature is raised these vibrations increase in amplitude and exchange energy with the electrons through the "electron–phonon interaction." To understand how we describe these vibrations, let us start with a simple example, namely a hydrogen molecule.

As we discussed in Chapter 3, we could describe the vibrations of this molecule in terms of a mass and spring system (Fig. 10.4.1). The mass M is just that of the two hydrogen atoms, while the spring constant K is equal to the second derivative of the potential energy with respect to the interatomic distance. We know from freshman physics that such a system behaves like an oscillator with a resonant frequency $\omega = \sqrt{K/M}$. Experimentalists have measured this frequency to be approximately $\omega = 2\pi(10^{14}/\mathrm{s})$. Knowing M we could calculate K and compare against theory, but that is a different story. The question we wish to address is the following. As we raise the temperature we expect the molecule to vibrate with increasing amplitude due to the thermal energy that it takes up from the surroundings. At very high temperatures the vibrations could become so violent that the molecule dissociates. But we are assuming that the temperature is way below that so that the amplitude of this vibration is much smaller than the bond length a. What is this amplitude as a function of the temperature?

To make our discussion quantitative, let us define a variable $u(t)$ that measures the distance between the two atoms, relative to the equilibrium value of u_0. We expect the atoms to oscillate with a frequency ω such that

$$u(t) = A\cos(\omega t + \phi)$$

It is convenient to define a complex amplitude $\tilde{u} = (A/2)\exp[\mathrm{i}\phi]$ and write

$$u(t) = \tilde{u}\exp[-\mathrm{i}\omega t] + \tilde{u}^*\exp[+\mathrm{i}\omega t] \tag{10.4.1}$$

The kinetic energy associated with this vibration is written as

$$KE = \frac{M}{2}\left(\frac{\mathrm{d}u}{\mathrm{d}t}\right)^2 = \frac{M\omega^2 A^2}{2}\sin^2(\omega t + \phi)$$

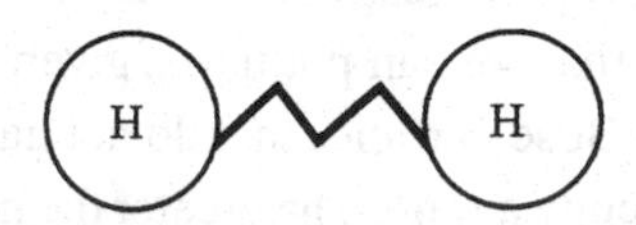

Fig. 10.4.1 A hydrogen molecule can be viewed as two masses connected by a spring.

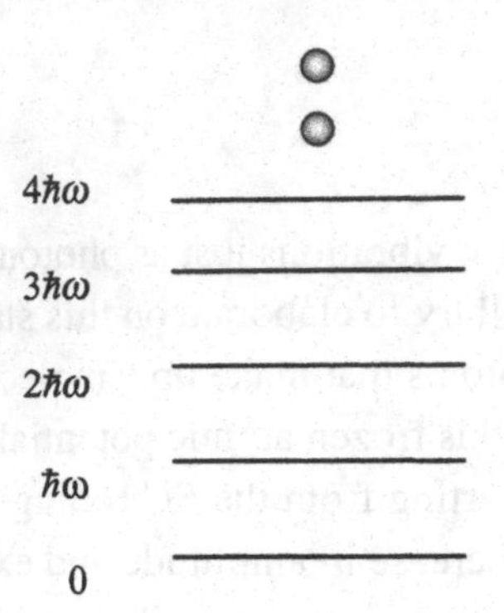

Fig. 10.4.2 Allowed energy levels of an oscillator with a resonant frequency ω.

while the potential energy is given by (note that $K = M\omega^2$)

$$PE = \frac{K}{2}u^2 = \frac{M\omega^2 A^2}{2}\cos^2(\omega t + \phi)$$

so that the total energy is independent of time as we might expect:

$$E = KE + PE = M\omega^2 A^2/2 = 2M\omega^2|\tilde{u}|^2 \tag{10.4.2}$$

Classical physicists believed that the energy E could have any value. Early in the twentieth century Planck showed that the experimentally observed radiation spectra could be explained by postulating that the energy of electromagnetic oscillators is quantized in units of $\hbar\omega$. It is now believed that the energies of all harmonic oscillators occur only in integer multiples of $\hbar\omega$ as shown in Fig. 10.4.2 ($n = 0, 1, 2, \ldots$):

$$E = 2M\omega^2|\tilde{u}|^2 = n\hbar\omega \rightarrow |\tilde{u}| = \sqrt{n\hbar/2M\omega} \tag{10.4.3}$$

Using Eq. (10.4.3) we can rewrite Eq. (10.4.1) as

$$u(t) = \sqrt{\frac{\hbar}{2M\omega}}(a\exp[-i\omega t] + a^*\exp[+i\omega t]) \tag{10.4.4}$$

where $|a|^2 = n = 0, 1, 2, \ldots$

Since the energy of the oscillator is an integer multiple of $\hbar\omega$, we can alternatively visualize it as a single energy level of energy $\hbar\omega$ into which we can put an integer number $\nu = 0, 1, 2, \ldots$ of particles called phonons (for electromagnetic oscillators the particles are called photons). If the oscillator is in a state with an energy $4\hbar\omega$ we say that four phonons are present. Vibrations too thus acquire a particulate character like electrons. The difference is that electrons are Fermi particles, obeying the exclusion principle. Consequently the number of particles n_k that we can put into a given state k is either zero or one. Phonons on the other hand are Bose particles that do not have any exclusion principle. Any number of phonons can occupy a state. The greater the number of particles ν, the greater the energy of the vibrations and hence the greater the amplitude

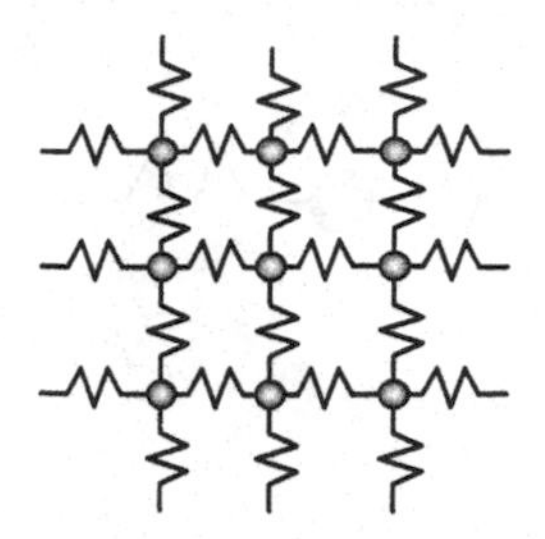

Fig. 10.4.3 A crystal lattice as a mass and spring system.

of the oscillation a in Eq. (10.4.4). For coherent vibrations, the amplitudes a are complex numbers with well-defined phases, but for thermal vibrations, there is no definite phase.

What is the average number of phonons at a given temperature? For electrons we know that the number of particles n_k in a state k can be either 0 or 1 and that the average value at equilibrium is given by the Fermi function (Eq. (1.1.1)). For phonons, the number of particles ν for a particular vibrational mode can be any positive number and the average value $\langle n \rangle$ at equilibrium is given by the Bose function:

$$\langle n \rangle = \frac{1}{\exp(\hbar\omega/k_B T) - 1} \tag{10.4.5}$$

I will not discuss the physical basis for this function (we did not disuss the Fermi function either) but both the Bose and Fermi functions follow from the general principle of equilibrium statistical mechanics stated in Eq. (3.4.5).

So what is the mean square amplitude of vibration at a temperature T? From Eq. (10.4.3)

$$\langle |\tilde{u}|^2 \rangle = \langle n \rangle \, \hbar/2M\omega$$

showing that the vibration amplitude increases with the number of phonons and thus can be expected to increase with temperature. At temperatures far below the melting point, this amplitude should be a small fraction of the equilibrium bond length.

Phonons: We argued in Chapter 3 that a hydrogen molecule could be visualized as two masses connected by a spring whose equilibrium length is equal to the hydrogen–hydrogen bond length. Extending the same argument we could visualize a solid lattice as a periodic array of masses connected by springs (Fig. 10.4.3).

How do we describe the vibrations of such a mass and spring array? One could equate the total force exerted by the springs on an individual mass to its mass times its acceleration and write down an infinite set of equations, one for each mass. This sounds like a complicated insoluble (non-quantum-mechanical) problem, but we will show that it can be tackled in much the same way that we tackled the bandstructure problem in Chapter 5.

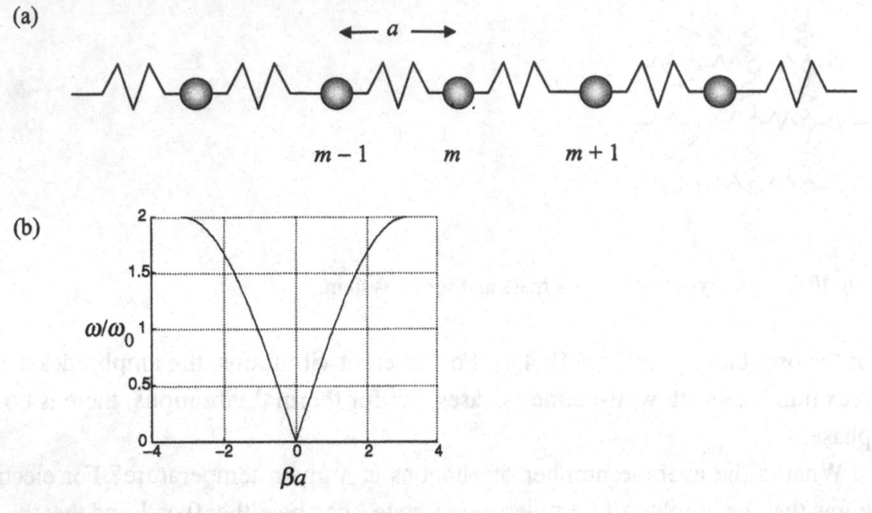

Fig. 10.4.4 (a) An infinite one-dimensional lattice of atoms. (b) Dispersion law ω/ω_0 vs. βa.

1D solid: For simplicity, let us consider a one-dimensional array of atoms represented by a mass and spring system (Fig. 10.4.4a). Assuming that u_m is the displacement of the mth atom from its equilibrium position in the lattice, the force exerted on the mth atom by the spring on the left is $K(u_m - u_{m-1})$, while that exerted by the spring on the right is $K(u_m - u_{m+1})$. From Newton's law,

$$M\frac{\mathrm{d}^2 u_m}{\mathrm{d}t^2} = K[u_{m+1} - 2u_m + u_{m-1}] \tag{10.4.6}$$

where M is the mass of the atom. Assuming sinusoidal vibrations with frequency $u_m = \tilde{u}_m \exp(-\mathrm{i}\omega t)$ we can write

$$M\omega^2 \tilde{u}_m = K[2\tilde{u}_m - \tilde{u}_{m-1} - \tilde{u}_{m+1}]$$

which can be written in the form of a matrix equation:

$$\omega^2\{\tilde{u}\} = [\Omega]\{\tilde{u}\} \tag{10.4.7}$$

where $\omega_0 \equiv \sqrt{K/M}$ and

$$\Omega = \begin{array}{c|cccccc} & 1 & 2 & \cdots & \cdots & N-1 & N \\ \hline 1 & 2\omega_0^2 & -\omega_0^2 & & & 0 & -\omega_0^2 \\ 2 & -\omega_0^2 & 2\omega_0^2 & & & 0 & 0 \\ & & \cdots & \cdots & & & \\ N-1 & 0 & 0 & & & 2\omega_0^2 & -\omega_0^2 \\ N & -\omega_0^2 & 0 & & & -\omega_0^2 & 2\omega_0^2 \end{array}$$

Here we have used periodic boundary conditions, that is, we have assumed the solid to be in the form of a ring with atom N connected back to atom 1. Note that this matrix has exactly the same form as the Hamiltonian matrix for a one-dimensional solid (Eq. (5.1.1)).

The similarity is of course purely mathematical. Equations (5.1.1) and (10.4.7) describe very different physics: the former describes the quantum mechanics of electrons in a periodic lattice, while the latter describes the classical dynamics of a periodic mass and spring system. But since the matrix *[H]* in Eq. (5.1.1) has exactly the same form as the matrix $[\Omega]$ in Eq. (10.4.7) they can be diagonalized in exactly the same way to obtain (cf. Eq. (5.1.2)):

$$\omega^2 = 2\omega_0^2(1 - \cos\beta a) \qquad \text{where } \beta a = n2\pi/N \tag{10.4.8}$$

The values of βa run from $-\pi$ to $+\pi$ and are spaced by $2\pi/N$, just like the values of ka in Fig. 5.1.2. The eigendisplacements $\tilde{u}_m$ corresponding to a given eigenmode β are given by

$$\sum_\beta \tilde{u}_\beta \exp[\mathrm{i}\beta ma] = \sum_\beta u_\beta \exp[\beta ma - \omega(\beta)t]$$

The actual instantaneous values of the displacement are obtained by taking the real parts of these phasor amplitudes (just as we do with voltages and currents in ac circuits):

$$u_m(t) = 2\mathrm{Re}\left\{\sum_\beta u_\beta \exp[\mathrm{i}(\beta ma - \omega(\beta)t)]\right\}$$

$$= \sum_\beta u_\beta \exp[\mathrm{i}(\beta ma - \omega(\beta)t)] + u_\beta^* \exp[-\mathrm{i}(\beta ma - \omega(\beta)t)] \tag{10.4.9}$$

What exactly the normal mode amplitudes u_β are depends on the history of how the lattice was excited. But the point is that any arbitrary displacement pattern $u_m(t)$ in the lattice can be expressed as a superposition of normal modes as expressed in Eq. (10.4.9) with an appropriate choice of the normal mode amplitudes u_β.

Optical phonons: We saw earlier that with two atoms per unit cell (see Fig. 5.1.4), the $E(k)$ plot acquires two branches (see Fig. 5.1.5). A similar effect is obtained for the dispersion $\omega(\beta)$ of vibrational modes too. Consider a lattice with two distinct masses $M_{1,2}$ and/or spring constants $K_{1,2}$ per unit cell as shown in Fig. 10.4.5a. Application of Newton's law now yields

$$\omega^2\{\tilde{u}\} = [\Omega]\{\tilde{u}\} \tag{10.4.10}$$

where $\omega_1 \equiv \sqrt{K_1/M_1}$, $\omega_2 \equiv \sqrt{K_2/M_2}$, $M_{1,2}$ and $K_{1,2}$ being the two masses and the

(a)

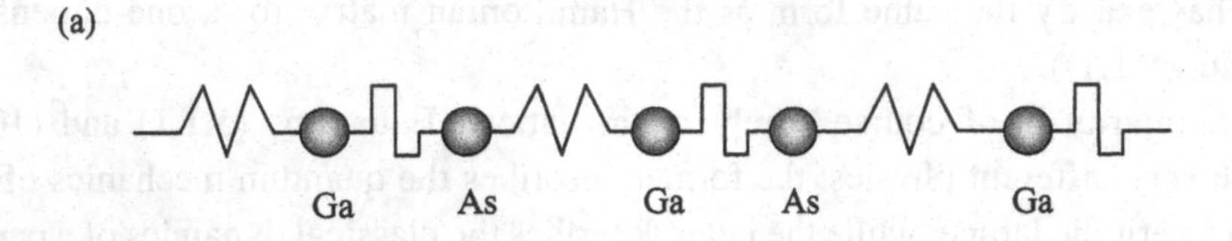

(b)

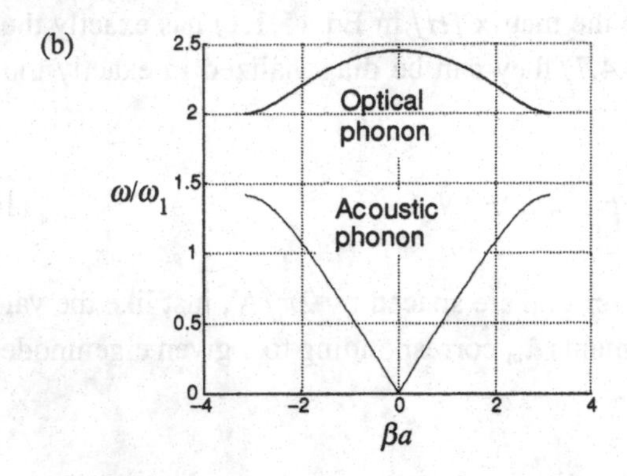

Fig. 10.4.5 (a) Snapshot of a GaAs lattice viewed along the (111) direction. (b) Dispersion law ω/ω_1 vs. βa showing acoustic and optical phonons assuming $\omega_2 = \sqrt{2}\omega_1$.

two spring constants, and

$$\Omega = \begin{array}{c|ccccc} & 1 & 2 & 3 & 4 & \\ 1 & \omega_1^2+\omega_2^2 & -\omega_1^2 & 0 & 0 & \cdots \\ 2 & -\omega_1^2 & \omega_1^2+\omega_2^2 & -\omega_2^2 & 0 & \cdots \\ 3 & 0 & -\omega_2^2 & \omega_1^2+\omega_2^2 & -\omega_1^2 & \cdots \\ 4 & 0 & 0 & -\omega_1^2 & \omega_1^2+\omega_2^2 & \cdots \\ & \cdots & \cdots & & & \\ & \cdots & \cdots & & & \end{array}$$

Note that this looks just like Eq. (5.1.6) and its eigenvalues can be obtained using the same method to yield (cf. Eq. (5.1.10)):

$$\begin{bmatrix} \omega_1^2+\omega_2^2 & \omega_1^2+\omega_2^2 e^{-i\beta a} \\ \omega_1^2+\omega_2^2 e^{+i\beta a} & \omega_1^2+\omega_2^2 \end{bmatrix} \begin{Bmatrix} u_1(t) \\ u_2(t) \end{Bmatrix} = \omega^2(\beta) \begin{Bmatrix} u_1(t) \\ u_2(t) \end{Bmatrix} \tag{10.4.11}$$

Setting the determinant to zero

$$\begin{bmatrix} \omega_1^2+\omega_2^2-\omega^2 & \omega_1^2+\omega_2^2 e^{-i\beta a} \\ \omega_1^2+\omega_2^2 e^{+i\beta a} & \omega_1^2+\omega_2^2-\omega^2 \end{bmatrix} = 0$$

we obtain the dispersion relation

$$\omega^2(\beta) = \omega_1^2+\omega_2^2 \pm \left(\omega_1^4+\omega_2^4+2\omega_1^2\omega_2^2\cos\beta a\right)^{1/2} \tag{10.4.12}$$

Like Eq. (5.1.10), Eq. (10.4.12) leads to two branches in the phonon spectrum as shown in Fig. 10.4.5b (cf. Fig. 5.1.5). The lower branch is called the acoustic phonon branch since it corresponds to ordinary acoustic waves at low frequencies. The displacement of both atoms in a unit cell (Ga and As) have the same sign for the acoustic branch. By contrast their displacements have opposite signs for the optical branch. The name "optical" comes from the fact that these vibrations can be excited by an incident electromagnetic radiation whose electric field sets the Ga and the As atoms moving in opposite directions, since one is negatively charged and one is positively charged.

It is easy to see why ω increases with β for the acoustic branch, but is nearly constant for the optical branch. We know that the resonant frequency of a mass and spring system increases as the spring gets stiffer. An acoustic mode with a small β results in very little stretching of the springs because all the atoms tend to move together. So the springs appear less stiff for smaller β, leading to a lower frequency ω. But with optical modes, the distortion of the springs is nearly independent of β. Even if β is zero the Ga and As atoms move against each other, distorting the springs significantly. The frequency ω thus changes very little with β.

Three-dimensional solids: In real three-dimensional solids, the displacement is actually a vector with three components: u_x, u_y, u_z for each of the two atoms in a unit cell. This leads to six branches in the dispersion curves, three acoustic and three optical. One of the three is a longitudinal mode with a displacement along the vector $\vec{\beta}$, while the other two are transverse modes with displacements perpendicular to $\vec{\beta}$. We thus have a longitudinal acoustic (LA) mode, two transverse acoustic (TA) modes, one longitudinal optical (LO) mode, and two transverse optical (TO) modes. The overall displacement at a point $(\vec{r}, t)$ can be written in the form ($\vec{r}$ denotes the position vector of the different points of the lattice)

$$\{\vec{u}(\vec{r}, t)\} = \sum_{\nu,\vec{\beta}} \{u_{\nu,\vec{\beta}}\} \exp[\mathrm{i}(\vec{\beta}\cdot\vec{r} - \omega_\nu(\beta)t)] + \{u_{\nu,\vec{\beta}}\}^* \exp[-\mathrm{i}(\vec{\beta}\cdot\vec{r} - \omega_\nu(\beta)t)] \tag{10.4.13}$$

Here the index ν runs over the six branches of the dispersion curve (LA, TA, etc.) and $\{u_{\nu,\vec{\beta}}\}$ are the (6×1) eigenvectors of the (6×6) eigenvalue equation obtained by generalizing Eq. (10.4.9) to include three components for each of the two displacement vectors. We could write the displacements as (2×1) eigenvectors one for each the three polarizations $\hat{\nu}$:

$$\{\vec{u}(\vec{r}, t)\}_\nu = \hat{\nu} \sum_{\vec{\beta}} \{u_{\vec{\beta}}\}_\nu \exp[\mathrm{i}(\vec{\beta}\cdot\vec{r} - \omega_\nu(\beta)t)] + \{u_{\vec{\beta}}\}_\nu^* \exp[-\mathrm{i}(\vec{\beta}\cdot\vec{r} - \omega_\nu(\beta)t)] \tag{10.4.14}$$

The two components represent the displacements of the two atoms in unit cell.

Following the same arguments as we used to arrive at Eq. (10.4.4) we can write

$$\{u_{\vec{\beta}}\}_\nu = \hat{\nu}\sqrt{\frac{\hbar}{2\rho\Omega\omega}}a_{\nu,\vec{\beta}}\begin{pmatrix}u_1\\u_2\end{pmatrix} \tag{10.4.15}$$

where (u_1, u_2) is equal to $(1, 1)/\sqrt{2}$ for acoustic phonons and to $(1, -1)/\sqrt{2}$ for optical phonons. The square of the amplitude is the number of phonons occupying the mode, whose average value is given by the Bose–Einstein factor:

$$|a_{\nu,\vec{\beta}}|^2 = n_{\nu,\vec{\beta}} \qquad \langle n_{\nu,\vec{\beta}}\rangle = \frac{1}{\exp(\hbar\omega_\nu(\beta)/k_BT) - 1} \tag{10.4.16}$$

From Eqs. (10.4.14)–(10.4.16), the mean squared displacement of an atom due to phonons can be written as

$$\langle u^2\rangle = \sum_{\vec{\beta}} \frac{\hbar}{2\rho\omega\Omega}\frac{1}{\exp(\hbar\omega/k_BT) - 1} \tag{10.4.17}$$

Strain due to a single phonon: Let us obtain the result we stated earlier in Eq. (10.1.16). The longitudinal strain is defined as the divergence of the displacement:

$$S = \vec{\nabla}\cdot\vec{u} = \sum_{\vec{\beta}} \vec{\beta}\cdot\hat{\nu}\sqrt{\frac{\hbar}{2\rho\omega\Omega}}\left(a_{\vec{\beta},\hat{\nu}}\exp[\mathrm{i}(\vec{\beta}\cdot\vec{r} - \omega_\nu(\beta)t)] + a^*_{\vec{\beta},\hat{\nu}}\exp[-\mathrm{i}(\vec{\beta}\cdot\vec{r} - \omega_\nu(\beta)t)]\right)$$

For a single phonon with wavevector $\vec{\beta}$ and polarization $\hat{\nu}$, we can set $a_{\vec{\beta},\hat{\nu}} = 1$, so that

$$S = (\hat{\nu}\cdot\vec{\beta})\sqrt{2\hbar/\rho\omega\Omega}\,\cos(\vec{\beta}\cdot\vec{r} - \omega(\beta)t)$$

as stated in Eq. (10.1.16).

EXERCISES

E.10.1. Assume that that the electron velocity is $\hbar k/m$ and the sound velocity is c_s. Evaluate Eqs. (10.2.16a, b) by: (a) converting the summation into an integral; (b) expressing the argument of the delta function in the form

$$\delta\left(\frac{\hbar^2 k\beta}{m}\left[\cos\theta - \frac{\bar{c}}{\hbar k/m} \pm \frac{\beta}{2k}\right]\right)$$

(c) performing the integral over $\cos\theta$ to get rid of the delta function and set a finite range to the limits of the integral over β: $\beta_{\min} < \beta < \beta_{\max}$; (d) performing the integral over β. Show that the lifetime due to acoustic phonon absorption and emission is given by

$$\frac{1}{\tau(k)} = \frac{mD^2k_BT}{\pi\hbar^3\rho c_s^2}k$$

(e) What is the angular distribution of the emitted phonons?

E.10.2. Consider an electron in a state $\vec{k}$ in a parabolic band with mass m having an energy E that exceeds the optical phonon energy $\hbar\omega_0$. Equating the argument of the delta function in Eq. (10.1.13) to zero, obtain an expression relating the magnitude of the wavevector β of an *emitted* optical phonon to the angle θ at which it is emitted (measured from $\vec{k}$ as shown in Fig. E.10.2). What is the range of values of θ outside which no optical phonons are emitted?

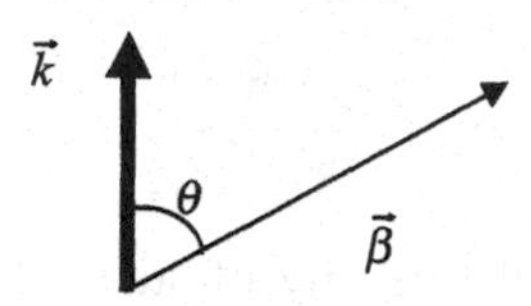

Fig. E.10.2

E.10.3. The mean squared displacement of an atom due to phonons is given in Eq. (10.4.17) as

$$\langle u^2 \rangle = \sum_{\vec{\beta}} \frac{\hbar}{2\rho\omega\Omega} \frac{1}{\exp(\hbar\omega/k_B T) - 1}$$

Convert the summation into an integral using periodic boundary conditions and evaluate the integral numerically to plot $\sqrt{\langle u^2 \rangle}$ vs. T over the range $0\,\text{K} < T < 1000\,\text{K}$. Assume acoustic phonons with $\omega_\beta = c_s\beta$ ($c_s = 5 \times 10^3$ m/s) and $\rho_s = 5 \times 10^3$ kg/m^3.

E.10.4. This book is focused on quantum transport and as such we have not discussed optical processes, except in passing in terms of its effect on electron transport. However, it is interesting to see how optical processes can be mapped onto corresponding transport problems so that we can make use of the ideas developed in this book (this is not commonly done, to my knowledge). Consider for example, an optical absorption process that takes electrons from the valence to the conduction band. Conceptually we could view this as an elastic transport process from a valence band state with N photons to a conduction band state with (N–1) photons as shown below (K is the matrix element for absorption given in Eq. (10.1.20a)):

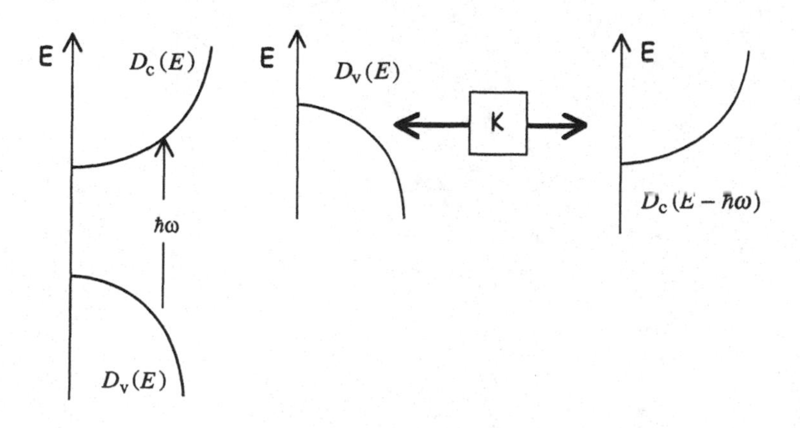

Explain why the rate at which photons are absorbed per second can be written as

$$R_{\text{abs}} = \frac{1}{2\pi h}\int \mathrm{d}E\,\overline{T}(E)[f_{\text{v}}(E) - f_{\text{c}}(E)]$$

where the "transmission" can be conveniently calculated from the alternative expression given in Exercise E.9.3:

$$\overline{T}(E) = \text{Trace}[A_1 M A_2 M^+]$$

by identifying the spectral functions A_1 and A_2 with the density of states $2\pi D_{\text{v}}(E)$ and $2\pi D_{\text{c}}(E - \hbar\omega)$ in the two "contacts" and the matrix element M with K given by Eq. (10.1.20a) to obtain

$$R_{\text{abs}} = \frac{2\pi}{h}\int \mathrm{d}E|K|^2 D_{\text{v}}(E) D_{\text{c}}(E - \hbar\omega)[f_{\text{v}}(E) - f_{\text{c}}(E)]$$

E.10.5. Illustrative exercises on dissipative quantum transport can be found in Chapter 11.

11 Atom to transistor

In Chapter 1, I used the generic structure shown in Fig. 11.1.1a (see Section 11.1.1) to focus and motivate this book. We spent Chapters 2 through 7 understanding how to write down a Hamiltonian matrix $[H_0]$ for the active region of the transistor structure whose eigenvalues describe the allowed energy levels (see Fig. 11.1.1b). In Chapter 8, I introduced the broadening $[\Gamma_1]$ and $[\Gamma_2]$ arising from the connection to the source and drain contacts. In Chapter 9, I introduced the concepts needed to model the flow of electrons, *neglecting* phase-breaking processes. In Chapter 10 we discussed the nature and meaning of phase-breaking processes, and how the resulting inflow and outflow of electrons is incorporated into a transport model. We now have the full "machinery" needed to describe dissipative quantum transport within the self-consistent field model (discussed in Chapter 3) which treats each electron as an independent particle moving in an average potential U due to the other electrons. I should mention, however, that this independent electron model misses what are referred to as "strong correlation effects" (see Section 1.5) which are still poorly understood. To what extent such effects can be incorporated into this model remains to be explored (see Appendix, Section A.5).

My purpose in this chapter is to summarize the machinery we have developed (Section 11.1) and illustrate how it is applied to concrete problems. I believe these examples will be useful as a starting point for readers who wish to use it to solve other problems of their own. At the same time, I have chosen these examples in order to illustrate conceptual issues that are of great importance in understanding the nature of electrical resistance on an atomic scale, namely the emergence of Ohm's law $V = IR$, with R proportional to the length of a conductor (Section 11.2), the spatial distribution of the reversible and irreversible heat associated with current flow (Section 11.3), and finally the spatial distribution of the voltage drop (Section 11.4).

11.1 Quantum transport equations

Let me quickly summarize the general model for dissipative quantum transport that we have discussed so far.

$$G^n = G\,\Sigma^{in}\,G^+ \tag{11.1.1}$$

$$G = [EI - H_0 - U - \Sigma]^{-1} \tag{11.1.2}$$

$$A = i[G - G^+] \qquad \Gamma = i[\Sigma - \Sigma^+] \tag{11.1.3}$$

where

$$\Sigma^{in} = \Sigma_1^{in} + \Sigma_2^{in} + \Sigma_s^{in}$$
$$\Sigma = \Sigma_1 + \Sigma_2 + \Sigma_s \tag{11.1.4}$$

These equations can be used to calculate the correlation function G^n and hence the density matrix ρ whose diagonal elements give us the electron density:

$$\rho = \int dE\, G^n(E)/2\pi \tag{11.1.5}$$

The current (per spin) at any terminal i can be calculated from

$$I_i = (q/\hbar) \int_{-\infty}^{+\infty} dE\, \tilde{I}_i(E)/2\pi \tag{11.1.6}$$

with

$$\tilde{I}_i = \text{Trace}\left[\Sigma_i^{in} A\right] - \text{Trace}[\Gamma_i G^n] \tag{11.1.7}$$

which is depicted in Fig. 11.1.1b in terms of an inflow ($\Sigma_i^{in} A$) and an outflow ($\Gamma_i G^n$) for the generic structure shown in Fig. 11.1.1a.

Input parameters: To use these equations, we need a channel Hamiltonian $[H_0]$ and the inscattering $[\Sigma^{in}]$ and broadening $[\Gamma]$ functions. For the two contacts, these are related:

$$\Sigma_1^{in} = \Gamma_1 f_1 \quad \text{and} \quad \Sigma_2^{in} = \Gamma_2 f_2 \tag{11.1.8}$$

and the broadening/self-energy for each contact can be determined from a knowledge of the surface spectral function a, the surface Green's function g of the contact, and the matrices $[\tau]$ describing the channel contact coupling:

$$\Gamma = \tau\, a\, \tau^+ \quad \text{and} \quad \Sigma = \tau\, g\, \tau^+ \tag{11.1.9}$$

For all the numerical results presented in this chapter we will use the simple one-band effective mass (equal to the free electron mass) model, both for the channel Hamiltonian $[H_0]$ and for the contact self-energy functions Σ_1, Σ_2.

For the scattering "terminal," unlike the contacts, there is no simple connection between Σ_s^{in} and Σ_s (or Γ_s). If the scattering process is essentially elastic ($E \approx E \pm \hbar\omega$),

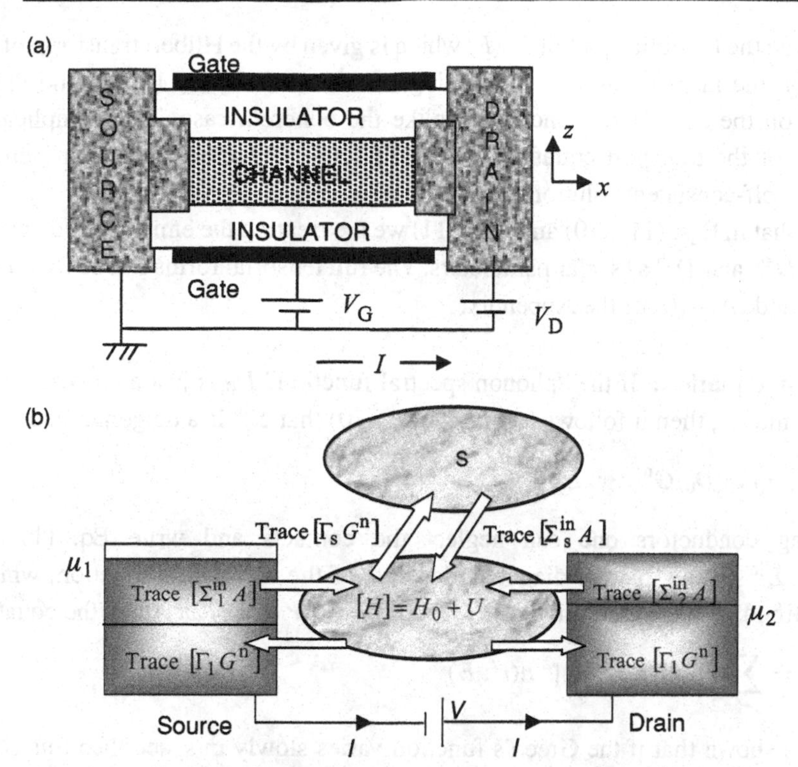

Fig. 11.1.1 (a) The generic transistor structure we used in the introduction to motivate this book. (b) Inflow and outflow of electrons for the generic structure in (a).

then (see Eqs. (10.3.12)–(10.3.14))

$$\Sigma_s^{in}(E) = D_0 G^n(E)$$
$$\Gamma_s(E) = D_0 A(E)$$
$$\Sigma_s(E) = D_0 G(E) \tag{11.1.10}$$

For general inelastic scattering processes (see Eqs. (10.3.8), (10.3.9))

$$\Sigma_s^{in}(E) = \int_0^\infty \frac{d(\hbar\omega)}{2\pi} \begin{pmatrix} D^{em}(\hbar\omega) \cdot G^n(E + \hbar\omega) \\ + D^{ab}(\hbar\omega) \cdot G^n(E - \hbar\omega) \end{pmatrix}$$

$$\Gamma_s(E) = \int_0^\infty \frac{d(\hbar\omega)}{2\pi} \begin{pmatrix} D^{em}(\hbar\omega) \cdot [G^p(E - \hbar\omega) + G^n(E + \hbar\omega)] \\ + D^{ab}(\hbar\omega) \cdot [G^n(E - \hbar\omega) + G^p(E + \hbar\omega)] \end{pmatrix}$$

but for our examples we will consider phonons with a single frequency ω_0

$$\Sigma_s^{in}(E) = D_0^{em} G^n(E + \hbar\omega_0) + D_0^{ab} G^n(E - \hbar\omega_0)$$
$$\Gamma_s(E) = D_0^{em}[G^p(E - \hbar\omega_0) + G^n(E + \hbar\omega_0)]$$
$$+ D_0^{ab}[G^n(E - \hbar\omega_0) + G^p(E + \hbar\omega_0)] \tag{11.1.11}$$

and ignore the hermitian part of $\Sigma_s(E)$ which is given by the Hilbert transform of $\Gamma_s(E)$. Note that the inscattering and broadening functions in Eqs. (11.1.10) and (11.1.11) depend on the correlation functions, unlike the coherent case. This complicates the solution of the transport equations (Eqs. (11.1.1)–(11.1.7)), requiring in general an iterative self-consistent solution.

Note that in Eqs. (11.1.10) and (11.1.11) we are treating the emission and absorption tensors D^{em} and D^{ab} as scalar parameters. The full tensorial forms can be found in Eqs. (A.4.9) and (A.4.10) of the Appendix.

Diffusion equation: If the "phonon spectral function" D_0 is just a constant times an identity matrix, then it follows from Eq. (11.1.10) that Σ_s^{in} is a diagonal matrix with

$$\Sigma_s^{in}(r, r; E) = D_0\, G^n(r, r; E)$$

For long conductors one can neglect the contacts and write Eq. (11.1.1) as $G^n = G\, \Sigma_s^{in}\, G^+$ so that the diagonal elements of the correlation function, which can be identified with the electron density $n(r; E) = G^n(r, r; E)/2\pi$, obey the equation

$$n(r; E) = \sum_{r'} D_0 |G(r, r'; E)|^2\, n(r'; E)$$

It can be shown that if the Green's function varies slowly in space then this equation reduces to the diffusion equation: $\nabla^2 n(r; E) = 0$ (Datta, 1990).

Transmission: In Chapter 9 we saw that the concept of transmission is a very useful one and developed an expression for the current in terms of the transmission function (see Eq. (9.1.9)). For quantum transport with dissipation, the concept is still useful as a qualitative heuristic tool, but in general it is not possible to write down a simple quantitative expression for the current in terms of the transmission function, because there is no simple connection between Σ_s^{in} and Γ_s, unlike the contacts where $\Sigma_i^{in} = \Gamma_i f_i$. It is more convenient to calculate the current directly from Eq. (11.1.6) instead. We can define an effective transmission by comparing Eq. (11.1.6) with Eq. (9.1.9)

$$\overline{T}_{\text{eff}}(E) = \frac{\tilde{I}_i(E)}{f_1(E) - f_2(E)} = \frac{\text{Trace}\left[\Sigma_i^{in} A\right] - \text{Trace}[\Gamma_i G^n]}{f_1 - f_2} \qquad (11.1.12)$$

which can be a useful parameter to compare with coherent transport.

Self-consistent calculation: In general it is necessary to perform a self-consistent solution of the transport equations with the "Poisson" equation, which accounts for electron–electron interactions through a potential $U(\vec{r})$ (see Fig. 11.1.2). We write Poisson within quotes as a reminder that this part of the problem could include corrections for correlation effects (see Chapter 3) in addition to standard electrostatics. This aspect is commonly ignored (as we do in Sections 11.2 and 11.3) when calculating

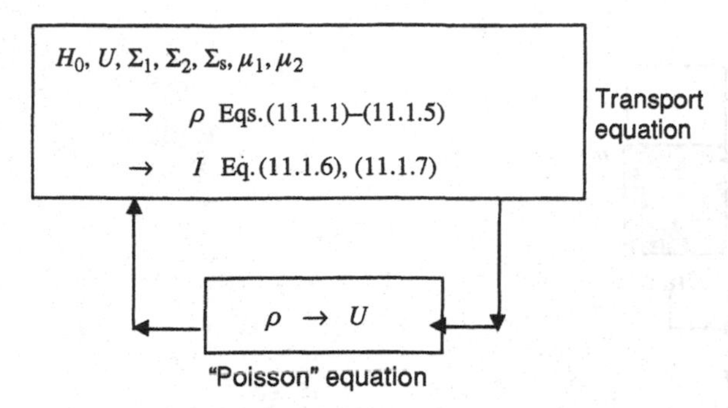

Fig. 11.1.2 In general, the transport problem has to be solved self-consistently with the "Poisson" equation, which accounts for electron–electron interactions through the potential U.

"linear response" for bias voltages that are small compared to the thermal energy $k_B T$ and/or the energy scale on which the density of states changes significantly. The current is then independent of the precise spatial profile $U(\vec{r})$ arising from the applied drain voltage. But if we are interested in the shape of the current–voltage (I–V) characteristics over a wide range of bias values as we are, for example, when calculating the ON-current of a transistor (Section 11.4), then the potential profile is of crucial importance as explained in Section 1.4.

A simple one-level example: In Chapter 1, we went through an example with just one level so that the electron density and current could all be calculated from a rate equation with a simple model for broadening. I then indicated that in general we need a matrix version of this "scalar model" and that is what the rest of the book is about (see Fig. 1.6.5).

Now that we have the full "matrix model" we have the machinery to do elaborate calculations as we will illustrate in the rest of the chapter. But before getting into such details, it is instructive to specialize to a one-level system with elastic phase-breaking (Fig. 11.1.3) so that all the matrices reduce to pure numbers and the results are easily worked out analytically. From Eqs. (11.1.1)–(11.1.4):

$$G(E) = (E - \varepsilon + (\mathrm{i}\Gamma/2))^{-1}$$

$$A(E) = \frac{\Gamma}{(E-\varepsilon)^2 + (\Gamma/2)^2}$$

$$\Gamma = \Gamma_1 + \Gamma_2 + D_0 A$$

$$G^{\mathrm{n}}(E) = \Sigma^{\mathrm{in}}(E)\, A(E)/\Gamma$$

$$\Sigma^{\mathrm{in}} = \Gamma_1 f_1 + \Gamma_2 f_2 + D_0 G^{\mathrm{n}}$$

Hence

$$\frac{G^{\mathrm{n}}}{A} = \frac{\Gamma_1 f_1 + \Gamma_2 f_2 + D_0 G^{\mathrm{n}}}{\Gamma_1 + \Gamma_2 + D_0 A} = \frac{\Gamma_1 f_1 + \Gamma_2 f_2}{\Gamma_1 + \Gamma_2}$$

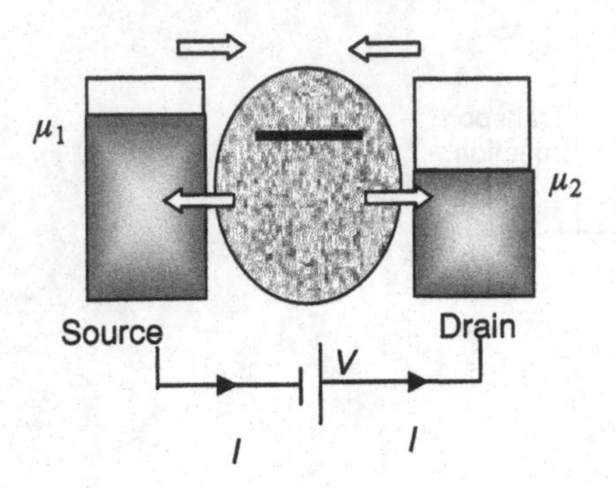

Fig. 11.1.3

independent of the scattering strength D_0. From Eqs. (11.1.6) and (11.1.7),

$$\tilde{I}_s = D_0 G^n A - D_0 A G^n = 0$$

and

$$\tilde{I}_1 = \Gamma_1 A(f_1 - (G^n/A)) = \frac{\Gamma_1 \Gamma_2 A}{\Gamma_1 + \Gamma_2}(f_1 - f_2) = -\tilde{I}_2$$

so that

$$I = \frac{q}{h}\int \mathrm{d}E \, \frac{\Gamma_1 \Gamma_2 A}{\Gamma_1 + \Gamma_2}(f_1 - f_2)$$

showing that elastic phase-breaking of this sort in a one-level system has no effect on the current, which is independent of D_0.

Even for a one-level system, inelastic phase-breaking is a little more complicated since the inscattering function Σ_s^{in} at energy E depends on the correlation function G^n at other energies $E + \hbar\omega$ (see Exercise E.11.6).

11.2 Physics of Ohm's law

My objective in this section is to show how the general quantum transport equations can be used to model conductors with phase-breaking processes. At the same time these examples will help illustrate how an ultrashort ballistic conductor evolves into a familiar macroscopic one obeying Ohm's law, which states that the conductance is directly proportional to the cross-sectional area S and inversely proportional to the length L. In chapter 1 I noted that for a ballistic conductor it is easy to see why the conductance should increase with S using elementary arguments. Now that we have discussed the concept of subbands (see Chapter 6) we can make the argument more precise. The conductance of a conductor increases with cross-sectional area because the number of subbands available for conduction increases and for large conductors this number is directly proportional to S.

But why should the conductance decrease with length L? Indeed a ballistic conductor without scattering has a conductance that is independent of its length. But for long conductors with scattering, the conductance decreases because the average transmission probability of electrons from the source to the drain decreases with the length of the conductor. We saw in Chapter 9 that the conductance is proportional to the total transmission $\overline{T}$ around the Fermi energy (see Eq. (9.4.4)), which can be expressed as the product of the number of modes M and the average transmission probability per mode T:

$$G = (2q^2/h)\overline{T} = (2q^2/h)MT \quad \textit{(Landauer formula)} \tag{11.2.1}$$

For large conductors, $M \sim S$ and $T \sim 1/L$, leading to Ohm's law: $G \sim S/L$. And that brings us to the question: why does the transmission probability decrease with length?

11.2.1 Classical transport

If we think of the electrons as classical particles, then it is easy to see why the transmission probability $T \sim 1/L$. Consider a conductor consisting of two sections in series as shown in Fig. 11.2.1. The first section has a transmission probability T_1; the second has a transmission probability T_2. What is the probability T that an electron will transmit through both? It is tempting to say that the answer is obviously $T = T_1T_2$, but that is wrong. That is the probability that the electron will get through both sections in its first attempt. But an electron turned back from section 2 on its first attempt has a probability of $T_1T_2R_1R_2$ of getting through after two reflections as shown in the figure ($R_1 = 1 - T_1$ and $R_2 = 1 - T_2$).

We can sum up the probabilities for all the paths analytically to obtain

$$T = T_1T_2((R_1R_2) + (R_1R_2)^2 + (R_1R_2)^3 + \cdots)$$

$$= \frac{T_1T_2}{1 - R_1R_2} = \frac{T_1T_2}{T_1 + T_2 - T_1T_2}$$

so that

$$\frac{1}{T} = \frac{1}{T_1} + \frac{1}{T_2} - 1 \tag{11.2.2}$$

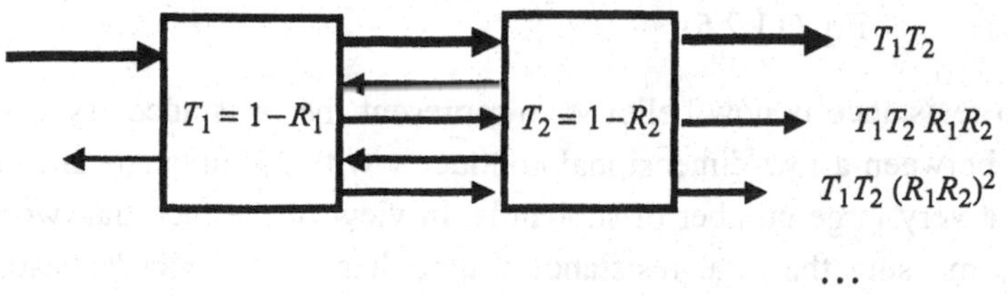

$T_1 = 1 - R_1$ | $T_2 = 1 - R_2$ → T_1T_2, $T_1T_2R_1R_2$, $T_1T_2(R_1R_2)^2$, ...

Fig. 11.2.1 Classical "addition" of the transmission probabilities for two successive sections to obtain an overall transmission probability.

This relation tells us the resulting transmission T if we cascade two sections with transmission T_1 and T_2 respectively. From this relation we can derive a general expression for the transmission $T(L)$ of a section of length L by deducing what function satisfies the relation:

$$\frac{1}{T(L_1+L_2)} = \frac{1}{T(L_1)} + \frac{1}{T(L_2)} - 1$$

It is easy to check that the following function fits the bill

$$T = \frac{\Lambda}{L+\Lambda} \tag{11.2.3}$$

where Λ is a constant of the order of a mean free path, representing the length for which the transmission probability is 0.5.

Equation (11.2.3) represents the transmission probability for classical particles as a function of the length of a conductor. Combining with Eq. (11.2.1) we obtain

$$G = (2q^2M/h)\frac{\Lambda}{L+\Lambda}$$

so that the resistance can be written as a constant interface resistance in series with a "device" resistance that increases linearly with length as required by Ohm's law:

$$\frac{1}{G} = \underbrace{\frac{h}{2q^2M}}_{\textit{interface resistance}} + \underbrace{\frac{h}{2q^2M}\frac{L}{\Lambda}}_{\textit{"device" resistance}} \tag{11.2.4}$$

Equation (11.2.4) suggests that the device conductance itself should be written as

$$G = \frac{2q^2M}{h}\,\frac{T}{1-T} \tag{11.2.5}$$

which was the original form advocated by Landauer. Equation (11.2.5) yields a resistance of zero as one might expect for a ballistic conductor (with $T = 1$), while Eq. (11.2.1) yields a non-zero resistance whose physical meaning caused extensive debate in the 1980s:

$$\underset{\text{Eq. (11.2.1)}}{1/G = (h/2q^2M)} \qquad \underset{\text{Eq. (11.2.5)}}{1/G = 0}$$

This non-zero resistance is now believed to represent the resistance associated with the interfaces between a low-dimensional conductor with M subbands and two large contacts with a very large number of subbands. In view of the fact that two-terminal measurements measure the total resistance rather than the "device" resistance, the present trend is to use Eq. (11.2.1) with the understanding that it includes the interface resistance along with the device resistance. Four-terminal measurements, on the other hand, are typically interpreted on the basis of the multi-terminal Büttiker formula discussed in Chapter 9.

11.2.2 Coherent transport (one subband)

We have seen that if we view electrons as classical particles we recover Ohm's law, together with a constant interface resistance in series. But is this true if we treat electrons as quantum mechanical particles obeying the Schrödinger wave equation? The answer is no! This is easy to check numerically if we calculate the transmission through a device with one scatterer (A or B) and a device with two scatterers (A and B) as shown in Fig. 11.2.2.

The transmission was calculated using the equations stated at the beginning of Section 11.2 with the phase-breaking terms (Σ_s and Σ_s^{in}) set equal to zero. Since we are dealing with coherent transport, we could calculate the transmission directly from Eq. (9.1.10). But it is better to use Eq. (11.1.12) since it is applicable to non-coherent transport and can be used in our later examples as well.

The important message from the example shown in Fig. 11.2.2 is that the quantum transmission through two scatterers does not follow a simple rule like the one we obtained for the transmission for classical particles (Eq. (11.2.2)). The basic reason is the interference between the two scatterers. If they are spaced by half a wavelength the reflections from the two scatterers interfere constructively, leading to a dip in the transmission. But if they are spaced by a quarter of a wavelength, the reflections interfere

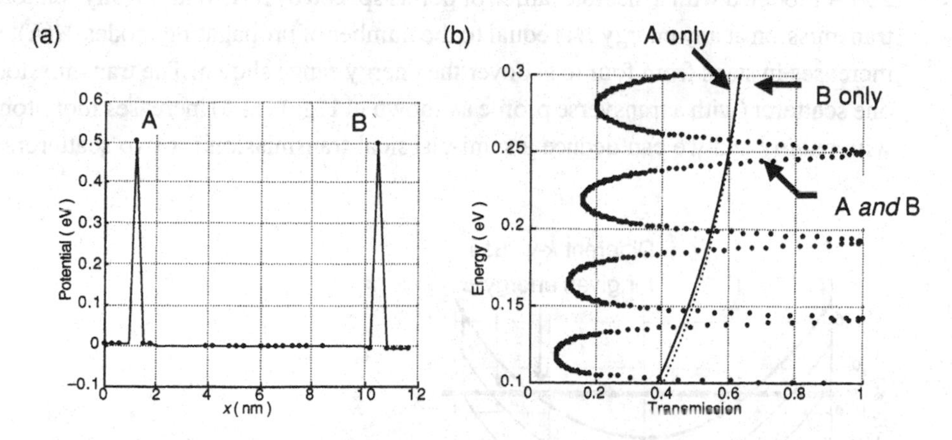

Fig. 11.2.2 (a) A short device with two scatterers, A and B. A discrete lattice with 40 sites separated by $a = 3$ Å was used in the calculation along with an effective mass equal to the free electron mass m. A small bias potential was assumed varying linearly from +5 meV to –5 meV across the device. Each scatterer is represented by a potential of 0.5 eV at one lattice site.
(b) Transmission versus energy calculated with only one scatterer (A or B) and with both scatterers (A and B).

destructively, leading to a peak in the transmission. This shows up as large oscillations in the transmission as a function of energy (which determines the de Broglie wavelength of the electrons). Clearly then if we cascade two sections, we do not expect the composite transmission to have an additive property as required by Ohm's law. Indeed, depending on the location of the Fermi energy, the transmission through two scatterers could even *exceed* that through the individual scatterers. This could never happen with classical particles that cannot have a higher probability of getting through two sections than of getting through one. But this is a well-known phenomenon with waves due to interference effects: light transmits better into a lens if we put an extra anti-reflection coating on top. Similarly with electrons: one section could act as an anti-reflection coating for the next section, leading to greater transmission and hence a lower resistance for two sections than for one!

11.2.3 Coherent transport (multiple subbands)

One could argue that this is really an artifact of a one-dimensional model whereby electrons of a given energy have a single wavelength. By contrast, in a multi-moded conductor, electrons of a given energy have many different values of k and hence many different wavelengths (in the longitudinal direction), one for each mode, as shown in Fig. 11.2.3.

As a result, we can expect interference effects to be diluted by the superposition of many oscillations with multiple wavelengths. This is indeed true. Figure 11.2.4 shows the transmission through a two-dimensional wire having a width of 75 Å and a length of 200 Å modeled with a discrete lattice of points spaced by 5 Å. Without any scatterer, the transmission at any energy E is equal to the number of propagating modes $M(E)$, which increases in steps from four to six over the energy range shown. The transmission with one scatterer (with a transverse profile as shown in Fig. 11.2.5) increases monotonically with energy and we can deduce a semi-classical transmission for two scatterers using

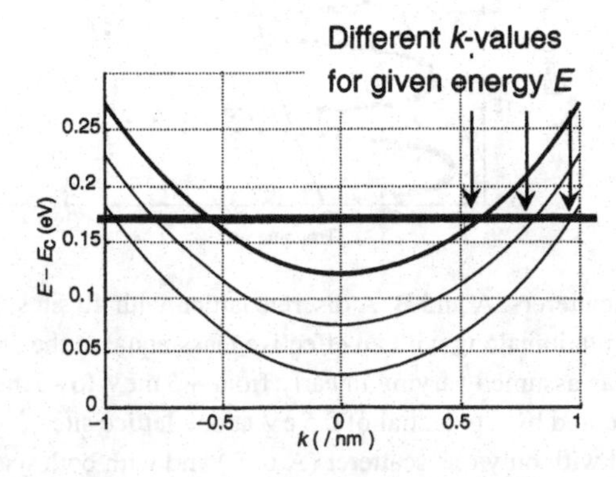

Fig. 11.2.3

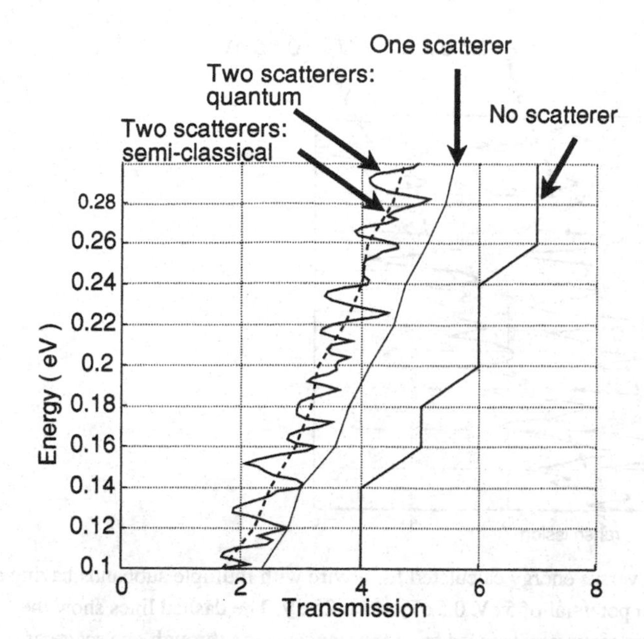

Fig. 11.2.4 Transmission versus energy calculated for a wire with multiple subbands having no scatterer, one scatterer, and two scatterers. Each scatterer has a maximum potential of 0.25 eV at the center of the wire.

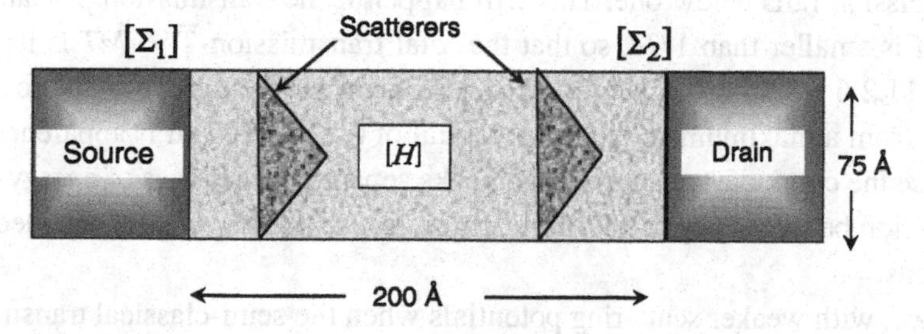

Fig. 11.2.5

Eq. (11.2.2) with $T = \overline{T}/M$ and noting that $T_2 \approx T_1$

$$\frac{M}{\overline{T}} = \frac{2M}{\overline{T}_1} - 1 \rightarrow \overline{T} = \frac{\overline{T}_1}{2 - (\overline{T}_1/M)}$$

It is apparent that the quantum transmission through two scatterers fluctuates around this semi-classical result, the size of the fluctuation being of order one. Such fluctuations in the conductance of narrow wires as a function of the gate voltage (which shifts the Fermi energy relative to the levels) have been observed experimentally and are often referred to as universal conductance fluctuations. Fluctuations have also been observed as a function of the magnetic field, which changes the effective wavelength at a given energy.

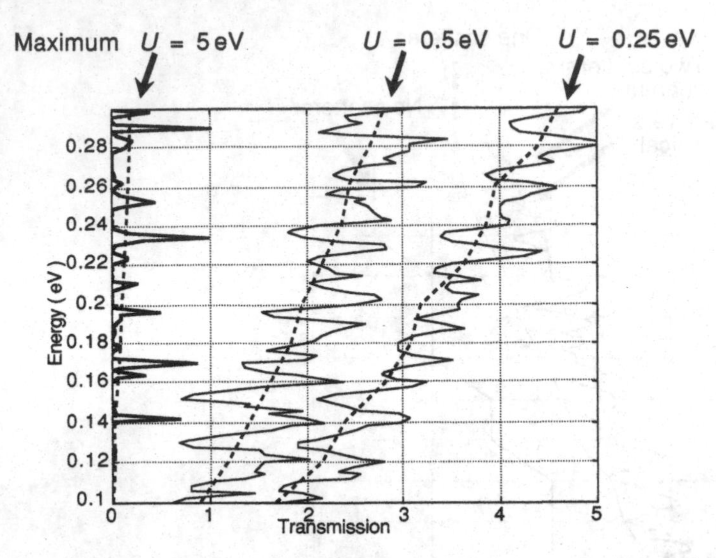

Fig. 11.2.6 Transmission versus energy calculated for a wire with multiple subbands having two scatterers with a maximum potential of 5 eV, 0.5 eV, and 0.25 eV. The dashed lines show the semi-classical result for two scatterers deduced from the transmission through one scatterer.

Since the size of the fluctuation is of order one, the reader might wonder what happens if the transmission falls below one. This will happen if the transmission probability T per mode M is smaller than $1/M$, so that the total transmission $\overline{T} = MT$ is less than one. In Fig. 11.2.6 we show the calculated transmission as the strength of the scatterers is increased from a maximum scattering potential of 0.25 to 5 eV. It is apparent that in the latter case the conductance shows large peaks separated by ranges of energy where the transmission becomes negligible, indicating a *strong localization* of the electronic wavefunctions.

By contrast, with weaker scattering potentials when the semi-classical transmission is larger than one, the quantum transmission shows fluctuations of order one around the semi-classical result. With multiple scatterers, the average quantum transmission turns out to be approximately one *less* than the semi-classical result. This is not at all evident from Fig. 11.2.6 which only involves two scatterers. With a large number of independent scatterers, it turns out that the backscattering per mode is enhanced by $(1/M)$ due to constructive interference leading to an equal decrease in the transmission per mode and hence to a decrease of one in the total transmission. A magnetic field destroys the constructive interference, causing an increase in the transmission which has been experimentally observed as a negative magnetoresistance (reduction in the resistance in a magnetic field) and is ascribed to this so-called *weak localization* effect. The basic phenomenon involving a coherent increase in backscattering has also been observed with electromagnetic waves in a number of different contexts unrelated to electron transport.

11.2.4 Quantum transport with dephasing

To summarize, as we make a wire longer the semi-classical transmission will decrease in accordance with Ohm's law and the quantum transmission will exhibit fluctuations of the order of one around the semi-classical result. If we make any wire long enough, we will reach a length for which the semi-classical transmission will be less than one (see Eq. (11.2.3)):

$$\overline{T} \approx \frac{M\Lambda}{L+\Lambda} < 1 \text{ if } L > M\Lambda$$

The quantum transmission for a wire longer than the localization length ($\sim M\Lambda$) will show large fluctuations characteristic of the strong localization regime. It would seem that even a copper wire, if it is long enough, will eventually enter this regime and cease to obey anything resembling Ohm's law! However, that is not what happens in real life. Why?

The reason is that our observations are valid for phase-coherent conductors where we do not have significant phase-breaking processes to dilute the quantum interference effects. A wire will exhibit strong localization only if the localization length $M\Lambda$ is shorter than the phase-breaking length. Since this length is typically quite short especially at room temperature, there is little chance of a copper wire (with its enormous number of modes M) ever entering this regime. Figure 11.2.7 shows the effective transmission for a one-dimensional wire having two coherent scatterers, with and without phase-breaking scattering. It is apparent that the interference effects are effectively washed out by the presence of phase-breaking processes. Figure 11.2.8 shows that a one-dimensional wire with only phase-breaking scatterers leads to Ohm's-law-like behavior as a function of length. Of course in this limit a full quantum transport model is unnecessary. We could probably use a semi-classical model that neglects interference effects altogether and treats electrons as particles. However, in general, we have both coherent and phase-breaking scattering and the quantum transport model described in Section 11.1 allows us to include both.

In these calculations we have assumed that the phase-breaking scatterers carry negligible energy away ($\hbar\omega \to 0$), so that we can use the simplified equations (cf. Eqs. (11.1.10))

$$\Sigma_s^{in}(E) = D_0[G^n(E)] \quad \text{and} \quad \Sigma_s(E) = D_0[G(E)] \tag{11.2.6}$$

to evaluate the self-energy and inscattering functions. We are also assuming the scattering to be diagonal in real space and uniformly distributed so that D_0 is a constant that we have set equal to 0.01 eV^2 in Fig. 11.2.7 and to 0.05 eV^2 in Fig. 11.2.8.

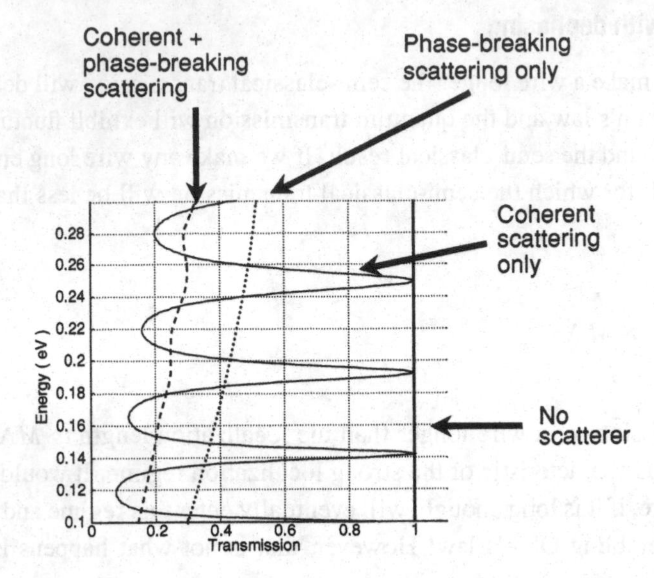

Fig. 11.2.7 Transmission versus energy for a one-dimensional wire with different types of scattering as indicated.

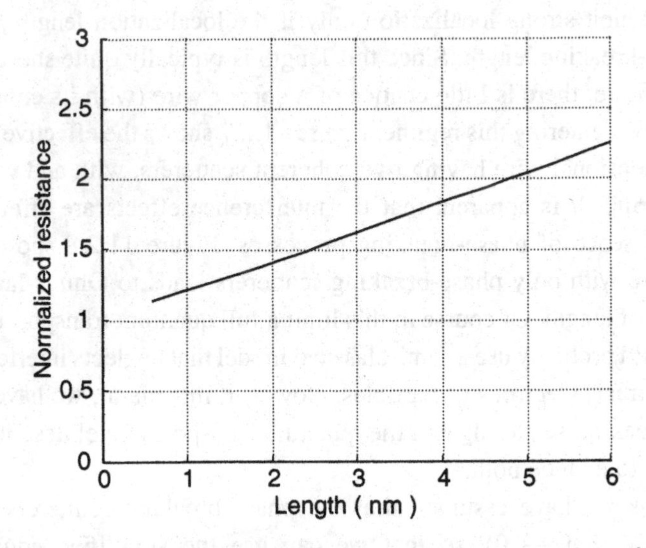

Fig. 11.2.8 Normalized resistance (inverse transmission) as a function of length for a one-dimensional wire with phase-breaking scattering only.

11.3 Where is the heat dissipated?

We have seen that with adequate amounts of phase-breaking scattering, the resistance of a conductor increases linearly with length (Fig. 11.2.8) in accordance with Ohm's law. However, as the length tends to zero (ballistic conductor), the resistance tends to a constant representing the interface resistance associated with the interfaces between the low-dimensional conductor and the three-dimensional contacts. But where does the associated I^2R loss (or the Joule heating) occur?

The answer to this question depends on the nature of the scattering process. The point is that resistance comes from the loss of momentum associated with scattering while the associated Joule heating comes from the loss of energy. For example, in Figs. 11.2.7 and 11.2.8 we have modeled the phase-breaking scattering as an elastic process, neglecting any associated energy loss. This means that, in this model, no energy is dissipated inside the device at all. Nonetheless this elastic scattering does give rise to a resistance that obeys Ohm's law because of the associated loss of momentum, not the loss of phase. Indeed in small conductors, a significant fraction of the Joule heating I^2R associated with a resistor R could be dissipated in the contacts rather than in the conductor itself. There is concrete evidence that this is true, allowing experimentalists to pump far more current through small conductors than what would be needed to destroy them if all the heat were dissipated inside them.

The fact that elastic scattering is not associated with any energy loss can be seen by noting at the normalized current per unit energy $\tilde{I}_i(E)$ (see Eq. (11.1.6)) is identical at each of the two terminals (source and drain) as shown in Fig. 11.3.1. The point is that the *energy current* at any terminal is given by (cf. Eq. (11.1.6))

$$I_{E,i} = \int_{-\infty}^{+\infty} \mathrm{d}E\, E \tilde{I}_i\,(E)/2\pi\hbar \tag{11.3.1}$$

and if there is power (P_d) dissipated inside the device then it must be reflected as a difference in the energy currents at the source and drain terminals:

$$P_\mathrm{d} = I_{E,\mathrm{drain}} - I_{E,\mathrm{source}} \tag{11.3.2}$$

Since current conservation requires the current to be the same at the source and the drain, the energy currents can be different only if they are distributed differently as a function of energy. Clearly there is no power dissipated in the device shown in Fig. 11.3.1, since the current has the same energy distribution at the source and drain. But if we model the same device assuming that the scatterers have an associated phonon energy of $\hbar\omega = 20$ meV, the energy distribution of the current is different at the source and drain, showing that some fraction of the I^2R loss occurs inside the device. Electrons, on the average, enter the source at a higher energy than the energy at which they exit the drain (Fig. 11.3.2).

11.3.1 "Peltier" effect

It is interesting to note that power need not be dissipated everywhere. In an inhomogeneous device there could be local regions where energy is absorbed by the electrons and the solid is locally *cooled*. Consider, for example, a one-dimensional current with a

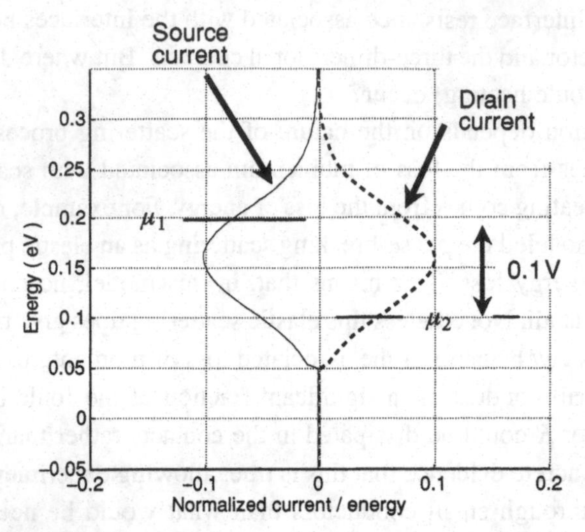

Fig. 11.3.1 Normalized current per unit energy $\tilde{I}_i(E)$ in a one-dimensional wire with phase-breaking elastic scattering.

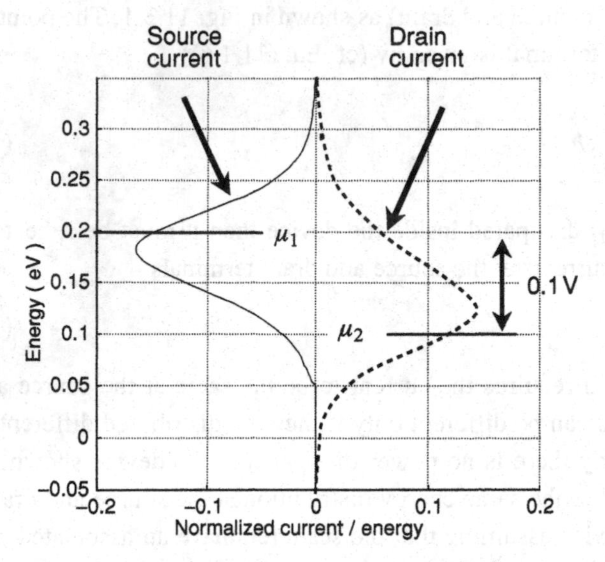

Fig. 11.3.2 Normalized current per unit energy $\tilde{I}_i(E)$ in a one-dimensional wire with inelastic scattering by phonons with energy $\hbar\omega = 20\,\text{meV}$ ($D_0 = 0.1\,\text{eV}^2$). The drain current flows at a lower energy than the source current due to the energy relaxation inside the device.

potential step in the center that forces the current to flow at a higher energy at the drain terminal than at the source terminal (Fig. 11.3.3). It is easy to see from Eq. (11.3.2) that the junction is cooled by the flow of current, which can be considered a microscopic version of the well-known Peltier effect where an electrical current cools one junction at the expense of another. Indeed if we had a potential barrier with an upstep followed by a downstep, then the upstep would be cooled while the downstep would be heated.

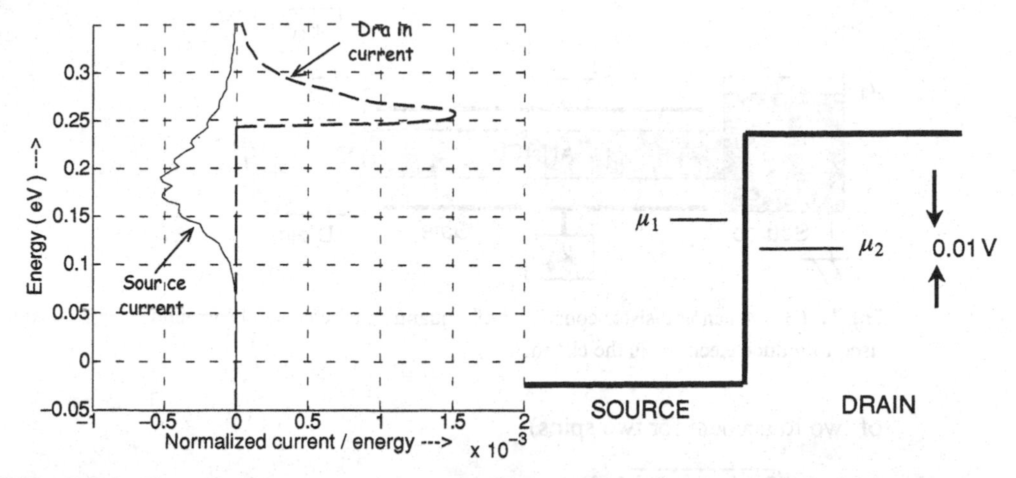

Fig. 11.3.3 Normalized current per unit energy E in a one-dimensional wire with inelastic scattering by phonons with energy $\hbar\omega = 20$ meV ($D_0 = 0.1$ eV^2), having a potential step in the middle as shown. The drain current flows at a higher energy than the source current, indicating that the device is cooled by the flow of current.

This aspect of the energy exchange is reversible and is proportional to the current, unlike the irreversible Joule heating which is proportional to the square of the current (Lake and Datta, 1992b).

11.4 Where is the voltage drop?

In Section 11.1 I stressed the importance of doing quantum transport calculations self-consistently with the "Poisson" equation (see Fig. 11.1.2) for the self-consistent potential U representing the effect of the other electrons. This is particularly important when calculating the current under a "large" applied voltage: the shape of the current–voltage characteristic can sometimes be significantly different depending on the potential profile (or the "voltage drop") inside the channel. For example, in determining the maximum (or the ON) current of a transistor it is important to know where the voltage drops.

Consider a "nanotransistor" composed of a narrow quantum wire labeled the "channel" (see Fig. 11.4.1) of radius a surrounded by a coaxial gate of radius b which is used to induce electrons in the channel as we discussed in Chapter 7. Assume that the electrons in the channel belong to a single subband with a parabolic dispersion relation

$$E = E_c + (\hbar^2 k^2/2m) \tag{11.4.1}$$

At equilibrium, with $\mu_2 = \mu_1$ and low temperatures ($T \to 0$ K) the density of electrons (per unit length) in the channel can be written as (see Table 6.2.1 with an extra factor

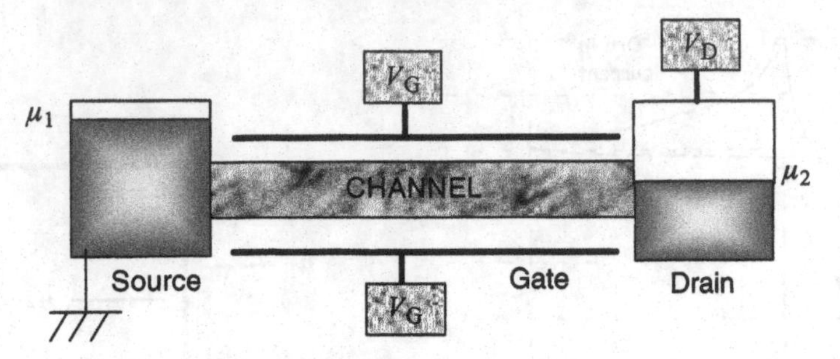

Fig. 11.4.1 A nanotransistor consisting of a quantum wire channel surrounded by a coaxial gate used to induce electrons in the channel.

of two to account for two spins)

$$n_L = 2\sqrt{2m(\mu_1 - E_c)}/\pi\hbar \tag{11.4.2}$$

If we make the gate voltage V_G more positive, it will induce more electrons in the channel, while if we make it more negative it will deplete the channel of electrons, in much the same way that we discussed in Section 7.3, except that in Chapter 7 we were talking primarily about a flat two-dimensional conductor, while now we are talking about a cylindrical one-dimensional conductor. In Chapter 7 we discussed only the equilibrium problem with $\mu_2 = \mu_1$. The problem I wish to discuss now is a non-equilibrium one. A voltage V_D is applied to the drain relative to the source making, $\mu_2 = \mu_1 - qV_D$. What is the current I? Formally we can calculate by following the self-consistent procedure depicted in Fig. 11.1.2 and numerical results are shown later (Fig. 11.4.4). But first let us try to understand the physics in simple terms.

Ballistic nanotransistor: We will start with a ballistic transistor (no scattering) having perfect contacts. If the contacts are good and there is no scattering we would expect the low-bias conductance to be equal to twice (for spin) the conductance quantum:

$$I = (2q^2/h)V_D \tag{11.4.3a}$$

The $+k$ states are filled up from the left contact with an electrochemical potential μ_1 while the $-k$ states are filled from the right contact with an electrochemical potential μ_2. In the energy range between μ_1 and μ_2 (plus a few k_BT on either side) the $+k$ states are nearly filled and carry current, but this current is not balanced by the $-k$ states since they are nearly empty. Since a 1D wire carries a current of $(2q/h)$ per unit energy, there is a net current given by $(2q/h)$ times the energy range $(\mu_1 - \mu_2)$ which is equal to $(2q^2/h)V_D$ as stated above.

Once μ_2 has dropped below the bottom of the band (E_c), the current cannot increase any further and we expect the ON-current to be given by

$$I_{ON}^{(L)} = (2q^2/h)(\mu_1 - E_c) \tag{11.4.3b}$$

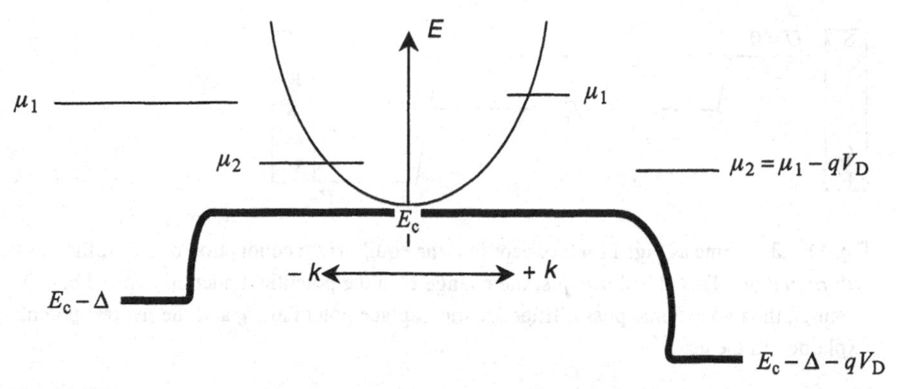

Fig. 11.4.2 An applied voltage lowers the energy levels and the electrochemical potential in the drain region. The $+k$ states are occupied from the source up to μ_1 while the $-k$ states are occupied from the drain up to μ_2, causing a net current to flow as discussed in Chapter 6 (see Section 6.3).

Equations (11.4.3a, b) suggest that the current should increase linearly with voltage and then level off as the voltage approaches $(\mu_1 - E_c)$. It is indeed true that the current increases linearly and then saturates but, depending on the electrostatics, the ON-current could be much larger, up to *four* times as large as that given by Eq. (11.4.3b), which we have labeled with (L) to denote the Laplace limit. Let me explain what I mean by that and what the other limit is.

The result given in Eq. (11.4.3b) is based on the picture shown in Fig. 11.4.2, which assumes that the only effect of the increasing drain voltage is to lower μ_2, *while the energy levels in the channel remain fixed* relative to the source. However, depending on the electrostatics, it is possible that the potential energy U in the channel would drop by some fraction of the drain potential $-qV_D$, thus lowering the bottom of the band and increasing the ON-current to (note that U is a negative quantity for positive drain voltages)

$$I_{ON} = (2q^2/h)(\mu_1 - E_c - U) \tag{11.4.4}$$

(I am assuming that U remains less than Δ, see Fig. 11.4.2.) So in estimating the ON-current, the all-important question is "How does the voltage drop?" The source is at zero, the drain is at $-qV_D$: what is the potential energy U inside the channel?

In general we determine the channel potential from a self-consistent solution of the electrostatics and the transport problems. For our present problem we can write the electrostatics in the form

$$U = U_L + (q^2/C_E)\delta n_L \tag{11.4.5}$$

where C_E is the capacitance per unit length of a coaxial capacitor with inner and outer

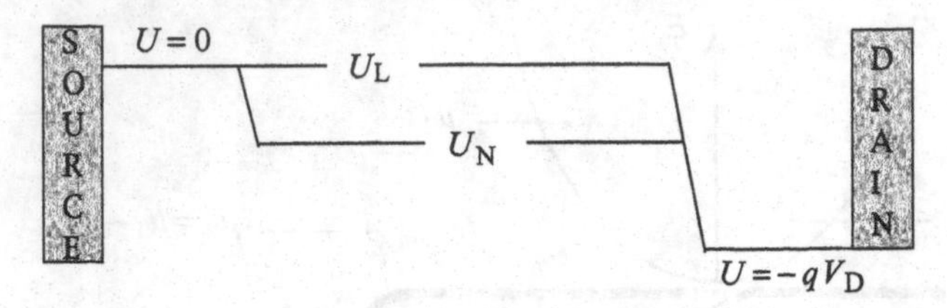

Fig. 11.4.3 Same as Fig. 11.4.2 except that the equilibrium conduction band profile has been *subtracted* off. This plot shows just the change U in the potential under an applied bias. Inside the channel, the two extreme possibilities are the Laplace potential U_L and the neutral potential U_N as explained in the text.

radii equal to a and b respectively:

$$C_E = 2\pi \varepsilon_r \varepsilon_0 / \ln(b/a) \tag{11.4.6}$$

δn_L is the change in the electron density per unit length and U_L is the solution to corresponding to $\delta n_L = 0$, sometimes called the Laplace solution, which is shown in Fig. 11.4.3. Since the potential is applied only to the drain and not the gate the Laplace potential has the shape shown in Fig. 11.4.3. It is essentially equal to the source potential throughout the channel and rapidly changes to the drain potential at the other end; how rapidly depends on the closeness of the gate to the channel.

The actual potential inside the channel is close to the Laplace limit if the electrostatic capacitance C_E is large: the second term in Eq. (11.4.5) then is negligible. This is the case when we assume a very "high-K" dielectric with $\varepsilon_r = 100$ (see Eq. (11.4.6) with b set equal to $2a$). But when we use a smaller $\varepsilon_r = 2$, the current increases significantly by nearly a factor of two (Fig. 11.4.4).

If the capacitance is even smaller, then in principle we could be in the other limit where $\delta n_L \to 0$, but the second term is finite. We call this the neutral limit and the corresponding potential the neutral potential U_N (see Section 7.3). What is U_N? We can write the electron density in the ON-state as

$$[n_L]_{ON} = \sqrt{2m(\mu_1 - E_c - U)}/\pi\hbar \tag{11.4.7}$$

since the potential energy U (which is negative) moves the bottom of the band down to $E_c + U$, but we lose a factor of two because only half the states (having $+k$) are occupied. Subtracting Eq. (11.4.7) from (11.4.2) we write the change in the electron density as

$$\delta n_L = (\sqrt{2m}/\pi\hbar)(2\sqrt{\mu_1 - E_c} - \sqrt{\mu_1 - E_c - U}) \tag{11.4.8}$$

Setting $\delta n_L = 0$, we obtain the neutral potential:

$$2\sqrt{\mu_1 - E_c} = \sqrt{\mu_1 - E_c - U_N} \rightarrow \mu_1 - E_c - U_N = 4(\mu_1 - E_c) \tag{11.4.9}$$

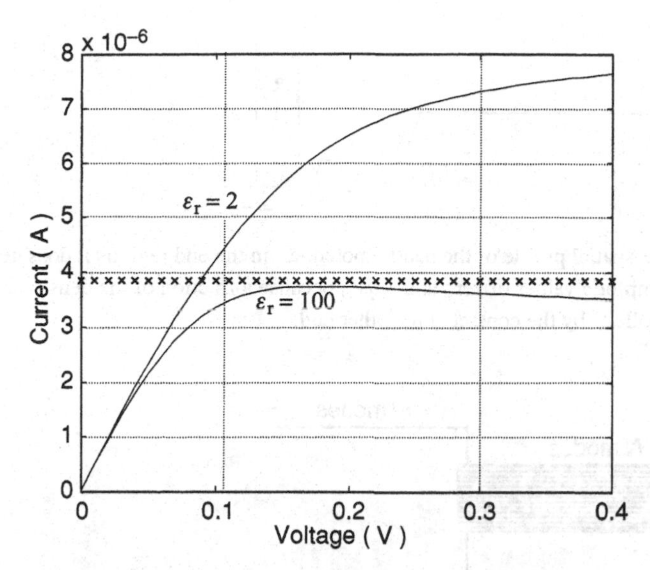

Fig. 11.4.4 Current (I) vs. drain voltage (V_D) for a ballistic quantum wire transistor, for different values of the capacitance C_E. Also shown (crosses) is the Laplace limit with $U = U_L$. (Courtesy of S. Salahuddin.)

which means that in the neutral limit the ON-current from Eq. (11.4.4) is four times what we expect in the Laplace limit (cf. Eq.(11.4.3b))

$$I_{ON}^{(N)} = (2q^2/h)(\mu_1 - E_c - U_N) = 4I_{ON}^{(L)} \tag{11.4.10}$$

I should note that the neutral limit of the ON-current need not always be four times the Laplace limit. The factor of four is specific to the one-dimensional example considered here arising from the fact that the electron density is proportional to the square root of the energy (see Eq. (11.4.2)). For a two-dimensional sheet conductor, the electron density increases linearly with energy and we can show that the neutral limit is two times the Laplace limit. The important point is that there is a Laplace limit and a neutral limit and the actual value could lie anywhere in between depending on the capacitance C_E.

Electrostatic boundary conditions: In Fig. 11.4.3 we have shown the potential U approaching the asymptotic values of zero and $-qV_D$ set by the external voltage in the source and drain regions respectively. It is common to assume that this statement will be true if we make these regions long enough. However, it is important to note that if the end regions are assumed ballistic then the potential may not reach the asymptotic values, no matter how long we make these regions.

The reason is that in these conductive end regions the potential U will approach the neutral value U_N needed to make $\delta n_L = 0$. Consider the region near the source, for example. If the potential U were zero, δn_L would be negative because a fraction of

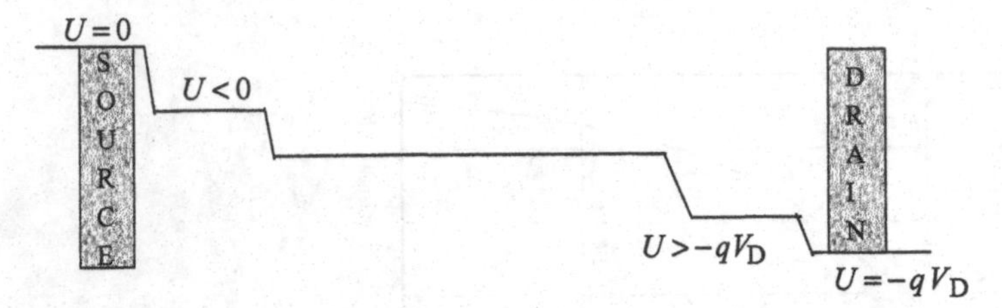

Fig. 11.4.5 Sketch of the spatial profile of the neutral potential: in the end regions it does not approach the correct asymptotic values of zero and $-qV_D$ because a fraction of the density of states in these regions is "controlled" by the contact at the other end.

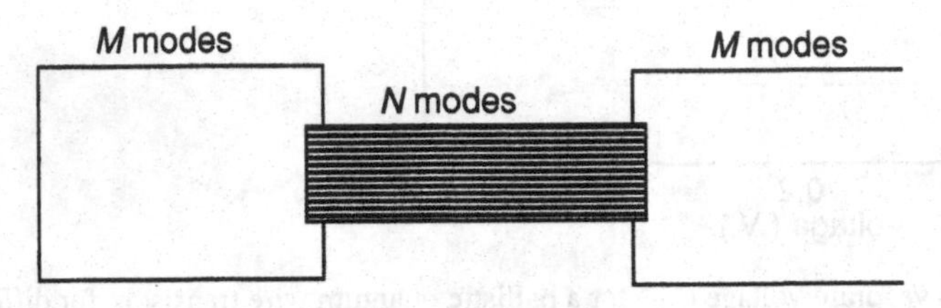

Fig. 11.4.6

the density of states in this region is now occupied according to the electrochemical potential μ_2 in the drain. This fraction is described by the partial spectral function $[A_2]$ that we discussed in Chapter 9. To keep $\delta n_L = 0$, the neutral potential in this region takes on a negative value (Fig. 11.4.5). This situation will not change if we simply make the end regions longer. As long as they are ballistic, it can be shown that there will be no change in the fraction of the density of states at one end that is controlled by the contact at the other end. Consequently the neutral potential in the source region will be less than the asymptotic value of zero and using a similar argument we can show that in the drain region it will be more than the asymptotic value of $-qV_D$.

The potential U will revert to the correct asymptotic values $\pm qV/2$ only if a negligible fraction of the density of states (or spectral function) at one end is controlled by the contact at the other end. This can happen if there is enough scattering within the device or if there is strong geometrical dilution at the contacts ($M \gg N$) (Fig. 11.4.6).

This means that in modeling near-ballistic devices without significant geometric dilution at the contacts we should not fix the potential at the two ends to the usual asymptotic values as we did in solving the capacitance problem (see Eq. (7.2.17)). One solution is to use a zero-field boundary condition for the Poisson equation and let the potential U develop self-consistently.

From a conceptual point of view, we could view the spatial profile of the neutral potential U_N (which may be different from the profile of the actual potential U) as an indicator of the spatial distribution of the resistance. The neutral potential across any ballistic region remains flat indicating zero resistance as we might intuitively expect. Figure 11.4.5 shows that only a fraction of the applied voltage V_D actually appears

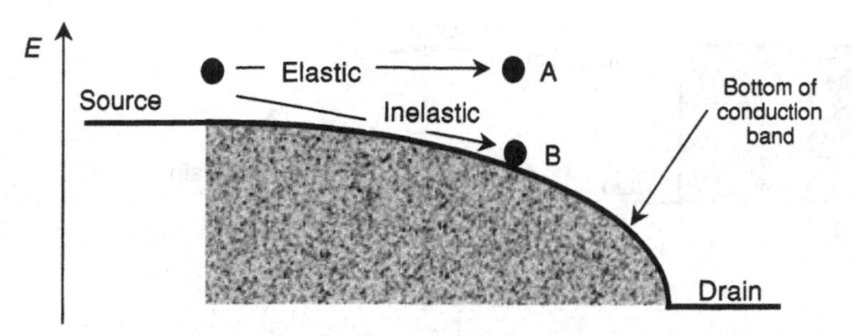

Fig. 11.4.7 Once an electron has lost sufficient energy through inelastic processes (point B) it cannot be turned back towards the source and current is not reduced by further scattering. But if it has not lost enough energy (point A) then backscattering to the source can occur.

between the two ends of the device, indicating that the resistance we calculate is only partly due to the channel and the rest should be associated with the interface between the narrow channel regions and the wide contact regions.

Nanotransistor with scattering: We expect the ON-current to be reduced by the presence of scattering, since the transmission is now less than one by a factor $T = \Lambda/(L + \Lambda)$ (see Eq. (11.2.3)), which depends on the length of the channel relative to a mean free path. In practice, however, the ON-current is higher that what Eq. (11.4.10) suggests, if the scattering processes are inelastic rather than elastic.

To understand why, we note that inelastic processes cause the electrons coming in from the source to lose energy as they propagate towards the drain (see Fig. 11.4.7). Once they have lost sufficient energy (indicated by point B) they cannot easily be turned back towards the source any more, since there are no allowed states in the source at this lower energy. Consequently, the electron proceeds to the drain and the current is not reduced. But if the scattering processes are elastic then electrons do not relax in energy (indicated by point A) and can be turned back towards the source with a reduction in the current. This physics can be described approximately by replacing the device length L in Eq. (11.2.3) with the inelastic energy relaxation length L_i that an electron traverses before it loses a few $k_B T$ of energy (large enough that coming back is near impossible)

$$T = \Lambda/(L_i + \Lambda) \tag{11.4.11}$$

This can make the actual ON-current much larger than what one might expect otherwise. Even purely elastic scattering causes a similar increase in the ON-current in two-dimensional conductors since it relaxes the *longitudinal* (directed from the source to the drain) kinetic energy, although the total energy (longitudinal + transverse) remains the same. Indeed, commercial transistors for many years now have shown ON-currents that are within 50% of their ballistic value even though they may be over ten mean free paths in length. The reason is that the energy relaxation length L_i tends to be of the same

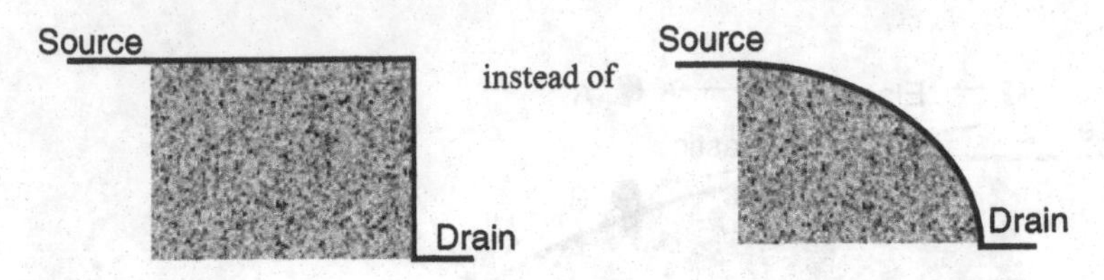

Fig. 11.4.8

order as the mean free path Λ, making the transmission probability approximately 0.5 regardless of the actual length of the channel.

My reason for bringing up this issue here is that this is another example of the importance of the self-consistent potential profile in determining the current at large bias voltages. For example, the effect just described would not arise if the potential profile looked as shown in Fig. 11.4.8 since there would then be little room for energy relaxation. It seems reasonable to expect that the actual magnitude of the effect in a real device will depend on the potential profile that has to be calculated self-consistently as indicated in Fig. 11.1.2. This is a point that is often overlooked because the transport block in Fig. 11.1.2 is intellectually more demanding than the "Poisson" block and tends to overshadow it in our mind. So it is worth remembering that in many problems, like the ON-current of a nanotransistor, the Poisson block could well represent the key "physics," making the transport block just a "detail"!

EXERCISES

E.11.1. (a) Consider a short 1D channel with two scatterers, A and B, modeled with a discrete lattice of 40 sites separated by $a = 3$ Å. Assume a small bias potential varying linearly from +5 meV to –5 meV across the device. Each scatterer is represented by a potential of 0.5 eV at one lattice site. Calculate the transmission versus energy with one scatterer only (A or B) and with both scatterers (A and B) and compare with Fig. 11.2.2. (b) Repeat including elastic phase-breaking scattering processes as indicated in Eq. (11.2.6) and compare with Fig. 11.2.7. (c) Plot the inverse transmission (at a fixed energy of 0.1 eV) versus length for a 1D wire with elastic phase-breaking scattering only and compare with Fig.11.2.8.

E.11.2. (a) Consider a multi-moded channel 75 Å wide and calculate the transmission through one scatterer and through two scatterers assuming each scatterer to be represented by a triangular potential (with a maximum of 0.25 eV at the center) in the transverse direction localized at one lattice plane in the longitudinal direction. Also, plot the semi-classical result obtained for two scatterers using the result for one scatterer. Compare with Fig. 11.2.4. (b) Repeat for different strengths of the scattering potential and compare with Fig. 11.2.6.

E.11.3. (a) Consider a 1D channel with phase-breaking elastic scattering and plot the current per unit energy as shown in Fig. 11.3.1. (b) Repeat for a channel with inelastic scattering and compare with Fig. 11.3.2. (c) Repeat for a channel with a potential step in the middle with inelastic scattering and compare with Fig. 11.3.3.

E.11.4. Calculate the current (I) versus drain voltage (V_D) self-consistently for a ballistic quantum wire nanotransistor and compare with Fig. 11.4.4.

E.11.5. Tunneling in the presence of phase-breaking: Calculate the inverse transmission at low drain voltage at $E = \mu$ for a conductor having its equilibrium electrochemical potential μ located at 0.1 eV, with the conduction band edge in the contact at 0 eV and that in the channel at 0.5 eV as shown in Fig. E.11.5a.

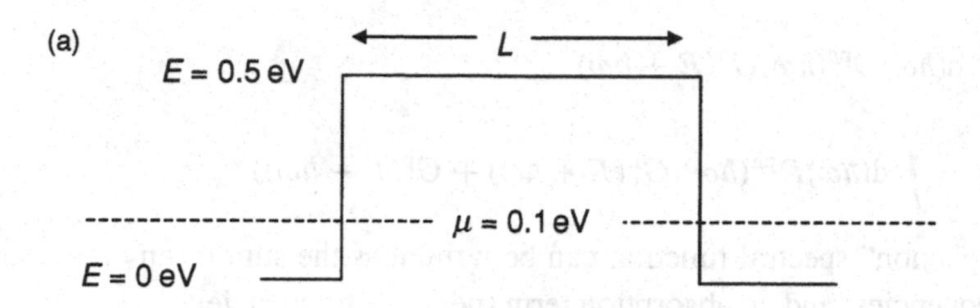

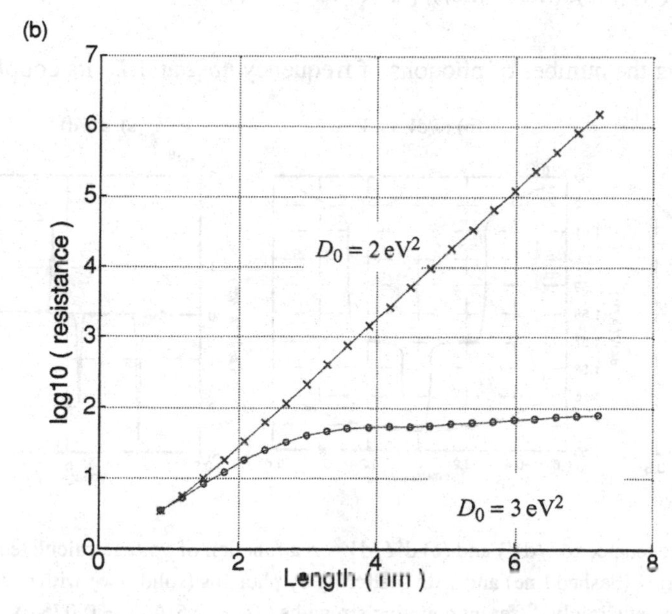

Fig. E.11.5

Transmission through the channel is by tunneling, so that the inverse transmission varies exponentially with the length L of the barrier in the absence of phase-breaking processes. Plot the logarithm of the inverse transmission (normalized resistance) as a function of the length for 1 nm < L < 7 nm with two different scattering strengths, $D_0 = 2\,\text{eV}^2$ and $3\,\text{eV}^2$. You should obtain a plot like the one shown in Fig. E.11.5b. Note that the expected exponential dependence of the resistance (linear dependence of the logarithm of the resistance) does not hold at higher scattering strengths. See, for example, Neofotistos *et al.* (1991).

E.11.6. Inelastic electron tunneling spectroscopy (IETS): At the end of Section 11.1, we looked at a simple one-level example with elastic phase-breaking scattering. It is instructive to do the one-level example with inelastic scattering, which is more complicated because different energies are coupled together. The basic equations are just the ones listed in Section 11.1 where the various matrices are just scalar quantities (or (1 × 1) matrices). The inscattering and broadening matrices are given by

$$\gamma_s^{\text{in}}(E) = \int \text{d}(\hbar\omega) D^{\text{ph}}(\hbar\omega) G^{\text{n}}(E + \hbar\omega)$$

$$\text{and} \quad \Gamma_s(E) = \int \text{d}(\hbar\omega) D^{\text{ph}}(\hbar\omega)(G^{\text{n}}(E + \hbar\omega) + G^{\text{p}}(E - \hbar\omega))$$

where the "phonon" spectral function can be written as the sum of an emission term (positive frequencies) and an absorption term (negative frequencies)

$$D^{\text{ph}}(\hbar\omega) = \sum_i D_i[(N_i + 1)\delta(\hbar\omega - \hbar\omega_i) + N_i\delta(\hbar\omega + \hbar\omega_i)]$$

with N_i representing the number of phonons of frequency $\hbar\omega_i$, and D_i its coupling.

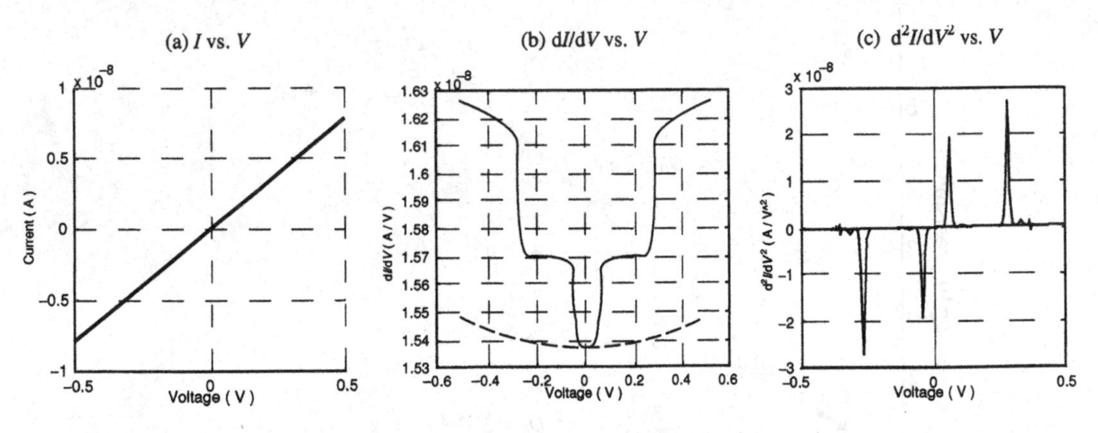

(a) Current (I), (b) conductance ($\text{d}I/\text{d}V$) and (c) $\text{d}^2I/\text{d}V^2$ as a function of voltage calculated without phonon scattering (dashed line) and with scattering by phonons (solid line) with two distinct frequencies having slightly different coupling strengths ($D_1 = 0.5, \hbar\omega_1 = 0.075$ eV and $D_2 = 0.7, \hbar\omega_2 = 0.275$ eV). Reproduced with permission from Datta (2004).

Calculate the current versus voltage (ignoring the self-consistent potential), assuming that the energy level $\varepsilon = 5$ eV lies much above the equilibrium electrochemical potential $\mu = 0$, so that current flows by tunneling. You should obtain a plot as shown above.

The current calculated without any phonon scattering (all $D_i = 0$) and with phonon scattering ($D_1 = 0.5$, $\hbar\omega_1 = 0.075$ eV and $D_2 = 0.7$, $\hbar\omega_2 = 0.275$ eV) shows no discernible difference. The difference, however, shows up in the conductance dI/dV where there is a discontinuity proportional to D_i when the applied voltage equals the phonon frequency $\hbar\omega_1$. This discontinuity shows up as peaks in d^2I/dV^2 whose location along the voltage axis corresponds to molecular vibration quanta, and this is the basis of the field of inelastic electron tunneling spectroscopy (IETS) (see for example, Wolf (1989)).

E.11.7. Spin-flip scattering: We saw at the end of Section 11.1 that elastic phase-breaking has little effect on the current in a one-level system. But it can have significant effect if we have two levels with very different couplings to the two contacts. Consider an up-spin level connected to the left contact and a down-spin level connected to the right contact so that

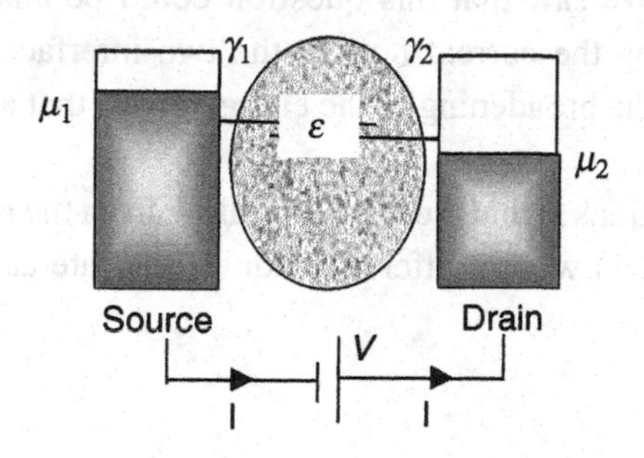

$$[H] = \begin{bmatrix} \varepsilon & \Delta \\ \Delta^* & \varepsilon \end{bmatrix}, \qquad [\Gamma_1] = \begin{bmatrix} \gamma_1 & 0 \\ 0 & 0 \end{bmatrix}, \qquad [\Gamma_2] = \begin{bmatrix} 0 & 0 \\ 0 & \gamma_2 \end{bmatrix}$$

(a) Obtain an expression for the transmission $\bar{T}(E)$. Explain physically why it is zero when $\Delta = 0$.

(b) Assume $\Delta = 0$ so that no current will flow if there is no scattering. Now assume that there are elastic phase-breaking processes described by $\Sigma_{ij} = \sum_{k,l} D_{ijkl} G_{kl}$, $\Sigma_{ij}^{in} = \sum_{k,l} D_{ijkl} G_{kl}^{n}$. Calculate the current for a fixed small drain voltage as a function of D_0, assuming that the only non-zero elements of 'D' are (a) $D_{1122} = D_{2211} = D_0$, and (b) $D_{1111} = D_{2222} = D_0$ (for further discussion see Datta (2005)).

12 Epilogue

I started this book with a "simple" problem: two contacts are made to a really small object having just one energy level in the energy range of interest. A voltage V is applied between the contacts so that their electrochemical potentials separate, $\mu_1 - \mu_2 = qV$ (see Fig. 1.2.2). What is the current? We saw that this question could be answered on the basis of two equations describing the current flow at the two interfaces (See Eqs. (1.2.3a, b)), provided we included the broadening of the energy levels that accompanies the process of coupling.

The essential point behind these equations is that there is an outflow from the contact and an inflow from the contact (Fig. 12.1) whose difference equals the rate at which the number of electrons changes:

$$\hbar \frac{dN}{dt} + \underbrace{\gamma N}_{\text{Outflow}} = \underbrace{\gamma f}_{\text{Inflow}} \tag{12.1}$$

In Chapters 8 and 9 we discussed a quantum mechanical treatment of this problem based on the one-electron Schrödinger equation

$$i\hbar \frac{d\psi}{dt} - H\psi - \underbrace{\Sigma\psi}_{\text{Outflow}} = \underbrace{S}_{\text{Inflow}} \tag{12.2}$$

with an additional self-energy term $\Sigma\psi$ and a source term S that give rise to outflow and inflow respectively ((Eq. (8.2) in Chapter 8 can be viewed as the Fourier transform of this equation). Traditional quantum mechanics courses focus on the Hamiltonian $[H]$ describing the internal dynamics of electrons in a closed system, just as we did in Chapters 2 through 7 of this book. But from Chapter 8 onwards we have been discussing different aspects of inflow and outflow, since it is so central to the problem of electron transport. The quantum transport equations summarized in Section 11.1 express the inflow and outflow in terms of the correlation functions of the quantities appearing in Eq. (12.2): $G^n \sim \psi\psi^+$, $\Sigma^{in} \sim SS^+$, etc., which are represented by matrices.

These equations provide the bridge from the atom to the transistor: both a conceptual bridge and a quantitative bridge that connects the reversible microscopic world

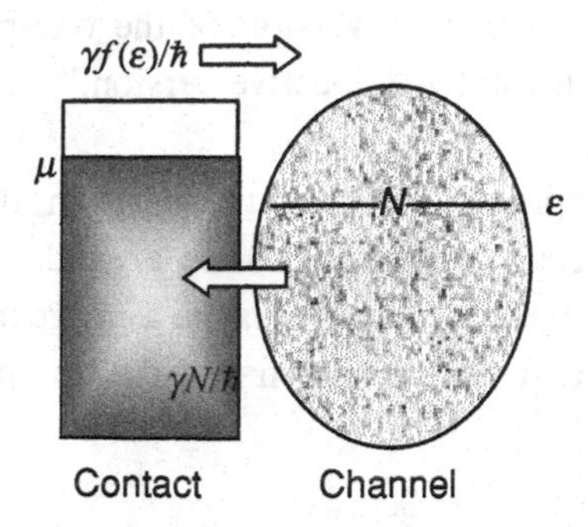

Fig. 12.1

described by the Hamiltonian [*H*] to the irreversible macroscopic world with its currents and dissipative flows. The problem of describing irreversible processes starting from reversible laws is indeed one of the deeper recurring problems of physics which has been widely discussed in the literature. A detailed discussion would take us too far afield but let me try to say a few words about classical transport that may help put the problem of quantum transport in perspective. I will write the relevant equations in one dimension to make them look less forbidding to beginners, but it is straightforward to generalize them to three dimensions.

Langevin equation: Classical mechanics is based on Newton's law $m(\mathrm{d}v/\mathrm{d}t) = F(t)$, which describes frictionless motion (like that of the planets around the Sun) very well but it has to be supplemented with a frictional force γv in order to describe the motion of everyday objects (like cars)

$$m\frac{\mathrm{d}v}{\mathrm{d}t} + \gamma v = F(t)$$

This relation suggests that the velocity approaches zero once all external forces $F(t)$ are removed, which is not exactly true since microscopic objects in equilibrium want to jiggle around with an average kinetic energy of $k_\mathrm{B}T/2$ per degree of freedom. This jiggling is not noticeable for cars, but quite significant for electrons or even for large molecules (so-called Brownian motion, for example). To include it in the equation of motion we need to add a random force $N(t)$

$$m\frac{\mathrm{d}v}{\mathrm{d}t} + \gamma v = F(t) + N(t) \tag{12.3}$$

thereby turning Newton's law into a stochastic differential equation, sometimes called the Langevin equation. With an appropriate choice of the statistical properties of $N(t)$ and the frictional coefficient γ, one can describe irreversible transport processes like that of electrons in solids. Indeed, the drift–diffusion equations widely used to model semiconductor devices can be viewed as the collective version of the one-particle equation (12.3). Let me briefly explain what I mean by "collective version."

Boltzmann equation: In the one-particle approach we track the position $x(t)$ and velocity $v(t)$ of individual particles. In the collective approach, x and v are independent variables and we define a distribution function $f(x, v, t)$ that tells us the average number of particles having a velocity v at the point x. It can be shown that the one-particle relations

$$\mathrm{d}x/\mathrm{d}t = v(t) \quad \text{and} \quad m(\mathrm{d}v/\mathrm{d}t) = F(t)$$

can equivalently be written in collective terms as (see, for example, Lundstrom (2000))

$$\frac{\partial f}{\partial t} + \frac{F}{m}\frac{\partial f}{\partial v} + v\frac{\partial f}{\partial x} = 0$$

The stochastic force $N(t)$ and the frictional term γv can both be included by treating these as processes that take electrons from one value of v to another, thus resulting in an inscattering S^{in} and an outscattering S^{out} at each point in phase space (x, v):

$$\frac{\partial f}{\partial t} + \frac{F}{m}\frac{\partial f}{\partial v} + v\frac{\partial f}{\partial x} = S^{\text{in}} - S^{\text{out}} \tag{12.4}$$

This is the celebrated Boltzmann equation (1D version, readily generalized to 3D), which is widely used today to describe the transport of diverse entities including neutrons and electrons. The drift–diffusion equations are obtained from Eq. (12.4) by taking its "first moment," that is by multiplying by v and summing over all v.

Modern versions of this approach, often referred to as semi-classical transport theory, take bandstructure into account by modifying Newton's laws to read

$$\mathrm{d}x/\mathrm{d}t = v \rightarrow \mathrm{d}x/\mathrm{d}t = (1/\hbar)\partial E/\partial k \equiv v$$

$$m(\mathrm{d}v/\mathrm{d}t) = F \rightarrow \hbar \mathrm{d}k/\mathrm{d}t = F$$

For parabolic bands with $E = \hbar^2k^2/2m^*$, the two forms are equivalent with the free electron mass m replaced by the effective mass m^*. For non-parabolic bands the two versions are not equivalent and the modified version (in terms of k) can be shown to be the correct one. The corresponding modified Boltzmann equation for $f(x, k, t)$ (rather than $f(x, v, t)$) is given by

$$\frac{\partial f}{\partial t} + \frac{F}{h}\frac{\partial f}{\partial k} + v\frac{\partial f}{\partial x} = S^{\text{in}} - S^{\text{out}} \tag{12.5}$$

"Arrow" of time: The functional dependence of S^{in} and S^{out} on the distribution function f can be selected to reflect different kinds of scattering, such as impurity scattering or phonon scattering or electron–electron scattering. In general, however, they have the property that the equilibrium distribution function $f_0(\text{E})$ with $E = U(x) + \varepsilon(k)$ gives no net inscattering or outscattering: $S^{\text{in}} - S^{\text{out}} = 0$. This ensures that $f_{\text{eq}} = f_0(E)$ with $E = U(x) + \varepsilon(k)$ is a solution to the Boltzmann equation since (note that $F = -\mathrm{d}U/\mathrm{d}x$)

$$\partial f_{\text{eq}}/\partial t = 0$$

$$\partial f_{\text{eq}}/\partial k = (\mathrm{d}f_0/\mathrm{d}E)(\partial E/\partial k) = (\mathrm{d}f_0/\mathrm{d}E)\hbar v$$

$$\partial f_{\text{eq}}/\partial x = (\mathrm{d}f_0/\mathrm{d}E)(\partial E/\partial x) = -(\mathrm{d}f_0/\mathrm{d}E)F$$

so that $$\frac{\partial f_{\text{eq}}}{\partial t} + \frac{F}{\hbar}\frac{\partial f_{\text{eq}}}{\partial k} + v\frac{\partial f_{\text{eq}}}{\partial x} = 0$$

It is believed that arbitrary initial distributions will eventually relax to the state $f_{\text{eq}}(x, k)$ in agreement with what is generally observed. This property defines a preferred direction (or "arrow") in time: non-equilibrium distributions always evolve into equilibrium distributions and never the other way around.

By contrast, if we were to write a set of N coupled equations for a set of N interacting particles

$$m\frac{\mathrm{d}^2 x_n}{\mathrm{d}t^2} = m\frac{\mathrm{d}v_n}{\mathrm{d}t} = -\frac{\partial}{\partial x_n}U_{\text{int}}(x_1, \ldots, x_N) \tag{12.6}$$

(U_{int} represents the interaction energy so that its negative gradient with respect to x_n gives the force felt by particle n), the solutions would have no preferred direction in time. If we were to videotape the time evolution of any solution to Eq. (12.6) and play it backwards, it would not look wrong, for t can be replaced by $-t$ without changing the equation. But a videotaped solution of the Boltzmann equation would look absurd if played backwards: instead of seeing a highly non-equilibrium distribution evolve towards an equilibrium distribution, we would see the reverse, which is completely at odds with experience.

Boltzmann was strongly criticized by many in his day who argued that an equation like Eq. (12.5) with a preferred direction in time cannot be equivalent to a time-reversible one like Eq. (12.6). It is now believed that with a sufficiently large number of particles Eq. (12.6) too will exhibit behavior similar to that predicted by the Boltzmann equation, at least on practical time scales shorter than what is called the recurrence time (see, for example McQuarrie (1976) and references therein). But this is by no means obvious and many profound papers continue to be written on the subject, which I will not go into. The point I am trying to make is that it is difficult to have the equilibrium solution emerge purely out of Newton's law (Eq. (12.6)). Boltzmann bypassed these complexities simply by a proper choice of the scattering functions, often called the "Stosszahlansatz" which ignores subtle correlations in the velocities of different particles. In the field of quantum transport too, one would expect significant difficulty in making the correct equilibrium solutions emerge purely from the Schrödinger equation (or even a multiparticle Schrödinger equation like Eq. (3.2.1)). It seems natural then to look for ways to do for the Schrödinger equation what Boltzmann did for Newton's law.

Quantum Boltzmann equation: Indeed this is exactly what the non-equilibrium Green's function (NEGF) formalism initiated by the works of Schwinger, Baym, Kadanoff, and Keldysh (see Kadanoff and Baym, 1962; Keldysh, 1965; Martin and Schwinger, 1959), sought to do. The basic quantity is the correlation function (generally written in the literature as $-iG^{<}(x, x'; t, t')$)

$$G^{n}(x, x'; t, t') \equiv \langle \psi^{+}(x', t')\psi(x, t)\rangle$$

which plays a role analogous to the distribution function $f(x, k, t)$ in Boltzmann theory. For steady-state transport the correlation function depends only on the difference time coordinate $(t - t')$ whose Fourier transform is the energy coordinate E that we have used throughout the book. Note that the quantum formalism has one extra dimension (x, x', E) relative to the Boltzmann formalism (x, k), in addition to the time coordinate. This is because the Boltzmann formalism assumes "k" states to be eigenstates, so that "k" and "E" are uniquely related. In the general quantum formalism such an assumption is not warranted. For further discussion of this point and related issues such as the Wigner transformation, see for example, section 8.8 of Datta (1995).

Landauer model: Work on electron transport (quantum or otherwise) before the 1980s was focused largely on the problem of electron–electron / electron–phonon / electron–impurity interactions and how they introduce irreversibility, dissipation, and ultimately electrical resistance. Contacts back then were viewed as minor experimental distractions. But this changed in the 1980s when experiments in mesoscopic physics revealed the important role played by contacts and a different model introduced by Landauer gained increasing popularity. In this model, the conductor itself is assumed to be a wire free of all interactions: irreversibility and dissipation arise from the connection to the contacts. This model seems relevant to the modeling of electronic devices as they scale down to atomic dimensions. At the same time it serves to illustrate the microscopic nature of resistance in its simplest context, one that we can call coherent transport (Chapter 9).

In this book I have used an even simpler version of this model as our starting point: a really small object (rather than a quantum wire) connected to two large contacts. Indeed Eq. (12.1), introduced in Chapter 1, can be viewed as a special version of the Boltzmann equation applied to a system so small that the distribution function is a single number N. The contacts are held in local equilibrium by irreversible interactions whose complex dynamics is completely bypassed simply by legislating that the inscattering and outscattering terms have the particularly simple forms: $S^{\text{in}} = \gamma f_1$ and $S^{\text{out}} = \gamma N$ which ensure that the equilibrium solution is the correct one with $N = f_1$. This is somewhat in the spirit of Boltzmann, who bypassed complex Newtonian dynamics with his inspired ansatz regarding the nature of S^{in} and S^{out}. In Chapter 9 we developed this model into a full coherent transport model (equivalent to Landauer's) and then introduced dissipation in Chapter 10 to obtain the complete equations for dissipative quantum transport (for weak interactions), which are summarized and illustrated with examples in Chapter 11.

Beyond coherent transport: Overall, I have tried to combine the physical insights from the Landauer approach with the power of the NEGF formalism, which I believe is needed to extend quantum transport models beyond the coherent regime. For example, I have stated earlier that the self-energy term $\Sigma\psi$ in Eq. (12.2) represents the outflow of the electrons and it is natural to ask if Σ (whose imaginary part gives the broadening or the inverse lifetime) should depend on whether the final state (to which outflow occurs) is empty or full. As I stated in Chapter 1 and later in Chapter 9, such exclusion principle factors do not appear as long as purely coherent processes are involved. But as we saw in Chapter 10, they do arise for non-coherent interactions in a non-obvious way that is hard to rationalize from the one-electron picture that we have been using. A proper description requires us to go beyond this picture.

In the one-electron picture, individual electrons are described by a one-electron wavefunction ψ and the electron density is obtained by summing $\psi^*\psi$ from different electrons. A more comprehensive viewpoint describes the electrons in terms of field operators c such that c^+c is the number operator which can take on one of two values, 0 or 1, indicating whether a state is empty or full. These field operators are often referred to as "second quantized" operators, implying that the passage from a particulate Newtonian viewpoint to a wavelike Schrödinger viewpoint is the "first quantization" (a term that is seldom used). Second quantization completes the formalism by incorporating both the particulate view and the wavelike view.

It can be shown that these operators obey differential equations

$$i\hbar\frac{d}{dt}c - Hc - \Sigma c = S \qquad (12.7)$$

that look much like the ones we have been using to describe one-electron wavefunctions (see Eq. (12.2)). But unlike $\psi^*\psi$, which can take on any value, operators like c^+c can only take on one of two values, 0 or 1, thereby reflecting a particulate aspect that is missing from the Schrödinger equation. This advanced formalism is needed to progress beyond coherent quantum transport to inelastic interactions and onto more subtle many-electron phenomena like the Kondo effect, which is largely outside the scope of this book. Indeed in strongly interacting systems it may be desirable to use similar approaches to write transport equations, not for the bare electron described by c, but for composite or dressed particles like polarons obtained through appropriate unitary transformations. In the Appendix I have tried to introduce the reader to these advanced concepts by rederiving our results from Chapters 8–10 (including the full time-dependent form not discussed earlier) using the second quantized formalism and showing how it can be used to to treat Coulomb blockade and the Kondo resonance arising from strong electron–electron interactions. This last phenomenon has been experimentally observed, but is inaccessible on the basis of the simple one-particle picture. As the field of nanoelectronics continues to progress, we can expect to discover many such phenomena where contacts get modified in non-trivial ways through their interactions with the channel and no longer function like simple reservoirs. As I mentioned in Section 8.4, this includes not only the metallurgical contacts, but also phonons and other "contact-like" entities represented by various self-energies $[\Sigma]$. It may then be necessary to write separate equations to determine the state of each "contact." Or perhaps it will prove more useful to write a transport equation not for individual electrons but for composite objects that include part of the "contact." At this time we can only speculate.

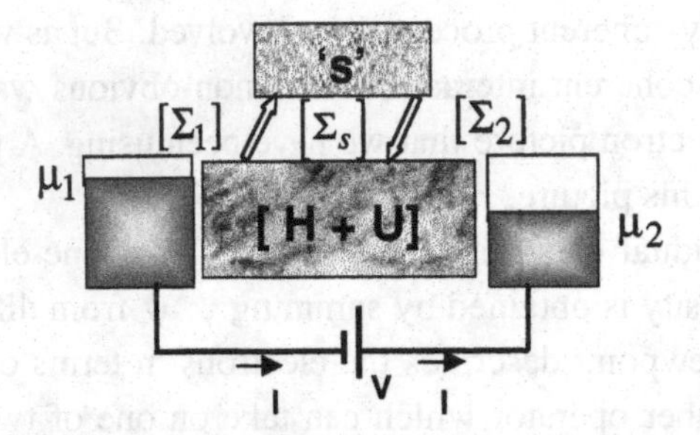

That is why I believe this field is currently at a very exciting stage where important advances can be expected from both applied and basic points of view. On the one hand we will continue to acquire a better quantitative understanding of nanoscale devices based on different materials and geometry. Although many of the observations appear to be described well within the basic self-consistent field model we have discussed, much remains to be done in terms of discovering better basis functions for representing the Hamiltonian $[H]$ and self-energy $[\Sigma]$, including inelastic scattering processes and implementing more efficient algorithms for the solution of the quantum transport equations. At the same time we can hope to discover new quantum transport phenomena (both steady-state and time-dependent) involving strong electron–phonon and electron–electron interactions, which are largely unexplored. In this respect I believe what we have seen so far represents only the "tip of the iceberg." As experimentalists acquire greater control of the degree of coupling between localized "channels" and delocalized "contacts," many more subtle quantum transport phenomena will be discovered, some of which may even be useful! I hope the broad conceptual framework I have tried to describe here will help the reader join this small but active subset of what Feynman called the "greatest adventure the human mind has ever begun" (Feynman, 1965).

Appendix: advanced formalism

In developing the general "matrix" model depicted in Fig. 1.6.5b, our starting point was the one-electron Schrüdinger equation

$$i\hbar\frac{d\psi}{dt} - H\psi = 0 \tag{A.1}$$

describing the time evolution of the one-electron wavefunction ψ from which the electron density is obtained by summing $\psi^*\psi$ for different electrons. Although it is a little tricky explaining exactly what one means by "different electrons," this procedure is adequate for dealing with simple problems involving coherent interactions where the background remains unaffected by the flow of electrons. But to go beyond such phenomena onto more complex processes involving phase-breaking interactions or strong electron–electron interactions it is desirable to use a more comprehensive viewpoint that describes the electrons in terms of field operators c. For non-interacting electrons these second quantized operators obey differential equations

$$i\hbar\frac{d}{dt}c - Hc = 0 \tag{A.2}$$

that look much like the one describing the one-electron wavefunction (see Eq. (A.1)). But unlike $\psi^*\psi$, which can take on any value, the number operator c^+c can only take on one of two values, 0 or 1, thereby reflecting a particulate aspect that is missing from the Schrüdinger equation. The two values of c^+c indicate whether a state is full or empty. At equilibrium, the average value of c^+c for a one-electron state with energy ε is given by the corresponding Fermi function $\langle c^+c\rangle = f_0(\varepsilon - \mu)$. However, this is true only if our channel consists of non-interacting electrons (perhaps with interactions described by a self-consistent field) in equilibrium. For non-equilibrium problems, a transport equation is needed that allows us to calculate c^+c based on our knowledge of the source terms from the contacts, which are assumed to remain locally in equilibrium and hence described by the equilibrium relations.

In this Appendix, I will: introduce the field operators and define the correlation functions in Section A.1; use these advanced concepts to rederive the expression (Eq. (11.1.1)) for the non-equilibrium density matrix in Section A.2; rederive the expression for the current (Eq. (11.1.6)) in Section A.3; derive the inscattering and broadening functions for incoherent processes (Eq. (11.1.11)) in Section A.4; and show how the approach is extended to include strong electron–electron interactions leading to Coulomb blockade and the Kondo resonance in Section A.5.

A.1 Correlation functions

Creation and annihilation operators: Consider a set of one-electron states labeled by j. In the multi-electron picture, we have two states 0_j and 1_j for each such one-electron state j. The creation operator c_j^+ inserts an electron in state j taking us from 0_j to 1_j and can be represented in the form

$$c_j(t) = \begin{array}{c} \begin{matrix} 0_j & 1_j \end{matrix} \\ \begin{bmatrix} 0 & 1 \\ 0 & 0 \end{bmatrix} \end{array} \exp(-i\varepsilon_j t/\hbar) \tag{A.1.1a}$$

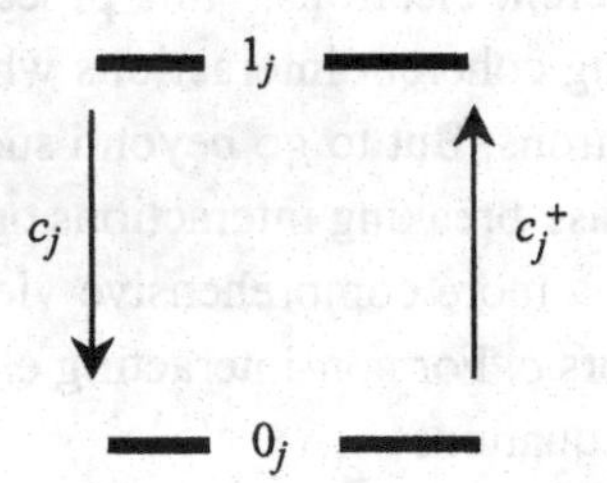

while the annihilation operator c_j represented by

$$c_j^+(t) = \begin{bmatrix} 0 & 0 \\ 1 & 0 \end{bmatrix} \exp(+i\varepsilon_j t/\hbar) \tag{A.1.1b}$$

takes an electron out of state j taking us from 1_j to 0_j. It is straightforward to show that

$$c_j^+(t)\, c_j(t) = \begin{array}{c} \begin{matrix} 0_j & 1_j \end{matrix} \\ \begin{bmatrix} 0 & 0 \\ 0 & 1 \end{bmatrix} \end{array} \qquad c_j(t) c_j^+(t) = \begin{array}{c} \begin{matrix} 0_j & 1_j \end{matrix} \\ \begin{bmatrix} 1 & 0 \\ 0 & 0 \end{bmatrix} \end{array}$$

independent of the time t. The former is called the number operator since its eigenvalues 0 and 1 represent the number of electrons in the corresponding state, while the latter tells us the number of empty states. Their expectation values are interpreted as the number of electrons and the number of holes in state j:

$$\langle c_j^+(t)\, c_j(t)\rangle = f_j \quad \text{and} \quad \langle c_j(t)\, c_j^+(t)\rangle = 1 - f_j \tag{A.1.2}$$

It can be checked that

$$c_j(t)\,c_j^+(t) + c_j^+(t)\,c_j(t) = 1$$
$$c_j(t)\,c_j(t) = 0$$
$$c_j^+(t)\,c_j^+(t) = 0$$

What is much less intuitive is the relation between the field operators for two different states i and j:

$$c_i(t)\,c_j^+(t) + c_j^+(t)\,c_i(t) = \delta_{ij} \tag{A.1.3a}$$
$$c_i(t)\,c_j(t) + c_j(t)\,c_i(t) = 0 \tag{A.1.3b}$$
$$c_i^+(t)\,c_j^+(t) + c_j^+(t)\,c_i^+(t) = 0 \tag{A.1.3c}$$

Ordinarily we would expect the operators for two distinct states to be independent of each other, so that for $i \neq j$, $c_i(t)\,c_j^+(t) = c_j^+(t)\,c_i(t)$. Equation (A.1.3a) would then imply that each is equal to zero. However, due to the exclusion principle, two distinct states are not really independent and $c_i(t)\,c_j^+(t) = -c_j^+(t)\,c_i(t)$. We can show that this is ensured if we modify Eqs. (A.1.1) to read

$$c_j(t) = \begin{bmatrix} 0 & (-1)^\nu \\ 0 & 0 \end{bmatrix} \exp(-i\varepsilon_j t/\hbar)$$

and

$$c_j^+(t) = \begin{bmatrix} 0 & 0 \\ (-1)^\nu & 0 \end{bmatrix} \exp(+i\varepsilon_j t/\hbar)$$

ν being the number of occupied states to the "left" of state j. This means that when dealing with a number of one-electron states, we need to agree on a specific order (it does not matter what order we choose) and stick to it, so that "left" has a well-defined meaning throughout the calculation. In practice, however, we do not need to worry about this, since we will be manipulating the operators making use of the algebra described by Eqs. (A.1.3). I just want to point out that this algebra, which is an expression of the exclusion principle, implies that putting an electron in state j is affected by the presence or absence of another electron in state i, even before we have included any interactions between electrons.

Correlation and spectral functions: In general we can define a two-time electron correlation function

$$G_{ij}^{\mathrm{n}}(t, t') \equiv \langle c_j^+(t')\,c_i(t) \rangle \equiv -iG_{ij}^{<}(t, t') \tag{A.1.4a}$$

and a two-time hole correlation function

$$G_{ij}^{\mathrm{p}}(t, t') \equiv \langle c_i(t)\,c_j^+(t') \rangle \equiv +iG_{ij}^{>}(t, t') \tag{A.1.4b}$$

whose values for equal time $t' = t$ give us the number operators in Eq. (A.1.2). Their sum is defined as the spectral function:

$$A_{ij}(t,t') \equiv G^{\mathrm{p}}_{ij}(t,t') + G^{\mathrm{n}}_{ij}(t,t') = \mathrm{i}\big[G^{>}_{ij}(t,t') - G^{<}_{ij}(t,t')\big] \tag{A.1.4c}$$

Fourier transformed functions: Under steady-state conditions the correlation functions depend only on the difference between the two time coordinates and it is convenient to work with the Fourier transformed functions: $(t - t') \rightarrow E$:

$$G^{\mathrm{n}}_{ij}(E) \equiv \int_{-\infty}^{+\infty} (\mathrm{d}\tau/\hbar)\exp(\mathrm{i}E\tau/\hbar)G^{\mathrm{n}}_{ij}(t, t-\tau) \tag{A.1.5a}$$

The inverse transform is given by

$$G^{\mathrm{n}}_{ij}(t, t-\tau) \equiv \langle c_j^{+}(t-\tau)\, c_i(t)\rangle = \int_{-\infty}^{+\infty} (\mathrm{d}E/2\pi)\, G^{\mathrm{n}}_{ij}(E)\, \exp(-\mathrm{i}E\tau/\hbar) \tag{A.1.5b}$$

so that the equal-time correlation function can be written as

$$G^{\mathrm{n}}_{ij}(t,t) \equiv \langle c_j^{+}(t)\, c_i(t)\rangle = \int_{-\infty}^{+\infty} (\mathrm{d}E/2\pi) G^{\mathrm{n}}_{ij}(E) \tag{A.1.6}$$

Similar relations hold for the hole correlation function G^{p} and the spectral function A.

Equilibrium: In general the electron and hole correlation functions can take on any value so long as they add up to give the spectral function as indicated in Eq. (A.1.4c). But at equilibrium, the former is proportional to the Fermi function so that (I am using lower case symbols to indicate equilibrium quantities)

$$g^{\mathrm{n}}_{ij}(E) = a_{ij}(E) f_0(E-\mu) \tag{A.1.7a}$$

$$g^{\mathrm{p}}_{ij}(E) = a_{ij}(E)(1 - f_0(E-\mu)) \tag{A.1.7b}$$

For an isolated system described by a Hamiltonian $[h]$, the spectral function is written down easily in the eigenstate representation

$$a_{rs}(E) = \delta_{rs}\delta(E-\varepsilon_{\mathrm{r}}) \tag{A.1.7c}$$

where r, s are the eigenstates of $[h]$, so that

$$g^{\mathrm{n}}_{rs}(E) = \delta_{rs}\delta(E-\varepsilon_{\mathrm{r}}) f_{\mathrm{r}}, \quad \text{with} \quad f_{\mathrm{r}} \equiv f_0(\varepsilon_{\mathrm{r}} - \mu) \tag{A.1.8a}$$

$$g^{\mathrm{n}}_{rs}(t,t') = \delta_{rs} f_{\mathrm{r}} \exp[-\mathrm{i}\varepsilon_{\mathrm{r}}(t-t')/\hbar]\exp(-\eta\,|t-t'|\,/\hbar) \tag{A.1.8b}$$

η being a positive infinitesimal. Our general approach is to use these relations with the appropriate Fermi functions for the contacts and then calculate the *resulting* correlation functions in the region of interest, namely the channel.

Boson operators: In discussing problems that involve phonon or photon emission, we will need to include operators b_α^+, b_α describing the phonon/photon fields. These operators obey a somewhat different algebra:

$$b_\beta(t)\, b_\alpha^+(t) - b_\alpha^+(t)\, b_\beta(t) = \delta_{\alpha\beta} \quad \text{(A.1.9a)}$$

$$b_\alpha(t)\, b_\beta(t) - b_\beta(t)\, b_\alpha(t) = 0 \quad \text{(A.1.9b)}$$

$$b_\alpha^+(t)\, b_\beta^+(t) - b_\beta^+(t)\, b_\alpha^+(t) = 0 \quad \text{(A.1.9c)}$$

where α, β are the different phonon/photon modes. Comparing with Eqs. (A.1.3) for the electron operators, it is easy to see that the difference lies in replacing a positive sign with a negative one. However, this "minor" change makes these operators far more intuitive since distinct modes $\alpha \neq \beta$ now function independently

$$b_\beta(t)\, b_\alpha^+(t) = b_\alpha^+(t)\, b_\beta(t), \quad \alpha \neq \beta$$

as "common sense" would dictate. Indeed, readers who have taken a course in quantum mechanics will recognize that in an operator treatment of the harmonic oscillator problem, one defines creation and annihilation operators that are linear combinations of the position (x) and momentum (p) operators that obey precisely the same algebra as in Eqs. (A.1.9). What this means is that, unlike electron operators, we could represent the phonon/photon operators using ordinary differential operators. However, in this Appendix we will not really use any representation. We will simply manipulate these operators making use of the algebra described in Eqs. (A.1.9). One consequence of this change in the algebra is that instead of Eq. (A.1.2) we have

$$\langle b_\alpha^+(t)\, b_\alpha(t)\rangle = N_\alpha \quad \text{and} \quad \langle b_\alpha(t)\, b_\alpha^+(t)\rangle = 1 + N_\alpha \quad \text{(A.1.10)}$$

where N_α is the number of phonons. As with the electron operators (see Eqs. (A.1.4)), we can define two-time correlation functions

$$G_{\alpha\beta}^{\text{ab}}(t, t') \equiv \langle b_\beta^+(t')\, b_\alpha(t)\rangle \quad \text{(A.1.11a)}$$

and

$$G_{\alpha\beta}^{\text{em}}(t, t') \equiv \langle b_\alpha(t) b_\beta^+(t')\rangle \quad \text{(A.1.11b)}$$

which we have labeled with "ab" and "em" (instead of "n" and "p" for electrons), which stand for absorption and emission respectively. Under steady-state conditions these functions depend only on $(t - t')$, which can be Fourier transformed to yield frequency domain correlation functions. At equilibrium (cf. Eqs. (A.1.7))

$$g^{\text{ab}}_{\alpha\beta}(\hbar\omega) = a^{\text{ph}}_{\alpha\beta}(\hbar\omega)N(\hbar\omega)$$
$$g^{\text{em}}_{\alpha\beta}(\hbar\omega) = a^{\text{ph}}_{\alpha\beta}(\hbar\omega)(1 + N(\hbar\omega)) \tag{A.1.12}$$

where the Fermi function has been replaced by the Bose function:

$$N(\hbar\omega) = [\exp(\hbar\omega/k_{\text{B}}T) - 1]^{-1} \tag{A.1.13}$$

and the phonon spectral function in the eigenmode representation is given by

$$a^{\text{ph}}_{\alpha\beta}(\hbar\omega) = \delta_{\alpha\beta}\delta(\hbar\omega - \hbar\omega_\alpha) \tag{A.1.14}$$

so that

$$g^{\text{ab}}_{\alpha\beta}(E) = \delta_{\alpha\beta}\delta(\hbar\omega - \hbar\omega_\alpha)\, N_\alpha \quad \text{with } N_\alpha \equiv N(\hbar\omega_\alpha)$$
$$g^{\text{ab}}_{\alpha\beta}(t, t') = \delta_{\alpha\beta} N_\alpha \exp[-\text{i}\omega(t - t')] \exp(-\eta |\, t - t' \,| /\hbar) \tag{A.1.15}$$

and

$$g^{\text{em}}_{\alpha\beta}(E) = \delta_{\alpha\beta}\delta(\hbar\omega - \hbar\omega_\alpha)(N_\alpha + 1)$$
$$g^{\text{em}}_{\alpha\beta}(t, t') = \delta_{\alpha\beta}(N_\alpha + 1) \exp[-\text{i}\omega(t - t')] \exp(-\eta \,|\, t - t' \,| /\hbar)$$

Energy–time relationship: The next point to note is that the wavefunction $\psi(t)$ and the field operator $c(t)$ are not observable quantities. What is observable are correlation functions like $\psi^*(t')\,\psi(t)$ or $c^+(t')\,c(t)$, in somewhat the same way that the noise voltage $V(t)$ across a resistor is described by its correlation function $V(t')V(t)$. Under steady-state conditions, such two-time correlation functions depend only on time differences $(t - t')$ and the corresponding Fourier transform variable is the energy E. For example, an electron with a wavefunction $\psi(t) = \psi_0 \exp(-\text{i}\varepsilon t/\hbar)$ has a correlation function of the form

$$\psi^*(t')\,\psi(t) = \psi_0^*\,\psi_0 \exp[-\text{i}\varepsilon(t - t')/\hbar]$$

and the Fourier transform with respect to $(t - t')$ is proportional to the delta function $\delta(E - \varepsilon)$ at $E = \varepsilon$. Steady-state phenomena can be described in terms of such Fourier transformed quantities with E as an independent variable, as we have been doing following the introduction of broadening in Section 1.3. In each of the following sections, I will first derive expressions in the time domain and then Fourier transform with respect to $(t - t')$ to obtain energy domain expressions suitable for steady-state analysis.

A.2 Non-equilibrium density matrix

Partitioning: Operators obeying Eqs. (A.1.9) are generally referred to as Boson operators while those obeying Eqs. (A.1.3) are referred to as Fermion operators. At

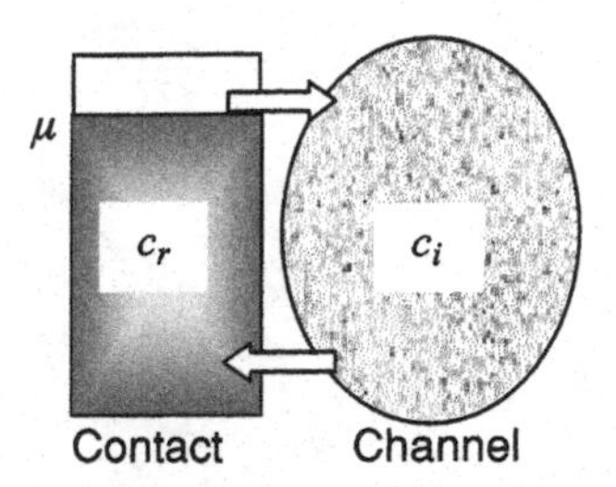

Fig. A.2.1

equilibrium, the corresponding correlation functions are given by Eqs. (A.1.8) and (A.1.15) respectively. The problem is to calculate them in the channel driven away from equilibrium and our strategy will be to assume that the channel is driven by the contacts and by the phonon/photon baths that are maintained in their respective equilibrium states by the environment.

To implement this procedure, we partition the overall structure into a part that is of interest (labeled channel) and a reservoir (labeled contact) having a large continuous density of states (Fig. A.2.1). We start from the equations describing the composite system:

$$i\hbar \frac{\mathrm{d}}{\mathrm{d}t}\begin{Bmatrix} c_i \\ c_r \end{Bmatrix} = \begin{bmatrix} \varepsilon_i & \tau_{ir} \\ [\tau^+]_{ri} & \varepsilon_r \end{bmatrix} \begin{Bmatrix} c_i \\ c_r \end{Bmatrix} \tag{A.2.1}$$

and obtain an effective equation for the channel of the form stated in Eq. (12.7), eliminating the reservoir variables r (i and r are assumed to represent eigenstates of the isolated channel and reservoir respectively) by assuming that they are maintained in equilibrium by the environment so that the corresponding fields are described by Eqs. (A.1.7). In this section let me illustrate the approach assuming that the coupling with the contact involves purely elastic interactions, just as we assumed in our discussion of coherent transport in Chapters 8 and 9. As we will see, the mathematics in this case will look just as if the quantities c are ordinary complex numbers like the wavefunctions we have been using. We will not really need to make use of the fact that these are Fermion operators. However, in subsequent sections we will discuss more general problems involving electron–phonon and electron–electron interactions, where we will make use of the properties of Fermion and Boson operators to obtain results that would be hard to rationalize from a one-electron picture.

Eliminating the reservoir variables: The contact subset of Eq. (A.2.1) yields (note: $[\tau^+]_{rj} = \tau^*_{jr}$)

$$i\hbar \frac{\mathrm{d}}{\mathrm{d}t} c_r = (\varepsilon_r - i\eta)c_r + \sum_j \tau^*_{jr}\, c_j + S_r \tag{A.2.2}$$

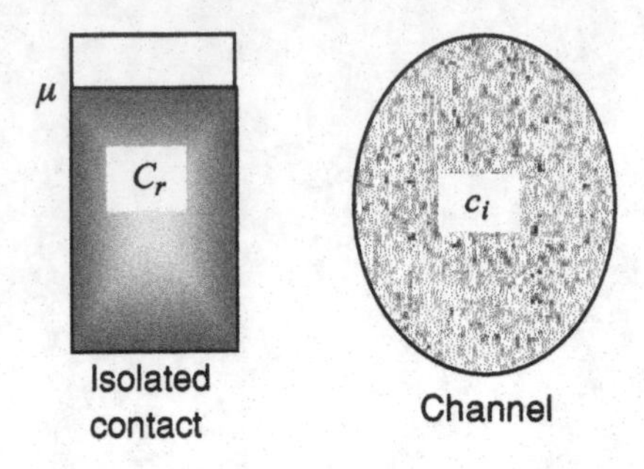

Fig. A.2.2

after adding an infinitesimal imaginary part η to the energy and a source term S_r following the same arguments that we used in Eq. (8.1.8a). We can write the solution for the reservoir field in the form

$$c_r(t) = \underbrace{C_r(t)}_{\text{Isolated contact}} + \underbrace{\sum_j \int_{-\infty}^{+\infty} dt_1 g_{rr}(t, t_1)\tau_{jr}^* c_j(t_1)}_{\text{Channel-induced}} \tag{A.2.3}$$

where

$$g_{rr}(t, t') = (1/i\hbar)\vartheta\,(t - t')\exp(-i\varepsilon_r - \eta)(t - t') \tag{A.2.4}$$

represents the Green's function (or "impulse response") of the differential operator appearing in Eq. (A.2.2):

$$L_r g_{rr}(t, t') = \delta(t - t')$$

with

$$L_r \equiv i\hbar\frac{\mathrm{d}}{\mathrm{d}t} - (\varepsilon_r - i\eta) \tag{A.2.5}$$

and $C_r(t)$ is the solution to the homogeneous equation: $L_r C_r(t) = S_r$.

Physically, $C_r(t)$ represents the field operator in the *isolated* contact (Fig. A.2.2) before connecting to the channel at $t = 0$. This allows us to use the law of equilibrium (see Eqs. (A.1.8a, b)) to write down the corresponding correlation function ($f_r \equiv f_0(\varepsilon_r - \mu)$):

$$g_{rs}^{\mathrm{n}}(E) = \delta_{rs}\delta\,(E - \varepsilon_r) f_r \tag{A.2.6a}$$

$$g_{rs}^{\mathrm{n}}(t, t') \equiv \langle C_s^+(t')\,C_r(t)\rangle$$
$$= \delta_{rs} f_r \exp[-i\varepsilon_r(t - t')]\exp(-\eta|t - t'|) \tag{A.2.6b}$$

The channel itself is assumed to be completely empty prior to $t = 0$ and is filled by the electrons "spilling over" from the contact after the connection is established at $t = 0$.

This process is described by starting from the channel subset of Eq. (A.2.1):

$$i\hbar \frac{d}{dt} c_i = \varepsilon_i \, c_i + \sum_j \tau_{ir} \, c_r$$

and substituting for $c_r(t)$ from Eq. (A.2.3):

$$i\hbar \frac{d}{dt} c_i - \varepsilon_i \, c_i - \sum_j \int_{-\infty}^{+\infty} dt_1 \, \Sigma_{ij}(t, t_1) \, c_j(t_1) = S_i(t) \tag{A.2.7}$$

where

$$\Sigma_{ij}(t, t') \equiv \sum_r \tau_{ir} g_{rr}(t, t') \tau^*_{jr} \tag{A.2.8}$$

and

$$S_i(t) \equiv \sum_r \tau_{ir} C_r(t) \tag{A.2.9}$$

Equation relating correlation functions: Defining a Green's function for the integro-differential operator appearing in Eq. (A.2.7):

$$i\hbar \frac{d}{dt} G_{ik}(t, t') - \varepsilon_i \, G_{ik}(t, t') - \sum_j \int_{-\infty}^{+\infty} dt_1 \, \Sigma_{ij}(t, t_1) \, G_{jk}(t_1, t') = \delta_{ik} \delta \, (t - t')$$

we can write the solution to Eq. (A.2.7) in the form

$$c_i(t) = \sum_k \int_{-\infty}^{+\infty} dt_1 \; G_{ik}(t, t_1) S_k(t_1) \tag{A.2.10}$$

so that the correlation function is given by

$$G^n_{ij}(t, t') \equiv \langle c^+_j(t') \, c_i(t) \rangle$$

$$= \sum_{k,l} \int_{-\infty}^{+\infty} dt_1 \int_{-\infty}^{+\infty} dt_2 \; G_{ik}(t, t_1) G^*_{jl}(t', t_2) \langle S^+_l(t_2) \, S_k(t_1) \rangle$$

Defining

$$\Sigma^{in}_{kl}(t_1, t_2) \equiv \langle S^+_l(t_2) \, S_k(t_1) \rangle \tag{A.2.11}$$

and

$$[G^+]_{lj}(t_2, t') \equiv [G]^*_{jl}(t', t_2) \tag{A.2.12}$$

we can write in matrix notation:

$$[G^n(t, t')] = \int_{-\infty}^{+\infty} dt_1 \int_{-\infty}^{+\infty} dt_2 \, [G(t, t_1)][\Sigma^{in}(t_1, t_2)][G^+(t_2, t')] \tag{A.2.13}$$

where

$$L[G(t,t')] = [I]\delta(t-t')$$

with

$$L \equiv i\hbar\frac{\mathrm{d}}{\mathrm{d}t}[I] - [h] - \int_{-\infty}^{+\infty} \mathrm{d}t_1\, [\Sigma(t,t_1)] \tag{A.2.14}$$

$$[\Sigma(t,t')] \equiv [\tau][g(t,t')][\tau^+] \tag{A.2.15}$$

$[h]$ being the Hamiltonian describing the isolated channel whose eigenstates are labeled by i. Equation (A.2.13) relates the correlation function in the channel to the inscattering function Σ^{in} describing the correlation of the source term.

Under steady-state conditions, all functions depend only on the difference between the two time coordinates so that the time integrals in Eq. (A.2.13) represent convolutions that turn into ordinary products if we Fourier transform with respect to the difference coordinate:

$$[G^{\mathrm{n}}(E)] \equiv [G(E)][\Sigma^{\mathrm{in}}(E)][G^+(E)] \tag{A.2.16}$$

with

$$L[G(E)] = [I]$$

$$L \equiv E[I] - [h] - [\Sigma(E)] \tag{A.2.17}$$

$$[\Sigma(E)] \equiv [\tau][G(E)][\tau^+] \tag{A.2.18}$$

Inscattering function: To evaluate the inscattering function

$$\Sigma_{ij}^{\mathrm{in}}(t,t') \equiv \langle S_j^+(t')\, S_i(t)\rangle$$

we substitute for the source term from Eq. (A.2.9)

$$\Sigma_{ij}^{\mathrm{in}}(t,t') = \sum_s \tau_{js}^* \sum_r \tau_{ir} \langle C_s^+(t')\, C_r(t)\rangle$$

$$= \sum_{r,s} \tau_{ir}\, g_{rs}^{\mathrm{n}}(t,t')\tau_{js}^*$$

where $g_{rs}^{\mathrm{n}}(t,t')$ is given by Eq. (A.2.6) since it represents the correlation function for the isolated contact in equilibrium if it were not connected to the channel. In matrix notation

$$[\Sigma^{\mathrm{in}}(t,t')] = [\tau][g^{\mathrm{n}}(t,t')][\tau^+] \tag{A.2.19}$$

Once again, since this correlation function depends only on the difference between the two time arguments, we can Fourier transform with respect to the difference coordinate to write

$$[\Sigma^{\text{in}}(E)] = [\tau][g^{\text{n}}(E)][\tau^{+}] \tag{A.2.20}$$

Inscattering vs. broadening: It is interesting to note that the anti-hermitian component of the self-energy function (see Eq. (A.2.18)), also called the broadening function, is given by

$$[\Gamma(E)] = \mathrm{i}[\Sigma(E) - \Sigma^{+}(E)] = [\tau][a(E)][\tau^{+}]$$

where $[a(E)]$ is the spectral function for the isolated contact: $[a] = \mathrm{i}[g - g^{+}]$. If the contact is assumed to be in equilibrium with the Fermi function $f(E)$, then the inscattering function from Eq. (A.2.20) can be written as

$$[\Sigma^{\text{in}}(E)] = [\tau][a(E)][\tau^{+}]f(E)$$

so that the inscattering and the corresponding broadening are related:

$$[\Sigma^{\text{in}}(E)] = [\Gamma(E)]f(E) \tag{A.2.21}$$

A.3 Inflow and outflow

We will now obtain an expression for the inflow and outflow, starting from the expression for the current (note that $\langle c_i{}^{+}c_i\rangle$ tells us the number of electrons in state i)

$$\begin{aligned} I(t) &\equiv \sum_i \frac{\mathrm{d}}{\mathrm{d}t}\langle c_i^{+}(t)\, c_i(t)\rangle \\ &= \frac{1}{\mathrm{i}\hbar}\sum_i \left\langle c_i^{+}(t)\left(\mathrm{i}\hbar\frac{\mathrm{d}}{\mathrm{d}t}c_i(t)\right)\right\rangle - \left\langle\left(-\mathrm{i}\hbar\frac{\mathrm{d}}{\mathrm{d}t}c_i^{+}(t)\right)c_i(t)\right\rangle \end{aligned}$$

and substitute from Eq. (A.2.7) to obtain explicit expressions for the inflow and outflow. More generally we could define a two-time version $I(t, t')$ as

$$I(t,t') \equiv \sum_i \left(\frac{\mathrm{d}}{\mathrm{d}t} + \frac{\mathrm{d}}{\mathrm{d}t'}\right)\langle c_i^{+}(t')\, c_i(t)\rangle \tag{A.3.1}$$

whose "diagonal" elements ($t = t'$) give us the total current. The advantage of the two-time version is that for steady-state transport we can Fourier transform with respect to $(t - t')$ to obtain the energy spectrum of the current. Substituting from Eq. (A.2.7) into Eq. (A.3.1) we obtain

$$I(t,t') = I^{\text{in}}(t,t') - I^{\text{out}}(t,t')$$

where

$$I^{\text{out}}(t,t')$$

$$= (-1/i\hbar)\sum_{i,j}\int_{-\infty}^{+\infty} dt_1(\Sigma_{ij}(t,t_1)\langle c_i^+(t')\,c_j(t_1)\rangle - \Sigma_{ij}^*(t',t_1)\langle c_j^+(t_1)\,c_i(t)\rangle)$$

$$= (-1/i\hbar)\sum_{i,j}\int_{-\infty}^{+\infty} dt_1(\Sigma_{ij}(t,t_1)\,G_{ji}^{\text{n}}(t_1,t') - \Sigma_{ij}^*(t',t_1)\,G_{ij}^{\text{n}}(t,t_1))$$

and

$$I^{\text{in}}(t,t') = (1/i\hbar)\sum_i (c_i^+(t')\,S_i(t) - S_i^+(t')\,c_i(t))$$

$$= (1/i\hbar)\sum_{i,j}\int_{-\infty}^{+\infty} dt_1 G_{ij}^*(t',t_1)\,\langle S_j^+(t_1)\,S_i(t)\rangle - G_{ij}(t,t_1)\,\langle S_i^+(t')\,S_j(t_1)\rangle$$

(making use of Eq. (A.2.10))

$$= (1/i\hbar)\sum_{i,j}\int_{-\infty}^{+\infty} dt_1 G_{ij}^*(t',t_1)\,\Sigma_{ij}^{\text{in}}(t,t_1) - G_{ij}(t,t_1)\,\Sigma_{ji}^{\text{in}}(t_1,t')$$

(making use of Eq. (A.2.11))

In matrix notation we can write

$$I^{\text{in}}(t,t') = (1/i\hbar)\,\text{Trace}\int_{-\infty}^{+\infty} dt_1[\Sigma^{\text{in}}(t,t_1)][G^+(t_1,t')] - [G(t,t_1)][\Sigma^{\text{in}}(t_1,t')] \tag{A.3.2}$$

$$I^{\text{out}}(t,t') = (1/i\hbar)\,\text{Trace}\int_{-\infty}^{+\infty} dt_1[\Sigma(t,t_1)][G^{\text{n}}(t_1,t')] - [G^{\text{n}}(t,t_1)][\Sigma^+(t_1,t')] \tag{A.3.3}$$

Note that these are just the conduction currents to which one should add the displacement currents to obtain the net terminal current.

Once again, under steady-state conditions each of the quantities depends only on the difference between the two time arguments and on Fourier transforming with respect to the difference coordinate the convolution turns into a normal product:

$$I^{\text{in}}(E) = \text{Trace}[\Sigma^{\text{in}}(E)\,G^+(E) - G(E)\,\Sigma^{\text{in}}(E)]/i\hbar$$
$$= \text{Trace}[\Sigma^{\text{in}}(E)\,A(E)]/\hbar \tag{A.3.4}$$

$$I^{\text{out}}(E) = \text{Trace}[G^{\text{n}}(E)\,\Sigma^+(E) - \Sigma(E)\,G^{\text{n}}(E)]/i\hbar$$
$$= \text{Trace}[\Gamma(E)\,G^{\text{n}}(E)]/\hbar \tag{A.3.5}$$

Multi-terminal devices: We have now obtained the basic expressions for inflow (Eq. (A.3.4)) and outflow (Eq. (A.3.5)) that we advertised in Fig. 1.6.5 along with an equation for G^n (Eq. (A.2.16)). For simplicity we considered a channel connected to just one contact, but the results can be readily extended to real devices with two or more contacts labeled by indices p, q. For example, Eqs. (A.2.13–A.2.15) are modified to read

$$[G^n(t,t')] = \sum_p \int_{-\infty}^{+\infty} dt_1 \int_{-\infty}^{+\infty} dt_2 \, [G(t,t_1)][\Sigma^{in}(t_1,t_2)]^{(p)}[G^+(t_2,t')] \tag{A.3.6}$$

where

$$L[G(t,t')] = [I]\delta(t-t')$$

with

$$L \equiv i\hbar\frac{d}{dt}[I] - [h] - \sum_p \int_{-\infty}^{+\infty} dt_1 [\Sigma(t,t_1)]^{(p)} \tag{A.3.7}$$

$$[\Sigma(t,t')]^{(p)} \equiv [\tau]^{(p)}[g(t,t')]^{(p)}[\tau^+]^{(p)} \tag{A.3.8}$$

while Eqs. (A.2.19), (A.3.2) and (A.3.3) become

$$[\Sigma^{in}(t,t')]^{(p)} = [\tau]^{(p)}[g^n(t,t')]^{(p)}[\tau^+]^{(p)} \tag{A.3.9}$$

$$I^{in}(t,t')^{(p)} = (1/i\hbar)\,\text{Trace} \int_{-\infty}^{+\infty} dt_1 [\Sigma^{in}(t,t_1)]^{(p)}[G^+(t_1,t')] - [G(t,t_1)][\Sigma^{in}(t_1,t')]^{(p)} \tag{A.3.10}$$

$$I^{out}(t,t')^{(p)} = (-1/i\hbar)\,\text{Trace} \int_{-\infty}^{+\infty} dt_1 [\Sigma(t,t_1)]^{(p)}[G^n(t_1,t')] - [G^n(t,t_1)][\Sigma^+(t_1,t')]^{(p)}$$

Under steady-state conditions the multi-terminal versions take the form

$$[G^n(E)] = \sum_p [G(E)][\Sigma^{in}(E)]^{(p)}[G^+(E)] \tag{A.3.11}$$

$$L[G] = [I] \qquad L \equiv E[I] - [h] - \sum_p [\Sigma(E)]^{(p)} \tag{A.3.12}$$

$$[\Sigma(E)]^{(p)} \equiv [\tau]^{(p)}[g(E)]^{(p)}[\tau^+]^{(p)} \tag{A.3.13}$$

$$[\Sigma^{in}(E)]^{(p)} = [\tau]^{(p)}[g^n(E)]^{(p)}[\tau^+]^{(p)} \tag{A.3.14}$$

$$I^{in}(E)^{(p)} = \text{Trace}[\Sigma^{in}(E)]^{(p)}[A(E)]/\hbar \tag{A.3.15}$$

$$I^{out}(E)^{(p)} = \text{Trace}[\Gamma(E)]^{(p)}[G^n(E)] \tag{A.3.16}$$

The terminal current is given by the difference between the inflow and outflow:

$$I(E)^{(p)} = (1/\hbar)\,\text{Trace}([\Sigma^{in}(E)]^{(p)}[A(E)] - [\Gamma(E)]^{(p)}[G^n(E)])$$

Making use of Eq. (A.3.10) and the relation

$$A = \sum_q G\Gamma^{(q)}G^+ = \sum_q G^+\Gamma^{(q)}G$$

(which follows from Eqs. (A.3.11) and (A.3.12)), we can write

$$I(E)^{(p)} = (1/\hbar)\sum_q \text{Trace}(\Sigma^{\text{in}(p)}G\Gamma^{(q)}G^+ - \Gamma^{(p)}G\Sigma^{\text{in}(q)}G^+)$$
$$= (1/\hbar)\sum_q \text{Trace}[\Gamma^{(p)}G\Gamma^{(q)}G^+](f_p - f_q)$$

if $\Sigma^{\text{in}(p)} = \Gamma^{(p)} f_p$. We can then write the current as

$$I(E)^{(p)} = (1/\hbar)\sum_q T^{(pq)}(f_p - f_q) \tag{A.3.17}$$

in terms of the transmission function:

$$T^{(pq)} \equiv \text{Trace}[\Gamma^{(p)}G\Gamma^{(q)}G^+] \tag{A.3.18}$$

Linearizing Eq. (A.3.17) about the equilibrium electrochemical potential we obtain the standard Büttiker equations (using δ to denote the change from the equilibrium value)

$$\delta I(E)^{(p)} = (1/\hbar)\sum_q T^{(pq)}(\delta\mu_p - \delta\mu_q) \tag{A.3.19}$$

widely used in mesoscopic physics (see Büttiker (1988) and Chapter 2 of Datta (1995)).

A.4 Inelastic flow

We have seen earlier (Fig. 10.3.1) that inelastic processes within the channel can be understood by regarding the *channel as its own contact.* With this in mind, let us consider a problem in which an electron can enter the channel from the contact by emitting (creating) or absorbing (annihilating) a phonon (Fig. A.4.1).

Note that in general we cannot assume the "contact" to be in equilibrium since it represents the channel itself, in other words, we should not make use of Eqs. (A.2.6). We start with the equation of motion for a channel field operator:

$$i\hbar\frac{\mathrm{d}}{\mathrm{d}t}c_i - \varepsilon_i c_i = s_i$$

where

$$s_i = \sum_{r,\alpha} \tau_{ir\alpha}c_r b_\alpha + \tau^*_{ri\alpha}c_r b^+_\alpha \tag{A.4.1}$$

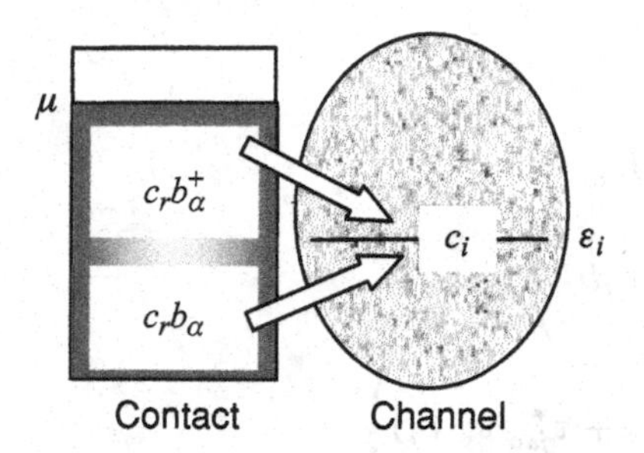

Fig. A.4.1

Equations of motion of this type describing the time evolution of the field operators (in the Heisenberg picture) are obtained using the Heisenberg equations of motion for any operator P

$$i\hbar \mathrm{d}P/\mathrm{d}t = PH - HP$$

where H is the second quantized Hamiltonian:

$$\begin{aligned} H = &\sum_i \varepsilon_i c_i^+ c_i + \sum_r \varepsilon_r c_r^+ c_r + \sum_\alpha \hbar\omega_\alpha (b_\alpha^+ b_\alpha + 1/2) \\ &+ \sum_{i,r,\alpha} (\tau_{ir\alpha} c_i^+ c_r b_\alpha + \tau_{ir\alpha}^* c_r^+ c_i b_\alpha^+ + \tau_{ri\alpha} c_r^+ c_i b_\alpha + \tau_{ri\alpha}^* c_i^+ c_r b_\alpha^+) \end{aligned} \tag{A.4.2}$$

The electron–phonon interaction term (the last term in Eq. (A.4.2)) is obtained by writing the potential $U(x)$ felt by one electron due to the phonon mode $\alpha(\tau_\alpha(\vec{x})$ is similar to $U_\beta^{\mathrm{ab}}(\vec{r})$ in Eq. (10.1.17)):

$$U(\vec{x}) = \sum_\alpha b_\alpha \tau_\alpha(\vec{x}) + b_\alpha^+ \tau_\alpha(\vec{x})^+$$

and then writing the second quantized version as we would for any ordinary potential $U(\vec{x})$:

$$H_{\text{el-ph}} = \sum_{i,r} U_{ir} c_i{}^+ c_r + U_{ri} c_r^+ c_i$$

We will now evaluate the two source terms in Eq. (A.4.1) one by one.

Evaluating $c_r b_\alpha$: We start from the equations of motion for the individual electron and phonon operators:

$$i\hbar \frac{\mathrm{d}}{\mathrm{d}t} c_r = \varepsilon_r c_r + \sum_{i,\alpha} \tau_{ri\alpha} c_i b_\alpha + \tau_{ir\alpha}^* c_i b_\alpha^+$$

$$i\hbar \frac{\mathrm{d}}{\mathrm{d}t} b_\alpha = \hbar\omega_\alpha b_\alpha + \sum_{i,r} \tau_{ri\alpha}^* c_i^+ c_r + \tau_{ir\alpha}^* c_r^+ c_i$$

and write an equation for the product

$$i\hbar\frac{\mathrm{d}}{\mathrm{d}t}\,c_r b_\alpha = \left(i\hbar\frac{\mathrm{d}}{\mathrm{d}t}\,c_r\right)b_\alpha + c_r\left(i\hbar\frac{\mathrm{d}}{\mathrm{d}t}\,b_\alpha\right)$$

$$= \left(\varepsilon_r c_r + \sum_{j,\beta}\tau_{rj\beta}\,c_j b_\beta + \tau^*_{jr\beta}\,c_j b^+_\beta\right)b_\alpha$$

$$+ c_r\left(\hbar\omega_\alpha b_\alpha + \sum_{j,s}(\tau^*_{sj\alpha}\,c^+_j c_s + \tau^*_{js\alpha}\,c^+_s\,c_j)\right)$$

which is simplified by replacing the products involving reservoir variables with their average values (which are time-independent numbers) to obtain (note that $\langle b_\beta b_\alpha\rangle = 0$ and $\langle c_r c_s\rangle = 0$)

$$i\hbar\frac{\mathrm{d}}{\mathrm{d}t}\,c_r b_\alpha - (\varepsilon_r + \hbar\omega_\alpha)c_r b_\alpha$$

$$= \sum_j\sum_{s,\beta}\tau^*_{js\beta}(\langle b^+_\beta\,b_\alpha\rangle\delta_{rs} + \langle c_r c^+_s\rangle\delta_{\alpha\beta})c_j$$

$$= \sum_j\sum_{s,\beta}\tau^*_{js\beta}(\langle b^+_\beta\,b_\alpha\rangle\langle c^+_s c_r\rangle + \langle c_r c^+_s\rangle\langle b_\alpha b^+_\beta\rangle)c_j$$

For the last step we have made use of the relations $\delta_{rs} = c^+_r c_s + c_s c^+_r$ (Eq. (A.1.3a)) and $\delta_{\alpha\beta} = b_\alpha b^+_\beta - b^+_\beta b_\alpha$ (Eq. (A.1.9a)).

As we did in Eq. (A.2.3), we can write the total solution as a sum of an isolated solution and a response induced by the source term:

$$c_r(t)\,b_\alpha(t) = C_r(t)\,B_\alpha(t) + \int_{-\infty}^{+\infty}\mathrm{d}t_1\sum_j c_j(t_1)\sum_{s,\beta}\tau^*_{js\beta}g_{r\alpha}(t,t_1)$$

$$\times\,(\langle b^+_\beta(t_1)\,b_\alpha(t_1)\rangle\langle c^+_s(t_1)c_r(t_1)\rangle + \langle c_r(t_1)c^+_s(t_1)\rangle\langle b_\alpha(t_1)b^+_\beta(t_1)\rangle)$$

where

$$i\hbar\frac{\mathrm{d}}{\mathrm{d}t}\,g_{r\alpha}(t,t_1) - (\varepsilon_r - i\eta + \hbar\omega_\alpha)g_{r\alpha}(t,t_1) = \delta(t-t_1)$$

To lowest order, we can write $c_r(t)\,b_\alpha(t) \approx g_{r\alpha}(t,t_1)\,c_r(t_1)\,b_\alpha(t_1)$, so that

$$c_r(t)\,b_\alpha(t) = C_r(t)B_\alpha(t) + \int_{-\infty}^{+\infty}\mathrm{d}t_1\sum_j c_j(t_1)\sum_{s,\beta}\tau^*_{js\beta}$$

$$\times\,(\langle b^+_\beta(t_1)\,b_\alpha(t)\rangle\langle c^+_s(t_1)\,c_r(t)\rangle + \langle c_r(t)\,c^+_s(t_1)\rangle\langle b_\alpha(t)\,b^+_\beta(t_1)\rangle)$$

$$= C_r(t)B_\alpha(t) + \int_{-\infty}^{+\infty}\mathrm{d}t_1\sum_j c_j(t_1)\sum_{s,\beta}\tau^*_{js\beta}$$

$$\times\left(G^{\mathrm{ab}}_{\alpha\beta}(t,t_1)G^{\mathrm{n}}_{rs}(t,t_1) + G^{\mathrm{p}}_{rs}(t,t_1)G^{\mathrm{em}}_{\alpha\beta}(t,t_1)\right) \tag{A.4.3a}$$

Evaluating $c_r b_\alpha^+$: Now that we have evaluated $c_r b_\alpha$, the next step is to evaluate $c_r b_\alpha^+$. With this in mind, we start from

$$i\hbar \frac{d}{dt} c_r = \varepsilon_r c_r + \sum_{i,\alpha} \tau_{ri\alpha}\, c_i b_\alpha + \tau_{ir\alpha}^*\, c_i b_\alpha^+$$

and

$$-i\hbar \frac{d}{dt} b_\alpha^+ = \hbar\omega_\alpha b_\alpha^+ + \sum_{i,r} \tau_{ir\alpha}\, c_i^+ c_r + \tau_{ri\alpha} c_{r^+} c_i$$

and write an equation for the product

$$i\hbar \frac{d}{dt} c_r b_\alpha^+ = \left(i\hbar \frac{d}{dt} c_r\right) b_\alpha^+ - c_r \left(-i\hbar \frac{d}{dt} b_\alpha^+\right)$$

$$= \left((\varepsilon_r - i\eta) c_r + \sum_{j,\beta} \tau_{rj\beta}\, c_j b_\beta + \tau_{jr\beta}^*\, c_j b_\beta^+\right) b_\alpha^+$$

$$- c_r \left(\hbar\omega_\alpha b_\alpha^+ + \sum_{j,s} (\tau_{js\alpha}\, c_j^+ c_s + \tau_{sj\alpha} c_s^+\, c_j)\right)$$

which is simplified by replacing the products involving reservoir variables with their average values as before:

$$i\hbar \frac{d}{dt} c_r b_\alpha^+ - (\varepsilon_r - \hbar\omega_\alpha) c_r b_\alpha^+ = \sum_j \sum_{s,\beta} \tau_{sj\beta} (\langle b_\beta b_\alpha^+ \rangle \delta_{rs} - \langle c_r c_s^+ \rangle \delta_{\alpha\beta}) c_j$$

$$= \sum_j \sum_{s,\beta} \tau_{sj\beta} (\langle b_\beta b_\alpha^+ \rangle \langle c_s^+ c_r \rangle + \langle c_r c_s^+ \rangle \langle b_\alpha^+ b_\beta \rangle) c_j$$

Once again we can write the solution as a sum of an isolated solution and a response induced by the source term:

$$c_r(t)\, b_\alpha^+(t) = C_r(t) B_\alpha^+(t) + \int_{-\infty}^{+\infty} dt_1 \sum_j c_j(t_1) \sum_{s,\beta} \tau_{sj\beta} \overline{g}_{r\alpha}(t, t_1)$$

$$\times \left(\langle b_\beta(t_1)\, b_\alpha^+(t_1) \rangle \langle c_s^+(t_1)\, c_r(t_1) \rangle + \langle c_r(t_1)\, c_s^+(t_1) \rangle \langle b_\alpha^+(t_1)\, b_\beta(t_1) \rangle\right)$$

where

$$i\hbar \frac{d}{dt} \overline{g}_{r\alpha}(t, t_1) - (\varepsilon_r - i\eta - \hbar\omega_\alpha) \overline{g}_{r\alpha}(t, t_1) = \delta(t - t_1)$$

Proceeding as before we write

$$c_r(t) b_\alpha^+(t) \approx C_r(t)\, B_\alpha^+(t) + \int_{-\infty}^{+\infty} dt_1 \sum_j c_j(t_1) \sum_{s,\beta} \tau_{sj\beta}$$

$$\times \left(\langle b_\beta(t_1)\, b_\alpha^+(t) \rangle \langle c_s^+(t_1)\, c_r(t) \rangle + \langle c_r(t)\, c_s^+(t_1) \rangle \langle b_\alpha^+(t)\, b_\beta(t_1) \rangle\right)$$

$$= C_r(t)B_\alpha^+(t) + \int_{-\infty}^{+\infty} \mathrm{d}t_1 \sum_j c_j(t_1) \sum_{s,\beta} \tau_{sj\beta}$$
$$\times \left(G_{\beta\alpha}^{\mathrm{em}}(t_1,t)G_{rs}^{\mathrm{n}}(t,t_1) + G_{rs}^{\mathrm{p}}(t,t_1)G_{\beta\alpha}^{\mathrm{ab}}(t_1,t)\right) \quad \text{(A.4.3b)}$$

Eliminating reservoir variables: Substituting Eqs. (A.4.3a, b) into Eq. (A.4.1) we obtain an equation for the channel field (cf. Eqs. (A.2.7)–(A.2.9)):

$$\mathrm{i}\hbar\frac{\mathrm{d}}{\mathrm{d}t}c_i - \varepsilon_i c_i - \sum_j \int_{-\infty}^{+\infty} \mathrm{d}t_1\, \Sigma_{ij}(t,t_1)\, c_j(t_1) = S_i(t) \quad \text{(A.4.4)}$$

where

$$S_i(t) \equiv \sum_{r,\alpha} \tau_{ir\alpha} C_r(t)\, B_\alpha(t) + \tau_{ri\alpha}^* C_r(t)\, B_\alpha^+(t) \quad \text{(A.4.5)}$$

$$\Sigma_{ij}(t,t_1) \equiv \vartheta(t-t_1)\Gamma_{ij}(t,t_1) \quad \text{(A.4.6)}$$

and

$$\Gamma_{ij}(t,t_1) = \sum_{r,s,\alpha,\beta} \tau_{ir\alpha}\, \tau_{js\beta}^* \left(G_{\alpha\beta}^{\mathrm{ab}}(t,t_1)\, G_{rs}^{\mathrm{n}}(t,t_1) + G_{rs}^{\mathrm{p}}(t,t_1)\, G_{\alpha\beta}^{\mathrm{em}}(t,t_1)\right)$$
$$+ \tau_{sj\beta}\, \tau_{ri\alpha}^* \left(G_{\beta\alpha}^{\mathrm{em}}(t_1,t)\, G_{rs}^{\mathrm{n}}(t,t_1) + G_{rs}^{\mathrm{p}}(t,t_1)\, G_{\beta\alpha}^{\mathrm{ab}}(t_1,t)\right) \quad \text{(A.4.7)}$$

Inscattering function: We evaluate the inscattering function

$$\Sigma_{ij}^{\mathrm{in}}(t,t') \equiv \langle S_j^+(t')\, S_i(t)\rangle$$

as before by substituting for the source term from Eq. (A.4.5)

$$\Sigma_{ij}^{\mathrm{in}}(t,t') = \sum_{r,s,\alpha,\beta} \tau_{ir\alpha}\, \tau_{js\beta}^* \langle C_s^+(t')\, C_r(t)\rangle \langle B_\beta^+(t')B_\alpha(t)\rangle$$
$$+ \tau_{sj\beta}\, \tau_{ri\alpha}^* \langle C_s^+(t')\, C_r(t)\rangle \langle B_\beta(t')B_\alpha^+(t)\rangle$$

which yields

$$\Sigma_{ij}^{\mathrm{in}}(t,t') = \sum_{r,s,\alpha,\beta} \tau_{ir\alpha}\, \tau_{js\beta}^* G_{rs}^{\mathrm{n}}(t,t')\, G_{\alpha\beta}^{\mathrm{ab}}(t,t') + \tau_{sj\beta}\, \tau_{ri\alpha}^* G_{rs}^{\mathrm{n}}(t,t')\, G_{\beta\alpha}^{\mathrm{em}}(t',t) \quad \text{(A.4.8)}$$

Steady state: Under steady-state conditions we Fourier transform with respect to the difference time coordinate to obtain:

$$\Gamma_{ij}(E) = \int_0^\infty \mathrm{d}(\hbar\omega)/2\pi$$
$$\times \sum_{r,s,\alpha,\beta} \tau_{ir\alpha}\, \tau_{js\beta}^* \left(G_{rs}^{\mathrm{n}}(E-\hbar\omega)G_{\alpha\beta}^{\mathrm{ab}}(\hbar\omega) + G_{rs}^{\mathrm{p}}(E-\hbar\omega)G_{\alpha\beta}^{\mathrm{em}}(\hbar\omega)\right)$$
$$+ \tau_{sj\beta}\, \tau_{ri\alpha}^* \left(G_{rs}^{\mathrm{n}}(E+\hbar\omega)G_{\beta\alpha}^{\mathrm{em}}(\hbar\omega) + G_{rs}^{\mathrm{p}}(E+\hbar\omega)G_{\beta\alpha}^{\mathrm{ab}}(\hbar\omega)\right) \quad \text{(A.4.9)}$$

from Eq. (A.4.7) and

$$\Sigma^{\text{in}}_{ij}(E) = \int_0^\infty \mathrm{d}(\hbar\omega)/2\pi$$

$$\times \sum_{r,s,\alpha,\beta} \tau_{ir\alpha}\,\tau^*_{js\beta} G^{\text{n}}_{rs}(E-\hbar\omega)\,G^{\text{ab}}_{\alpha\beta}(\hbar\omega) + \tau_{sj\beta}\,\tau^*_{ri\alpha} G^{\text{n}}_{rs}(E+\hbar\omega)\,G^{\text{em}}_{\beta\alpha}(\hbar\omega) \qquad \text{(A.4.10)}$$

from Eq. (A.4.8). If we assume the phonons to remain in equilibrium we can use Eqs. (A.1.15) to obtain Eq. (10.3.5) from Eq. (A.4.10), noting that the present $\tau_\alpha(\vec{x})$ is similar to $U^{\text{ab}}_\beta(\vec{r})$ in Section 10.3. Eq. (A.4.9) gives the full tensorial version of Eq. (10.3.9).

Current: Note that there is no need to rederive the expressions for the inflow and the outflow (see Eqs. (A.3.3)–(A.5.5)) since these were derived making use of Eq. (A.2.7) which is still valid (see Eq. (A.4.4)). What has changed is the expression for the self-energy $[\Sigma]$ (cf. Eqs. (A.2.8) and (A.4.7)) and the inscattering $[\Sigma^{\text{in}}]$ (cf. Eqs. (A.2.19) and (A.4.8)). But the basic expressions for the current in terms of these quantities remains the same.

A.5 Coulomb blockade/Kondo resonance

Next we consider a problem involving purely coherent coupling to the contact, but we now take into account the Coulomb interaction between the two spin levels inside the channel denoted by c and d (Fig. A.5.1). For simplicity, we will only consider an equilibrium problem (with one contact) and include one level of each type in this discussion:

$$i\hbar\frac{\mathrm{d}}{\mathrm{d}t}c = \varepsilon c + \sum_r \tau_r c_r + U d^+ d c \qquad \text{(A.5.1a)}$$

$$i\hbar\frac{\mathrm{d}}{\mathrm{d}t}c_r = \varepsilon_r c_r + \tau^*_r c \qquad \text{(A.5.1b)}$$

We can proceed as before to write Eq. (A.2.7) from Eq. (A.2.1). We now obtain the same result except for an additional term Ud^+dc:

$$i\hbar\frac{\mathrm{d}}{\mathrm{d}t}c - \varepsilon c - \int_{-\infty}^{+\infty} \mathrm{d}t_1\,\Sigma_0(t,t_1)\,c(t_1) = S^c(t) + U d^+(t)\,d(t)\,c(t) \qquad \text{(A.5.2)}$$

Fourier transforming we obtain

$$(E-\varepsilon-\Sigma_0)\,c = S^c + U\{d^+dc\}(E) \qquad \text{(A.5.3)}$$

$$\Sigma_0 \equiv \sum_r \frac{|\tau_r|^2}{E-\varepsilon_r+i\eta} \qquad S^c \equiv \sum_r \tau_r C_r$$

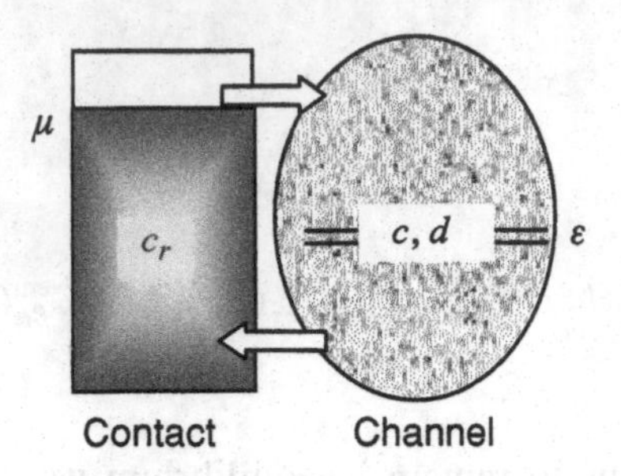

Fig. A.5.1

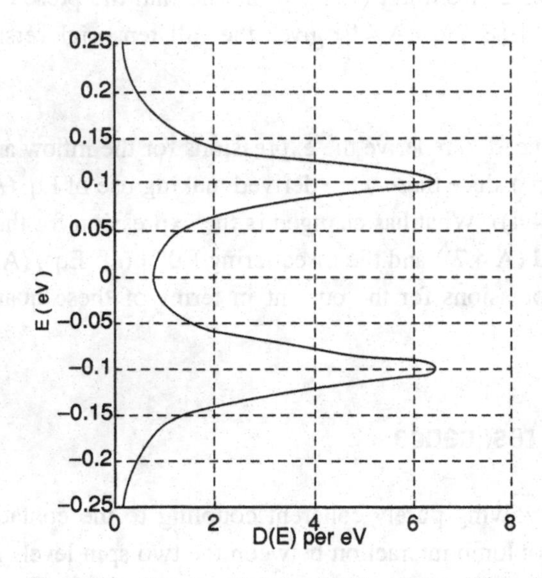

Fig. A.5.2 Density of states, $D(E) = \mathrm{i}[G - G^+]$ calculated using $G(E)$ from Eq. (A.5.12) with $\varepsilon = -0.1$ eV, $U = +0.2$ eV, $\langle d^+ d\rangle = 0.5$ and Σ_0 calculated from Eq. (A.5.3) assuming that the reservoir consists of a set of levels uniformly spaced by 0.4 meV over a range of energies from -1 eV to $+1$ eV, with $\eta = 1$ meV and $\tau_r = 1.8$ meV.

where we have added the braces { } around d^+dc to indicate that we need the Fourier transform of the product (which is different from the product of the individual Fourier transforms).

Similarly, we can work with the equations for the down-spin components:

$$\mathrm{i}\hbar\frac{\mathrm{d}}{\mathrm{d}t}d = \varepsilon d + \sum_r \tau_r d_r + Uc^+cd \tag{A.5.4a}$$

and

$$\mathrm{i}\hbar\frac{\mathrm{d}}{\mathrm{d}t}d_r = \varepsilon_r d_r + \tau_r^* d \tag{A.5.4b}$$

to obtain

$$(E - \varepsilon - \Sigma_0)d = S^d + U\{c^+cd\}(E) \tag{A.5.5}$$

where $S^d \equiv \Sigma_r \tau_r D_r$.

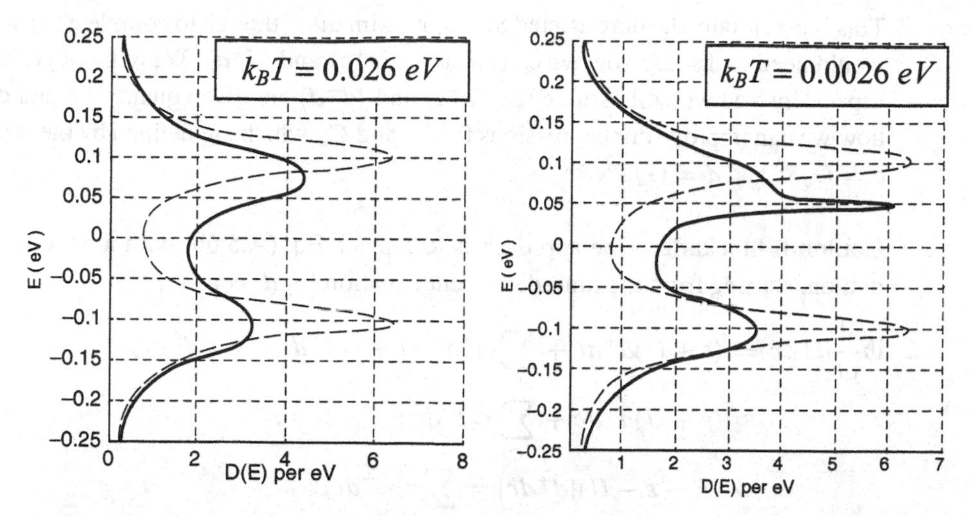

Fig. A.5.3 Density of states, $D(E) = \mathrm{i}[G - G^+]$ calculated using $G(E)$ from Eq. (A.5.20) with $\mu = 0.05\,\text{eV}$ and the same parameters as in Fig. A.5.2 (also shown for reference is the result using $G(E)$ from Eq. (A.5.12)). Note how the "Kondo peak" appears at $\mu = 0.05\,\text{eV}$ at the lower temperature (the author is grateful to A. W. Ghosh and D. Sen for their help).

What we propose to show in this section is that if we treat the new terms Ud^+dc and Uc^+cd using a self-consistent field method, we will get a single peak for the spectral function as shown in Fig. 1.5.1, but if we use a better approximation for these terms we will obtain two peaks as shown in Fig. A.5.2, which can be viewed as a combination of the two possibilities sketched in Fig. 1.5.2. If we take our approximation a step further we will obtain the central peak around $E = \mu$ as shown in Fig. A.5.3. This peak is responsible for the increased resistivity in bulk metals with magnetic impurities at low temperatures as explained in the 1960s by Kondo (see Kouwenhoven and Glazman, 2001). More recently, experiments on single quantum dots and molecules in the 1990s have revealed an enhanced conductance believed to arise from the same Kondo peak in the spectral function (Meir *et al.*, 1991).

Self-consistent field (SCF) approximation: The simplest approximation is to write

$$\{c^+cd\}(E) \approx \langle c^+c\rangle d(E) \tag{A.5.6a}$$

$$\{d^+dc\}(E) \approx \langle d^+d\rangle c(E) \tag{A.5.6b}$$

so that from Eqs. (A.5.3) and (A.5.5)

$$c = \underbrace{\frac{1}{E - \varepsilon - \Sigma_0 - U\langle d^+d\rangle}}_{G_c(E)} S^c \tag{A.5.7a}$$

$$d = \underbrace{\frac{1}{E - \varepsilon - \Sigma_0 - U\langle c^+c\rangle}}_{G_d(E)} S^d \tag{A.5.7b}$$

This is essentially the unrestricted SCF approximation, though to complete the story we would need to discuss how we can calculate $\langle c^+c\rangle$ and $\langle d^+d\rangle$. We will not go into this aspect. Instead we will assume that $\langle c^+c\rangle$ and $\langle d^+d\rangle$ are given quantities, and discuss how we can improve our expressions for G_c and G_d which are defined by the relations: $c = G_c S^c$ and $d = G_d S^d$.

Coulomb blockade: One approach is to replace Eq. (A.5.6b) with a better approximation, starting from the time-dependent equation for d^+dc,

$$i\hbar\frac{\mathrm{d}}{\mathrm{d}t}d^+dc = (\varepsilon + U)d^+dc + \sum_r(\tau_r^* d_r^+ cd + \tau_r d^+ d_r, c + \tau_r d^+ dc_r) \tag{A.5.8}$$

$$\approx (\varepsilon + U)d^+dc + \sum_r \tau_r d^+ dc_r \tag{A.5.9}$$

$$\rightarrow (E - \varepsilon - U)\{d^+dc\} = \sum_r \tau_r d^+ dc_r$$

and making use of Eq. (A.5.2) to write

$$\{d^+dc\}(E) = \sum_r \frac{\tau_r\langle d^+d\rangle C_r}{E - \varepsilon - U - \Sigma_0} = \frac{\langle d^+d\rangle}{E - \varepsilon - U - \Sigma_0}S^c \tag{A.5.10}$$

Substituting into Eq. (A.5.3) we obtain

$$(E - \varepsilon - \Sigma_0)c = \left(1 + \frac{U\langle d^+d\rangle}{E - \varepsilon - U - \Sigma_0}\right)S^c \tag{A.5.11}$$

from which we can write the Green's function, $G(E)$

$$c = \left(\frac{1}{E - \varepsilon - \Sigma_0} + \frac{U\langle d^+d\rangle}{(E - \varepsilon - \Sigma_0)(E - \varepsilon - U - \Sigma_0)}\right)S^c$$

$$= \underbrace{\left(\frac{1 - \langle d^+d\rangle}{E - \varepsilon - \Sigma_0} + \frac{\langle d^+d\rangle}{E - \varepsilon - U - \Sigma_0}\right)}_{G(E)}S^c \tag{A.5.12}$$

The density of states, $D(E) = i[G - G^+]$, calculated using the above $G(E)$ for a typical set of parameters is shown in Fig. A.5.2. This is similar to the picture we expect under Coulomb blockade conditions (see Fig. 1.5.2).

Kondo resonance: To go further we step back to Eq. (A.5.8)

$$i\hbar\frac{\mathrm{d}}{\mathrm{d}t}d^+dc = (\varepsilon + U)d^+dc + \sum_r(\tau_r^* d_r^+ cd + \tau_r d^+ d_r c + \tau_r d^+ dc_r)$$

$$\rightarrow (E - \varepsilon - U)\{d^+dc\} = \sum_r(\tau_r^*\{d_r^+ cd\} + \tau_r\{d^+ d_r c\} + \tau_r\{d^+ dc_r\}) \tag{A.5.13}$$

and try to do better than what we did in going to Eq. (A.5.9) by starting from the time-dependent equations for each of the three quantities appearing in Eqs. (A.5.13):

$$i\hbar \frac{d}{dt} d^+ dc_r = (\varepsilon_r - i\eta) d^+ dc_r + \sum_r (\tau_r^* d^+ dc + \tau_r d^+ d_r c_r + \tau_r^* d_r^+ c_r d)$$

$$\approx (\varepsilon_r - i\eta) d^+ dc_r + \sum_r \tau_r^* d^+ dc$$

$$\{d^+ dc_r\} \approx \langle d^+ d \rangle C_r + \sum_r \frac{\tau_r^*}{E - \varepsilon_r + i\eta} \{d^+ dc\} \quad \text{(A.5.14)}$$

$$i\hbar \frac{d}{dt} d^+ d_r c = \varepsilon_r d^+ d_r c + \sum_r (\tau_r^* d^+ dc + \tau_r d^+ d_r c_r - \tau_r^* d_r^+ d_r c)$$

$$\approx \varepsilon_r d^+ d_r c + \sum_r \tau_r^* d^+ dc - \tau_r^* f_r c$$

$$\{d^+ d_r c\} \approx \sum_r \frac{\tau_r^*}{E - \varepsilon_r + i\eta} \{d^+ dc\} - \frac{\tau_r^* f_r}{E - \varepsilon_r + i\eta} c \quad \text{(A.5.15)}$$

$$i\hbar \frac{d}{dt} d_r^+ cd = (2\varepsilon + U - \varepsilon_r - i\eta) d_r^+ cd + \sum_r (\tau_r d^+ dc - \tau_r d_r^+ d_r c - \tau_r d_r^+ c_r c)$$

$$\approx (2\varepsilon + U - \varepsilon_r - i\eta) d_r^+ cd + \sum_r (\tau_r d^+ dc - \tau_r f_r c)$$

$$\{d_r^+ cd\} \approx \sum_r \frac{\tau_r}{E - 2\varepsilon - U + \varepsilon_r + i\eta} \{d^+ dc\} - \frac{\tau_r f_r}{E - 2\varepsilon - U + \varepsilon_r + i\eta} c \quad \text{(A.5.16)}$$

Substituting Eqs. (A.5.14), (A.5.15) and (A.5.16) into Eq. (A.5.13) we obtain a better expression for $\{d^+ dc\}$ than previously (see Eq. (A.5.10))

$$(E - \varepsilon - U - 2\Sigma_0 - \Sigma_1)\{d^+ dc\} = \langle d^+ d \rangle S^c - (\Sigma_2 + \Sigma_3) c$$

$$\rightarrow \{d^+ dc\} = \frac{\langle d^+ d \rangle}{E - \varepsilon\ U - 2\Sigma_0 - \Sigma_1} S^c - \frac{\Sigma_2 + \Sigma_3}{E - \varepsilon - U - 2\Sigma_0 - \Sigma_1} c \quad \text{(A.5.17)}$$

where

$$\Sigma_1 \equiv \sum_r \frac{|\tau_r|^2}{E - 2\varepsilon - U + \varepsilon_r + i\eta}$$

$$\Sigma_2 \equiv \sum_r \frac{|\tau_r|^2 f_r}{E - \varepsilon_r + i\eta} \quad \text{(A.5.18)}$$

$$\Sigma_3 \equiv \sum_r \frac{|\tau_r|^2 f_r}{E - 2\varepsilon - U + \varepsilon_r + i\eta}$$

Substituting Eq. (A.5.17) back into Eq. (A.5.13) we obtain

$$\left(E - \varepsilon - \Sigma_0 + \frac{U(\Sigma_2 + \Sigma_3)}{E - \varepsilon - U - 2\Sigma_0 - \Sigma_1} \right) c$$

$$= \left(1 + \frac{U \langle d^+ d \rangle}{E - \varepsilon - U - 2\Sigma_0 - \Sigma_1} \right) S^c \quad \text{(A.5.19)}$$

so that the Green's function $G(E)$ (such that $c = GS^c$) is given by

$$G = \left[1 - \langle d^+d\rangle \middle/ \left(E - \varepsilon - \Sigma_0 + \frac{U(\Sigma_2 + \Sigma_3)}{E - \varepsilon - U - 2\Sigma_0 - \Sigma_1}\right)\right] + \left[\langle d^+d\rangle \middle/ \left(E - \varepsilon - \Sigma_0 - U - \frac{U(\Sigma_0 + \Sigma_1 - \Sigma_2 - \Sigma_3)}{E - \varepsilon - 2\Sigma_0 - \Sigma_1}\right)\right] \tag{A.5.20}$$

The density of states $D(E) = \mathrm{i}[G - G^+]$ calculated using $G(E)$ from Eq. (A.5.20) with the same parameters as in Fig. A.5.2 is shown in Fig. A.5.3 (also shown for reference is the result using $G(E)$ from Eq. (A.5.12)). Note how the central "Kondo peak" grows in amplitude as the temperature is lowered. Indeed at low temperatures and also for stronger coupling to the reservoir, the treatment presented above may be inadequate. It may be necessary to go to higher orders (perhaps infinite!) in perturbation theory and for many years this has been one of the major challenges in many-electron theory.

Summary: We have derived the expressions for the current starting from the second quantized description of the composite channel–contact system and eliminating the contact variables to obtain an effective equation for the channel having the form

$$\mathrm{i}\hbar\frac{\mathrm{d}}{\mathrm{d}t}c - Hc - \underbrace{\Sigma c}_{\text{Outflow}} = \underbrace{S}_{\text{Inflow}}$$

It is possible to base the quantum mechanical treatment on the one-electron wavefunction instead of second quantized operators, but this does not provide satisfactory answers to subtle issues like the presence or absence of Pauli blocking. For example, it is natural to ask if Σ in the outflow term should depend on whether the final state (to which outflow occurs) is empty or full. The answer is usually "no" for coherent processes, but not always, as for example in the Kondo effect. Such questions are difficult to answer from the one-electron picture, but the answers come out naturally in a second quantized treatment. In strongly interacting systems it may be desirable to use similar approaches to write transport equations not for the bare electron described by c, but for "composite" or dressed particles obtained through appropriate unitary transformations that include part of the "contact."

MATLAB codes used to generate text figures

 It is planned to make "soft copies" of these codes available through my website and also through the nanohub where they can be executed without installation using a web browser.

Chapter 1

% Fig.1.1.1

```
clear all

%Constants (all MKS, except energy which is in eV)
hbar=1.055e-34;q=1.602e-19;eps0=8.854E-12;epsr=4;m=0.25*9.11e-31;%Effective mass
I0=q*q/hbar;

%Parameters
W=1e-6;L=10e-9;t=1.5e-9;%W=Width,L=Length of active region,t=oxide thickness
Cg=epsr*eps0*W*L/t;Cs=0.05*Cg;Cd=0.05*Cg;CE=Cg+Cs+Cd;U0=q/CE;
alphag=Cg/CE,alphad=Cd/CE
    %alphag=1;alphad=0.5;U0=0.25;

kT=0.025;mu=0;ep=0.2;
    v=1e5;%Escape velocity
        g1=hbar*v/(q*L);g2=g1;g=g1+g2;
            %g1=0.005;g2=0.005;g=g1+g2;

%Energy grid
NE=501;E=linspace(-1,1,NE);dE=E(2)-E(1);
    D0=m*q*W*L/(pi*hbar*hbar);% Step Density of states per eV
    D=D0*[zeros(1,251) ones(1,250)];
    %D=(2*g/(2*pi))./((E.^2)+((g/2)^2));% Lorentzian Density of states per eV
        %D=D./(dE*sum(D));%Normalizing to one

%Reference number of electrons
f0=1./(1+exp((E+ep-mu)./kT));N0=2*dE*sum(D.*f0);ns=N0/(L*W*1e4),%/cm^2

%Bias
IV=61;VV=linspace(0,0.6,IV);
for iV=1:IV
    Vg=0.5;Vd=VV(iV);
```

```
    %Vd=0.5;Vg=VV(iV);
        mu1=mu;mu2=mu1-Vd;UL=-(alphag*Vg)-(alphad*Vd);

U=0;%Self-consistent field
dU=1;
while dU>1e-6
    f1=1./(1+exp((E+UL+U+ep-mu1)./kT));
        f2=1./(1+exp((E+UL+U+ep-mu2)./kT));
    N(iV)=dE*sum(D.*((f1.*g1/g)+(f2.*g2/g)));
        Unew=U0*(N(iV)-N0);dU=abs(U-Unew);
            U=U+0.1*(Unew-U);
end
I(iV)=dE*I0*(sum(D.*(f1-f2)))*g1*g2/g;
end

hold on
h=plot(VV,I,'b');
set(h,'linewidth',[2.0])
set(gca,'Fontsize',[25])
xlabel(' Voltage (V) --->')
ylabel(' Current (A) ---> ')
grid on
```

% Fig.1.1.3

```
clear all

E=linspace(-.25,.25,501);dE=E(2)-E(1);kT=0.025;Ef=0;
V=0;mu1=Ef+(V/2);mu2=Ef-(V/2);
f1=1./(1+exp((E-mu1)./kT));f2=1./(1+exp((E-mu2)./kT));
%dE*(sum(f1-f2))/V

hold on
h=plot(f1,E,'g');
set(h,'linewidth',[2.0])
set(gca,'Fontsize',[25])
xlabel(' Fermi function --->')
ylabel(' E - mu (eV) ---> ')
grid on
```

% Fig.1.3.3, 1.5.1

```
clear all

E=linspace(-.5,.5,50001);dE=E(2)-E(1);gam=0.05;
D=(gam/(2*pi))./((E.^2)+((gam/2)^2));
%D=(gam/(2*pi))./(((E-0.25).^2)+((gam/2)^2));%Use for Fig.1.5.2
%D=D+((gam/(2*pi))./(((E+0.25).^2)+((gam/2)^2)));%Use for Fig.1.5.2
dE*sum(D)

hold on
h=plot(D,E,'g');
```

```
set(h,'linewidth',[2.0])
set(gca,'Fontsize',[25])
xlabel(' D (E) --->')
ylabel(' E (eV) ---> ')
grid on
```

% Fig.1.4.6

```
clear all

%Constants (all MKS, except energy which is in eV)
hbar=1.055e-34;q=1.602e-19;I0=q*q/hbar;

%Parameters
U0=0.025;kT=0.025;mu=0;ep=0.2;
g1=0.005;g2=0.005;g=g1+g2;
alphag=1;alphad=0.5;

%Energy grid
NE=501;E=linspace(-1,1,NE);dE=E(2)-E(1);
D=(g/(2*pi))./((E.^2)+((g/2)^2));% Lorentzian Density of states per eV
D=D./(dE*sum(D));%Normalizing to one

%Bias
IV=101;VV=linspace(0,1,IV);
for iV=1:IV
    Vg=0;Vd=VV(iV);
    %Vd=0;Vg=VV(iV);
        mu1=mu;mu2=mu1-Vd;UL=-(alphag*Vg)-(alphad*Vd);

U=0;%Self-consistent field
dU=1;
while dU>1e-6
    f1=1./(1+exp((E+ep+UL+U-mu1)./kT));
        f2=1./(1+exp((E+ep+UL+U-mu2)./kT));
    N(iV)=dE*sum(D.*((f1.*g1/g)+(f2.*g2/g)));
        Unew=U0*N(iV);dU=abs(U-Unew);
            U=U+0.1*(Unew-U);
end
I(iV)=dE*I0*(sum(D.*(f1-f2)))*(g1*g2/g);
end

hold on
h=plot(VV,N,'b');
%h=plot(VV,I,'b');
set(h,'linewidth',[2.0])
set(gca,'Fontsize',[25])
xlabel(' Voltage ( V ) --->')
%ylabel(' Current ( A ) ---> ')
ylabel(' Number of electrons ---> ')
grid on
```

% Fig.E.1.3

```
clear all

%Constants (all MKS, except energy which is in eV)
hbar=1.055e-34;q=1.602e-19;I0=q*q/hbar;

%Parameters
U0=0.025;kT1=0.026;kT2=0.025;ep=0.2;
g1=0.005;g2=0.005;g=g1+g2;
alphag=1;alphad=0.5;

%Energy grid
NE=501;E=linspace(-1,1,NE);dE=E(2)-E(1);
g1=0.005*(E+abs(E))./(E+E+1e-6);% zero for negative E
g2=0.005*ones(1,NE);g1=g2;
g=g1+g2;

%Bias
IV=101;VV=linspace(-0.25,0.25,IV);
for iV=1:IV
    mu1=ep+VV(iV);mu2=mu1;
        f1=1./(1+exp((E-mu1)./kT1));
        f2=1./(1+exp((E-mu2)./kT2));
        D=(g./(2*pi))./(((E-ep).^2)+((g./2).^2));
            D=D./(dE*sum(D));
I(iV)=dE*2*I0*(sum(D.*(f1-f2).*g1.*g2./g));
end

hold on
%h=plot(VV,N/2,'b');%Part (a)
h=plot(VV,I,'b');
set(h,'linewidth',[2.0])
set(gca,'Fontsize',[25])
xlabel(' Voltage ( V ) --->')
ylabel(' Current ( A ) ---> ')
%ylabel(' Number of electrons ---> ')
grid on
```

% Fig.E.1.4

```
clear all

%Constants (all MKS, except energy which is in eV)
hbar=1.055e-34;q=1.602e-19;I0=q*q/hbar;

%Parameters
U0=0.025;kT=0.025;mu=0;ep=0.2;N0=0;
g1=0.005;g2=0.005;g=g1+g2;
alphag=1;alphad=0.5;

%Energy grid
NE=501;E=linspace(-1,1,NE);dE=E(2)-E(1);
```

```
g1=0.005*(E+abs(E))./(E+E+1e-6);% zero for negative E
g2=0.005*ones(1,NE);
g=g1+g2;

%Bias
IV=101;VV=linspace(-.6,.6,IV);
for iV=1:IV
    Vg=0;Vd=VV(iV);
    %Vd=0;Vg=VV(iV);
        mu1=mu;mu2=mu1-Vd;UL=-(alphag*Vg)-(alphad*Vd);

U=0;%Self-consistent field
dU=1;
while dU>1e-6
    f1=1./(1+exp((E-mu1)./kT));
        f2=1./(1+exp((E-mu2)./kT));
        D=(g./(2*pi))./(((E-ep-UL-U).^2)+((g./2).^2));
            D=D./(dE*sum(D));
    N(iV)=dE*2*sum(D.*((f1.*g1./g)+(f2.*g2./g)));
        Unew=U0*(N(iV)-N0);dU=abs(U-Unew);
            U=U+0.1*(Unew-U);
end
I(iV)=dE*2*I0*(sum(D.*(f1-f2).*g1.*g2./g));
end

hold on
%h=plot(VV,N/2,'b');%Part (a)
h=plot(VV,I,'b');
set(h,'linewidth',[2.0])
set(gca,'Fontsize',[25])
xlabel(' Voltage (V) --->')
ylabel(' Current (A) ---> ')
%ylabel(' Number of electrons ---> ')
grid on
```

Chapter 2

% Fig.2.3.2a, b

```
clear all

%Constants (all MKS, except energy which is in eV)
hbar=1.055e-34;m=9.110e-31;epsil=8.854e-12;q=1.602e-19;

%Lattice
Np=100;a=1e-10;X=a*[1:1:Np];t0=(hbar^2)/(2*m*(a^2))/q;L=a*(Np+1);
T=(2*t0*diag(ones(1,Np)))-(t0*diag(ones(1,Np-1),1))-(t0*diag(ones(1,Np-1),-1));

[V,D]=eig(T);D=diag(D);[Enum,ind]=sort(D);

E1=D(ind(1));psi1=abs(V(:,ind(1)));P1=psi1.*conj(psi1);
E2=D(ind(25));psi2=abs(V(:,ind(25)));P2=psi2.*conj(psi2);
```

```
%analytical eigenvalues
Ean=(((hbar*pi)^2)/(2*m*(L^2))/q)*[1:Np].*[1:Np];

hold on
%h=plot(Enum,'bx');% Part (a)
%h=plot(Ean,'b');% Part (a)
h=plot(P1,'b');% Part (b)
h1=plot(P2,'b');% Part (b)
set(h,'linewidth',[3.0])
set(h1,'linewidth',[1.0])
set(gca,'Fontsize',[25])

%xlabel(' Eigenvalue Number , alpha --->');% Part (a)
%ylabel(' E (eV) ---> ');% Part (a)
xlabel(' Lattice site # --->');% Part (b)
ylabel(' Probability ---> ');% Part (b)
grid on
```

% Fig.2.3.5

```
clear all

%Constants (all MKS, except energy which is in eV)
hbar=1.055e-34;m=9.110e-31;epsil=8.854e-12;q=1.602e-19;

%Lattice
Np=100;a=1e-10;X=a*[1:1:Np];t0=(hbar^2)/(2*m*(a^2))/q;L=a*(Np+1);
T=(2*t0*diag(ones(1,Np)))-(t0*diag(ones(1,Np-1),1))-(t0*diag(ones(1,Np-1),-1));
T(1,Np)=-t0;T(Np,1)=-t0;

[V,D]=eig(T);D=diag(D);[Enum,ind]=sort(D);

E1=D(ind(1));psi1=abs(V(:,ind(1)));P1=psi1.*conj(psi1);
E2=D(ind(50));psi2=abs(V(:,ind(50)));P2=psi2.*conj(psi2);

%analytical eigenvalues
Ean=(((hbar*pi)^2)/(2*m*(L^2))/q)*[1:Np].*[1:Np];

hold on
h=plot(Enum,'bx');
set(h,'linewidth',[3.0])
set(gca,'Fontsize',[25])
xlabel(' Eigenvalue Number, alpha --->');
ylabel(' E (eV) ---> ');
grid on
```

% Fig.2.3.6, 2.3.7

```
clear all

%Constants (all MKS, except energy which is in eV)
hbar=1.055e-34;m=9.110e-31;epsil=8.854e-12;q=1.602e-19;
a0=4*pi*epsil*hbar*hbar/(m*q*q),E0=q/(8*pi*epsil*a0)
```

```
%Lattice
Np=100;a=(5e-10*2/Np);% *1 for Fig.1.3.6 and *2 for Fig.1.3.7
R=a*[1:1:Np];t0=(hbar^2)/(2*m*(a^2))/q;

%Quantum numbers
n=1;l=0;% for 1s, n=1 and for 2s, n=2

%Hamiltonian,H = Kinetic,K + Potential,U
K=(2*t0*diag(ones(1,Np)))-(t0*diag(ones(1,Np-1),1))-(t0*diag(ones(1,Np-1),-1));
U=((-q/(4*pi*epsil)./R)+(l*(l+1)*hbar*hbar/(2*m*q))./(R.*R));U=diag(U);
[V,D]=eig(K+U);D=diag(D);[DD,ind]=sort(D);
E=D(ind(n-l));psi=V(:,ind(n-l));
P=psi.*conj(psi);[-E0/(n^2) E]

%analytical solutions
P1s=(4*a/(a0^3))*R.*R.*exp(-2*R./a0);
P2s=(4*a/(2*4*4*(a0^3)))*R.*R.*((2-(R./a0)).^2).*exp(-2*R./(2*a0));
P3s=(4*a/(3*81*81*(a0^3)))*R.*R.*((27-(18*R./a0)+(2*(R./a0).^2)).^2).*exp(-2*R./(3*a0));
P2p=(4*a/(3*32*(a0^3)))*R.*R.*((R./a0).^2).*exp(-2*R./(2*a0));
P3p=(8*a/(3*81*81*(a0^3)))*R.*R.*((6-(R./a0)).^2).*((R./a0).^2).*exp(-2*R./(3*a0));

hold on
h=plot(R,P,'b');
h=plot(R,P1s,'bx');% use P1s for '1s' and P2s for '2s'
set(h,'linewidth',[2.0])
set(gca,'Fontsize',[25])
xlabel(' x (m ) --->');
ylabel(' Probability ---> ');
grid on
```

Chapter 3

% Fig.3.1.4

```
clear all

%Constants (all MKS, except energy which is in eV)
hbar=1.055e-34;m=9.110e-31;epsil=8.854e-12;q=1.602e-19;
%Lattice
Np=200;a=(10e-10/Np);R=a*[1:1:Np];t0=(hbar^2)/(2*m*(a^2))/q;

%Hamiltonian,H = Kinetic,T + Potential,U + Uscf
T=(2*t0*diag(ones(1,Np)))-(t0*diag(ones(1,Np-1),1))-(t0*diag(ones(1,Np-1),-1));
UN=(-q*2/(4*pi*epsil))./R;% Z=2 for Helium

Uscf=zeros(1,Np);change=1;
while change>0.01
    [V,D]=eig(T+diag(UN+Uscf));D=diag(D);[DD,ind]=sort(D);
    E=D(ind(1));psi=V(:,ind(1));P=psi.*conj(psi);P=P';
        Unew=(q/(4*pi*epsil))*((sum(P./R)-cumsum(P./R))+(cumsum(P)./R));
        change=sum(abs(Unew-Uscf))/Np,Uscf=Unew;
end
```

```
%analytical solutions for 1s hydrogen
a0=4*pi*epsil*hbar*hbar/(m*q*q);
P0=(4*a/(a0^3))*R.*R.*exp(-2*R./a0);

hold on
%h=plot(R,UN,'b');% Part (a)
%h=plot(R,Uscf,'b');% Part(a)
h=plot(R,P,'b');% Part (b)
h=plot(R,P0,'bx');% Part (b)
set(h,'linewidth',[2.0])
set(gca,'Fontsize',[25])
xlabel(' R ( m ) --->');
%ylabel(' U ( eV ) ---> ');% Part (a)
%axis([0 1e-9 -100 20]);% Part (a)
ylabel(' Probability ---> ');% Part (b)
axis([0 1e-9 0 0.1]);% Part (b)
grid on
```

% Fig.3.1.5

```
clear all

%Constants (all MKS, except energy which is in eV)
hbar=1.055e-34;m=9.110e-31;epsil=8.854e-12;q=1.602e-19;

%Lattice
Np=200;a=(10e-10/Np);R=a*[1:1:Np];t0=(hbar^2)/(2*m*(a^2))/q;

%Hamiltonian,H = Kinetic,T + Potential,U + Ul + Uscf
T=(2*t0*diag(ones(1,Np)))-(t0*diag(ones(1,Np-1),1))-(t0*diag(ones(1,Np-1),-1));
UN=(-q*14/(4*pi*epsil))./R;% Z=14 for silicon
l=1;Ul=(l*(l+1)*hbar*hbar/(2*m*q))./(R.*R);

Uscf=zeros(1,Np);change=1;
while change>0.1
    [V,D]=eig(T+diag(UN+Uscf));D=diag(D);[DD,ind]=sort(D);
    E1s=D(ind(1));psi=V(:,ind(1));P1s=psi.*conj(psi);P1s=P1s';
    E2s=D(ind(2));psi=V(:,ind(2));P2s=psi.*conj(psi);P2s=P2s';
    E3s=D(ind(3));psi=V(:,ind(3));P3s=psi.*conj(psi);P3s=P3s';

    [V,D]=eig(T+diag(UN+Ul+Uscf));D=diag(D);[DD,ind]=sort(D);
    E2p=D(ind(1));psi=V(:,ind(1));P2p=psi.*conj(psi);P2p=P2p';
    E3p=D(ind(2));psi=V(:,ind(2));P3p=psi.*conj(psi);P3p=P3p';
    n0=(2*(P1s+P2s+P3s))+(6*P2p)+(2*P3p);

    n=n0*(13/14);
    Unew=(q/(4*pi*epsil))*((sum(n./R)-cumsum(n./R))+(cumsum(n)./R));
    %Uex=(-q/(4*pi*epsil))*((n./(4*pi*a*R.*R)).^(1/3));%Unew=Unew+Uex;
    change=sum(abs(Unew-Uscf))/Np,Uscf=Unew;
end

[E1s E2s E2p E3s E3p]
```

```
%analytical solution for 1s hydrogen
a0=4*pi*epsil*hbar*hbar/(m*q*q);
P0=(4*a/(a0^3))*R.*R.*exp(-2*R./a0);

hold on
h=plot(R,P1s,'b');
h=plot(R,P0,'bx');
h=plot(R,P3p,'bo');
set(h,'linewidth',[2.0])
set(gca,'Fontsize',[25])
xlabel(' R ( m ) --->');
ylabel(' Probability ---> ');
axis([0 5e-10 0 0.08]);
grid on
```

% Fig.3.3.4

```
clear all

%Constants (all MKS, except energy which is in eV)
hbar=1.055e-34;m=9.110e-31;epsil=8.854e-12;q=1.602e-19;
a0=4*pi*epsil*hbar*hbar/(m*q*q);E0=q/(8*pi*epsil*a0);

R0=.05*[1:200];
a=(-2*E0)*(1-(exp(-2*R0).*(1+R0)))./R0;
b=(-2*E0)*exp(-R0).*(1+R0);
s=(1+R0+((R0.^2)/3)).*exp(-R0);
Uee=(2*E0)./sqrt(1+(R0.^2));UNN=(2*E0)./R0;

EB0=(a+b)./(1+s);R=a0*R0;

hold on
h=plot(R,EB0,'b--');
h=plot(R,Uee,'bx');
h=plot(R,UNN,'b');
h=plot(R,(2*EB0)+UNN+Uee,'b+');
set(h,'linewidth',[2.0])
set(gca,'Fontsize',[25])
grid on
xlabel(' R ( m ) --->')
ylabel(' Energy (eV) ---> ')
axis([0 4e-10 -25 25])
```

% Fig.3.4.2

```
clear all

%Constants (all MKS, except energy which is in eV)
hbar=1.055e-34;q=1.602e-19;I0=q*q/hbar;

%Parameters
U0=0.5;% U0 is 0.25 for part(a), 0.1 for part (b)
```

```
kT=0.025;mu=0;ep=0.2;
g1=0.005;g2=0.005;g=g1+g2;
alphag=1;alphad=0.5;

%Bias
IV=101;VV=linspace(0,1,IV);
for iV=1:IV
    Vd=0;Vg=VV(iV);
        mu1=mu;mu2=mu1-Vd;UL=-(alphag*Vg)-(alphad*Vd);
f1=1/(1+exp((ep+UL-mu1)/kT));f2=1/(1+exp((ep+UL-mu2)/kT));
f1U=1/(1+exp((ep+UL+U0-mu1)/kT));f2U=1/(1+exp((ep+UL+U0-mu2)/kT));

P1=((g1*f1)+(g2*f2))/(1e-6+(g1*(1-f1))+(g2*(1-f2)));
P2=P1*((g1*f1U)+(g2*f2U))/(1e-6+(g1*(1-f1U))+(g2*(1-f2U)));
P0=1/(1+P1+P1+P2);P1=P1*P0;P2=P2*P0;
p0(iV)=P0;p1(iV)=P1;p2(iV)=P2;
end

hold on
h=plot(VV,p0,'bo');
h=plot(VV,p1,'b');
h=plot(VV,p2,'bx');
set(h,'linewidth',[2.0])
set(gca,'Fontsize',[25])
grid on
xlabel(' Gate voltage, VG ( volts ) --->')
ylabel(' Current ( Amperes ) ---> ')
axis([0 1 0 1])
```

% Fig.3.4.3

```
clear all

%Constants (all MKS, except energy which is in eV)
hbar=1.055e-34;q=1.602e-19;I0=q*q/hbar;

%Parameters
U0=0.1;% U0 is 0.25 for part(a), 0.025 for part (b)
kT=0.025;mu=0;ep=0.2;
g1=0.005;g2=0.005;g=g1+g2;
alphag=1;alphad=0.5;

%Bias
IV=101;VV=linspace(0,1.5,IV);
for iV=1:IV
    Vg=0;Vd=VV(iV);
    %Vd=0;Vg=VV(iV);
        mu1=mu;mu2=mu1-Vd;UL=-(alphag*Vg)-(alphad*Vd);

%Multielectron method
f1=1/(1+exp((ep+UL (-U0/2)-mu1)/kT));f2=1/(1+exp((ep+UL(-U0/2)-mu2)/kT));
f1U=1/(1+exp((ep+UL+(U0/2)-mu1)/kT));f2U=1/(1+exp((ep+UL+(U0/2)-mu2)/kT));
```

```
P1=((g1*f1)+(g2*f2))/(1e-6+(g1*(1-f1))+(g2*(1-f2)));
P2=P1*((g1*f1U)+(g2*f2U))/(1e-6+(g1*(1-f1U))+(g2*(1-f2U)));
P0=1/(1+P1+P1+P2);P1=P1*P0;P2=P2*P0;

I1(iV)=2*I0*((P0*g1*f1)-(P1*g1*(1-f1))+(P1*g1*f1U)-(P2*g1*(1-f1U)));
I2(iV)=2*I0*((P0*g2*f2)-(P1*g2*(1-f2))+(P1*g2*f2U)-(P2*g2*(1-f2U)));
end

%RSCF method (same as Fig.1.4.6 with added factor of two)
%Energy grid
NE=501;E=linspace(-1,1,NE);dE=E(2)-E(1);
D=(g/(2*pi))./((E.^2)+((g/2)^2));% Lorentzian Density of states per eV
D=D./(dE*sum(D));%Normalizing to one

%Bias
for iV=1:IV
    Vg=0;Vd=VV(iV);
    %Vd=0;Vg=VV(iV);
        mu1=mu;mu2=mu1-Vd;UL=-(alphag*Vg)-(alphad*Vd);

U=0;%Self-consistent field
dU=1;
while dU>1e-6
    F1=1./(1+exp((E+ep+UL+U-mu1)./kT));
        F2=1./(1+exp((E+ep+UL+U-mu2)./kT));
    N(iV)=dE*2*sum(D.*((F1.*g1/g)+(F2.*g2/g)));
        Unew=U0*N(iV);
        dU=abs(U-Unew);U=U+0.1*(Unew-U);
end
I(iV)=dE*2*I0*(sum(D.*(F1-F2)))*(g1*g2/g);
end

hold on
h=plot(VV,I1,'b');
h=plot(VV,I,'b--');
set(h,'linewidth',[2.0])
set(gca,'Fontsize',[25])
grid on
xlabel(' Drain Voltage, VD ( volts ) --->')
ylabel(' Current ( Amperes ) ---> ')
axis([0 1.5 0 1.4e-6])

%E.3.5c: Unrestricted scf
clear all

%Constants (all MKS, except energy which is in eV)
hbar=1.055e-34;q=1.602e-19;I0=q*q/hbar;

%Parameters
U0=0.25;kT=0.025;mu=0;ep=0.2;
g1=0.005;g2=0.005;g=g1+g2;
alphag=1;alphad=0.5;
```

```
%Energy grid
NE=501;E=linspace(-1,1,NE);dE=E(2)-E(1);
D=(g/(2*pi))./((E.^2)+((g/2)^2));% Lorentzian Density of states per eV
D=D./(dE*sum(D));%Normalizing to one

%Bias
IV=101;VV=linspace(0,1,IV);
for iV=1:IV
    Vg=0;Vd=VV(iV);
    %Vd=0;Vg=VV(iV);
        mu1=mu;mu2=mu1-Vd;UL=-(alphag*Vg)-(alphad*Vd);

Uup=0;Udn=0.1;%Unrestricted self-consistent field
dU=1;while dU>.001
    f1up=1./(1+exp((E+ep+UL+Uup-mu1)./kT));
        f2up=1./(1+exp((E+ep+UL+Uup-mu2)./kT));
            Nup(iV)=dE*sum(D.*((f1up.*g1)+(f2up.*g2))./(g1+g2));
    f1dn=1./(1+exp((E+ep+UL+Udn-mu1)./kT));
        f2dn=1./(1+exp((E+ep+UL+Udn-mu2)./kT));
            Ndn(iV)=dE*sum(D.*((f1dn.*g1)+(f2dn.*g2))./(g1+g2));
            Udnnew=2*U0*(Nup(iV)-0.5);Udn=Udn+0.1*(Udnnew-Udn);
            Uupnew=2*U0*(Ndn(iV)-0.5);Uup=Uup+0.1*(Uupnew-Uup);
                dU=abs(Uup-Uupnew)+abs(Udn-Udnnew);
end
Iup(iV)=dE*I0*sum(D.*(f1up-f2up))*(g1*g2/g);
Idn(iV)=dE*I0*sum(D.*(f1dn-f2dn))*(g1*g2/g);
end

hold on
%h=plot(VV,Nup,'bo');%Part (b)
%h=plot(VV,Ndn,'bx');%Part (b)
h=plot(VV,Iup+Idn,'b');
set(h,'linewidth',[2.0])
set(gca,'Fontsize',[25])
xlabel(' Voltage (V) --->')
ylabel(' Current (A) ---> ')
%ylabel(' Number of electrons ---> ');%Part (b)
grid on
```

Chapter 4

% Fig.4.1.4

```
clear all

%Constants (all MKS, except energy which is in eV)
hbar=1.055e-34;m=9.110e-31;epsil=8.854e-12;q=1.602e-19;
a0=4*pi*epsil*hbar*hbar/(m*q*q);E0=q/(8*pi*epsil*a0);
```

```
%Basis
L=.074e-9/a0;s=exp(-L)*(1+L+((L^2)/3));
r=linspace(-2e-10,+2e-10,101);r0=r/a0;
psi=sqrt(1/(pi*(a0^3)))*(exp(-abs(r0-(L/2)))+exp(-abs(r0+(L/2))));
n=2*psi.*conj(psi)./(2*(1+s));

a=-2*E0*(1-((1+L)*exp(-2*L)))/L;
b=-2*E0*(1+L)*exp(-L);
EB0=-E0+((a+b)/(1+s));
[a b s EB0]

hold on
h=plot(r,n,'b');
set(h,'linewidth',[2.0])
set(gca,'Fontsize',[25])
grid on
xlabel(' R ( m ) --->')
ylabel(' Electron density ( /m^3 ) ---> ')
axis([-2e-10 2e-10 0 2e30])
```

% Fig.4.3.1

```
clear all

%Constants (all MKS, except energy which is in eV)
hbar=1.055e-34;m=9.110e-31;q=1.602e-19;mu=0.25;
kT=0.025;% 0.025 for Part (c),(e) and 0.0025 for Part (d),(f)

%Lattice
Np=100;a=2e-10;X=a*[1:1:Np];t0=(hbar^2)/(2*m*(a^2))/q;U=linspace(0,0,Np);
T=(2*t0*diag(ones(1,Np)))-(t0*diag(ones(1,Np-1),1))-(t0*diag(ones(1,Np-1),-1));
T(1,Np)=-t0;T(Np,1)=-t0;%Periodic boundary conditions for Parts (d), (f)
U(Np/2)=U(Np/2)+10;%Impurity potential with Parts (d), (f)

[V,D]=eig(T+diag(U));E=sort(diag(D)');
D=diag(D)-mu;
rho=1./(1+exp(D./kT));rho=V*diag(rho)*V';rho=diag(rho)/a;

hold on
grid on
%h=plot(E,'b');h=plot(mu*ones(Np/2,1),'b');% Part (b)
h=plot(X,rho,'b');% Part (c)-(f)
set(h,'linewidth',[2.0])
set(gca,'Fontsize',[25])
grid on
%xlabel(' Eigenvalues number --->');% Part (b)
%ylabel(' Energy ( eV ) ---> ');% Part (b)
xlabel(' x ( m ) --->');% Part (c)-(f)
ylabel(' Electron density ( /m^3 ) ---> ');% Part (c)-(f)
```

```
%axis([0 100 0 4]);% Part (b)
axis([0 2e-8 0 1e9]);% Part (c)-(f)
```

% Fig.4.4.1, 4.4.2

```
clear all

%Constants (all MKS, except energy which is in eV)
hbar=1.055e-34;m=9.110e-31;epsil=8.854e-12;q=1.602e-19;
a0=4*pi*epsil*hbar*hbar/(m*q*q);E0=q/(8*pi*epsil*a0);

F=linspace(0,1e9,11);A=(a0*128*sqrt(2)/243)*F;B=(-3*a0)*F;
for kF=1:11
M=[-E0 0 A(kF);0 -E0/4 B(kF);A(kF) B(kF) -E0/4];
[V,D]=eig(M);D=diag(D);[DD,ind]=sort(D);
E1(kF)=D(ind(1));E2(kF)=D(ind(2));E3(kF)=D(ind(3));
end

%perturbation theory results
E1s=-E0-((A.^2)/(3*E0/4));
E2s=-(E0/4)+B;
E2p=-(E0/4)-B;

hold on
%h=plot(F,E1,'b');% Fig.3.4.1
%h=plot(F,E1s,'bx');% Fig.3.4.1
h=plot(F,E2,'b');% Fig.3.4.2
h-plot(F,E3,'b');% Fig.3.4.2
h=plot(F,E2s,'bx');% Fig.3.4.2
h=plot(F,E2p,'bo');% Fig.3.4.2
set(h,'linewidth',[2.0])
set(gca,'Fontsize',[25])
grid on
xlabel(' Field ( V/m ) --->');
ylabel(' Energy ( eV ) ---> ');
%axis([0 2e-8 0 1e9]);
```

Chapter 5

% Fig.5.1.5

```
clear all

k=linspace(-1,1,21);a=2;b=1;
E1=sqrt((a^2)+(b^2)+(2*a*b.*cos(pi*k)));

hold on
h=plot(k,E1,'b');
h=plot(k,-E1,'b');
set(h,'linewidth',[2.0])
set(gca,'Fontsize',[25])
xlabel('k (in units of pi/a)--->')
```

```
ylabel('Energy (eV) ---> ')
grid on
```

% Fig.5.3.2

```
clear all

Esa=-8.3431;Epa=1.0414;Esc=-2.6569;Epc=3.6686;Esea=8.5914;Esec=6.7386;
Vss=-6.4513;Vxx=1.9546;Vxy=5.0779;Vsapc=4.4800;Vpasc=5.7839;Vseapc=4.8422;
Vpasec=4.8077;

%Either of the following choices for d1,d2,d3 and d4 should give the same result.
d1=[1 1 1]/4;d2=[1 -1 -1]/4;d3=[-1 1 -1]/4;d4=[-1 -1 1]/4;
d1=[0 0 0]/2;d2=[0 -1 -1]/2;d3=[-1 0 -1]/2;d4=[-1 -1 0]/2;

l=1;m=1;n=1;kmax=pi;Nt=21;%L-direction
%l=1;m=0;n=0;kmax=2*pi;Nt=21;%X-direction

for Nk=1:Nt
k=[l m n]*kmax*(Nk-1)/(Nt-1);
    p1=exp(i*sum(k.*d1));p2=exp(i*sum(k.*d2));
    p3=exp(i*sum(k.*d3));p4=exp(i*sum(k.*d4));
        g0=(p1+p2+p3+p4)/4;g1=(p1+p2-p3-p4)/4;
        g2=(p1-p2+p3-p4)/4;g3=(p1-p2-p3+p4)/4;

h=[Esa/2 Vss*g0 0 0 0 Vsapc*g1 Vsapc*g2 Vsapc*g3 0 0;
   0  Esc/2  -Vpasc*conj(g1)  -Vpasc*conj(g2)  -Vpasc*conj(g3)  0 0 0 0 0;
   0  0   Epa/2  0      0      Vxx*g0   Vxy*g3 Vxy*g2  0 -Vpasec*g1;
   0  0   0      Epa/2  0      Vxy*g3   Vxx*g0 Vxy*g1  0 -Vpasec*g2;
   0  0   0      0      Epa/2  Vxy*g2   Vxy*g1 Vxx*g0  0 -Vpasec*g3;
   0  0   0      0      0      Epc/2    0      0       Vseapc*(g1)   0;
   0  0   0      0      0      0        Epc/2  0       Vseapc*(g2)   0;
   0  0   0      0      0      0        0      Epc/2   Vseapc*(g3)   0;
   0  0   0      0      0      0        0      0       Esea/2   0;
   0  0   0      0      0      0        0      0       0        Esec/2];
H=h+h';
[V,D]=eig(H);
    eigst = sum(D);
    E(Nk,:) = sort(real(eigst));
        X(Nk)=-(Nk-1)/(Nt-1);%L-direction
        X1(Nk)=(Nk-1)/(Nt-1);%X-direction
end

hold on
h=plot(X,E,'b');
%h=plot(X1,E,'b');
set(h,'linewidth',[2.0])
set(gca,'Fontsize',[25])
xlabel('k (as fraction of maximum value)---> ')
ylabel('Energy (eV) ---> ')
grid on
```

```
%Note: X-axis from 0 to +1 represents the -X direction
%while the section from 0 to -1 represents the -L direction
```

% Fig.5.4.1a

```
clear all
soa=.3787/3;soc=.0129/3;Esa=-8.3431;Epa=1.0414;Esc=-2.6569;Epc=3.6686;Esea=8.5914;
Esec=6.7386;Vss=-6.4513;Vxx=1.9546;Vxy=5.0779;Vsapc=4.4800;
Vpasc=5.7839;Vseapc=4.8422;Vpasec=4.8077;
d1=[1 1 1]/4;d2=[1 -1 -1]/4;d3=[-1 1 -1]/4;d4=[-1 -1 1]/4;
d1=[0 0 0]/2;d2=[0 -1 -1]/2;d3=[-1 0 -1]/2;d4=[-1 -1 0]/2;

l=1;m=1;n=1;kmax=pi;Nt=101;%L-direction
l=1;m=0;n=0;kmax=2*pi;Nt=101;%X-direction

for Nk=1:Nt
k=[l m n]*kmax*(Nk-1)/(Nt-1);
    p1=exp(i*sum(k.*d1));p2=exp(i*sum(k.*d2));
    p3=exp(i*sum(k.*d3));p4=exp(i*sum(k.*d4));
        g0=(p1+p2+p3+p4)/4;g1=(p1+p2-p3-p4)/4;
        g2=(p1-p2+p3-p4)/4;g3=(p1-p2-p3+p4)/4;
h=[Esa/2 Vss*g0 0 0 0 Vsapc*g1 Vsapc*g2 Vsapc*g3 0 0;
   0  Esc/2  -Vpasc*conj(g1)  -Vpasc*conj(g2)  -Vpasc*conj(g3)  0 0 0 0 0;
   0  0      Epa/2  0         0       Vxx*g0   Vxy*g3 Vxy*g2  0 -Vpasec*g1;
   0  0      0      Epa/2     0       Vxy*g3   Vxx*g0 Vxy*g1  0 -Vpasec*g2;
   0  0      0      0         Epa/2   Vxy*g2   Vxy*g1 Vxx*g0  0 -Vpasec*g3;
   0  0      0      0         0       Epc/2    0      0       Vseapc*(g1)   0;
   0  0      0      0         0       0        Epc/2  0       Vseapc*(g2)   0;
   0  0      0      0         0       0        0      Epc/2   Vseapc*(g3)   0;
   0  0      0      0         0       0        0      0       Esea/2   0;
   0  0      0      0         0       0        0      0       0        Esec/2];
H=[h+h' zeros(10);
    zeros(10)    h+h'];

hso=zeros(20);
hso(3,4)=-i*soa;hso(3,15)=soa;
hso(4,15)=-i*soa;
hso(5,13)=-soa;hso(5,14)=i*soa;
hso(6,7)=-i*soc;hso(6,18)=soc;
hso(7,18)=-i*soc;
hso(8,16)=-soc;hso(8,17)=i*soc;
hso(13,14)=i*soa;
hso(16,17)=i*soc;
Hso=hso+hso';

[V,D]=eig(H+Hso);
    eigst = sum(D);
    E(Nk,:) = sort(real(eigst));
```

```
        X(Nk)=-(Nk-1)/(Nt-1);%L-direction
        X1(Nk)=(Nk-1)/(Nt-1);%X-direction
end

hold on
%h=plot(X,E,'b');
h=plot(X1,E,'b');
axis([-1 1 -3 3])
set(h,'linewidth',[2.0])
set(gca,'Fontsize',[25])
xlabel('k (as fraction of maximum value)--->')
ylabel('Energy (eV) ---> ')
grid on
```

Chapter 6

% Fig.6.1.2

```
clear all

z=zeros(5);Z=zeros(10);
%Constants (all MKS, except energy which is in eV)
hbar=1.055e-34;q=1.602e-19;a=2.45e-10*4/sqrt(3);m=9.110e-31;
d1=[1 1 1]/4;d2=[1 -1 -1]/4;d3=[-1 1 -1]/4;d4=[-1 -1 1]/4;

%sp3s* model parameters
soa=.3787/3;soc=.0129/3;
Esa=-8.3431;Epa=1.0414;Esc=-2.6569;Epc=3.6686;Esea=8.5914;Esec=6.7386;
Vss=-6.4513;Vpasc=-5.7839;Vpasec=-4.8077;
Vsapc=4.4800;Vseapc=4.8422;Vxx=1.9546;Vxy=5.0779;

%Conduction band effective mass model parameters
Ec=1.55;meff=.12*m;

Nt=101;kk=1*linspace(0,1,Nt);
l=0.5;m=0.5;n=0.5;%L-direction
%l=1;m=0;n=0;%X-direction

for Nk=1:Nt
k=2*pi*kk(Nk)*[l m n];

%sp3s* model
    p1=exp(i*sum(k.*d1));p2=exp(i*sum(k.*d2));
    p3=exp(i*sum(k.*d3));p4=exp(i*sum(k.*d4));
        g0=(p1+p2+p3+p4)/4;g1=(p1+p2-p3-p4)/4;
        g2=(p1-p2+p3-p4)/4;g3=(p1-p2-p3+p4)/4;

a1=diag([Esa Epa Epa Epa Esea]);A1=[a1 z;z a1];
a2=diag([Esc Epc Epc Epc Esec]);A2=[a2 z;z a2];
b=[Vss*g0 Vsapc*g1 Vsapc*g2 Vsapc*g3 0;
    Vpasc*g1     Vxx*g0 Vxy*g3 Vxy*g2 Vpasec*g1;
    Vpasc*g2 Vxy*g3 Vxx*g0 Vxy*g1 Vpasec*g2;
```

```
    Vpasc*g3 Vxy*g2 Vxy*g1 Vxx*g0 Vpasec*g3;
    0 Vseapc*conj(g1) Vseapc*conj(g2) Vseapc*conj(g3) 0];B=[b z;z b];
    h=[a1 b;b' a2];H=[A1 B;B' A2];

aso=soa*[0 0 0 0 0 0 0 0 0 0;
         0 0 -i 0 0 0 0 0 1 0;
         0 i 0 0 0 0 0 0 -i 0;
         0 0 0 0 0 0 -1 i 0 0;
         0 0 0 0 0 0 0 0 0 0;
         0 0 0 0 0 0 0 0 0 0;
         0 0 0 -1 0 0 0 i 0 0;
         0 0 0 -i 0 0 -i 0 0 0;
         0 1 i 0 0 0 0 0 0 0;
         0 0 0 0 0 0 0 0 0 0];
    cso=soc*[0 0 0 0 0 0 0 0 0 0;
         0 0 -i 0 0 0 0 0 1 0;
         0 i 0 0 0 0 0 0 -i 0;
         0 0 0 0 0 0 -1 i 0 0;
         0 0 0 0 0 0 0 0 0 0;
         0 0 0 0 0 0 0 0 0 0;
         0 0 0 -1 0 0 0 i 0 0;
         0 0 0 -i 0 0 -i 0 0 0;
         0 1 i 0 0 0 0 0 0 0;
         0 0 0 0 0 0 0 0 0 0];H=H+[aso Z;Z cso];

[V,D]=eig(H);
    eiglst = sum(D);
    E(Nk,:) = sort(real(eiglst));

%Conduction band effective mass model
Em(Nk)=Ec+((hbar^2)*sum(k.*k)/(2*meff*q*(a^2)));
end

kk=-kk;%L-direction

hold on
h1=plot(kk,E, 'b');
h2=plot(kk,Em,'b--');
axis([-1 1 -3 3])
set(h1,'linewidth',[1.0])
set(h2,'linewidth',[2.0])
set(gca,'Fontsize',[24])
xlabel(' ka (fraction of maximum value ---> ')
ylabel(' Energy ( eV ) ---> ')
grid on
```

% Fig.6.1.7

```
clear all

t=3;m=65;%66 for (a), 65 for (b)
D=2*m*0.14*sqrt(3)/(2*pi);
```

```
Eg=2*t*0.14/D;nu=round(2*m/3)+0;% +1 is used for higher mode
kyb=2*pi*nu/(2*m);
kxa=0.05*linspace(-pi,pi,101);

E1=(3*t/2)*sqrt(((kxa*2/3).^2)+(((abs(kyb)-(2*pi/3))*2/sqrt(3)).^2));%a0=b*2/sqrt(3)=a*2/3;
E2=t*sqrt(1+(4*cos(kyb).*cos(kxa))+(4*cos(kyb).^2));

k=kxa./pi;[D Eg nu min(E1)]

hold on
h=plot(k,E1, 'b');
h=plot(k,-E1, 'b');
axis([-0.05 0.05 -0.6 0.6])
set(h,'linewidth',[1.0])
set(h,'linewidth',[2.0])
set(gca,'Fontsize',[24])
xlabel(' kxa/pi (fraction of maximum value ---> ')
ylabel(' Energy ( eV ) ---> ')
grid on
```

% Fig.6.1.9

```
clear all

t=3;kxa=0;
kyb=linspace(-pi,pi,101);
E1=(3*t/2)*sqrt(((kxa*2/3).^2)+(((abs(kyb)-(2*pi/3))*2/sqrt(3)).^2));%a0=b*2/sqrt(3)=a*2/3;
E2=t*sqrt(1+(4*cos(kyb).*cos(kxa))+(4*cos(kyb).^2));

k=kyb./pi;

hold on
h=plot(k,E1,'b');
h=plot(k,-E1,'b');
h=plot(k,E2,'bx');
h=plot(k,-E2,'bx');
axis([-1 1 -15 15])
set(h,'linewidth',[1.0])
set(h,'linewidth',[2.0])
set(gca,'Fontsize',[24])
xlabel(' kyb/pi ---> ')
ylabel(' Energy ( eV ) ---> ')
grid on
```

% Fig.6.2.1

```
clear all

%Constants (all MKS, except energy which is in eV)
hbar=1.055e-34;m=9.110e-31;q=1.602e-19;L=1e-9;
D2=zeros(1,101);
```

```
Lz=20e-9;%5e-9 for (a),20e-9 for (b)
E0=(hbar^2)*(pi^2)/(2*q*m*Lz^2);
for p=1:25
E=linspace(-0.1,0.25,101);thet=(E+abs(E))./(2*E);
EE=E-(p*p*E0);theta=(EE+abs(EE))./(2*EE);
D1=(L)*q*m*thet.*real((2*m*E*q).^(-0.5))./(pi*hbar);
D2=D2+((L^2)*q*m*theta./(2*pi*hbar*hbar));
D3=(L^3)*q*m*thet.*real((2*m*E*q).^0.5)./(2*pi*pi*hbar*hbar*hbar);
end

hold on
h=plot(D2,E,'b');
h=plot(D3.*Lz/L,E,'b');
%axis([0 10 -0.1 0.25]);%Part (a)
axis([0 40 -0.1 0.25]);%Part (b)
set(h,'linewidth',[1.0])
set(h,'linewidth',[2.0])
set(gca,'Fontsize',[24])
xlabel(' D(E) (per eV per nm^2) ---> ')
ylabel(' Energy ( eV ) ---> ')
grid on
```

% Fig.6.2.2

```
clear all
t=3;m=800;% Use 200 and 800 for two plots
a0=0.14;D=2*m*a0*sqrt(3)/(2*pi);Eg=2*t*0.14/D;c=pi*D;L=1;D
nu0=round(2*m/3);a=3*a0/2;

E=linspace(0,0.25,101);
DG=(2*c*L/(2*pi*a*a*t*t))*E;

DN=zeros(1,101);
for nu=nu0-100:nu0+100
Ek=((t*2*pi/sqrt(3))*((3*nu/(2*m))-1))+(i*1e-12);
    DN=DN+((2*L/(pi*a*t))*abs(real(E./(sqrt((E.^2)-(Ek^2))))));
end

hold on
h1=plot(DG,E,'bx');
h2=plot(DN,E,'b');
hold on
axis([0 50 0 0.25]);
set(h1,'linewidth',[1.0])
set(h2,'linewidth',[2.0])
set(gca,'Fontsize',[24])
xlabel(' D(E) (per eV per nm) ---> ')
ylabel(' Energy ( eV ) ---> ')
grid on
```

% Fig.6.3.3

```
clear all

%Constants (all MKS, except energy which is in eV)
hbar=1.055e-34;m=9.110e-31;q=1.602e-19;a=5e-10;L=10e-9;

k=0.5*linspace(-1,1,201)/a;
Ek=(hbar^2)*(k.^2)/(2*0.25*m*q);
EE=linspace(0,0.2,201);

%Subband (1,1)
E1=2*(hbar^2)*(pi^2)/(2*0.25*m*q*L^2);
M=((EE-E1)+abs(EE-E1))./(2*abs(EE-E1));

%Subbands (1,2) and (2,1)
E2=5*(hbar^2)*(pi^2)/(2*0.25*m*q*L^2);
M=M+(((EE-E2)+abs(EE-E2))./(abs(EE-E2)));

%Subband (2,2)
E3=8*(hbar^2)*(pi^2)/(2*0.25*m*q*L^2);
M=M+(((EE-E3)+abs(EE-E3))./(2*abs(EE-E3)));

hold on
h=plot(k,E1+Ek,'b');%Part (a)
h=plot(k,E2+Ek,'b');%Part (a)
h=plot(k,E3+Ek,'b');%Part (a)
%h=plot(M,EE,'b');%Part (b)
set(h,'linewidth',[2.0])
set(gca,'Fontsize',[24])
xlabel('k ( / m )');%Part (a)
%xlabel(' M ( E ) ');%Part (b)
ylabel('E - Ec ( eV ) -->');
axis([-1e9 1e9 0 0.3]);%Part (a)
%axis([0 5 0 0.3]);%Part (b)
grid on
```

% Fig.6.3.4

```
clear all

%Constants (all MKS, except energy which is in eV)
hbar=1.055e-34;m=9.110e-31;q=1.602e-19;a=5e-10;L=10e-9;

k=0.5*linspace(-1,1,201)/a;
Ek=-(hbar^2)*(k.^2)/(2*0.25*m*q);
EE=linspace(0,-0.2,201);

%Subband (1,1)
E1=-2*(hbar^2)*(pi^2)/(2*0.25*m*q*L^2);
M=((E1-EE)+abs(E1-EE))./(2*abs(E1-EE));
```

```
%Subbands (1,2) and (2,1)
E2=-5*(hbar^2)*(pi^2)/(2*0.25*m*q*L^2);
M=M+(((E2-EE)+abs(E2-EE))./(abs(E2-EE)));

%Subband (2,2)
E3=-8*(hbar^2)*(pi^2)/(2*0.25*m*q*L^2);
M=M+(((E3-EE)+abs(E3-EE))./(2*abs(E3-EE)));

hold on
%h=plot(k,E1+Ek,'b');%Part (a)
%h=plot(k,E2+Ek,'b');%Part (a)
%h=plot(k,E3+Ek,'b');%Part (a)
h=plot(M,EE,'b');%Part (b)
set(h,'linewidth',[2.0])
set(gca,'Fontsize',[24])
%xlabel('k ( / m )');%Part (a)
xlabel(' M ( E ) ');%Part (b)
ylabel('E - Ev ( eV ) -->');
%axis([-1e9 1e9 -0.3 0]);%Part (a)
axis([0 5 -0.3 0]);%Part (b)
grid on
```

Chapter 7

% Fig.7.1.5

```
clear all

%Constants (all MKS, except energy which is in eV)
hbar=1.055e-34;q=1.602e-19;a=3e-10;m=9.110e-31;

%Conduction band parameters
mw=.07*m;ma=.22*m;mb=(.7*mw)+(.3*ma);kk=0*.1*pi;
Ec=0;Eb=(.7*0)+(.3*1.25);

for nk=1:24
Nw=nk+10;Nb=2*Nw;Np=Nb+Nw+Nb;W(nk)=(Nw-1)*a*1e9;
tb=(hbar^2)/(2*mb*(a^2)*q);tw=(hbar^2)/(2*mw*(a^2)*q);
t=[tb*ones(1,Nb) tw*ones(1,Nw-1) tb*ones(1,Nb)];
tt=[0 t]+[t 0];
Ebk=Eb+(tb*(kk^2));Ewk=tw*(kk^2);Ebwk=(Eb/2)+((tb+tw)*(kk^2)/2);
U=Ec+[Ebk*ones(1,Nb) Ebwk Ewk*ones(1,Nw-2) Ebwk Ebk*ones(1,Nb)];
H=-diag(t,1)-diag(t,-1)+diag(tt)+diag(U);
[V,D]=eig(H);D=diag(D);D=(sort(real(D)))';
E1(nk)=D(1);E2(nk)=D(2);

end

hold on
h1=plot(W,E1,'b');
h1=plot(W,E2,'b--');
```

```
set(h1,'linewidth',[2.0])
set(gca,'Fontsize',[24])
xlabel(' W ( nm ) ---> ')
ylabel(' Energy ( eV ) ---> ')
axis([2 10 0 .4])
grid on
```

% Fig.7.1.6

```
clear all

%Constants (all MKS, except energy which is in eV)
hbar=1.055e-34;q=1.602e-19;a=3e-10;m=9.110e-31;

%Conduction band parameters
mw=.07*m;ma=.22*m;mb=(.7*mw)+(.3*ma);
Nw=24;Nb=2*Nw;Np=Nb+Nw+Nb;W=(Nw-1)*a*1e9
Ec=0;Eb=(.7*0)+(.3*1.25);

for nk=1:26
kk=(nk-1)*a*1e10/500;k(nk)=kk/(a*1e9);
tb=(hbar^2)/(2*mb*(a^2)*q);tw=(hbar^2)/(2*mw*(a^2)*q);
t=[tb*ones(1,Nb) tw*ones(1,Nw-1) tb*ones(1,Nb)];
tt=[0 t]+[t 0];
Ebk=Eb+(tb*(kk^2));Ewk=tw*(kk^2);Ebwk=(Eb/2)+((tb+tw)*(kk^2)/2);
U=Ec+[Ebk*ones(1,Nb) Ebwk Ewk*ones(1,Nw-2) Ebwk Ebk*ones(1,Nb)];
H=-diag(t,1)-diag(t,-1)+diag(tt)+diag(U);
[V,D]=eig(H);D=diag(D);D=(sort(real(D)))';
E1(nk)=D(1);E2(nk)=D(2);
end

E1w=E1(1)+(hbar^2)*(k.^2)./(2*mw*1e-18*q);
E2w=E2(1)+(hbar^2)*(k.^2)./(2*mw*1e-18*q);
E1b=E1(1)+(hbar^2)*(k.^2)./(2*mb*1e-18*q);
E2b=E2(1)+(hbar^2)*(k.^2)./(2*mb*1e-18*q);

hold on
h=plot(k,E1,'b');
h=plot(k,E2,'b');
h=plot(k,E1w,'b:');
h=plot(k,E2w,'b:');
h=plot(k,E1b,'b--');
h=plot(k,E2b,'b--');
set(h,'linewidth',[1.0])
set(gca,'Fontsize',[24])
xlabel(' k ( / nm ) ---> ')
ylabel(' Energy ( eV ) ---> ')
axis([0 .5 0 0.4])
grid on
```

% Fig.7.2.5

```
clear all

%Constants (all MKS, except energy which is in eV)
hbar=1.06e-34;q=1.6e-19;eps0=8.85E-12;epsr=4;m=.25*9.1e-31;
mu=0;kT=.025;n0=m*kT*q/(2*pi*(hbar^2));n0

%inputs
a=3e-10;t0=(hbar^2)/(2*m*(a^2)*q);e0=q*a/eps0;
Nox=7;Nc=10;%use Nc=10,30 for 3,9nm channel respectively
Np=Nox+Nc+Nox;XX=a*1e9*[1:1:Np];
Ec=[3*ones(Nox,1);0*ones(Nc,1);3*ones(Nox,1)];

%Hamiltonian matrix
T=(2*t0*diag(ones(1,Np)))-(t0*diag(ones(1,Np-1),1))-(t0*diag(ones(1,Np-1),-1));

%dielectric constant matrix
D2=epsr*((2*diag(ones(1,Np)))-(diag(ones(1,Np-1),1))-(diag(ones(1,Np-1),-1)));
iD2=inv(D2);

Vg=.25;Ubdy=-epsr*[Vg;zeros(Np-2,1);Vg];
%Ubdy=-epsr*[0;zeros(Np-2,1);Vg];;%for asymmetric bias
U0=iD2*Ubdy;

%self-consistent calculation
    U1=1e-6*ones(Np,1);UU=U1;change=1;
    while change>1e-3
        U1=U1+(0.1*(UU-U1));
        [P,D]=eig(T+diag(Ec)+diag(U1));D=diag(D);
        rho=log(1+exp((mu-D)./kT));rho=P*diag(rho)*P';
        n=2*n0*diag(rho);
            for kp=1:Np
                ncl(kp)=a*2*(n0^1.5)*Fhalf((mu-Ec(kp)-U1(kp))/kT);
            end
            %n=ncl';%use for semiclassical calculation
            UU=U0+(iD2*e0*n);
            change=max(max((abs(UU-U1))));
        U=Ec+U1;%self-consistent band profile
end

%electron density in channel per cm2
    ns=1e-4*sum(sum(n.*[zeros(Nox,1);ones(Nc,1);zeros(Nox,1)]));Vg,ns
        nn=1e-6*n./a;%electron density per cm3
        Fn=mu*ones(Nc+Nox+Nox,1);

hold on
h=plot(XX,nn,'g');
%h=plot(XX,Ec,'g--');
%h=plot(XX,Ec+U1,'g');
%h=plot(XX,Fn,'g:');
set(h,'linewidth',[2.0])
```

```
set(gca,'Fontsize',[24])
xlabel(' z (nm) ---> ')
%ylabel(' E (eV) ---> ')
ylabel(' n (/cm3) --->')
%axis([0 8 -.5 3])
axis([0 8 0 15e18])
grid on
```

% Fig.7.3.1

```
clear all

%Constants (all MKS, except energy which is in eV)
hbar=1.06e-34;q=1.6e-19;eps0=8.85E-12;epsr=4;m=.25*9.1e-31;
mu=0;kT=.025;n0=m*kT*q/(2*pi*(hbar^2));

%inputs
a=3e-10;t0=(hbar^2)/(2*m*(a^2)*q);e0=q*a/eps0;
Nox=7;Nc=10;%use Nc=10,30 for 3,9nm channel respectively
Np=Nox+Nc+Nox;XX=a*1e9*[1:1:Np];
Ec=[3*ones(Nox,1);0*ones(Nc,1);3*ones(Nox,1)];

%Hamiltonian matrix
T=(2*t0*diag(ones(1,Np)))-(t0*diag(ones(1,Np-1),1))-(t0*diag(ones(1,Np-1),-1));

%dielectric constant matrix
D2=epsr*((2*diag(ones(1,Np)))-(diag(ones(1,Np-1),1))-(diag(ones(1,Np-1),-1)));
iD2=inv(D2);

Vg=linspace(-.25,.25,26);
for kg=1:26
Ubdy=-epsr*[Vg(kg);zeros(Np-2,1);Vg(kg)];kg;
%Ubdy=-epsr*[0;zeros(Np-2,1);Vg(kg)];;%for asymmetric bias
U0=iD2*Ubdy;

%self-consistent calculation
    U1=1e-6*ones(Np,1);UU=U1;change=1;
    while change>1e-3
        U1=U1+(0.1*(UU-U1));
        [P,D]=eig(T+diag(Ec)+diag(U1));D=diag(D);
        rho=log(1+exp((mu-D)./kT));rho=P*diag(rho)*P';
        n=2*n0*diag(rho);
            for kp=1:Np
                ncl(kp)=a*2*(n0^1.5)*Fhalf((mu-Ec(kp)-U1(kp))/kT);
            end
            %n=ncl';%use for semiclassical calculation
            UU=U0+(iD2*e0*n);
            change=max(max((abs(UU-U1))));
        U=Ec+U1;%self-consistent band profile
end
```

```
%electron density in channel per cm2
      ns(kg)=1e-4*sum(sum(n.*[zeros(Nox,1);ones(Nc,1);zeros(Nox,1)]));
            nn(:,kg)=1e-6*n./a;%electron density per cm3
            Fn(:,kg)=mu*ones(Nc+Nox+Nox,1);
end
C=q*(ns(26)-ns(25))/(Vg(26)-Vg(25));
d=1e-4*epsr*eps0*2/C;d,C
%ns=log10(ns)

hold on
h=plot(Vg,ns,'b');
set(h,'linewidth',[2.0])
set(gca,'Fontsize',[24])
xlabel(' Vg (V) ---> ')
ylabel(' ns (/cm2) ---> ')
%axis([0 .3 0 3.5e12])
grid on
```

% Fig.7.3.2

```
clear all

E=linspace(-.5,1,1001);
D=sqrt(E);

hold on
h=plot(D,E,'b');
set(h,'linewidth',[2.0])
set(gca,'Fontsize',[24])
xlabel(' D ( E ) (arb. units) ')
ylabel(' E ( eV ) ')
grid on
```

% Fig.7.3.4

```
clear all

E=linspace(-.25,.25,501);dE=E(2)-E(1);kT=0.025;Ef=0;
V=0;mu1=Ef+(V/2);mu2=Ef-(V/2);
f1=1./(1+exp((E-mu1)./kT));f2=1./(1+exp((E-mu2)./kT));
FT=[0 diff(f1)];FT=FT.*(-1/dE);
%dE*(sum(f1-f2))/V

hold on
h=plot(FT,E,'b');
set(h,'linewidth',[2.0])
set(gca,'Fontsize',[24])
grid on
```

% Fig.7.4.1, Fig.7.4.2

```
clear all

z=zeros(5);Z=zeros(10);
%Constants (all MKS, except energy which is in eV)
hbar=1.055e-34;q=1.602e-19;a=2.45e-10*4/sqrt(3);m=9.110e-31;
d1=[1 1 1]/4;d2=[1 -1 -1]/4;d3=[-1 1 -1]/4;d4=[-1 -1 1]/4;

%sp3s* model parameters
soa=.3787/3;soc=.0129/3;
Esa=-8.3431;Epa=1.0414;Esc=-2.6569;Epc=3.6686;Esea=8.5914;Esec=6.7386;
Vss=-6.4513;Vpasc=-5.7839;Vpasec=-4.8077;
Vsapc=4.4800;Vseapc=4.8422;Vxx=1.9546;Vxy=5.0779;

%Valence band Luttinger-Kohn parameters
Ev=-.1;del=.3;g1=6.85;g2=2.1;g3=2.9;
t1=(hbar^2)*g1/(2*m*q*(a^2));
t2=(hbar^2)*g2/(2*m*q*(a^2));
t3=(hbar^2)*g3/(2*m*q*(a^2));

Nt=101;kk=1*linspace(0,1,Nt);
l=1;m=0;n=0;%X-direction
l=0.5;m=0.5;n=0.5;%L-direction

for Nk=1:Nt
k=2*pi*kk(Nk)*[l m n];

%sp3s* model
    p1=exp(i*sum(k.*d1));p2=exp(i*sum(k.*d2));
    p3=exp(i*sum(k.*d3));p4=exp(i*sum(k.*d4));
        g0=(p1+p2+p3+p4)/4;g1=(p1+p2-p3-p4)/4;
        g2=(p1-p2+p3-p4)/4;g3=(p1-p2-p3+p4)/4;

a1=diag([Esa Epa Epa Epa Esea]);A1=[a1 z;z a1];
a2=diag([Esc Epc Epc Epc Esec]);A2=[a2 z;z a2];
b=[Vss*g0 Vsapc*g1 Vsapc*g2 Vsapc*g3 0;
    Vpasc*g1    Vxx*g0 Vxy*g3 Vxy*g2 Vpasec*g1;
    Vpasc*g2 Vxy*g3 Vxx*g0 Vxy*g1 Vpasec*g2;
    Vpasc*g3 Vxy*g2 Vxy*g1 Vxx*g0 Vpasec*g3;
    0 Vseapc*conj(g1) Vseapc*conj(g2) Vseapc*conj(g3) 0];B=[b z;z b];
    h=[a1 b;b' a2];H=[A1 B;B' A2];

aso=soa*[0 0 0 0 0 0 0 0 0 0;
             0 0 -i 0 0 0 0 0 1 0;
             0 i 0 0 0 0 0 0 -i 0;
             0 0 0 0 0 0 -1 i 0 0;
             0 0 0 0 0 0 0 0 0 0;
             0 0 0 0 0 0 0 0 0 0;
             0 0 0 -1 0 0 0 i 0 0;
             0 0 0 -i 0 0 -i 0 0 0;
```

```
                 0 1 i 0 0 0 0 0 0 0;
                 0 0 0 0 0 0 0 0 0 0];
    cso=soc*[0 0 0 0 0 0 0 0 0 0;
                 0 0 -i 0 0 0 0 0 1 0;
                 0 i 0 0 0 0 0 0 -i 0;
                 0 0 0 0 0 0 -1 i 0 0;
                 0 0 0 0 0 0 0 0 0 0;
                 0 0 0 0 0 0 0 0 0 0;
                 0 0 0 -1 0 0 0 i 0 0;
                 0 0 0 -i 0 0 -i 0 0 0;
                 0 1 i 0 0 0 0 0 0 0;
                 0 0 0 0 0 0 0 0 0 0];H=H+[aso Z;Z cso];

[V,D]=eig(H);
    eiglst = sum(D);
    E(Nk,:) = sort(real(eiglst));

%Valence band Luttinger-Kohn model
P=Ev+(t1*sum(k.*k));Q=t2*((k(1)^2)+(k(2)^2)-(2*(k(3)^2)));
R=-(sqrt(3)*t2*((k(1)^2)-(k(2)^2)))+(i*2*t3*sqrt(3)*k(1)*k(2));
S=2*t3*sqrt(3)*((k(1)-(i*k(2)))*k(3));

H4=-[P+Q -S R 0;
        -S' P-Q 0 R;
        R' 0 P-Q S;
        0 R' S' P+Q];[V,D]=eig(H4);
    eiglst = sum(D);
    ELK4(Nk,:) = sort(real(eiglst));

H6=-[P+Q -S R 0 -S/sqrt(2) sqrt(2)*R;
    -S' P-Q 0 R -sqrt(2)*Q sqrt(1.5)*S;
    R' 0 P-Q S sqrt(1.5)*S' sqrt(2)*Q;
    0 R' S' P+Q -sqrt(2)*R' -S'/sqrt(2);
    -S'/sqrt(2) -sqrt(2)*Q' sqrt(1.5)*S -sqrt(2)*R P+del 0;
    sqrt(2)*R' sqrt(1.5)*S' sqrt(2)*Q' -S/sqrt(2) 0 P+del];
        [V,D]=eig(H6);
    eiglst = sum(D);
    ELK6(Nk,:) = sort(real(eiglst));
end

kk=-kk;%L-direction

hold on
h1=plot(kk,E,'b');
%h2=plot(kk,ELK4,'b--');% Fig.6.4.1
h2=plot(kk,ELK6,'b--');% Fig.6.4.2
set(h1,'linewidth',[2.0])
set(h2,'linewidth',[3.0])
set(gca,'Fontsize',[24])
xlabel(' ka (fraction of maximum value) ---> ')
```

```
ylabel(' Energy ( eV ) ---> ')
axis([-1 1 -2 3])
grid on
```

% Fig.7.4.4

```
clear all

%Constants (all MKS, except energy which is in eV)
hbar=1.055e-34;q=1.602e-19;a=3e-10;m=9.110e-31;

Eb=.15;
%Luttinger-Kohn parameters
g1=6.85;g2=2.1;g3=2.9;%GaAs
w1=(hbar^2)*g1/(2*m*q*(a^2));
w2=(hbar^2)*g2/(2*m*q*(a^2));
w3=(hbar^2)*g3/(2*m*q*(a^2));
    g1=3.45;g2=0.68;g3=1.29;%AlAs
    a1=(hbar^2)*g1/(2*m*q*(a^2));b1=(.7*w1)+(.3*a1);
    a2=(hbar^2)*g2/(2*m*q*(a^2));b2=(.7*w2)+(.3*a2);
    a3=(hbar^2)*g3/(2*m*q*(a^2));b3=(.7*w3)+(.3*a3);
Ev=0;Evb=(0.7*0)+(0.3*0.75);kx=0*pi;ky=0*pi;k2=(kx^2)+(ky^2);

for nk=1:20
Nw=nk+10;Nb=Nw;Np=Nb+Nw+Nb;W(nk)=(Nw-1)*a*1e9;Z=zeros(Np);nk
X(nk)=Nw-1;
t=[b1*ones(1,Nb) w1*ones(1,Nw-1) b1*ones(1,Nb)];tt=[0 t]+[t 0];
Ebk=Evb+(b1*k2);Ewk=(w1*k2);Ebwk=(Ebk+Ewk)/2;
U=Ev+[Ebk*ones(1,Nb) Ebwk Ewk*ones(1,Nw-2) Ebwk Ebk*ones(1,Nb)];
P=-diag(t,1)-diag(t,-1)+diag(tt)+diag(U);

t=-2*[b2*ones(1,Nb) w2*ones(1,Nw-1) b2*ones(1,Nb)];tt=[0 t]+[t 0];
Ebk=b2*k2;Ewk=w2*k2;Ebwk=(Ebk+Ewk)/2;
U=[Ebk*ones(1,Nb) Ebwk Ewk*ones(1,Nw-2) Ebwk Ebk*ones(1,Nb)];
Q=-diag(t,1)-diag(t,-1)+diag(tt)+diag(U);

Ebk=-(sqrt(3)*b2*((kx^2)-(ky^2)))+(i*2*b3*sqrt(3)*kx*ky);
Ewk=-(sqrt(3)*w2*((kx^2)-(ky^2)))+(i*2*w3*sqrt(3)*kx*ky);
Ebwk=(Ebk+Ewk)/2;
U=[Ebk*ones(1,Nb) Ebwk Ewk*ones(1,Nw-2) Ebwk Ebk*ones(1,Nb)];
R=diag(U);

t=2*i*sqrt(3)*(kx-(i*ky))*[b3*ones(1,Nb) w3*ones(1,Nw-1) b3*ones(1,Nb)];
S=diag(t,1)-diag(t,-1);

H=[P+Q Z;Z P+Q];HL=[P-Q Z;Z P-Q];
HC=[-S R;R' S'];
    H=-[H HC;HC' HL];

[V,D]=eig(H);D=diag(D);D=-(sort(real(-D)))';
E1(nk)=D(1);E2(nk)=D(2);E3(nk)=D(3);E4(nk)=D(4);
```

```
E5(nk)=D(5);E6(nk)=D(6);E7(nk)=D(7);E8(nk)=D(8);
end

%Analytical results for infinite well
Ean1=-(w1-(2*w2))*(pi^2)./(X.^2);
Ean2=-(w1+(2*w2))*(pi^2)./(X.^2);

hold on
%h=plot(W,Ean1,'b');
%h=plot(W,Ean2,'b');
h=plot(W,E1,'b');
%h=plot(W,E2,'bx');
h=plot(W,E3,'b');
%h=plot(W,E4,'b+');
h=plot(W,E5,'b');
%h=plot(W,E6,'bx');
h=plot(W,E7,'b');
%h=plot(W,E8,'b+');
set(h,'linewidth',[2.0])
set(gca,'Fontsize',[24])
xlabel(' W ( nm ) ---> ')
ylabel(' Energy ( eV ) ---> ')
axis([3 9 -.25 0])
grid on
```

% Fig.7.4.5

```
clear all

%Constants (all MKS, except energy which is in eV)
hbar=1.055e-34;q=1.602e-19;a=3e-10;m=9.110e-31;

%Luttinger-Kohn parameters
g1=6.85;g2=2.1;g3=2.9;%GaAs
w1=(hbar^2)*g1/(2*m*q*(a^2));
w2=(hbar^2)*g2/(2*m*q*(a^2));
w3=(hbar^2)*g3/(2*m*q*(a^2));
    g1=3.45;g2=0.68;g3=1.29;%AlAs
    a1=(hbar^2)*g1/(2*m*q*(a^2));b1=(.7*w1)+(.3*a1);
    a2=(hbar^2)*g2/(2*m*q*(a^2));b2=(.7*w2)+(.3*a2);
    a3=(hbar^2)*g3/(2*m*q*(a^2));b3=(.7*w3)+(.3*a3);
Ev=0;Evb=(0.7*0)+(0.3*0.75);

Nw=18;Nb=Nw;Np=Nb+Nw+Nb;W=(Nw-1)*a*1e9,Z=zeros(Np);

for nk=1:26
k(nk)=(nk-1)/500;% in A^-1
l=0;m=1;lm=sqrt((l^2)+(m^2));
kx=(l/lm)*k(nk)*a*1e10;ky=(m/lm)*k(nk)*a*1e10;
k2=(kx^2)+(ky^2);
```

```
t=[b1*ones(1,Nb) w1*ones(1,Nw-1) b1*ones(1,Nb)];tt=[0 t]+[t 0];
Ebk=Evb+(b1*k2);Ewk=(w1*k2);Ebwk=(Ebk+Ewk)/2;
U=Ev+[Ebk*ones(1,Nb) Ebwk Ewk*ones(1,Nw-2) Ebwk Ebk*ones(1,Nb)];
P=-diag(t,1)-diag(t,-1)+diag(tt)+diag(U);

t=-2*[b2*ones(1,Nb) w2*ones(1,Nw-1) b2*ones(1,Nb)];tt=[0 t]+[t 0];
Ebk=b2*k2;Ewk=w2*k2;Ebwk=(Ebk+Ewk)/2;
U=[Ebk*ones(1,Nb) Ebwk Ewk*ones(1,Nw-2) Ebwk Ebk*ones(1,Nb)];
Q=-diag(t,1)-diag(t,-1)+diag(tt)+diag(U);

Ebk=-(sqrt(3)*b2*((kx^2)-(ky^2)))+(i*2*b3*sqrt(3)*kx*ky);
Ewk=-(sqrt(3)*w2*((kx^2)-(ky^2)))+(i*2*w3*sqrt(3)*kx*ky);
Ebwk=(Ebk+Ewk)/2;
U=[Ebk*ones(1,Nb) Ebwk Ewk*ones(1,Nw-2) Ebwk Ebk*ones(1,Nb)];
R=diag(U);

t=-2*i*sqrt(3)*(kx-(i*ky))*[b3*ones(1,Nb) w3*ones(1,Nw-1) b3*ones(1,Nb)]/2;
S=diag(t,1)-diag(t,-1);

H=[P+Q Z;Z P+Q];HL=[P-Q Z;Z P-Q];
HC=[-S R;R' S'];
    H=-[H HC;HC' HL];
[nk sum(sum(abs(H-H')))]

[V,D]=eig(H);D=diag(D);D=-(sort(real(-D)))';
E1(nk)=D(1);E2(nk)=D(2);E3(nk)=D(3);E4(nk)=D(4);
end

k=k*10;%per Angstrom to per nm
hold on
%h=plot(W,Ean1,'b');
%h=plot(W,Ean2,'b');
h=plot(k,E1,'b');
%h=plot(k,E2,'bx');
h=plot(k,E3,'b');
%h=plot(k,E4,'b+');
set(h,'linewidth',[2.0])
set(gca,'Fontsize',[24])
xlabel(' k ( /nm ) ---> ')
ylabel(' Energy ( eV ) ---> ')
axis([0 .5 -.1 0])
grid on
```

Chapter 8

% Fig.8.1

```
clear all

E=linspace(-.3,.3,50001);dE=E(2)-E(1);gam=0.05;
D=(gam/(2*pi))./(((E-.14).^2)+((gam/2)^2));
```

```
D=D+(gam/(2*pi))./(((E-.04).^2)+((gam/2)^2));%Use for Fig.P.5.2
D=D+((gam/(2*pi))./(((E+.06).^2)+((gam/2)^2)));%Use for Fig.P.5.2
D=D+(gam/(2*pi))./(((E+.15).^2)+((gam/2)^2));
dE*sum(D)

hold on
h=plot(D,E,'b');
set(h,'linewidth',[2.0])
set(gca,'Fontsize',[24])
xlabel(' D(E) (per eV) --->')
ylabel(' E (eV) ---> ')
grid on
```

% Fig.8.2.5

```
clear all

%Constants (all MKS, except energy which is in eV)
hbar=1.06e-34;q=1.6e-19;m=0.25*9.1e-31;mu=0.25;kT=0.025;

%inputs
a=2e-10;t0=(hbar^2)/(2*m*(a^2)*q);Np=50;t0
X=a*linspace(0,Np-1,Np);U=linspace(-0.05,0.05,Np);
H=(2*t0*diag(ones(1,Np)))-(t0*diag(ones(1,Np-1),1))-(t0*diag(ones(1,Np-1),-1));
H=H+diag(U);HP=H;
HP(1,Np)=-t0;HP(Np,1)=-t0;

[V,D]=eig(HP);D=diag(D);
rho=1./(1+exp((D-mu)./kT));
rho=V*diag(rho)*V';rho=diag(rho)/a;

%Energy grid for Green's function method
Emin=-0.1;Emax=0.4;NE=250;E=linspace(Emin,Emax,NE);dE=E(2)-E(1);zplus=i*1e-12;
f=1./(1+exp((E-mu)./kT));

%Green's function method
sig1=zeros(Np,Np);sig2=zeros(Np,Np);n=zeros(Np,1);
    for k=1:NE
        ck=(1-((E(k)+zplus-U(1))/(2*t0)));ka=acos(ck);
            sigma=-t0*exp(i*ka);sig1(1,1)=sigma;
        ck=(1-((E(k)+zplus-U(Np))/(2*t0)));ka=acos(ck);
            sigma=-t0*exp(i*ka);sig2(Np,Np)=sigma;
            G=inv(((E(k)+zplus)*eye(Np))-H-sig1-sig2);
            n=n+(f(k)*(dE*diag(i*(G-G'))/(2*pi*a)));
    end

hold on
h=plot(X,rho,'b');
h=plot(X,n,'bx');
grid on
set(h,'linewidth',[2.0])
```

```
set(gca,'Fontsize',[24])
xlabel(' X ( m ) --> ')
ylabel(' n ( / m ) --> ')
```

% Fig.8.2.6

```
clear all

%Constants (all MKS, except energy which is in eV)
hbar=1.06e-34;q=1.6e-19;m=0.25*9.1e-31;mu=0.25;kT=0.025;

%inputs
a=2e-10;t0=(hbar^2)/(2*m*(a^2)*q);Np=50;t0
X=a*linspace(0,Np-1,Np);U=linspace(-0.05,0.05,Np);
H=(2*t0*diag(ones(1,Np)))-(t0*diag(ones(1,Np-1),1))-(t0*diag(ones(1,Np-1),-1));
H=H+diag(U);

%Energy grid for Green's function method
Emin=-0.1;Emax=0.4;NE=250;E=linspace(Emin,Emax,NE);dE=E(2)-E(1);zplus=i*1e-12;
f=1./(1+exp((E-mu)./kT));

%Green's function method
sig1=zeros(Np,Np);sig2=zeros(Np,Np);
    for k=1:NE
        ck=(1-((E(k)+zplus-U(1))/(2*t0)));ka=acos(ck);
            sigma=-t0*exp(i*ka);sig1(1,1)=sigma;
        ck=(1-((E(k)+zplus-U(Np))/(2*t0)));ka=acos(ck);
            sigma=-t0*exp(i*ka);sig2(Np,Np)=sigma;
            G=inv(((E(k)+zplus)*eye(Np))-H-sig1-sig2);
            D0=diag(i*(G-G'))/(2*pi);D1(k)=D0(1);D2(k)=D0(Np);
    end

hold on
%h=plot(X,U,'b');
h=plot(D1,E,'b');
%h=plot(D2,E,'b');
grid on
set(h,'linewidth',[2.0])
set(gca,'Fontsize',[24])
xlabel(' X ( m ) --> ')
ylabel(' U ( eV ) --> ')
%axis([0 1e-8 -.1 .4])
axis([0 1.2 -.1 .4])
```

%Fig.8.4.1

```
ep=-0.25;ep1=0.25;t=0.5;eta=0.025;
H=[ep t;t ep1];

E=linspace(-1,1,201);
for kE=1:201
```

```
G=inv(((E(kE)+(i*eta))*eye(2,2))-H);
A=diag(i*(G-G'));D(kE)=A(1);
end

hold on
h=plot(D,E,'gx');
set(h,'linewidth',[3.0])
set(gca,'Fontsize',[24])
grid on
xlabel(' LDOS (/ eV) --> ')
ylabel(' Energy (eV) --> ')
```

%Fig.E.8.2

```
clear all
t0=1;zplus=1e-10;

NE=81;X=linspace(-1,3,NE);
for kE=1:NE
    E=2*X(kE);
    ck=1-((E+zplus)/(2*t0));ka=acos(ck);
    if imag(ka) < 0
        ka=ka';end
    sig(kE)=-t0*exp(i*ka);
end

hold on
h1=plot(real(sig),X,'g');
h2=plot(imag(sig),X,'g');
h1=plot(real(sig),X,'go');
set(h1,'linewidth',[2.0])
set(h2,'linewidth',[4.0])
set(gca,'Fontsize',[24])
grid on
xlabel(' --> ')
ylabel(' --> ')
```

%Fig. E.8.5

```
clear all

%Constants (all MKS, except energy which is in eV)
hbar=1.06e-34;q=1.6e-19;m=0.25*9.1e-31;zplus=i*5e-3;

%inputs
a=2.5e-10;t0=(hbar^2)/(2*m*(a^2)*q);Np=100;t0
X=a*linspace(0,Np-1,Np);U=zeros(1,Np);U(Np/2)=5/(a*1e10);
H=(2*t0*diag(ones(1,Np)))-(t0*diag(ones(1,Np-1),1))-(t0*diag(ones(1,Np-1),-1));
H=H+diag(U);
E=0.1;
```

```
%Green's function method
sig1=zeros(Np,Np);sig2=zeros(Np,Np);
    ck=(1-((E+zplus-U(1))/(2*t0)));ka=acos(ck);
        sigma=-t0*exp(i*ka);sig1(1,1)=sigma;
    ck=(1-((E+zplus-U(Np))/(2*t0)));ka=acos(ck);
        sigma=-t0*exp(i*ka);sig2(Np,Np)=sigma;
            G=inv(((E+zplus)*eye(Np))-H-sig1-sig2);
            D0=diag(i*(G-G'))/(2*pi);

hold on
h=plot(X,D0,'b');
grid on
set(h,'linewidth',[2.0])
set(gca,'Fontsize',[24])
xlabel(' X (m) --> ')
ylabel(' DOS ( / eV ) --> ')
```

Chapter 9

% Fig.9.4.2

```
clear all
%Constants (all MKS, except energy which is in eV)
hbar=1.055e-34;m=9.110e-31;q=1.602e-19;a=5e-10;L=10e-9;

k=0.5*linspace(-1,1,201)/a;
Ek=(hbar^2)*(k.^2)/(2*0.25*m*q);
EE=linspace(0,0.2,201);

%Subband (1,1)
E1=2*(hbar^2)*(pi^2)/(2*0.25*m*q*L^2);
M=((EE-E1)+abs(EE-E1))./(2*abs(EE-E1));

%Subbands (1,2) and (2,1)
E2=5*(hbar^2)*(pi^2)/(2*0.25*m*q*L^2);
M=M+(((EE-E2)+abs(EE-E2))./(abs(EE-E2)));

%Subband (2,2)
E3=8*(hbar^2)*(pi^2)/(2*0.25*m*q*L^2);
M=M+(((EE-E3)+abs(EE-E3))./(2*abs(EE-E3)));
k=k*1e-9;
hold on
h=plot(k,E1+Ek,'b');
h=plot(k,E2+Ek,'b');
h=plot(k,E3+Ek,'b');
set(h,'linewidth',[2.0])
set(gca,'Fontsize',[24])
xlabel('k ( / nm )');
ylabel('E - Ec ( eV ) -->');
axis([-1 1 0 0.3]);
grid on
```

% Fig.9.5.5

```
clear all

%Constants (all MKS, except energy which is in eV)
hbar=1.06e-34;q=1.6e-19;m=.25*9.1e-31;IE=(q*q)/(2*pi*hbar);
Ef=0.1;kT=.025;

%inputs
a=3e-10;t0=(hbar^2)/(2*m*(a^2)*q);
NS=15;NC=30;ND=15;Np=NS+NC+ND;

%Hamiltonian matrix
%NS=15;NC=20;ND=15;Np=NS+NC+ND;UB=0*ones(Np,1);%no barrier
%NS=23;NC=4;ND=23;Np=NS+NC+ND;
    %UB=[zeros(NS,1);0.4*ones(NC,1);zeros(ND,1);];%tunneling barrier
NS=15;NC=16;ND=15;Np=NS+NC+ND;
    UB=[zeros(NS,1);.4*ones(4,1);zeros(NC-8,1);.4*ones(4,1);zeros(ND,1)];%RT barrier
T=(2*t0*diag(ones(1,Np)))-(t0*diag(ones(1,Np-1),1))-(t0*diag(ones(1,Np-1),-1));
T=T+diag(UB);

%Bias
V=0;mu1=Ef+(V/2);mu2=Ef-(V/2);
U1=V*[.5*ones(1,NS) linspace(0.5,-0.5,NC) -.5*ones(1,ND)];
U1=U1';%Applied potential profile

%Energy grid for Green's function method
NE=501;E=linspace(-.2,.8,NE);zplus=i*1e-12;dE=E(2)-E(1);
    f1=1./(1+exp((E-mu1)./kT));
        f2=1./(1+exp((E-mu2)./kT));

%Transmission
I=0;%Current
for k=1:NE
    sig1=zeros(Np);sig2=zeros(Np);sig3=zeros(Np);
    ck=1-((E(k)+zplus-U1(1)-UB(1))/(2*t0));ka=acos(ck);
    sig1(1,1)=-t0*exp(i*ka);gam1=i*(sig1-sig1');
        ck=1-((E(k)+zplus-U1(Np)-UB(Np))/(2*t0));ka=acos(ck);
        sig2(Np,Np)=-t0*exp(i*ka);gam2=i*(sig2-sig2');
            G=inv(((E(k)+zplus)*eye(Np))-T-diag(U1)-sig1-sig2-sig3);
                TM(k)=real(trace(gam1*G*gam2*G'));
                    I=I+(dE*IE*TM(k)*(f1(k)-f2(k)));
end
V,I

XX=a*1e9*[1:1:Np];
XS=XX([1:NS-4]);XD=XX([NS+NC+5:Np]);

hold on
%h=plot(TM,E,'b');
h=plot(XX,U1+UB,'b');
h=plot(XS,mu1*ones(1,NS-4),'b--');
```

```
h=plot(XD,mu2*ones(1,ND-4),'b--');
%axis([0 1.1 -.2 .8])
axis([0 15 -.2 .8])
set(h,'linewidth',[2.0])
set(gca,'Fontsize',[24])
%xlabel(' Transmission ---> ')
xlabel(' z ( nm ) --->')
ylabel(' Energy ( eV ) ---> ')
grid on
```

% Fig.9.5.8

```
clear all

%Constants (all MKS, except energy which is in eV)
hbar=1.06e-34;q=1.6e-19;m=.25*9.1e-31;IE=(q*q)/(2*pi*hbar);
Ef=0.1;kT=.025;

%inputs
a=3e-10;t0=(hbar^2)/(2*m*(a^2)*q);
NS=15;NC=30;ND=15;Np=NS+NC+ND;

%Hamiltonian matrix
NS=15;NC=16;ND=15;Np=NS+NC+ND;
    UB=[zeros(NS,1);.4*ones(4,1);zeros(NC-8,1);.4*ones(4,1);zeros(ND,1)];%RT barrier
T=(2*t0*diag(ones(1,Np)))-(t0*diag(ones(1,Np-1),1))-(t0*diag(ones(1,Np-1),-1));
T=T+diag(UB);

%Bias
V=0;mu1=Ef+(V/2);mu2=Ef-(V/2);
U1=V*[.5*ones(1,NS) linspace(0.5,-0.5,NC) -.5*ones(1,ND)];
U1=U1';%Applied potential profile

%Energy grid for Green's function method
NE=501;E=linspace(-.2,.8,NE);zplus=i*1e-12;dE=E(2)-E(1);
    f1=1./(1+exp((E-mu1)./kT));
        f2=1./(1+exp((E-mu2)./kT));

%Transmission
I=0;%Current
for k=1:NE
    sig1=zeros(Np);sig2=zeros(Np);sig3=zeros(Np);
    ck=1-((E(k)+zplus-U1(1)-UB(1))/(2*t0));ka=acos(ck);
    sig1(1,1)=-t0*exp(i*ka);gam1=i*(sig1-sig1');
        ck=1-((E(k)+zplus-U1(Np)-UB(Np))/(2*t0));ka=acos(ck);
        sig2(Np,Np)=-t0*exp(i*ka);gam2=i*(sig2-sig2');
            sig3(Np/2,Np/2)=-i*0.25;gam3=i*(sig3-sig3');%Büttiker probe
            G=inv(((E(k)+zplus)*eye(Np))-T-diag(U1)-sig1-sig2-sig3);
                T12=real(trace(gam1*G*gam2*G'));
                T13=real(trace(gam1*G*gam3*G'));
```

```
                    T23=real(trace(gam2*G*gam3*G'));
                        TM(k)=T12+(T13*T23/(T12+T23));
                            I=I+(dE*IE*TM(k)*(f1(k)-f2(k)));
end
V,I

XX=a*1e9*[1:1:Np];
XS=XX([1:NS-4]);XD=XX([NS+NC+5:Np]);

hold on
h=plot(TM,E,'b');
axis([0 1.1 -.2 .8])
set(h,'linewidth',[2.0])
set(gca,'Fontsize',[24])
xlabel(' Transmission ---> ')
ylabel(' Energy ( eV ) ---> ')
grid on
```

% Fig.9.5.10

```
clear all

%Constants (all MKS, except energy which is in eV)
hbar=1.06e-34;q=1.6e-19;m=.25*9.1e-31;IE=(q*q)/(2*pi*hbar);
Ef=0.1;kT=.025;

%inputs
a=3e-10;t0=(hbar^2)/(2*m*(a^2)*q);
NS=15;NC=30;ND=15;Np=NS+NC+ND;
%Hamiltonian matrix
%NS=15;NC=20;ND=15;Np=NS+NC+ND;UB=0*ones(Np,1);%no barrier
%NS=23;NC=4;ND=23;Np=NS+NC+ND;
    %UB=[zeros(NS,1);0.4*ones(NC,1);zeros(ND,1);];%tunneling barrier
NS=15;NC=16;ND=15;Np=NS+NC+ND;
    UB=[zeros(NS,1);0.4*ones(4,1);zeros(NC-8,1);0.4*ones(4,1);zeros(ND,1)];%RT barrier
T=(2*t0*diag(ones(1,Np)))-(t0*diag(ones(1,Np-1),1))-(t0*diag(ones(1,Np-1),-1));
T=T+diag(UB);

%Bias
NV=26;VV=linspace(0,.5,NV);
for iV=1:NV
V=VV(iV);mu1=Ef+(V/2);mu2=Ef-(V/2);
U1=V*[.5*ones(1,NS) linspace(0.5,-0.5,NC) -.5*ones(1,ND)];
U1=U1';%Applied potential profile

%Energy grid for Green's function method
NE=101;E=linspace(-.2,.8,NE);zplus=i*1e-12;dE=E(2)-E(1);
    f1=1./(1+exp((E-mu1)./kT));
    f2=1./(1+exp((E-mu2)./kT));
        %For infinite 2-D cross-section
```

```
                %f1=(2*m*kT*q/(2*pi*hbar^2)).*log(1+exp((mu1-E)./kT));
                %f2=(2*m*kT*q/(2*pi*hbar^2)).*log(1+exp((mu2-E)./kT));

%Transmission
I=0;%Current
for k=1:NE
    sig1=zeros(Np);sig2=zeros(Np);sig3=zeros(Np);
    ck=1-((E(k)+zplus-U1(1)-UB(1))/(2*t0));ka=acos(ck);
    sig1(1,1)=-t0*exp(i*ka);gam1=i*(sig1-sig1');
        ck=1-((E(k)+zplus-U1(Np)-UB(Np))/(2*t0));ka=acos(ck);
        sig2(Np,Np)=-t0*exp(i*ka);gam2=i*(sig2-sig2');
            sig3(Np/2,Np/2)=-i*0.00025;gam3=i*(sig3-sig3');%Büttiker probe
            G=inv(((E(k)+zplus)*eye(Np))-T-diag(U1)-sig1-sig2-sig3);
                T12=real(trace(gam1*G*gam2*G'));
                T13=real(trace(gam1*G*gam3*G'));
                T23=real(trace(gam2*G*gam3*G'));
                    TM(k)=T12+(T13*T23/(T12+T23));
                        I=I+(dE*IE*TM(k)*(f1(k)-f2(k)));
end
II(iV)=I;V,I
end

XX=a*1e9*[1:1:Np];
XS=XX([1:NS-4]);XD=XX([NS+NC+5:Np]);

hold on
h=plot(VV,II,'b');
%h=plot(XX,U1+UB,'b');
%h=plot(XS,mu1*ones(1,NS-4),'b--');
%h=plot(XD,mu2*ones(1,ND-4),'b--');
axis([0 .5 0 3.5e-7])
%axis([0 15 -.3 .7])
set(h,'linewidth',[2.0])
set(gca,'Fontsize',[24])
xlabel(' Voltage ( V ) ---> ')
%xlabel(' z ( nm ) --->')
%ylabel(' Energy ( eV ) ---> ')
ylabel(' Current ( A ) ---> ')
grid on
```

Chapter 10

% Fig.10.4.4

```
clear all

beta=pi*linspace(-1,1,201);w0=1;
y=sqrt(2*w0*(1-cos(beta)));

hold on
h=plot(beta,y,'b');
```

```
set(h,'linewidth',[2.0])
set(gca,'Fontsize',[24])
grid on
```

% Fig.10.4.5

```
clear all

beta=pi*linspace(-1,1,201);w1=1;w2=2;
for n=1:201
    A=[w1+w2 w1+(w2*exp(-i*beta(n)));w1+(w2*exp(i*beta(n))) w1+w2];
    [V,D]=eig(A);D=sort(real(diag(D)));
    D1(n)=real(sqrt(D(1)));D2(n)=real(sqrt(D(2)));
end

hold on
h=plot(beta,D1,'b');
h=plot(beta,D2,'b');
set(h,'linewidth',[2.0])
set(gca,'Fontsize',[24])
grid on
```

Chapter 11

% Fig.11.2.2, 11.2.7

```
clear all

%1-D with elastic phase-breaking and/or coherent, T vs. E, fixed length

%Constants (all MKS, except energy which is in eV)
hbar=1.06e-34;q=1.6e-19;m=1*9.1e-31;IE=(q*q)/(2*pi*hbar);kT=.025;

%inputs
a=3e-10;t0=(hbar^2)/(2*m*(a^2)*q);D=0.01*0;

%Energy grid
NE=401;E=linspace(.1,.3,NE);zplus=i*1e-12;dE=E(2)-E(1);

%Bias
V=0.01;f1=1/(1+exp((-V/2)/kT));
        f2=1/(1+exp((V/2)/kT));

%Hamiltonian
    Np=40;UB=zeros(Np,1);UB(5)=0.5*1;UB(36)=0.5*1;
    U1=V*linspace(0.5,-0.5,Np)';XX=a*linspace(0,Np-1,Np);
    T=(2*t0*diag(ones(1,Np)))-(t0*diag(ones(1,Np-1),1))-(t0*diag(ones(1,Np-1),-1));

%Iterative solution
for k=1:NE
    sig1=zeros(Np);sig2=zeros(Np);
    ck=1-((E(k)+zplus-U1(1)-UB(1))/(2*t0));ka=acos(ck);
```

```
sig1(1,1)=-t0*exp(i*ka);gam1=i*(sig1-sig1');
        ck=1-((E(k)+zplus-U1(Np)-UB(Np))/(2*t0));ka=acos(ck);
        sig2(Np,Np)=-t0*exp(i*ka);gam2=i*(sig2-sig2');

    %calculating the Green function, G self-consistently
    G=inv(((E(k)+zplus)*eye(Np))-T-diag(U1+UB)-sig1-sig2);change=1;
        while(change>1e-4)
            sigp=diag(D*diag(G));
    S=inv(((E(k)+zplus)*eye(Np))-T-diag(U1+UB)-sig1-sig2-sigp);
    change=sum(sum(abs(G-S)))/(sum(sum(abs(G)+abs(S))));
        G=(0.5*G)+(0.5*S);
        end
        G=S;A=i*(G-G');
        M=D*(G.*conj(G));

    %calculating the electron density,n(r;E)
    gamp=i*(sigp-sigp');gamma=gam1+gam2+gamp;
        sigin1=f1*gam1;sigin2=f2*gam2;
        n=(inv(eye(Np)-M))*diag(G*(sigin1+sigin2)*G');
        siginp=D*diag(n);

    %calculating the correlation function Gn
        Gn=G*(sigin1+sigin2+siginp)*G';
    %calculating the effective transmission
        I1(k)=(1/(f2-f1))*real(trace(gam1*Gn)-trace(sigin1*A));
        I2(k)=(1/(f1-f2))*real(trace(gam2*Gn)-trace(sigin2*A));
end

hold on
h=plot(I1,E,'b');
%h=plot(I2,E,'bx');
%h=plot(1e9*XX,U1+UB,'b');
%h=plot(1e9*XX,U1+UB,'bo');
set(h,'linewidth',[2.0])
set(gca,'Fontsize',[24])
%xlabel(' x ( nm ) ---> ')
%ylabel(' Potential ( eV ) ---> ')
xlabel(' Transmission ---> ')
ylabel(' Energy ( eV ) ---> ')
grid on
axis([0 1.1 .1 .3])
```

% Fig.11.2.4, 11.2.6

```
clear all

%multi-moded coherent transport,T vs. E

%Constants (all MKS, except energy which is in eV)
hbar=1.06e-34;q=1.6e-19;m=9.1e-31;IE=(q*q)/(2*pi*hbar);kT=.025;
```

```
%inputs
a=5e-10;t0=(hbar^2)/(2*m*(a^2)*q);
%Energy grid
NE=11;% 11 for one scatterer, 101 for two
E=linspace(0.1,0.3,NE);zplus=i*1e-12;dE=E(2)-E(1);

%Bias
V=0.01;f1=1/(1+exp((-V/2)/kT));
        f2=1/(1+exp((V/2)/kT));

%Transverse modes
NW=15;NT=7;
alpha=(4*t0*diag(ones(1,NW)))-(t0*diag(ones(1,NW-1),1))-(t0*diag(ones(1,NW-1),-1));
[VT,D]=eig(alpha);[D ind]=sort(diag(D));
in=[];for k=1:NT
    in=[in ind(k)];end
VT=VT(:,in);D=diag(VT'*alpha*VT);

%Hamiltonian
    Np=40;UB=zeros(Np,1);UB(5)=0.25*1;UB(36)=0.25*0;
    impshape=[linspace(0,1,7) linspace(1,0,NW-7)];
    U1=V*linspace(0.5,-0.5,Np)';
    al=alpha+(U1(1)*eye(NW,NW));
    H=VT'*al*VT;H1=H;
    Z=zeros(NT,NT);bet=-t0*eye(NT,NT);
    for N=2:Np
        al=alpha+(U1(N)*eye(NW,NW));al1=al;
        al=al+(diag(UB(N)*impshape));
        al=VT'*al*VT;H=[H bet;bet' al];
        al1=VT'*al1*VT;H1=[H1 bet;bet' al1];%Use for one scatterer
        bet=[Z;bet];
    end
%calculating the transmission
for k=1:NE
    ck=(D-E(k)-zplus+U1(1))./(2*t0);ka=acos(ck);
    s1=-t0*exp(i.*ka);sig1=[diag(s1) zeros(NT,NT*(Np-1));zeros(NT*(Np-1),NT*Np)];
        ck=(D-E(k)-zplus+U1(Np))./(2*t0);ka=acos(ck);
        s2=-t0*exp(i.*ka);sig2=[zeros(NT*(Np-1),NT*Np);zeros(NT,NT*(Np-1)) diag(s2);];
        gam1=i*(sig1-sig1');gam2=i*(sig2-sig2');

    G=inv(((E(k)+zplus)*eye(NT*Np))-H-sig1-sig2);
        T(k)=real(trace(gam1*G*gam2*G'));
    G1=inv(((E(k)+zplus)*eye(NT*Np))-H1-sig1-sig2);%Use for one scatterer
        M(k)=real(trace(gam1*G1*gam2*G1'));[k T(k) M(k)],%use for one scatterer
end

Tsc=T./(2-(T./M));%semiclassical addition, use for one scatterer

%save condfluct2
XX=a*linspace(0,Np-1,Np);
```

```
hold on
%h=plot(T,E,'b');
h=plot(Tsc,E,'b--');
%h=plot(M,E,'b');
set(h,'linewidth',[2.0]);
set(gca,'Fontsize',[24]);
xlabel(' Transmission ---> ')
ylabel(' Energy ( eV ) ---> ')
axis([0 5 .1 .3])
grid on
```

% Fig.11.2.8

```
clear all

%1-D elastic coherent and/or phase-breaking, R vs. L, fixed E

%Constants (all MKS, except energy which is in eV)
hbar=1.06e-34;q=1.6e-19;m=.25*9.1e-31;IE=(q*q)/(2*pi*hbar);kT=.025;

%inputs
a=3e-10;t0=(hbar^2)/(2*m*(a^2)*q);D=0.05;V=0.01;

%Bias

%Energy grid
E=0.1;zplus=i*1e-12;
    f1=1/(1+exp((-V/2)/kT));
        f2=1/(1+exp((V/2)/kT));

%Current
for k=2:21
    Np=k;UB=zeros(Np,1);U1=V*linspace(0.5,-0.5,Np)';k
    T=(2*t0*diag(ones(1,Np)))-(t0*diag(ones(1,Np-1),1))-(t0*diag(ones(1,Np-1),-1));

    sig1=zeros(Np);sig2=zeros(Np);sig3=zeros(Np);
    ck=1-((E+zplus-U1(1)-UB(1))/(2*t0));ka=acos(ck);
    sig1(1,1)=-t0*exp(i*ka);gam1=i*(sig1-sig1');
        ck=1-((E+zplus-U1(Np)-UB(Np))/(2*t0));ka=acos(ck);
        sig2(Np,Np)=-t0*exp(i*ka);gam2=i*(sig2-sig2');

    %calculating the Green function, G self-consistently
    G=inv(((E+zplus)*eye(Np))-T-diag(U1+UB)-sig1-sig2);change=1;
        while(change>1e-4)
            sigp=diag(D*diag(G));
    S=inv(((E+zplus)*eye(Np))-T-diag(U1+UB)-sig1-sig2-sigp);
    change=sum(sum(abs(G-S)))/(sum(sum(abs(G)+abs(S))));
        G=(0.5*G)+(0.5*S);
        end
        G=S;A=i*(G-G');
        M=D*(G.*conj(G));
```

```
    %calculating the inscattering functions from the contacts F1,F2
    gam1=i*(sig1-sig1');gam2=i*(sig2-sig2');
    gamp=i*(sigp-sigp');gamma=gam1+gam2+gamp;
        sigin1=f1*gam1;sigin2=f2*gam2;
        n=(inv(eye(Np)-M))*diag(G*(sigin1+sigin2)*G');
        siginp=D*diag(n);

    %calculating the correlation function Gn
        Gn=G*(sigin1+sigin2+siginp)*G';

    %calculating the current
        I1(k-1)=(1/(f1-f2))*real(trace(gam1*Gn)-trace(sigin1*A));
        I2(k-1)=(1/(f1-f2))*real(trace(gam2*Gn)-trace(sigin2*A));
        L(k-1)=k*a*1e10;
end

L=L./10;% Angstrom to nm
hold on
h=plot(L,1./I2,'b');
%h=plot(I1+I2,'g--');
set(h,'linewidth',[2.0])
set(gca,'Fontsize',[24])
xlabel(' Length ( nm ) ---> ')
ylabel(' Normalized resistance ---> ')
axis([0 6 0 3])
grid on
```

% Fig.11.3.1, 11.3.2, 11.3.3

```
clear all

%1-D with inelastic scattering

%Constants (all MKS, except energy which is in eV)
hbar=1.06e-34;q=1.6e-19;m=.25*9.1e-31;Ef=0.15;kT=0.025;

%inputs
a=3e-10;t0=(hbar^2)/(2*m*(a^2)*q);

%Hamiltonian matrix
    Np=40;UB=0*[zeros(10,1);0.25*ones(Np-10,1)];
    T=(2*t0*diag(ones(1,Np)))-(t0*diag(ones(1,Np-1),1))-(t0*diag(ones(1,Np-1),-1));

%Bias
V=0.1;mu1=Ef+(V/2);mu2=Ef-(V/2);
U1=V*[.5*ones(1,1) linspace(0.5,-0.5,Np-2) -.5*ones(1,1)]';%Applied potential profile
    D=1e-1;%Scattering strength

%Energy grid
NE=101;E=linspace(-.05,.35,NE);zplus=i*1e-12;dE=E(2)-E(1);
    f1=1./(1+exp((E-mu1)./kT));
        f2=1./(1+exp((E-mu2)./kT));
```

```
%Initial guess
sigin=0*ones(Np,NE);sigout=0*ones(Np,NE);

%Iterative solution of transport equation
change=1;it=1;n=zeros(Np,NE);p=zeros(Np,NE);
while change>1e-3
for k=1:NE
    sig1=zeros(Np);sig2=zeros(Np);
        ck=1-((E(k)+zplus-U1(1)-UB(1))/(2*t0));ka=acos(ck);
        sig1(1,1)=-t0*exp(i*ka);gam1=i*(sig1-sig1');
            ck=1-((E(k)+zplus-U1(Np)-UB(Np))/(2*t0));ka=acos(ck);
            sig2(Np,Np)=-t0*exp(i*ka);gam2=i*(sig2-sig2');
        sigin1(:,k)=f1(k)*diag(gam1);sigin2(:,k)=f2(k)*diag(gam2);
        sigout1(:,k)=(1-f1(k))*diag(gam1);sigout2(:,k)=(1-f2(k))*diag(gam2);
        gamp=sigin(:,k)+sigout(:,k);

    G=inv(((E(k)+zplus)*eye(Np))-T-diag(U1+UB)-sig1-sig2+(i*0.5*diag(gamp)));
        A=diag(i*(G-G'));
        n(:,k)=real(diag(G*((f1(k)*gam1)+(f2(k)*gam2)+diag(sigin(:,k)))*G'));
        p(:,k)=A-n(:,k);
end

off=0;%less than NE-1, equal to 0 for elastic
C=exp(-dE*off/kT);
ne=n(:,[off+1:NE]);ne=[ne zeros(Np,off)];
na=n(:,[1:NE-off]);na=[zeros(Np,off) na];
pa=p(:,[off+1:NE]);pa=[pa zeros(Np,off)];
pe=p(:,[1:NE-off]);pe=[zeros(Np,off) pe];

siginnew=(D*ne)+(C*D*na);
sigoutnew=(D*pe)+(C*D*pa);
change=sum(sum(abs(siginnew-sigin)));
change=change+sum(sum(abs(sigoutnew-sigout)))
sigin=((1-it)*sigin)+(it*siginnew);
sigout=((1-it)*sigout)+(it*sigoutnew);
end

    I1=real((sigout1.*n)-(sigin1.*p));I1=sum(I1);
    I2=real((sigout2.*n)-(sigin2.*p));I2=sum(I2);
    I3=real((sigout.*n)-(sigin.*p));I3=sum(I3);

I123=(dE/V)*[sum(I1) sum(I2) sum(I3)],%Normalized Conductance
IE=(dE/(V*V))*[sum(E.*I1) sum(E.*I2) sum(E.*I3)],%Normalized Power
kirchoff=[sum(I123) sum(IE)],%checking for conservation of current and energy current

save inel0
hold on
h=plot(I1,E,'b');
h=plot(I2,E,'b--');
%h=plot(I3,E,'c');
set(h,'linewidth',[2.0])
```

```
set(gca,'Fontsize',[24])
xlabel(' Normalized current / energy ---> ')
ylabel(' Energy ( eV ) ---> ')
axis([-.2 .2 -.05 .35])
```

% Fig.11.4.4

```
clear all

%Ballistic self-consistent solution

%Constants (all MKS, except energy which is in eV)
hbar=1.06e-34;q=1.6e-19;m=.25*9.1e-31;
kT=0.0259;zplus=i*1e-12;eps0=8.854e-12;

%inputs
a=3e-10;t0=(hbar^2)/(2*m*(a^2)*q);N=40;
Ef=0.1;Ec=-0.5;Vg=0;
r=5e-9;tox=5e-9;K=2;%Use large value of permittivity K for Laplace limit
U0=q/2/pi/a/K/eps0.*log((r+tox)/r)

%Hamiltonian matrix
    Np=40;
    H0=(2*t0*diag(ones(1,Np)))-(t0*diag(ones(1,Np-1),1))-(t0*diag(ones(1,Np-1),-1));

%Energy grid
    NE=401;E=linspace(-0.5,0.3,NE);dE=E(2)-E(1);I0=(q^2)/hbar/2/pi

%Bias
iV=41;V=linspace(0,0.4,iV);n0=0;UL=-Vg*ones(Np,1);U=UL;
for kk=1:iV
    Vd=V(kk);mu1=Ef;mu2=Ef-Vd;
    sig1=zeros(Np);sig2=zeros(Np);
    epsilon=1;
    while (epsilon>0.001)
        rho=0;
        for k=1:NE
            f1=1/(1+exp((E(k)-mu1)/kT));f2=1/(1+exp((E(k)-mu2)/kT));

            cka1=1-(E(k)+zplus-Ec)/2/t0; ka1=acos(cka1);
            sig1(1,1)=-t0*exp(i*ka1);gam1=i*(sig1-sig1');
                cka2=1-(E(k)+zplus-Ec+Vd)/2/t0; ka2=acos(cka2);
        sig2(N,N)=-t0*exp(i*ka2);gam2=i*(sig2-sig2');
                G=inv((E(k)+zplus)*eye(N)-H0-diag(U)-sig1-sig2);A=i*(G-G');
                    sigin1=f1*gam1;sigin2=f2*gam2;
                        Gn=G*(sigin1+sigin2)*G';rho=rho+dE/2/pi*Gn;
                T(k)=trace(gam1*G*gam2*G');
        I1(k)=real(trace(sigin1*A)-trace(gam1*Gn));
        I2(k)=-real(trace(sigin2*A)-trace(gam2*Gn));
            end
        n=real(diag(rho));Unew=UL+(U0*(n-n0));
```

```
        dU=Unew-U;epsilon=max(abs(dU))
        U=U+0.25*dU;
            if Vd==0
            n0=n;epsilon=0;end
        end
        ID1=2*I0*dE*sum(I1);ID2=2*I0*dE*sum(I2);%2 for spin
        I(kk)=ID1;
end

save IV2
IonL=I0*Ef
hold on
h=plot(V,I,'b');
h=plot(V,IonL*ones(iV,1),'bx');
set(h,'linewidth',[2.0])
set(gca,'Fontsize',[24])
grid on
xlabel(' Voltage ( V ) --> ')
ylabel(' Current ( A ) --> ')
```

% Fig.E.11.5

```
clear all

%1-D tunneling and/or elastic phase-breaking, R vs. L, fixed E

%Constants (all MKS, except energy which is in eV)
hbar=1.06e-34;q=1.6e-19;m=.25*9.1e-31;IE=(q*q)/(2*pi*hbar);kT=.025;
zplus=i*1e-51;

%inputs
a=3e-10;t0=(hbar^2)/(2*m*(a^2)*q);
D=3e-1;% Scattering Strength: 2e-1 (x's) and 3e-1 (o's) eV^2
V=0.001;% Applied voltage
mu=0.1;% Fermi energy
    f1=1/(1+exp((-V/2)/kT));%Fermi function in contact 1 at E=mu
        f2=1/(1+exp((V/2)/kT));%Fermi function in contact 2 at E-mu

% Actual calculation
E=mu;
for k=5:26
    Np=k;%Length of barrier = (Np-2)*a
    UB=[0;0.5*ones(Np-2,1);0];% Barrier height
    U1=V*linspace(0.5,-0.5,Np)';% Applied potential profile
    T=(2*t0*diag(ones(1,Np)))-(t0*diag(ones(1,Np-1),1))-...
        (t0*diag(ones(1,Np-1),-1));%Tight-binding Hamiltonian

    sig1=zeros(Np);sig2=zeros(Np);sig3=zeros(Np);
    ck=1-((E+zplus-U1(1)-UB(1))/(2*t0));ka=acos(ck);
    sig1(1,1)=-t0*exp(i*ka);gam1=i*(sig1-sig1');%Self-energy for contact 1
```

```
        ck=1-((E+zplus-U1(Np)-UB(Np))/(2*t0));ka=acos(ck);
        sig2(Np,Np)=-t0*exp(i*ka);gam2=i*(sig2-sig2');%Self-energy for contact 2

    %calculating the Green function, G self-consistently
    G=inv(((E+zplus)*eye(Np))-T-diag(U1+UB)-sig1-sig2);change=1;
        while(change>1e-15)
        sigp=diag(D*diag(G));%Self-energy due to scattering
    S=inv(((E+zplus)*eye(Np))-T-diag(U1+UB)-sig1-sig2-sigp);
    change=sum(sum(abs(G-S)))/(sum(sum(abs(G)+abs(S))));
        G=(0.5*G)+(0.5*S);
        end
        G=S;A=i*(G-G');M=D*(G.*conj(G));

    %calculating the inscattering functions from the contacts F1,F2
    gam1=i*(sig1-sig1');gam2=i*(sig2-sig2');
    gamp=i*(sigp-sigp');gamma=gam1+gam2+gamp;
        sigin1=f1*gam1;sigin2=f2*gam2;
        n=(inv(eye(Np)-M))*diag(G*(sigin1+sigin2)*G');
        siginp=D*diag(n);%Inflow due to scattering

    %calculating the correlation function Gn
        Gn=G*(sigin1+sigin2+siginp)*G';

    %calculating the current
        I1(k-4)=(1/(f1-f2))*real(trace(gam1*Gn)-trace(sigin1*A));
        I2(k-4)=(1/(f1-f2))*real(trace(gam2*Gn)-trace(sigin2*A));
        L(k-4)=(k-2)*a*1e9;%in nanometers
end

hold on
h=plot(L,log10(-1./I1),'g');% Current at left end
h=plot(L,log10(1./I2),'go');% Current at right end
set(h,'linewidth',[2.0])
set(gca,'Fontsize',[24])
xlabel(' Length ( nm ) ---> ')
ylabel(' log10 ( resistance ) ---> ')
grid on
```

%Fig.E.11.6

```
clear all
%0-D with inelastic scattering

%Constants (all MKS, except energy which is in eV)
hbar=1.06e-34;q=1.6e-19;I0=q*q/(2*pi*hbar);

%Parameters
    H0=5;Ef=0;kT=0.0025;dE=0.0005;zplus=i*1e-12;gamma=0.1;
        D0=0;Dnu=0*[0.5 0.7];Nph=size(Dnu,2);
            hnu=[100 550];%Multiply by dE for actual hnu
                Nhnu=1./((exp(dE*hnu./kT))-1);
```

```
%Bias
NV=203;VV=linspace(-0.51,0.5,NV);dV=VV(2)-VV(1);
for iV=1:NV
V=VV(iV);mu1=Ef;mu2=Ef-V;U1=(-0.5)*V;

   %Energy grid
   E=[mu2-(10*kT)-(10*dE):dE:mu1+(10*kT)+(10*dE)];
      if V<0
         E=[mu1-(10*kT)-(10*dE):dE:mu2+(10*kT)+(10*dE)];
      end
      NE=size(E,2);[iV NE]
         f1=1./(1+exp((E-mu1)./kT));
         f2=1./(1+exp((E-mu2)./kT));

   %Initial guess
      n=zeros(1,NE);p=zeros(1,NE);
      sigin1=zeros(1,NE);sigout1=zeros(1,NE);
      sigin2=zeros(1,NE);sigout2=zeros(1,NE);
      sigin=0*ones(1,NE);sigout=0*ones(1,NE);
%Iterative solution of transport equation
change=1;it=1;
while change>1e-3
   for k=1:NE
   sig1=-i*gamma/2;gam1=i*(sig1-sig1');
   sig2=-i*gamma/2;gam2=i*(sig2-sig2');
      sigin1(k)=f1(k)*gam1;sigin2(k)=f2(k)*gam2;
      sigout1(k)=(1-f1(k))*gam1;sigout2(k)=(1-f2(k))*gam2;
      gamp=sigin(k)+sigout(k);
   G=inv((E(k)+zplus)-H0-U1-sig1-sig2+(i*0.5*gamp));
      A=i*(G-G');
         n(k)=real(G*((f1(k)*gam1)+(f2(k)*gam2)+sigin(k))*G');
         p(k)=A-n(k);
   end

   siginnew=D0*n;sigoutnew=D0*p;
   for iph=1:Nph
      inu=hnu(iph);
      if inu<NE
      ne=n([inu+1:NE]);ne=[ne zeros(1,inu)];
      na=n([1:NE-inu]);na=[zeros(1,inu) na];
      pe=p([inu+1:NE]);pe=[pe zeros(1,inu)];
      pa=p([1:NE-inu]);pa=[zeros(1,inu) pa];
         siginnew=siginnew+((Nhnu(iph)+1)*Dnu(iph)*ne)+(Nhnu(iph)*Dnu(iph)*na);
      sigoutnew=sigoutnew+(Nhnu(iph)*Dnu(iph)*pe)+((Nhnu(iph)+1)*Dnu(iph)*pa);
      end
   end

   change=sum(sum(abs(siginnew-sigin)));
   change=change+sum(sum(abs(sigoutnew-sigout)));
```

```
        sigin=((1-it)*sigin)+(it*siginnew);
        sigout=((1-it)*sigout)+(it*sigoutnew);
end

    I1=real((sigout1.*n)-(sigin1.*p));I1=sum(I1);
    I2=real((sigout2.*n)-(sigin2.*p));I2=sum(I2);
    I3=real((sigout.*n)-(sigin.*p));I3=sum(I3);

    I123=dE*[sum(I1) sum(I2) sum(I3)],%Normalized Conductance
    %IE=(dE/(V*V))*[sum(E.*I1) sum(E.*I2) sum(E.*I3)],%Normalized Power
    %kirchoff=[sum(I123) sum(IE)],%checking for conservation of current and energy current

II(iV)=sum(I2)*dE*I0;
end

G1=diff(II)./dV;VG=VV([2:NV]);
IETS=diff(G1)./dV;VETS=VV([3:NV]);

hold on
%h=plot(VV,II,'rx');
h=plot(VG,G1,'b–');
set(h,'linewidth',[2.0])
set(gca,'Fontsize',[24])
%xlabel(' Voltage (V) --> ')
%ylabel(' d2I/dV2 --> ')
```

Appendix

% Fig.A.5.2, A.5.3

```
clear all

NE=1001;E=linspace(-.25,.25,NE);zplus=i*1e-3;dE=E(2)-E(1);kT=.00026;
Nep=5001;ep=linspace(-1,1,Nep);tau=0.05;dep=ep(2)-ep(1);delta=3.117*tau*tau/2
ep0=-25*delta;U=50*delta;[U/pi abs(ep0)]/delta
D=ones(1,Nep);f=1./(1+exp(ep./kT));fK=1./(1+exp(E./kT));tau=0.06;

for kE=1:NE
s0(kE)=dep*tau*tau*sum(D./(E(kE)-ep+zplus));
s1(kE)=dep*tau*tau*sum(D./(E(kE)-ep0-ep0-U+ep+zplus));
s2(kE)=dep*tau*tau*sum(D.*f./(E(kE)-ep+zplus));
s3(kE)=dep*tau*tau*sum(D.*f./(E(kE)-ep0-ep0-U+ep+zplus));
end

g=U./(E-ep0-U-s0-s0-s1);
GK=(1+(0.5*g))./(E-ep0-s0+(g.*(s2+s3)));
G=(1+(0.5*U./(E-ep0-U-s0)))./(E-ep0-s0);
A=i*(G-conj(G))/(2*pi);dE*sum(A)
AK=i*(GK-conj(GK))/(2*pi);dE*sum(AK)
dE*sum(AK.*fK)
del=-dE*sum(imag(s0))
```

```
hold on
%h=plot(E,-imag(s0));
%h=plot(E,imag(s1));
%h=plot(E,imag(s2),'mx');
%h=plot(E,imag(s3),'m');
h=plot(A,E,'b--');
h=plot(AK,E,'b');
set(h,'linewidth',[2.0])
set(gca,'Fontsize',[24])
grid on
xlabel(' D(E) per eV --> ')
ylabel(' E ( eV ) --> ')
```

Further reading

This is not intended to be a representative bibliography, just a list of references that could help readers broaden their understanding of the material presented herein, which is a very small subset of the vast literature in the field.

General

This book has grown out of a graduate course (and recently its undergraduate version) that I have been teaching for a number of years. The reader may find it useful to view the videostreamed course lectures, which are publicly available through the web. Information should be available at my website

http://dynamo.ecn.purdue.edu/~datta

and through the "nanohub" at www.nanohub.org.

1 Prologue

This chapter and the Appendix have been adapted from Datta (2004). With corrections.

The example used to illustrate the importance of the potential profile in determining the current–voltage characteristics in Section 1.4 (Figs. 1.4.2, 1.4.3) is motivated by the experiments discussed in

Datta S., Tian W., Hong S., Reifenberger R., Henderson J., and Kubiak C. P. (1997). STM current–voltage characteristics of self-assembled monolayers (SAMs). *Phys. Rev. Lett.*, **79**, 2530.

We will not be discussing the Coulomb blockade regime in this book (other than Sections 1.5 and 3.4). For further reading, see for example, Bonet *et al.* (2002) and:

Kastner M. (1993). Artificial atoms. *Phys. Today*, **46**, 24.
Likharev K. (1999). Single-electron devices and their applications. *Proc. IEEE*, **87**, 606.
Kouwenhoven L. P. and McEuen P. L. (1997). Single electron transport through a quantum dot. In *Nano-Science and Technology*, ed. G. Timp. AIP Press.
Ferry D. K. and Goodnick S. M. (1997). *Transport in Nanostructures*. Cambridge University Press. Chapter 4.

The simple nanotransistor model briefly described in Section 1.6 is essentially the same as that described in:

Rahman A., Guo J., Datta S., and Lundstrom M. (2003). Theory of ballistic transistors. *IEEE Trans. Electron Dev.*, **50**, 1853.

I mentioned in Section 1.6 that the regime of transport with $t \leq U_0$ presents major theoretical challenges (see, for example, Georges (2004)). One of the paradigms widely investigated in this context is that of "Luttinger liquids," see, for example, Giamarchi (2004).

For an introduction to the Kondo effect, see for example, Kouwenhoven and Glazman (2001).

2 Schrödinger equation

For a summary of basic quantum mechanics and the hydrogen atom the reader could consult any beginning text, such as:

Eisberg R. and Resnick R. (1974, 1985). *Quantum Physics*. Wiley.
Cox P. A. (1996). *Introduction to Quantum Theory and Atomic Structure*. Oxford Chemistry Primers.

We have discussed only the finite difference method here. The reader may want to look up the finite element method. See for example Ramdas Ram-Mohan (2002) or White *et al.* (1989).

3 Self-consistent field

For a detailed discussion of the self-consistent field method, the reader is directed to Herman and Skillman (1963) and:

Slater, J. C. (1963–1974). *Quantum Theory of Molecules and Solids*, Vols. I–IV. McGraw-Hill.
Grant G. H. and Richards W. G. (1995). *Computational Chemistry*. Oxford Chemistry Primers.

For more discussion of the structural aspects:

Pettifor D. (1995). *Bonding and Structure of Molecules and Solids*. Oxford University Press.

The discussion in Section 3.4 is based on the approach described in:

Beenakker C. W. J. (1991). Theory of Coulomb blockade oscillations in the conductance of a quantum dot. *Phys. Rev.* B, **44**, 1646.

For a discussion of the interpretation of one-particle energy levels, see for example:

Brus L. E. (1983). A simple model for the ionization potential, electron affinity, and aqueous redox potentials of small semiconductor crystallites. *J. Chem. Phys.*, **79**, 5566.

More recent references include:

Bakkers E. P. A. M., Hens Z., Zunger A., Franceschetti A., Kouwenhoven L. P., Gurevich L., and Vanmaekelbergh D. (2001). Shell-tunneling spectroscopy of the single-particle energy levels of insulating quantum dots. *Nano Lett.*, **1**, 551.
Niquet Y. M., Delerue C., Allan G., and Lannoo M. (2002). Interpretation and theory of tunneling experiments on single nanostructures. *Phys. Rev.* B, **65**, 165334.

4 Basis functions

The reader may enjoy the discussion of "base states" in Feynman (1965), especially Chapter 8.

Orthogonal basis functions are widely discussed in all quantum mechanics texts such as:

Cohen-Tannoudji C., Diu B., and Laloe F. (1977). *Quantum Mechanics.* Wiley.
Schiff L. I. (1955). *Quantum Mechanics.* McGraw-Hill.

Non-orthogonal bases are discussed in more detail in quantum chemistry texts such as:

Szabo A. and Ostlund N. S. (1996). *Modern Quantum Chemistry.* Dover.

5 Bandstructure

Bandstructure-related concepts are discussed in solid-state physics texts like:

Ziman J. M. (1972). *Principles of the Theory of Solids.* Cambridge University Press.
Kittel C. (1976). *Introduction to Solid State Physics.* Wiley.

Section 5.3 closely follows the seminal work of Vogl *et al.* (1983). For more recent developments and references, see for example:

Boykin T. B., Klimeck G., Chris Bowen R., and Oyafuso F. (2002). Diagonal parameter shifts due to nearest-neighbor displacements in empirical tight-binding theory. *Phys. Rev.* B, **66**, 125207, and references therein.

We have only described the empirical tight-binding method. Many other approaches have been used, most notably the pseudopotential method and the $k \cdot p$ method. See for example Fischetti (1991) and Singh (1993).

6 Subbands

Detailed discussion of carbon nanotubes can be found in:

Dresselhaus M. S., Dresselhaus G., and Eklund P. C. (1996). *Science of Fullerenes and Carbon Nanotubes.* Academic.

7 Capacitance

The term "quantum capacitance" was probably first introduced by:

Luryi S. (1988). Quantum capacitance devices. *Appl. Phys. Lett.*, **52**, 501.

It has been used by other authors, see for example:

Katayama Y. and Tsui D. C. (1993). Lumped circuit model of two-dimensional tunneling transistors. *Appl. Phys. Lett.*, **62**, 2563.

For more on the multi-band effective mass model see Singh (1993).

8 Level broadening

For interesting discussions of the problem of describing irreversible phenomena starting from reversible laws, see for example:

Prigogine I. (1980). *From Being to Becoming.* Freeman.

Feynman R. P. (1965). The distinction of past and future. *Character of Physical Law*. MIT Press. Chapter 5.

9 Coherent transport

For discussions of the transmission formalism the reader can consult Datta (1995) and Büttiker (1988). Also:

Imry Y. (1997). *Introduction to Mesoscopic Physics*. Oxford University Press.
Ferry D. K. and Goodnick S. M. (1997). *Transport in Nanostructures*. Cambridge University Press.

To get a flavor of some of the more recent developments, see for example Beenakker (1997) and Büttiker (2001).

Coherent transport calculations are often carried out non-self-consistently as in Fig. 9.5.10, but this aspect can affect the current–voltage characteristics significantly. See for example Cahay *et al.* (1987), which uses the transmission formalism to calculate the electron density, but I believe it is more convenient to use the NEGF equations discussed in this chapter.

10 Non-coherent transport

For more on electron–photon interactions, see:

Feynman R. P. (1985). *QED: The Strange Theory of Light and Matter*. Princeton University Press.
Marcuse D. (1980). *Principles of Quantum Electronics*. Academic.

Semi-classical methods that ignore quantum interference but treat non-coherent scattering processes in great detail are widely used to model electronic devices. For a lucid introduction, see:

Lundstrom M. S. (2000). *Fundamentals of Carrier Transport*. Cambridge University Press.

The lowest order treatment of electron–phonon interaction described in Section 10.3 is similar to that described in Caroi *et al.* (1972).

I mention in Section 10.3 that the exclusion principle acts in a somewhat non-intuitive way. The reader may want to look at Section 4.4 of Kadanoff and Baym (1962) and Section 8.2 of *ETMS* (Datta, 1995).

11 Atom to transistor

For more detailed discussions of the physics of conduction the reader can consult any of those under "Coherent transport." Chapter 7 of *ETMS* (Datta, 1995) discusses the similarities and differences between the problem of electron transport and optical propagation.

The example on Peltier effect in Section 11.3.1 was suggested by Landauer and is based on Lake and Datta (1992b). The discussion in Section 11.4 is based on the approach described in Lundstrom (1997).

Quantum models for 1D devices based on the NEGF formalism have been extensively developed by the "NEMO" group. See, for example, Klimeck *et al.* (1995).

These are available for public use via their website (http://hpc.jpl.nasa.gov/PEP/gekco/nemo).

Quantum models for molecular electronic devices based on the NEGF formalism have also been developed. See for example, Zahid *et al.* (2003). This book also contains a brief discussion of the NEGF equations in *non-orthogonal* bases in the Appendix.

Appendix: advanced formalism

Despite the title, I believe the approach presented here is less "advanced" and more accessible to beginners than the treatments available in the literature that may be needed for a systematic approach to higher order calculations. A few standard references and review articles, which cover the classic work on the NEGF formalism as applied to infinite homogeneous media, are Martin and Schwinger (1959), Kadanoff and Baym (1962), Keldysh (1965), Danielewicz (1984), Rammer and Smith (1986), Mahan (1987), and Khan *et al.* (1987).

Many authors have applied the NEGF formalism to problems involving finite structures following Caroli *et al.* (1972). The basic picture presented in this book is based primarily on Datta (1989) and Meir and Wingreen (1992) as explained in Section 1.6. Chapter 8 of *ETMS* (Datta, 1995) has a discussion of the NEGF and its relation to the Landauer formalism.

The time-dependent equations described here are similar to those described in:

Jauho A. P., Wingreen N. S., and Meir Y. (1994). Time-dependent transport in interacting and non-interacting resonant tunneling systems. *Phys. Rev.* B, **50**, 5528.

Haug H. and Jauho A. P. (1996). *Quantum Kinetics in Transport and Optics of Semiconductors*. Springer-Verlag.

The Green's function equations described for the Kondo resonance are the same as those presented in Meir *et al.* (1991).

Electron transport in strongly correlated systems is a topic of much current research. A few references that may be useful are Georges (2004), Giamarchi (2004) and:

Fulde P. (1991). *Electron Correlations in Molecules and Solids*. Springer-Verlag.

Mahan G. D. (1991). *Many-Particle Physics*. Plenum.

Hewson A. C. (1993). *The Kondo Problem to Heavy Fermions*. Cambridge University Press.

References

Allen P. B. and Mitrovic B. (1982). Theory of superconducting T_c. In *Solid State Physics*, ed. H. Ehrenreich *et al.*, vol. 37. Academic Press.

Ashcroft N. W. and Mermin D. (1976). *Solid State Physics*. Holt, Rinehart and Winston.

Beenakker C. W. J. (1997). Random matrix theory of quantum transport. *Rev. Mod. Phys.*, **69**, 731.

Bonet E., Deshmukh M. M. and Ralph D. C. (2002). Solving rate equations for electron tunneling via discrete quantum states. *Phys. Rev.* B, **65**, 045317.

Buttiker M. (1988). Symmetry of electrical conduction. *IBM J. Res. Dev.*, **32**, 317.

(2001). Capacitance, charge fluctuations and dephasing in Coulomb coupled conductors. In *Interacting Electrons in Nanostructures*, ed. H. Schoeller and R. Haug. *Lecture Notes in Physics*, vol. 579. Springer-Verlag. p. 149.

Cahay M., McLennan M., Datta S., and Lundstrom M. S. (1987). Importance of space-charge effects in resonant tunneling devices. *Appl. Phys. Lett.*, **50**, 612.

Caroli C., Combescot R., Nozieres P., and Saint-James D. (1972). A direct calculation of the tunneling current: IV. Electron–phonon interaction effects. *J. Phys. C: Solid State Phys.*, **5**, 21.

Chen C. J. (1993). *Introduction to Scanning Tunneling Microscopy*. Oxford University Press.

Danielewicz, P. (1984). Quantum theory of non-equilibrium processes. *Ann. Phys., NY*, **152**, 239.

Datta S. (1989). Steady-state quantum kinetic equation. *Phys. Rev.* B, **40**, 5830.

(1990). A simple kinetic equation for steady-state quantum transport. *J. Phys. Cond. Matter*, **2**, 8023.

(1995). *Electronic Transport in Mesoscopic Systems*. Cambridge University Press.

(2000). Nanoscale Device Simulation: The Green's Function Formalism. *Superlattices and Microstructures*, **28**, 253.

(2004). Electrical resistance: an atomistic view. *Nanotechnology*, **15**, S433.

(2005). Spin Dephasing and "Hot Spins." *Proceedings of the International School of Physics "Enrico Fermi"*. Course CLX, 2004. Societa, Italiana di Fisica.

Feynman R. P. (1965). *Lectures on Physics*, vol. III. Addison-Wesley.

(1972). Statistical Mechanics. *Frontiers in Physics*. Addison-Wesley.

Fischetti M. V. (1991). Monte Carlo simulation of transport in significant semiconductors of the diamond and zinc-blende structures – Part I: homogeneous transport. *IEEE Trans. Electron Dev.*, **38**, 634.

Georges A. (2004). Strongly correlated electron materials: dynamical mean-field theory and electronic structure. *AIP Conference Proceedings*, 715. Also arxiv: cond-mat 0403123

Giamarchi T. (2004). *Quantum Physics in One Dimension*. Oxford University Press.

Harrison W. A. (1999). *Elementary Electronic Structure*. World Scientific.

Herman F. and Skillman S. (1963). *Atomic Structure Calculations*. Prentice-Hall. pp. 3–9.

Kadanoff L. P. and Baym G. (1962). Quantum Statistical Mechanics. *Frontiers in Physics Lecture Notes*. Benjamin/Cummings.

Keldysh L. V. (1965). Diagram technique for non-equilibrium processes. *Sov. Phys. JETP*, **20**, 1018.

Khan F. S., Davies J. H., and Wilkins J. W. (1987). Quantum transport equations for high electric fields. *Phys. Rev.* B, **36**, 2578.

Klimeck G., Lake R., Bowen R. C., Frensley W. R. and Moise T. S. (1995). Quantum device simulation with a generalized tunneling formula. *Appl. Phys. Lett.*, **67**, 2539.

Kouwenhoven L. and Glazman L. (2001). Revival of the Kondo effect. *Phys. World*, January, p. 33.

Lake R. and Datta S. (1992a). Non-equilibrium Green's function method applied to double-barrier resonant-tunneling diodes. *Phys. Rev.* B, **45**, 6670.

(1992b). Energy balance and heat exchange in mesoscopic systems. *Phys. Rev.* B, **46**, 4757.

Lawaetz P. (1971). Valence band parameters in cubic semiconductors. *Phys. Rev.* B, **4**, 3460.

Lindley D. (2001). *Boltzmann's Atom: The Great Debate that Launched a Revolution in Physics*. Free Press.

Lundstrom M. S. (1997). Elementary scattering theory of the MOSFET. *IEEE Electron Dev. Lett.*, **18**, 361.

Mahan G. D. (1987). Quantum transport equation for electric and magnetic fields. *Phys. Rep.*, **145**, 251.

Mann J. B. (1967). *Atomic Structure Calculations, 1: Hartree–Fock Energy Results for Elements Hydrogen to Lawrencium*. Distributed by Clearinghouse for Technical Information, Springfield, VA.

Martin P. C. and Schwinger J. (1959). Theory of many-particle systems. *Phys. Rev.*, **115**, 1342.

McQuarrie D. A. (1976). *Statistical Mechanics*. Harper & Row.

Meir Y. and Wingreen S. (1992). Landauer formula for the current through an interacting electron region. *Phys. Rev. Lett.*, **68**, 2512.

Meir Y., Wingreen N. S., and Lee P. A. (1991). Transport through a strongly interacting electron system: theory of periodic conductance oscillations. *Phys. Rev. Lett.*, **66**, 3048.

Neofotistos G, Lake R., and Datta S. (1991). Inelastic scattering effects on single barrier tunneling. *Phys. Rev.* B, **43**, 2242.

Paulsson M. and Datta S. (2003). Thermoelectric effects in molecular electronics. *Phys. Rev.* B, **67**, 241403.

Rakshit T., Liang G.-C., Ghosh A. and Datta S. (2004). Silicon-based molecular electronics. *Nano Lett.*, **4**, 1803.

Ramdas Ram-Mohan L. (2002). *Finite Element and Boundary Element Applications in Quantum Mechanics*. Oxford University Press.

Rammer J. and Smith H. (1986). Quantum field-theoretical methods in transport theory of metals. *Rev. Mod. Phys.*, **58**, 323.

Sakurai J. J. (1967). *Advanced Quantum Mechanics*. Addison-Wesley.

Salis G., Fuchs D. T., Kikkawa J. M., Awscalom D. D., Ohno Y. and Ohno H. (2001). Optical manipulation of nuclear spin by a two-dimensional electron gas. *Phys. Rev. Lett.*, **86**, 2677.

Singh J. (1993). *Physics of Semiconductors and Their Heterostructures*. McGraw-Hill.

Venugopal R., Goasguen S., Datta S. and Lundstrom M. S. 2004. A quantum mechanical analysis of channel access, geometry and series resistance in nanoscale transistors. *J. Appl. Phys.*, **95**, 292.

Vogl P., Hjalmarson H. P., and Dow J. (1983). A semi-empirical tight-binding theory of the electronic structure of semiconductors. *J. Phys. Chem. Solids*, **44**, 365.

White S. R., Wilkins J. W., and Teter M. P. (1989). Finite-element method for electronic structure. *Phys. Rev.* B, **39**, 5819.

Wolf E. I. (1989). *Principles of Electron Tunneling Spectroscopy*. Oxford Science Publications.

Zahid F., Paulsson M. and Datta S. (2003). Electrical conduction in molecules. In *Advanced Semiconductors and Organic Nanotechniques*, ed. H. Morkoc. Academic Press.

Index

Printed in the United States
By Bookmasters